ANCIENT PRELUDES

World Prehistory from the Perspectives of Archaeology, Geology, and Paleoecology

D. Bruce Dickson
Texas A&M University

and

David L. Carlson
Texas A&M University

eddie bowers publishing, inc.
2600 Jackson Street
Dubuque, Iowa 52001-3342 USA

Exclusive marketing and distributor rights for
U.K., Eire, and Continental Europe held by:

Gazelle Book Services Limited
Falcon House
Queen Square
Lancaster LA1 1RN U.K.

eddie bowers publishing, inc.
2600 Jackson Street
Dubuque, Iowa 52001-3342 USA
eddiebowerspub@hotmail.com

ISBN 1-57879-001-8

Printed in the United States of America.

1 2 3 4 5 6 7 8 9

Table of Contents

PART I:

HOW CAN WE KNOW OF PREHISTORY?

Major Points Raised in Chapter One

1. This chapter begins by contrasting the world view of contemporary people, especially our view of time, with that of our forefathers.

2. Scientific terms including **law**, **theory**, **model**, **hypothesis**, and **explanation** are defined.

3. The natural sciences of **prehistory, geology, paleoecology**, and **archaeology** and the social sciences of **anthropology, sociology, economics** and **cultural geography** are defined.

4. The remainder of the chapter is concerned with the emergence of scientific prehistory in Western thought. **Mythic**, **philosophic**, and **scientific** prehistories are defined and examples of each are provided.

5. The origin story of the ancient Sumerians of Mesopotamia is used as an example of a **mythic prehistory**.

6. Aristotle's account of the origin of the ancient Greek city-state is used as an example of a **philosophical prehistory**.

7. The modern **scientific prehistory** that developed out of archaeology, anthropology, geology, paleoecology and molecular biology over the last 200 years is discussed together with the contemporary **contextual** form of prehistory.

CHAPTER 1

INTRODUCTION TO THE STUDY OF THE PAST

Introduction

The world of our forefathers was very different from the one we inhabit. For them, history had begun a mere 6,000 years earlier with the Creation. What is more, the time between that Creation and the present was fully described beginning with the Old Testament account of Adam and Eve and the Biblical Patriarchs and continuing through the events described in the writings of the Greek, Roman, Medieval and Renaissance historians in more-or-less unbroken fashion. The world of our forefathers was **fixed** both in space and time and humankind was firmly at its center. There was no prehistory, only history.

In contrast, our world is neither finite, predictable nor young. It is a world set in an expanding universe of infinite dimension. It is a world of vast antiquity, the comprehension of which is barely within human capacity. It is a world "in process" in which sequences of dimly understood ages follow one another in seemingly endless ranks: Cambrian, Ordovician, and Silurian; or Paleolithic, Mesolithic, and Neolithic. Ours is no longer a Newtonian world. It has become, instead, a world of probabilities and uncertainties and the grandest uncertainty of all is the place of humanity in this vast progression of time. Far from stage center, humankind seems to have emerged rather breathlessly from the wings long after the play had begun. We are distinctly minor actors in the drama of time.

In large measure, our modern conception of the world is the product of the intellectual revolution of the last 250 or so years. It is a revolution in which the sciences of geology and archaeology have played a central—if not preeminent—part. In this book, therefore, we outline the contributions of the sciences of archaeology and geology to our modern understanding of the world and our place within it.

The purpose of this chapter is to introduce the general scientific framework and perspective of the book. In the first chapter of the book we define prehistory, geology, paleoecology, and archaeology, and distinguish between mythic, philosophic, scientific, and contextual prehistories. In the next chapter, we chronicle the rise of scientific study of the past and trace the impact of that study on Western thought.

Basic Concepts

We have adopted an explicitly scientific approach in this book and have tried to structure our accounts of the past in an orderly scientific fashion. As a consequence, we use a number of everyday words like explanation, law, theory, model and hypotheses in a very narrow manner. For example, in this narrow usage, a **scientific explanation** is an attempt at supplying unified and consistent answers to the questions "Why?" "How?" and "What is the Evidence?" An explanation might help us to understand why the earth travels around the sun, why children look like their parents, or why sandstone rock in the Arizona desert contains marine fossils. Explanations can also help us to understand how something came about. Protective coloration of some animals can be explained as an adaptation that would promote survival. Members of a species who are able camouflage themselves would survive longer and have more children. Over time there would be more and more camouflaged individuals.

A **law** is a generalization about the world that has survived repeated efforts to disprove it. A law often refers to causal forces, such as magnetism or gravity. Laws are not explanations. They are statements about the world around us that can be accepted as "fact," or at least as very reliable.

A **theory** employs laws in order to offer an explanation about the world-around-us. Theories are not trial laws. A theory can be well-accepted, provisionally accepted, or poorly-accepted. Theories gain acceptance when they account for observable phenomena better than any alternative theory, when they survive repeated efforts to disprove them, and when they predict unexpected consequences that turn out to be true.

A **model** is a provisional description or an explanation of something. As a description of something, a model tries to simplify reality by capturing the important characteristics of the phenomenon under investigation. A model of ceramic manufacture should help us to understand the major steps in making a pot, but minor steps might be left out of the model if they would complicate the model without adding to our understanding. As a provisional explanation, a model might represent a theory or part of a theory. As the detail of the model are filled in, the model might provide a relatively complete explanation of something and become a theory.

Theories and models can provide **hypotheses**. An hypothesis is a prediction or a conclusion that has been derived logically from a theory or model. As such, the hypothesis becomes a way to evaluate or test the law, theory, or model against evidence. If, in the face of this evidence, the hypothesis cannot be rejected, it is *provisionally* accepted as true pending further tests. If it is shown to be false, support for the theory or model from which it derived is weakened.

Individual sciences vary widely in their subject matter. Further, specific methods and techniques differs greatly from one area of science to another. Nonetheless, the sciences *in general* share two major characteristics: (1) the logical and systematic organization of knowledge into laws, theories, models and explanations and (2) the use of evidence to test hypotheses drawn from this organized body of knowledge. With this in mind let us turn to the definition of the major natural and social sciences concerned with the past.

Prehistory is that time before written records. In the broadest sense of the word, the prehistory of the Earth begins with the formation of our planet some 4.5 to 5 billion years ago. More narrowly, the term refers only to humankind's prehistory which begins a mere 4 million or more years ago with the appearance of the **hominids**, the upright-walking, large-brained members of the family of primates called Hominidae. This family includes us, *Homo sapiens sapiens,* as well as our now-extinct ancestors.

Prehistory ends with the invention of writing and the beginning of written history. The earliest writing appeared in Mesopotamia sometime after 3500 B.C. Prehistory did not end at the same time everywhere. Literacy and written history did not arrive in Western Europe until the Roman Conquest late in the First Millennium B.C. In North America prehistory lingered until the arrival of European explorers and colonists after 1492. The more remote parts of North America did not enter into history until well into the 19th century. The bulk of human time on Earth has been prehistoric time, beyond the range of written history.

Over this large chunk of time the environment has changed dramatically. Human capacities, such as intelligence, have increased greatly and human social organization has transformed family-based societies into complex nation-states. The aim of scientific prehistory is to understand prehistoric change. Prehistorians develop models of the past and compare those models with the material remains of ancient life, the archaeological record.

Geology is the scientific study of the composition, structure, dynamics, and history of the Earth. By studying rocks, rock formations, and the processes that generate and change them, geologists seek to reconstruct the physical and organic history of the Earth from its earliest formation to the present time. Geology is commonly divided into physical geology and historical geology but both branches of the field depend on the **uniformitarian assumption**. This is the premise that both the laws of physics and the geological processes active on the earth today (for example, wind and water erosion, volcanism, earthquakes, and sediment deposition) operated in the past as they do at present. Uniformitarian geologists assume that these physical laws and processes are sufficient to explain the formation of geological strata without recourse to extraordinary or supernatural causes.

Physical geology deals with the structure and chemical composition of the Earth and with the forces that have shaped them. Physical geology includes a variety of different specialized branches. Geophysics uses the principles of physics to study the dynamic changes in the earth and its atmosphere. Geochemistry studies the distribution of chemical elements of the earth. Geochemists are particularly interested in abundance of different elements and their distribution and movement. Mineralogy is the study of the growth, composition and structure of minerals. Geomorphology is the study of landforms and their formation. Physical geologists help archaeologists to describe the deposits in which archaeological sites are found.

Historical geology deals with the chronology of the Earth. Historical geology also includes a number of specialized fields. Stratigraphy is the study of rock layers, their composition, and their interrelationships. Sedimentology is the study of sediments and the depositional processes that form them. Geochronology is the dating of geological strata and the processes that formed them. Paleontology is the study of the fossil evidence of life forms. Paleogeography is the study of the pattern and location of ancient land masses and oceans. Paleogeographers use the theory of continental drift to show how the position and relationships of various land masses have changed over millions of years. Historical geologists help archaeologists identify the contexts in which fossils will be preserved and to estimate the age of those deposits.

Although their specific methods and subject matter contrast, history, prehistory, and historical geology are sister disciplines whose practical and theoretical goals are the similar. On the practical level, historians, prehistorians, and historical geologists alike are concerned with time, with the reconstruction of past events, and with the ordering of those events in proper chronological sequence. On the theoretical level, practitioners in all three disciplines seek to develop general theories describing the underlying causes of the specific sequences of past events.

Paleoecology. Ecology is the study of the relationships of organisms to their surroundings environments and to each other. Paleoecology is the study of such organisms and relationships in the past. The uniformitarian assumption provides the logical basis for paleoecology just as it does in geology. Paleoecologists attempt to reconstruct past changes in the Earth's environment in terms of processes that can be observed in the present.

Archaeology is the study of human behavior from material remains. It includes methods for systematically recovering, describing, and studying the material remains of human activity. Archaeologists recover these remains at archaeological sites. An **archaeological site** is a place that contains evidence of human activity. Archaeological sites range in size from small clusters of a few artifacts to whole cities and from surface scatters of stone tools in plowed fields to deep cave or river valley deposits. Material evidence includes artifacts, structures, features, human remains, floral and faunal remains. The nature of these materials and their position in the site are the combined result of human activity and of geomorphic processes that foster accumulation, preservation, and decay. Knowing the relationships between these materials and the sediments that contain them is the only way to infer the human activities that took place at the site.

Conventional archaeological methods include techniques for finding and excavating sites and for analyzing archaeological materials. Archaeologists construct theories and models of the past and test them against the patterns recovered through the location and excavation of archaeological sites.

The methods of archaeology are not limited to the study of prehistory. They are used to study material remains in historic and modern times. Thus, the methods of archaeology apply equally well to prehistoric caves, Classical Greek temples, abandoned medieval towns, 19th century factories—even the modern refuse dumps studied by archaeologists at the University of Arizona as part of the Tucson Garbage Project! While the excavation techniques used on each of these kinds of sites differ, the logic upon which the techniques are based remains the same.

Excavating a site also destroys it, so archaeological experiments are not repeatable. Although archaeology is closely related to natural sciences like geology or ecology, the data recovered by archaeologists are also of interest to the humanities and the social sciences as well as. Both in subject matter and in methodology, archaeology is a hybrid of the natural and social sciences and the humanities.

Since archaeology deals with the remains of cultural and social activity, the models which underlie and direct research are derived, in large measure, from other social sciences. **Anthropology** is the study of humankind and human culture, with a particular focus on non-Western, pre-industrial peoples and cultures. The concept of **culture** is central to anthropology. There are many ways to define the term. In 1952 two anthropologists wrote a book on the definition of culture (Kroeber and Kluckhohn 1952). They analyzed 161 different definitions of culture and 100 longer statements about culture. They created a composite definition:

> Culture consists of patterns, explicit and implicit, of and for behavior acquired and transmitted by symbols, constituting the distinctive achievement of human groups, including their embodiments in artifacts; the essential core of culture consists of traditional (i.e., historically derived and selected) ideas and especially their attached values; culture systems may, on the one hand, be considered as products of action, on the other as conditioning elements of further action (Kroeber and Kluckhohn 1952: 357).

Culture includes ideas about the world around us, ideas about how people should behave, as well as accepted sets of rules about how people interact with one another. Economic systems, concepts of leadership, technological knowledge, belief systems, and language are all part of culture. Culture is a uniquely human means of adapting to the environment.

Sociology is the study of society, that is of social groups, social institutions, and social interaction, especially that of the industrialized West. **Economics** is the study of production, distribution, and consumption. **Cultural geography** is the study of spatial location, distribution, and organization. All of these disciplines have provided the basis for models and theories about the past. However, unlike these disciplines, our studies deal with dead—rather than living—peoples and prehistoric—rather than contemporary—peoples, societies and cultures. Prehistorians must depend on shovel work rather than interviews or archives. As a result, theories in prehistory tend to be different than those of the other social sciences. Nonetheless, prehistorians have one advantage over other social scientists in the enormous time depth of prehistory. We have examples of many more societies than are now in existence or are known through historical records. Prehistorians have a huge comparative data base with which to work. They have a vast array of examples, cases, and types of social phenomenon stretching back millions of years. The scientific study of society and culture has been immensely enriched by the data of prehistory.

The Emergence of Scientific Prehistory

Archaeology is the product of the industrial and scientific transformation of life and thought that began in the late 18th century. During this time, archaeology developed into a truly scientific study of prehistory. This does not mean that people never thought about their origins in the "pre-archaeological period" before about A.D. 1750. They most surely did. An interest in the past and a willingness to speculate have always been features of human life.

Scientific prehistory is possible only after the archaeology emerges as a field of study. The transformation of prehistory from myth to science was a critical and seminal event in the history of Western thought. Western intellectual history can be viewed as a gradual organization of thought into systems of greater explanatory power. Copleston (1955: 22) believes this was due to an "eradicable tendency in the human mind to reduce multiplicity to unity, to seek for explanations and hypotheses which will cover an even wider range of facts and events." Human beings can accept neither chaos nor void and must seek some explanation of the world of events around them.

According to Auguste Comte (1798-1857), Western thought has moved through three historical stages. In each stage, thinkers try to answer different kinds of questions. The first was the theological or mythological stage. Thought in this stage was directed at answering the question "Why?" In the metaphysical or philosophical stage, the questions became "Why?" and "How?" In the final, scientific stage, the questions became "Why?" "How?" and "What is evidence?" Each stage transcended the one before it, but never wholly replaced it as a mode of intellectual organization (Lenzer 1975).

Mythical explanatory systems use speculation and intuition to organize and structure experience. The resulting structure is the myth. Myths provide explanations for human events and bring coherence to the chaos and tragedy of human history. Often they justify the basic institutions of society by relating them to a primordial act of foundation recorded by the myth. The myth's primary appeal is to emotion, not to logic.

In contrast, **philosophic explanatory systems** curtail or eliminate emotion. Instead, philosophical explanations use reason, observation, distinction, classification, and generalization as primary tools. Philosophical explanations seek to develop general propositions about things observed or experienced and hypotheses of wider generality. As Nussbaum puts it, "philosophy has traditionally been committed to an 'ascent' from the perception of particulars to the intellectual grasp of universals" (1986: 49) Philosophical proofs must be clear, logical—or at least internally consistent—and as simple as possible.

Scientific explanatory systems are also the result of searches for the nature of reality through reason, observation, classification and generalization. However, a scientific explanation also involves the rigorous testing of both general propositions about sense-data and hypotheses of wider generality. Such testing demands that a critical stance be taken as to the kinds and quality of evidence to be used in testing. Where possible, experimental methods are used to test hypotheses and repeat observations. When this is not possible, other, non-experimental means must be used. Whatever kinds of evidence are used to test a hypothesis (and ultimately, to reject or tentatively confirm a theory), the test must be objective. Commonly, such tests involve the comparison with a rival causal explanation. A choice is made on the basis of which better explains the data. That is, "data confirm a hypothesis if they make belief more reasonable than disbelief" (Miller 1987: 175). This process is never complete and does not produce absolute truth, merely a closer and closer approximation to it; "the strongest commitment that is reasonable to the sciences is tentative belief, with openness to revision in the face of further developments" (Miller 1987: 157).

The scientific mode replaced the mythic and

philosophic modes at different rates in the various areas of human thought and study. The use of myth to explain basic technological processes probably ended fairly early in the human career. For example, fire or iron-making have long been freed of mythical explanations as people came to understand them as physical processes governed by natural, rather than supernatural, causes. In contrast, replacing mythic modes of explanation took much longer in astronomy. The interpretation of the cosmos in mythological terms persisted well into the late Renaissance—a fact symbolized graphically by Galileo's 17th-century struggle with the Catholic church over the nature of the solar system. Like the great questions of astronomy, the prehistoric origins of humanity have been explained in terms of all three modes of thought. However, the scientific mode replaced the mythic mode in prehistoric studies only in the last two hundred years.

The following examples of a mythic, philosophic, and scientific prehistory should help to illustrate the process of intellectual growth and change.

Mythical prehistory: the origin myths of the ancient Sumerians of Mesopotamia. The earliest formal writing system thus far discovered is known as **cuneiform**. Cuneiform writing arose during the 4th millennium B.C. in ancient Mesopotamia, that area of southwestern Asia occupied today by Iraq and parts of Iran. The ancient Sumerians used wedge-shaped wooden sticks or styluses to inscribe the lines, dots and angular shapes that made up the characters of their writing system on clay tablets, that is, palm-sized pads of damp earth. These tablets could be preserved by baking them in pottery kilns. Such baking rendered the tablets virtually indestructible, an attribute that has delighted generations of archaeologists and students of ancient languages. A great number of cuneiform texts contain business, tax, and trade records. However, texts of a religious or literary nature also have been recovered and it is through the translations of these texts that we are able to gain insight into the earliest recorded concept of "prehistory." Such texts reveal that the Sumerians had developed a mythic explanation of their origins.

The following outline was compiled from various cuneiform tablets by Samuel Kramer (1959: 84). First was the primeval sea. Nothing is said of its origin or birth, and it is likely that the Sumerians thought of it as having always existed. The primeval sea begat the cosmic mountains which consisted of heaven and earth united. The Sumerians thought of heaven and earth as gods in human form, *An* (heaven) was male and *Ki* (earth) was female. The union of *An* and *Ki* begat the air-god, *Enlil*. *Enlil* split heaven from earth and carried off earth, his mother. The union of *Enlil* and

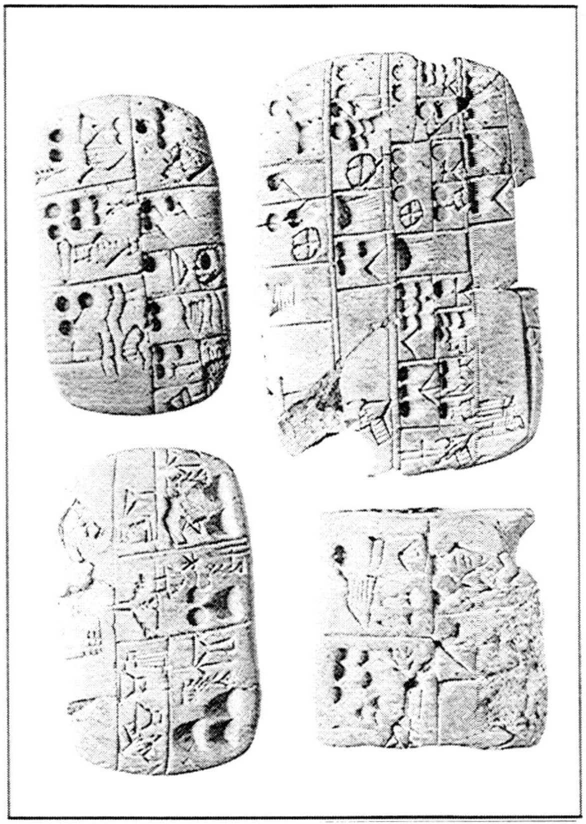

Figure 1. Cuneiform tablets (photograph courtesy, Field Museum of Natural History, Chicago, Illinois).

his mother set the stage for the organization of the universe–the creation of man, animals, and plants, and the establishment of civilization. *Enlil* as "cosmocrater" illustrates a common element in mythic prehistories. This account is a mythical one because it (1) takes a human-centered view of the universe; (2) willingly employs supernatural causes; and (3) is based entirely on speculation. In a mythic explanation of prehistory such as this one, no effort is made to portray logical action or internal consistency. For example, no one bothers to ask "Whence came the primeval sea?" We rely entirely on emotion and faith rather than reason and evidence.

Philosophical prehistory: Aristotle and the origin of the ancient Greek city-state. The ancient Greek's were not without their mythical explanations of nature, both secular and religious. For instance, the account of creation presented by Hesiod in his *Theogony* is no less mythic than the origin myths of the Sumerians.

But as early as the 6th century B.C., Greek philosophers were trying to determine the nature of the physical world and the place of humanity within it. They asked "What is the nature of the Good?" or "Who should rule?" or "What is the nature of humankind?" They used reason in their attempts to find answers to these questions and then judged their efforts in terms of logical consistency. In doing so, they introduced the philosophic mode of inquiry and explanation into Western thought.

Aristotle (384-322 B.C.) was arguably the most influential of the ancient Greek philosophers. He explored problems in realms as diverse as ethics, government, aesthetics, economics, natural history, and the origins of society. His work later became central to the philosophy of the Middle Ages. However, it is Aristotle's views on origins or prehistory of the Greek *polis* or "city-state" that concern us here. He provided this account in his general treatise on politics. Aristotle offered a rational, non-mythical theory of state origins that influenced Western thought for 2,000 years.

Aristotle defined the state as a self-sufficient political body, that is, one whose members could provide for all of their own needs. The purpose of the state is to organize society in such a way that the greatest number of people achieve the greatest happiness. His account of the origin of the state began with two assumptions: (1) the idea of **teleology**, that the development of a thing is caused by the purpose which it serves; and (2) that human beings are by nature political beings. In Aristotle's view, our political and social nature made the state natural and inevitable. Human beings naturally associate with one another but, prior to the rise of the city-state, the largest social group to which people could have belonged was the family. Over time, several families combined to make a village. A village would have been ruled by the eldest male as king. The state emerged when several villages combined to create a social entity large enough to be self-sufficient. According to Aristotle, "the state comes into existence, originating in the base needs of life, and continuing in existence for the sake of the good life" (Jowett 1943: 53-54).

Aristotle's explanation of the origin of the state is philosophical rather than mythical because it invokes no supernatural causes and is logical in approach. He first stated his assumptions, then reduced the state to its smallest social and political unit, the family. He then proposed a gradual development in stages from this simplest social formation to the city-state as it was known in the 5th century B.C. No empirical evidence was sought or thought to be needed. The only test of the adequacy of the account was its internal consistency: is it a more logical and parsimonious account of the rise of the state than competing theories?

Figure 2. In this detail from the great painting "The School of Athens" by the Italian Renaissance master, Raphael (1483-1520), the immortals of ancient Greek philosophy are shown in rapt conversation. Plato is pictured on the left, Aristotle on the right.

Scientific prehistory: human origins and the rise of the State. Science—whether physics, pathology, or economics—does not provide us with certainty. As noted earlier, the best stand for us to take towards the results of science "is *tentative belief, with openness to revision* in the face of further developments" (Miller 1987: 157, italics added).

Scientific prehistory begins with a "tentative belief" in what is referred to as **deep time** or the long chronology of the Earth. This is the view of the Earth as immensely old, 4.5 to 5 billion years, old enough for continents to drift apart, for mountain ranges to be thrust up and then eroded to sea level, for sediments to be compress to solid rock, and for self-replicating organisms to evolve from simple proteins and nucleic acids (Dyson 1988: 54-73). It is the view that humankind appeared very late in this vast time stream, a mere 4 to 6 million years ago in the Pliocene and subsequent Pleistocene geological epochs. It is the view that humankind, the family Hominidae, is firmly rooted in the animal kingdom with its closest relatives in the primate order, that is, the hominids are more similar to the other family of higher primates, the great apes or pongids, than to any other taxa in the animal kingdom.

Further, modern scientific prehistory embodies the notion that the physical nature of humankind was not fixed at some initial point but appears to have changed in a fairly plastic fashion largely as a result of selection of chance genetic variation. The Earth's environment has also not remained fixed during our evolutionary career. The last 4 million years have seen worldwide climatic and biogeographic change, shifts in sea level, long term cycles of glacial advance and retreat, and rapid mammalian evolution and extinction.

Within that changing environmental framework, the evolution of our line has several forms of hominids which are fairly well-documented through molecular biology and the fossil record. Comparative studies of the molecular biology of apes and humans suggests that the separation of the pongid and hominid lines occurred between 4 and 6 million years ago (mya). The **australopithecines,** fossil forms dated to between 4 to 2 mya, were the first hominids to possess bipedal locomotion, that is, to walk upright. Six different species of australopithecines have been identified. Larger-brained hominids, identified as the first members of the genus ***Homo***, appear about 2.2 mya. After about 1.6 mya, a new form of the genus, *Homo erectus,* migrated out of Africa into Asia and later, into Europe. Modern humans, *Homo sapiens sapiens* appear in the fossil record about 100,000 mya.

The physical evolution of humankind has been accompanied by a well-documented archaeological sequence of **cultural evolution**. For example, archaeologists have determined that hominids began making stone tools nearly 3 mya. Although it is still uncertain as to which form of hominid made these first implements, the course of their subsequent progressive change through time has been well documented with archaeological evidence. In fact, the trajectory of stone tool change in prehistory clearly illustrates what Robert Carneiro states are the two most important attributes of cultural evolution: growth and development.

> **Growth** consists of an increase in substance, while **development** consists of an increase in structure. Growth is usually manifested by a proliferation of structural elements already present, while development is marked by the rise of new structural elements. Growth, therefore, is essentially quantitative, while development is qualitative. One can occur without the other, but...evolution, in the full sense of the term, comprehends both (Carneiro 1969: 1020).

Prehistory shows that culture has both **grown**, meaning changed in degree, and **developed**, meaning changed in kind in dramatic fashion over the last three million years (Dole 1973: 248). These changes are measured by charting increases through time in human population size; the size and organizational complexity of human political units; the gross geographic distribution of humankind; the variety of environments settled or utilized for human purposes; and the number and variety of energy sources utilized.

Over the last three million years, cultural evolution has accelerated. It has moved from the glacial pace of the Lower Paleolithic to the breakneck speed of the present. During most of that time, people were hunters and gatherers who depended upon wild plants and animals. The nature of human subsistence began to change from food collection to food production around the close of the Pleistocene epoch. The emergence of agriculture during Holocene epoch seems to have occurred independently in a number of places on the globe shortly after 10,000 B.C. These changes coincided worldwide with various post-Pleistocene changes including glacial retreat, climate change, mammalian extinctions, a rise in worldwide sea level, and great changes in the distributions of plants and animals on earth.

The transformation from **food collecting** to **food producing** in southwestern Asia stimulated changes in many other areas of human culture. Population increased. There was an expansion trade as it became possible to store foodstuffs for longer periods. Problems of acquisition, control, the distribution of scarce resources like metal ores were prompted by population growth and technological changes. The need for administrative positions was spurred by irrigation agriculture. Religion began to emerge as a political force. Social and economic inequality, exploitation, and class differentiation followed from uneven distributions of wealth. Human impacts to the

Figure 3. Comparison of casts of skulls from three species of australopithecine with *Homo habilis* (upper right).

environment grew at an accelerating pace as deforestation and salinization increased and warfare was practiced on a greater and greater scale.

Several thousand years after ancient economies shifted from food collecting to food producing, cities, states, and civilization emerged independently in several parts of the world. A **city** is a "well-populated and unified center, many of whose inhabitants engaged in nonrural occupations, and one which came to control a larger or more productive area, and hence more wealth, than its mere subsistence required" (Hammond 1972: 8).

A **state** may be defined as "an autonomous political unit, encompassing many communities within its territory and having a centralized government with the power to collect taxes. draft men for work or war, and decree and enforce laws" (Carneiro 1970: 733).

Civilization is a level of social and cultural complexity characterized, according to V. Gordon Childe by the presence of cities, state political units, class stratified society, full-time occupational specialization, and long distance trade. In addition, civilizations develop writing, basic practical sciences like geometry, surveying and astronomy, monumental architecture and standardized art styles. At its base, a civilization depends upon subsistence systems capable of producing surplus foodstuffs and on political organization able to accumulate and redistribute that surplus.

Civilizations consist of networks of multiple states and other kinds of societies connected to one another through complex ties of trade, warfare and shared ideas. The growth of civilization has been witness to a sharp intensification of competition between societies in prehistory and history. This intensification is displayed most clearly in the violent rise of empires and other complex institutions of political control, domination, submission and injustice that often characterizes life among civilized peoples. Yet at the same time, civilization has fostered humankind's most noble achievements in the arts, the sciences, literature, law, philosophy and justice.

Archaeological evidence of the earliest cities, states and civilization come from the portion of the southwestern Asia known as Mesopotamia and date to sometime around 3,000 B.C. The rise of cities, states and civilization in Mesopotamia and elsewhere in the world appears to be a special instance of the general evolutionary trend from simpler to more complex forms that has occurred in human symbols, systems and institutions. Not long after the emergence of cities, methods of writing were developed. With writing comes the beginning of record keeping and the prehistoric period comes to a close.

Both scientific and philosophical prehistories are logical approaches to the past which invoke no supernatural causes. However, scientific prehistory does not rely exclusively on the strength and internal consistency of its logic; empirical evidence is essential to its construction and testing.

For example, the sequence of **human physical evolution** is based upon existing fossil evidence. Independent support comes from modern genetics and molecular biology. Our understanding of **cultural evolution** is based directly upon the archaeological and historical evidence. What we know about the nature of prehistoric cultural evolution has been assisted by studies of change in historic and modern times conducted by anthropologists and sociologists. Our current models to explain the rise of civilization, cities, and the state combine social science theory with archaeological data and historical documents.

Beyond science? Beginning the in 1970's some prehistorians have begun to question scientific explanatory systems. Following the lead of postmodernist thinking in literature, they view archaeological sites as "texts." Just as the interpretation of literature is often a matter of individual experience, **post-processual** or **contextual** archaeologists argue that the interpretation of archaeological sites is more like reading a book than an objective appraisal of the evidence in the context of a scientific hypothesis or theory (cf. Hodder 1992; Shanks and Tilley 1992). These archaeologists argue that objectivity is a mirage and that our understanding of prehistory is strongly affected by cultural beliefs and attitudes. The authors of this text would argue that objectivity is not an either/or condition, but a matter of degree. While we agree that no one is ever completely objective, that does not mean that we cannot strive to be more objective. We argue that the interplay between human beings trying to be as objective as possible and the materials recovered from archaeological sites, allow us to continually refine what we know about the past.

Of course, not every detail of the modern scientific picture of the past is supported up by evidence. Our knowledge of much of prehistory is tentative and partial. Some of the models we use to explain of the past are little more than speculation, but these are not the same as the speculations of mythic prehistory. Speculations in scientific prehistory are accepted only as hypotheses to be tested against future discoveries.

This summary of the scientific view of prehistory may not seem shocking to you. It is important to remember that the acceptance of the scientific approach to the past—and the corresponding replacement of mythic and philosophic explanations—represented a profound change in human thought. The slow pace of the general acceptance of a scientific approach to the past was due in part to a number of intellectual and institutional barriers. The major obstacles included: the belief in a universe centered around humankind and the Earth; the limited time span allotted by Biblical tradition for the creation of the world, humankind, and human society; and a willingness to invoke supernatural causes and worldwide catastrophes such as Noah's Flood to account for the past (Cameron 1969: 78-79).

The effects of these beliefs can be seen clearly in the attempts to explain stone tools and fossils. Scholars have been aware of both stone tools and fossils since the time of the Greeks and all three modes of explanation have been used to account for them. The enigmas of fossils and stone tools had to be solved before further progress could be made in several areas including estimating the age of the Earth, developing a scientific prehistory, geology, paleoecology, and archaeology, formulating the theory of organic evolution by natural selection, recognizing that human culture has evolved from simple to increasingly more complex over time, and developing theories to explain this evolution.

The solution of the stone tool and fossil enigmas—and the radical change in our thinking about the past that these solutions demanded—occupies us in the next chapter.

Major Points Raised in Chapter Two

1. This chapter deals with the history of modern scientific prehistory. The nature of this history is illustrated by examining three questions or "enigmas." The first was **the enigma of stone tools**. The second was **the enigma of fossils**. The third was **the enigma of species change**.

2. The answer to the stone tool enigma was provided by the **Three Age System** of Old World prehistory. In addition, Worsaae's **Law of Association** provided the basis organizing artifacts into contemporaneous groups.

3. The answer to the fossil enigma involved the recognition that fossils were once live plants and animals. **Nicolaus Steno's** work lead to the formulation of the **Law of Superposition** and the **Principle of the Initial Horizontality of Strata**, which form the basis of **stratigraphy**. The debate over **Uniformitarianism** in geology leads to the scientific study of the Earth.

4. The resolution to these two enigmas provided the tools needed to resolve the debate over the **association** between stone tools and extinct fossil animals. This opened the way to expanding the chronology of human prehistory.

5. The answer to the final enigma was the Darwin's **Theory of Natural Selection** as a way of understanding changes in species over time. Darwin's theory was substantially expanded and strengthened by modern **genetics** in creating the **Synthetic Theory of Evolution**.

6. Evolution and scientific archaeology produced the first unified theory of prehistory.

CHAPTER 2

ANCIENT TOOLS, FOSSILS AND THE RISE OF SCIENTIFIC PREHISTORY

The Stone-Tool Enigma

Today most people are aware that human technology has changed from simpler to increasingly more complex forms and that the Indian arrowheads or flaked or ground objects that commonly turn up today in plowed fields represent tools made long ago by ancient, technically-simpler peoples. However, the recognition that chipped stones are tools or implements and the understanding of their place in the evolution of human culture are relatively recent. Stone tools were replaced by metal tools in the Old World in a series of stages. The first metal tools were made by cold hammering native copper. Then copper and tine were smelted to make bronze. Next, iron was smelted. This sequence is called the Three-Age System after its key elements: stone, bronze, and iron. Recognizing the nature of stone tools and their role in human cultural development were important steps in the evolution of a scientific prehistory. This understanding emerged first in mythic, then in philosophic form. Only in the nineteenth century did it emerge in fully scientific form to provide a basis for a scientific prehistory.

Classical Period	ca 500 BC to AD 476
Early Middle Ages	AD 476 to 1050
High Middle Ages	AD 1050 to 1400
Renaissance	AD 1400 to 1600
Enlightenment	AD 1600 to 1799
19th Century	AD 1799 to 1914
Contemporary Era	AD 1914 to the present

Table 2-1. The Periods of European Intellectual History.

The Interpretation of Stone Tools in the Pre-Classic and Classic World. The Classical period of civilization begins in the 5th century B.C. and lasts until the fall of Rome in the west around A.D. 476. The first contribution to the solution of the stone-tool enigma was made by the ancient Greeks. Two of the greatest figures in Greek literature, the poets **Hesiodus** and Homer, provided mythic prehistories that referred to technologies based on stone.

Scholars generally date Hesiodus' works to the 7th century B.C. According to him, four historical ages preceded his own time. He called the five ages of mankind: Gold, Silver, Bronze, Heroic and Iron and asserted that the Age of Gold was preceded by an even earlier, more primitive age during which farming was unknown. Nonetheless, Hesiodus' account of technological change is a mythical one. The poet calls neither reason nor evidence to support his stages.

The Romans replaced the Greeks as the dominant people in the ancient world after about 150 B.C. Although they excelled as geographers, engineers, architects, soldiers, travelers, and compilers of knowledge, the Romans contributed far less to the sciences than did the Greeks. However Epicurean poet **Titus Lucretius Carus** (99-55 B.C.) wrote a lengthy scientific poem, *On the Nature of Things*. Lucretius' discussion of historical geology and prehistory is found in Book V. Here he deals with such questions as the origin of world, the origin and development of natural species, and the origin and growth of human institutions, language, art, religion, and technology. Lucretius accepted the bronze and iron ages of Hesiodus, but presents them as part of a philosophical explanation of technological development. According to Lucretius:

> Arms [i.e. weapons] of old were hands, nails and teeth and stones and boughs broken off from the forests, and flame and fire as soon as they had become known. Afterwards the force of iron and copper was discovered; and

> the use of copper was known before that of iron, as its nature is easier to work and it is found in greater quantity. With copper they would labor the soil of the earth, with copper stir up the billows of war and deal about wide-gaping wounds and seize cattle and lands; for every thing defenseless and unarmed would readily yield to them with arms in hand. Then by slow steps the sword of iron gained ground and the make of the copper sickle became a byword; and with iron they began to plough through the earth's soil, and the struggles of wavering war were rendered equal (Oates 1940: 188).

Lucretius' view of prehistory was favorably quoted and widely accepted. After Lucretius, few Romans contributed to the study of prehistory.

The Interpretation of Stone Tools in the Middle Ages. The rise of Christianity and the destruction of the Roman Empire in the West mark the onset of the Early Middle Ages (A.D. 476-1050). The 6th century closing of the schools in Athens by the Emperor of Constantinople is a dramatic symbol of the changes ushered in with the Middle Ages. In a stroke, a school that had been the dynamic center of philosophy, science and mathematics in the West for almost a millennium, ceased to exist (Hodgett 1972: 11). In Christian Europe, the philosophical and scientific achievements of the Classical world were preserved and passed on only imperfectly. Although the scholars of the Early Middle Ages were great compilers of Classical scientific knowledge, their interpretations of the natural world tended to be mixtures of Aristotle, practical knowledge, Biblical exegesis (close examination of the text of the Bible), peasant folklore, astrology, and myth. This combination of paganism with the Christian anthropocentric world view has come to be known as the **Medieval Synthesis**.

The Medieval world was a closed and relatively stable society in which "most of the essential truths about man, the universe, religion and philosophy were held to be fairly well known and permanently established" (Abrams, 1961: 78). Perhaps this overstates the case somewhat. According to Etienne Gilson (1936), there were two Medieval approaches to knowledge and two kinds of philosophers. One group lead by Tertullian and his successors believed that "since God has spoken to us it is no longer necessary for us to think"(Gilson 1936). The second group was lead by St. Augustine and his successors who believed "the divine law required man to seek God by the rational methods of philosophy" because it is man's first duty to use his God-given reason (Gilson 1936).

Medieval philosophers needed no prehistory. The Three-Age System of Lucretius was long forgotten and the Medieval world view, based on the *Old Testament* had a linear account of time which was complete in and of itself: Creation, the Fall, the Patriarchs, Noah, the Children of Israel, the life of Christ, the Last Judgment. The static qualities of the Medieval world view are powerfully expressed in illustrations of Biblical stories painted on the walls of cathedrals and churches during the period. The Patriarchs, Moses, Christ and the Apostles are all shown as Medieval European nobles, peasants, and soldiers. The nature and quality of human life and culture were assumed to be timeless and unchanging. Of course, ancient stone tools were commonly found in plowed fields by Medieval peasants in Europe just as they are today. Such objects were explained as celestial thunderbolts rather than man-made artifacts (Clarke 1978:4). And why not? There was no theoretical basis for people of the Middle Ages to interpret such artifacts as tools used by peoples in earlier times. "Earlier times" were already fully accounted for in Genesis and there is no mention of stone tools in the Garden of Eden! This uncritical approach to the world began to give way in the High Middle Ages (A.D. 1050 - 1450) as the New Science diffused into Western Christendom. Although the source was primarily Arabic nations, the New Science was founded on the works of the ancient Greeks. By the late 14th century, the rediscovery of Classical literature had begun challenge the Medieval approach to knowledge.

The Renaissance Recovery of the Past. The Renaissance (A.D. 1400 to 1600) began in the Italo-Byzantine cities of Genoa, Pisa, Venice, Rome, Florence and Milan in the late 14th century and continued through the 15th and 16th centuries. "In this period in Western Europe the arts of painting, sculpture, architecture and literature reached an eminence not exceeded by any civilization in any age" (Abrams 1961: 77). The beginning of the Renaissance seems closely related to two factors. First was the increase in trade and international contacts. The second was technological progress in navigation, sailing, and warfare. Ultimately this expanding trade and exploration lead to the discovery of the New World and to a new European world view. As Crosby says in *The Columbian Exchange*,

> ancient and medieval geographers had made fine maps of all the world, but the men of the Columbian generation discovered that 'Ptolemeus, and other knewe not the halfe.'

> [The Medieval Synthesis] . . . proved too cramped to accommodate the New World. Aristotle had quite logically supposed the equatorial zone of the Earth so hot that life could not exist there. Joseph Acosta crossed directly under the sun on his way to America in 1570 and 'felt so great cold, as I was forced to go into the sunne to warme me: what could I else doe then but laugh at Aristotle's Meterors and his philosophie . . . (Crosby 1972: 9).

The Medieval Synthesis was assaulted in other ways. The anthropocentric cosmos was forever shaken by the work of the Renaissance scholar **Nicolaus Copernicus** (1473-1543). Copernicus formulated the hypothesis that Earth was not the immobile center of the universe, but merely one of the planets in rotation both on its own axis and in an orbit around the sun. This hypothesis was not his most important contribution, however. Rather, it was his insistence "that old authorities were in error, that even observation and common sense were fallible, and that only reason supported by mathematical calculation could be trusted" (Dixon 1941: 279).

For the study of the past, the most significant event of the Renaissance was not the growth of science. Rather it was the rise of Humanism or the "New Learning" during the 14th and 15th centuries and the rediscovery of Classical Roman and Greek literature and civilization. By late Antiquity, the works of Classical Greek and Roman philosophers, mathematicians, and historians had been translated into Arabic. During the Middle Ages, Greek and Latin ideas mingled with Arabic and Indian thought to create a new philosophical and mathematical synthesis. At the same time, the Classical legacy had been partly preserved in the Byzantine Empire of the Greeks.

Another source of the Classics was the monastery libraries of Italy and France. Selected Greek and Latin manuscripts had been copied by generations of monks during the Middle Ages. By the 14th century, many of these libraries had begun to deteriorate. Humanists like Francesco Petrarca, better known as simply Petrarch, Dante Alighieri, Giovanni Boccaccio, and others, searched the decaying libraries and archives for copies of the writings of the ancients. Many of these were translated into Italian. In this way, the humanists of the Renaissance rediscovered a great number of Greek and Latin manuscripts, and added considerably to the number of Classical authors and works known to the Middle Ages. As a result of this work, Renaissance Italy became "Europe's first headquarters for exploring the past. What Portugal was for adventurers in geography, Italy was for history" (Boorstin, 1983: 574).

The rise of science, humanism, and the New Learning coincided with the invention of the movable-type printing press in the middle of the 15th century. Printed versions of the newly discovered Classical texts spread widely across Europe. But, "instead of perpetuating the ideology of feudalism, printing served to disseminate the spirit of skepticism, to interest man in his material surroundings, and to spread the knowledge of the expanding world disclosed by the bourgeois discoveries and explorations. Science was inestimably aided by Gutenberg's invention" (Dixon, 1941: 279).

Translations of the works of Lucretius re-introduced the idea of the Three-Age progression of stone to bronze to iron tools and weapons into Western thinking about the past.

Michel Mercati (1541-1593) combined these humanistic works and opinions with contemporary theories and empirical data. In doing so, he provided the first modern formulation of the Three-Age System. Mercati was Curator of the Vatican Botanical Gardens for Pope Clement VIII. During the Renaissance, these gardens played a role in the development of science analogous to that of MIT or Los Alamos in our own time. Mercati was charged with the oversight and curation of the growing Vatican collection of ethnological specimens that explorers and colonists were sending from the New World, Africa and Asia. As were most men of his class, he had been educated in the New Learning and was thus familiar with the Classical Greek and Roman authors, including the work of Lucretius on the Three-Age System. From his studies of the Book of Judges in the *Old Testament*, he learned that the ancient Israelites originally had only bronze weapons but acquired iron ones from their enemies, the Philistines.

In *Metalloteca*, Mercati combined the Greek and Roman classics, contemporary ethnological observation, and the historical accounts from the Old Testament to formulate the first **scientific** prehistory. Mercati proposed that European civilization had gone through the three ages outlined by Lucretius. As proof of a European stone age, he noted that stone artifacts found by peasants in plowed fields were similar to those known to be used by the native peoples in the New World. As proof of European metal ages, he cited the Biblical account of the transference of iron technology to the bronze-using Israelites. Thus, Mercati transformed the Classical Greek and Roman philosophical hypothesis about the development of technology into a scientific one by attempting to test it against evidence. Although the *Metalloteca* was not published until 1717, the manuscript circulated fairly widely during the 16th and 17th centuries, and influenced many scholars (Clarke 1978).

The Three-Age System in The Enlightenment. The Enlightenment or the Age of Reason (A.D. 1600 to 1800) refers to the developments in science and philosophy that began in the 17th century and became the mainstream of Western thought during the 18th century. The period was shaped by the physics of Isaac Newton, the rationalism of Rene Descartes, the empiricism of the Francis Bacon and John Locke, and the skepticism of Francois Voltaire, David Hume, and Adam Smith.

By the middle of the 17th century, the creative energies of the Italian Renaissance, obstructed by the Inquisition, had begun to diminish. Renaissance spirit passed to France under the patronage of the French court in the reign of Louis XIV. In its new setting, Italian learning helped foster the intellectual climate of the European Enlightenment. The writings of the Italian humanists were widely read in France during the Enlightenment and it appears that Mercati's views on the technological development of European civilization were widely accepted, especially in Jesuit circles. In the early 18th century, the Jesuit antiquarian scholar, Montfaucon, published an account of a Megalithic tomb in which stone implements were found. In 1721, the Jesuit Antoine de Jussieu compared French stone implements to those used by American Indians using Mercati's Three-Age System (Clarke 1978).

In 1725, the Italian historian **Giovanni Battista Vico** (1668-1744) published *The New Science*. Vico sought to discover laws of historical development by means of the scientific method. In this work Vico expanded the Three-Age System by asserting that all civilizations pass through the same stages from a stone to a bronze to an iron age.

By the end of the Enlightenment, the view that Western civilization had developed in stages from stone technology to metal was well accepted by thinking, educated people. In his celebrated account of his travels in Scotland, *Journey to the Western Islands* in 1773, Samuel Johnson remarked of the Island of Rassay that:

> A proof much stronger of the distance at which the first possessors of the island lived from the present time is afforded by the stone heads of arrows which are very frequently picked up. The people call them Elfbolts, and believe that the fairies shoot them at the cattle. They nearly resemble those which Mr. Banks lately brought here from the savage countries of the Pacific Ocean, and must have been made by a nation to which the use of metals was unknown (1990 [orig. 1775]: 48).

The late 17th and 18th centuries also witnessed the rise of a popular interest in ancient artifacts referred to as **Antiquarianism**. The discovery and excavation of the Roman sites of Pompeii and Herculaneum in 1714 had profound effect on public imagination. They stimulated demand for ancient objects and led collectors to locate and excavate ancient archaeological sites, tombs, and mounds all over Europe. Unfortunately, the spirit of collecting, with its lust for possession and its jealous secrecy about sources, is more often hostile to science. Antiquarianism eventually did give way to scientific archaeology in the 19th century. Interestingly, the transformation took place first in Denmark.

The Verification of the Three-Age System. Denmark had been under French political and cultural influence as early as the reign of Louis XIV. As a consequence, Danish scholars were familiar with the humanistic traditions of the Renaissance and the Enlightenment. The writings of Mercati and others had introduced the Three-Age System and this idea had long been accepted by Danish historians and scholars. As elsewhere in Europe, the upper class of Denmark shared the 18th century passion for antiquities. However, Danish antiquarians did not realize their potential.

The systematic application of the Three-Age System lead to a scientific archaeology in Denmark. Public interest in archaeology was stimulated by new discoveries and concern that they would be lost forever. According to Jensen:

> In the wake of the radical agricultural reforms of the 1780's terrible damage was wrought to thousands of prehistoric monuments in Denmark. New forms of land management led to the cultivation of regions which had not been ploughed for more than a thousand years. A profusion of buried treasures from heathen times was ploughed up from the soil—all the while their hiding-places, the prehistoric monuments, were damaged (1982: 1-2).

Further destruction was produced by the Napoleonic wars in the first decade of the 19th century when the British army bombarded and burned Copenhagen.

The Royal Antiquities Commission was formed in 1807 to save what sites they could from immediate destruction. Some nine years later in 1816, the commission appointed **Christian Jurgensen Thomsen** (1788-1865) as the director of its new Museum of Northern Antiquities (Trigger 1989: 75).

Thomsen's background in business had taught him the principles of warehousing and inventorying, two invaluable skills in museum work. In 1819 the museum collections were presented to the public. In an effort to introduce order into the chaos of the museum's prehistoric holdings, Thomsen arranged the exhibits of ancient stone and metal antiquities in terms of the Three-Age System: stone, bronze or iron. At first Thomsen merely used the Three-Age scheme as a way of organizing the materials. However, he quickly came to see it as a conceptual model of great predictive value. In 1836 he published the *Guide to Scandinavian Antiquities* which proposed that the sequential order of stone-bronze-iron tools had an important economic and technical basis. The made the Three-age system the first technological model of European prehistory. Thomsen also asserted that these groupings represented ancient tribal or racial groups (a point not now accepted).

Thomsen's efforts represent the first use of the Three-Age System to actually classify and date archaeological materials. The coherence of this approach was apparent to everyone who visited the museum or read Thomsen's *Guide*. In short order, the Three-Age System was applied to the prehistoric record all over Europe. While the museum display illustrated the utility of the Three-Age System, it did not **prove** that the theory itself was correct. Perhaps Thomsen's greatest contribution to scientific archaeology is the fact that he advanced the Three-Age System in a manner that allowed it to be tested directly against archaeological evidence.

Jens Jacob Asmussen Worsaae (1831-1885) was hired by Thomsen and charged with expanding the museum collections. He is generally regarded as the first full-time professional archaeologist. Later, he succeeded Thomsen as director of the museum. Worsaae, who was only nineteen when appointed, took his charge very seriously. Between 1839 and 1841 he performed a series of highly competent and creative field excavations in the **barrows** or burial mounds that dotted the Danish landscape.

During this fieldwork, he developed a number of excavation techniques and proposed the **Law of Association** or **Worsaae's Principle**. This law states that objects placed in a grave were in use at the time of the burial. That is, burials provide a kind of time capsule of prehistoric material culture. The concept of association provides the basis for making two kinds of inferences about the past. The first is that artifacts recovered from a grave (or other sealed deposit) were in use at the same time. This is the basis for the archaeological concept of **assemblage**. Secondly, if the age of one artifact is known, the date can reasonably be extended to the other artifacts in the assemblage. Unfortunately referring to Worsaae's Principle as a law is misleading because the inferences from it are not always valid. Most of the great debates in prehistory center on whether or not Worsaae's Principle applies to a particular case, that is whether or not two or more artifacts can be claimed to be **associated**. Worsaae's principle provides a way to test the key implication of the Three-Age theory. If stone, bronze and iron objects were in use at different times, then burials should never (or at least, seldom) contain mixtures of stone, bronze, and iron artifacts. Worsaae's excavations did reveal that artifacts of these materials were consistently found in **different** tombs (Rowe 1962: 130).

Worsaae published the results of this field work in 1843 in *Denmark's Past Elucidated by Means of Antiquities and Burial Mounds*. The work remains a landmark in archaeology (Daniels 1971: 141). By the time Worsaae died in 1885, he had established the basic theoretical framework that prevailed in Danish archaeology until the 1960's (Jensen 1982: 2). By the end of the century, the Three-Age System had itself come of age.

Over 2500 years, this rough chronology changed from a mythical to a philosophical and finally to a scientific, prehistory of great explanatory and predictive power. It had become a model from which hypotheses could be drawn and tested by excavating archaeological sites. Finally, the Three-Age System had become a major conceptual model which was accepted by archaeologists for nearly one hundred years following the publication of Thomsen's *Guide*.

The Three-Age System did not remain a static idea for long. The Stone Age was subdivided into two sub-stages in 1865 by Sir John Lubbock. In his book *Prehistoric Times*, Lubbock coined the terms **Paleolithic** and **Neolithic** and thus created a "five-age system." The recognition of the **Mesolithic** in the 1890s added still another age to the system (Daniel, 1971: 140). Nonetheless, it is with the work of Thomsen and Worsaae in the first half of the 19th century that modern archaeology truly began.

The Fossil Enigma

Just as we are aware that flaked stone arrowheads are products of human workmanship, so too do we realize that petrified casts of animals and plants found in sedimentary rocks are **fossils**, that is, remnants of living organisms whose organic parts have been replaced by minerals. The recognition of the organic nature of fossils was crucial to the development of a scientific understanding of prehistory. At least since the time of the Classical Greeks, people have wondered how shells and other marine fossils came to rest on top of mountains far from any ocean. Efforts to answer this question were first speculative and mythic, then philosophic, and only much later, scientific. Solving the fossil enigma lead to an understanding of the principles of stratigraphy and an awareness of the true magnitude of the Earth's age. In turn, these two achievements set the stage for the acceptance of Darwin's theory of evolution.

The Interpretation of Fossils in the Classical World. Despite their brilliance in other areas of human thought and science, the Classical Greeks never developed a geology worthy of them. Nonetheless, some Greek philosophers made startlingly accurate interpretations of fossils, among them two early pre-Socratic philosophers. Xenophanes of Colophon recognized fossils as the imprints of plants and animals in mud which later dried. Xanthus of Lydia saw sea shell fossils as evidence that sea and land were constantly undergoing changes of position and that great catastrophic events had overtaken the Earth in the past (Haber 1959a: 10).

In contrast, Aristotle expressed a gradualist view of geology. Unfortunately, he countered the catastrophist views of Xanthus with the mythical notion that fossils were formed in the Earth by a plastic or mineralizing force under the influence of the heavens. Even more incredibly, Aristotle stated that "a great many fishes live in the earth motionless and are found when excavations are made" (Haber 1959a:10).

Fossils in the Medieval World. Aristotle's view of fossils survived among some scholars during the Middle Ages. Other equally speculative explanations were advanced as alternatives from time to time. The most influential one was the view that fossils were the remains of creatures drowned in Noah's Flood. This idea was first advanced by Tertullian in the 2nd century A.D. Ristoro D'Arezzo, an Italian natural philosopher provided the clearest formulation of this **Deluge Theory** of fossils in 1282. The theory persisted throughout the Middle Ages and the Renaissance. It was widely accepted by scholars as it enabled them to link the Greek explanations of the origin of fossils with Biblical tradition. A variety of folk views and superstitions about fossils were also common (Haber 1959a: 12).

The Renaissance View of Fossils. The first essentially modern interpretation of fossils was made by **Leonardo da Vinci** (1452-1519). He deduced that fossils were formed at the bottom of rivers and oceans where mud replaced the soft parts of their bodies and then over time turned to stone. Leonardo observed that geological strata or layers were laid down in succession and he correctly interpreted this as evidence that they were formed by many depositions. He concluded therefore, that Noah's Flood could not account for all of the fossils that the deposits contained. He further concluded that the succession of layers indicated that the process had occurred over a long period of time. Noting that the Noah's Flood reportedly lasted only forty days and nights, he expressed doubt that the mussels, whose shells he found in the mountains of Lombardy, could have traveled fast enough to get to those elevations so quickly.

However, Leonardo used an elaborate philosophic explanation to account for the presence of fossils on mountain tops. Further, he never circulated these views and the notebooks containing them did not appear in print until 18th century. Consequently, Da Vinci's view of fossils had little influence on his or subsequent ages (Haber 1959a: 14-17).

Finally, **Conrad Gesner** (1516-1565), perhaps the greatest naturalist of the Renaissance, published an encyclopedic treatment of fossils entitled *On Fossil Objects* in 1565. Although Gesner's treatment of the subject was both comprehensive and well-documented, it illustrates the confusion in the minds of scholars of the period. In addition to the mineralized remains of plants and animals, stone tools, gemstones, and other objects excavated from the earth are treated as fossils in the book (Rudwick 1985: 1).

Fossils and Geology during the Early Enlightenment. The 17th century was a period of great scientific genius and remarkable accomplishment. Not surprisingly, throughout the century, theological authority found itself challenged by the spirit and method of the emerging critical rationalism. However, the rise of science at this time coincided with the intense Biblical literalism which had its roots in the Protestant Reformation (1517-1564) of the previous century. Practically, this meant that explanations of the natural world had to be squared with the revealed world found in the Bible. For example, the story of Noah and the Deluge had to be considered in any attempt to account for fossils and Earth's history.

In the emerging science of the late 16th and early 17th centuries, scholars harmonized natural history with sacred history in a literary manner. They treated both Nature (the work of God) and Scripture (the word of God) as "texts" and concluded that one could learn of God by "reading" either text (Haber 1959a: 21). However, since these two supposedly parallel "texts" in fact contained very different information, literalist theology was presented with numerous challenges as knowledge of the natural world expanded. Regarding the prehistory of the Earth, most scientists of the day accepted the account found in the first five books of the Bible. These books contain the account of the **Hexameron**, the six days of Creation, together with accounts of Adam and Eve, Noah and Deluge, Cain and Able and so forth. Scholars and philosophers as diverse as Rene Decartes in *Principle of Philosophy* (1644) and Thomas Burnet in his *Sacred Theory of the Earth* (1681 in Latin, 1684 in English) sought to provide scientific accounts of the Earth's origins using the Book of Genesis. A close reading of the *Old Testament* reveals that it contains a good deal of chronological information. For example, in Genesis 7: 18-20 we find the following:

> 18. And Jared lived 160 and 2 years, and he begat E'noch.
> 19. And Jared lived after he begot E'noch 800 years and begat sons and daughters.
> 20. And E'noch lived 65, and begat Methuselah.

The natural philosophers of the 17th century, intent on reading the Biblical text as a guide to the "text" of Nature, thought that this genealogical information was the key to discovering a chronology of Earth. They discovered serious discrepancies between the Hebrew Massoretic text, the Samaritan Pentateuch, and the Greek Septuagint.

A convincing reconciliation of these conflicting sources was proposed by the Anglican **Archbishop James Ussher** (1581-1656). Ussher published his celebrated book, *Annals of the Ancient and New Testaments*, in two parts between 1650 and 1654. In it he provided a scholarly and convincing argument that Creation had occurred in 4004 B.C. and that Noah's Flood had followed some 1,656 years later in 2319 B.C. Ussher's chronology represented an important refinement of the literal Judeo-Christian view of Biblical time, at least among Protestants, his date rapidly became orthodoxy. Before the end of the century, the year 4004 B.C. as the date of Creation was printed in most editions of the King James' version of the Bible.

By Ussher's calculations, there had been less than 6,000 years of human history since the Creation. This meant that human history was both short and fully accounted for in the first five books of the *Old Testament* and the histories of ancient writers like Herodotus, Thucydides, Livy, and Caesar. First came the Hebrew Patriarchs, followed by the Assyrians, the Persians, the Greeks, the Romans, the Middle Ages and modern times. There was no prehistory.

Such a world picture is not much improved over the closed view of time and the universe held during the Middle Ages. It inhibited the development of an adequate explanation of fossils and made developmental or evolutionary theories of life impossible. On the other hand, by the rigor of his scholarship and the strength of his logic, Archbishop Ussher had transformed the Biblical account of Earth history into a philosophical one. By structuring the Biblical chronology in terms of actual time, Ussher exposed it to objective testing and eventual rejection. Only a few years after the publication of Ussher's *Annals*, the basic tool for making that test, the Law of Superposition, was forged in Tuscany by a Danish physician called Nicolaus Steno.

Nicolaus Steno (1638-1686). Nils Stensen, later to Latinize his name to Nicolaus Stenoius or Steno, was born into a wealthy family of goldsmiths in Copenhagen in 1638. As a young man he studied medicine, physiology, anatomy and other sciences at Copenhagen, Amsterdam, Leiden, and Montpellier. In Paris, he conducted research in anatomy and established reputation in scientific circles by his dissection and insightful description of a sheep's brain. In 1665, Steno left Paris for Italy where, shortly after his arrival in Florence, he became physician to the Grand Duke Ferdinand II of Tuscany. Ferdinand, who had studied physics under Galileo, was a avid supporter of the sciences. He regarded Steno's arrival at his court as a great coup. Steno did not disappoint his noble patron. While in Florence he became an active and influential member of the Florentine Academy.

In 1666 Steno dissected a pregnant shark and became the first anatomist to provide an accurate description of the process of ovulation. Steno published an expanded and generalized description of the process some years later and this work assured him a place in the history of genetics (Cochrane 1973: 248). Steno also noted that the teeth of the shark corresponded to flat, triangular stones that he had collected during his wanderings about Tuscany on a commission from the Grand Duke to explore the mineral resources of the country (Cochrane 1973: 249). These triangular stones had been well known to the peoples of the Mediterranean world since Roman times as *glossopetrae* or tongue-stones. A host of fabulous folktales—that they had fallen from heaven or were the petrified tongues of snakes and so forth—had been concocted to explain them.

Steno proposed instead that, based on their morphology, the *glossopetrae* were shark's teeth. But such an claim raised other problems. How, after all, was he to explain "how the teeth got from the shark's head into the solid rock that encloses them. Or, in more general terms, how has one solid body come to be inside another solid body?" (Albritton 1980: 24-27).

To answer these questions, Steno used reason and observation. He devoted the next several years to the study of the natural landscape of Tuscany and central Italy. As part of this study, he described both the fossils and the geological strata in which they were recovered, relying particularly on exposures of rock visible in rock quarries. A **stratum** (pl. strata) is a sheet-like layer or bed of rock, earth, or other material usually found between other beds or layers of different composition. The order or arrangement of strata from lowest to highest is called its **stratification**.

Steno's studies of the stratification in numerous rock exposures convinced him that the strata of Tuscany had been deposited under water and that the fossils found within the strata had been buried during this process of sedimentation. From this premise, Steno deduced the deceptively simple conclusion that "at the time when any given stratum was being formed all the matter resting upon it was fluid, and, therefore, at the time when the lowest stratum was being formed, none of the upper strata existed" (cited in Albritton 1980: 33-34). In other words, any stratum is **younger** than the one upon which it was deposited and **older** than the one which was deposited upon it. This is the earliest formulation of the **Law of Superposition** or **Steno's Principle**. The principle is simple, but it identifies the logical relationship that underlies the complexity of the real world and leads to a logical basis for ordering the Earth's geological history.

Steno further reasoned that the process of deposition would have resulted in the formation of horizontal layers or beds of sediments. This assumption was later to be called the **Principle of Initial Horizontality of Strata**. "Strata either perpendicular to the horizon or inclined toward it were at one time parallel to the horizon" (Albritton 1980: 34-35). Thus, when he examined Tuscan rock exposures in which the strata were folded, inclined, or overturned, Steno recognized that these were changes of position that had occurred after the strata were formed and could infer their original horizontal position.

Steno hypothesized that the current disordered state of the strata evident in the rock exposures resulted either from the turbulence of Noah's Flood or from the collapse of caves in the interior of the Earth. He referred to both of these causes in his reconstruction of the geological history of Tuscany. Steno seems to have accepted the Biblical account of Earth's. His orthodoxy did not prevent him from observing that the beds of rock and the fossils they contained were laid down sequentially. That is, they were not due to a single deposition and therefore could not have been the result of the Deluge.

Steno presented these views with great clarity in 1669 in a work rather curiously titled *Podromus to a Dissertation Concerning a Solid Naturally Enclosed Within a Solid*. Unfortunately, Steno never completed the intended grand sequel. In the years that followed, he abandoned his scientific studies altogether. Having converted to Roman Catholicism in 1667, Steno took holy orders in 1675 and devoted the remaining eleven years of his life entirely to the church. His ecclesiastical career was marked by rapid professional advancement, rigorous austerity and fasting. In fact, relentless fasting may have contributed to his death in Germany in 1686.

Steno's influence on the geological thought during the 17th century spread rapidly. In 1671, only two years after its original appearance in Italy, Steno's *Podromus* was translated into English. In the years that followed, the work appears to have been widely read among scholars in England and on the continent. Through his dissection and field observation, Steno laid down basic laws of stratigraphy and superposition and in doing so made both a scientific geology and a scientific archaeology possible. Despite his retention of Genesis as the historical frame for his work, Steno began the transformation of the explanation of fossils and the study of Earth from myth to science (Fenton and Fenton 1952: 25).

Two other natural philosophers also contributed to the solution of the fossil enigma during the Scientific Revolution of the 17th century: Baron Gottfried Wilhelm von Leibniz (1646-1716) and Robert Hooke. Leibniz is perhaps best known for his contribution to philosophy—and for the ridicule heaped upon it by Voltaire in *Candide*. But Leibniz was also deeply interested in mining and geology; he recognized the organic nature of fossils and proposed that the presence of marine fossils on land indicated that the Earth's oceans were gradually shrinking (Rudwick 1985: 126).

Robert Hooke (1635--1703), Curator of Experiments for the British Royal Society, was a celebrated contributor to optics, biology and physics, as well as geology. Independently of Steno, Hooke arrived at the conclusion that fossils were organic in origin and his work seems to have been influential in shaping English geological thinking. He was the first scientist to examine petrified wood and other fossils under the microscope. These studies provided an empirical basis for his conclusions. Hooke also suggested that fossils might be used to make chronological comparisons between different layers of rock, although he did not actually attempt to test this idea. The work of Steno, Leibniz, and Hooke shaped and fueled the debate on the origins of fossils. By the end of the century, the arguments of these men, combined with the comparison between fossils and modern animals and plants, had led most natural philosophers to recognize the organic origins of fossils (Haber 1959a: 29).

Steno's and Hooke's accounts of fossils and their studies of superposition of layers of rock on the Earth's surface were not seen as being in conflict with the Judeo-Christian account found in Genesis. Steno concluded from his Tuscan evidence that the world was created about 4,000 B.C. Hooke also remained quite orthodox in his religious views. Nonetheless, general developments in the study of paleontology and historical geology were leading to a conflict between science and religion.

Fossils in the Later Enlightenment. Natural philosophers of the 17th century came to accept the view that fossils were the result of natural processes. But they avoided the related question of the length of time involved in producing such "petrifactions." However, scholars of the 18th century proved to be more willing to face the radical implications that the solution of the fossil enigma had for the question of the age of the Earth. By the later Enlightenment, natural philosophers were not afraid of rejecting the *Old Testament*. However, the interpretation of fossils as evidence that the Earth had experienced periodic sea-land transformations and other great changes appeared to run contrary to the 18th century view of the world as a vast, harmonic machine. This view of the world as a machine had been inspired by Newtonian physics. Voltaire, as an influential proponent of this view, opposed both the Church's view of fossils as relics of the Deluge and the notion that they represented any kind of massive, catastrophic reversals of natural laws.

Georges Louis Le Clere, Comte de Buffon (1707-1788), Curator of the Royal Garden in Paris, laid the foundation for the scientific interpretation of fossils in his comprehensive treatment of Earth science, *Theory of the Earth* published in 1749. In this book, he reviewed work of previous paleontologists and concluded that their analysis failed due to their reliance on the Deluge as the explanation of fossils. Buffon claimed to believe in the scriptures in matters of religion, but rejected them as a basis for explanation in science. As a result, he had trouble with the administration of the Sorbonne, and eventually was forced to recant of his views publicly in 1751. This did not stop him from continuing to publish and defend his theories until the end of his life in 1788.

In 1778, Buffon published his influential natural history of the Earth entitled *Epochs of Nature*. In it he divided the history of the Earth into seven epochs. The number was derived from Genesis but the time assigned to each of the seven was measured in thousands of years rather than hours. He defended this on the grounds that no one knows the length of one of God's "days." The first epoch in his history was the emergence and cooling of the Earth from a molten mass. On the basis of experiments that he had conducted cooling rocks, Buffon estimated that this epoch would have lasted 3,000 years. Land animals did not appear until Buffon's fifth epoch some 67,000 years later. Humans appeared 15,000 years afterwards in the seventh and final epoch. Buffon estimated that the development of humankind took about 5,000.

Altogether Buffon assigned an age of 85,000 years to the Earth and was among the first scientists to provide a non-Biblical estimate of the time needed to explain the fossil strata (Haber 1959b: 234). The significance of his *Epochs* is not due to the accuracy of his estimate; modern calculations place the age of the Earth at more than 4.5 **billion** years. Rather, it is his willingness to project a long, rather than a short, chronology for the Earth and to use experimental evidence to find it. Buffon also pointed to the changes in plant and animal life contained in the fossil record. He tentatively advanced an evolutionary hypothesis to explain these changes.

Buffon's *Theory* and his *Epochs* were among the most widely acclaimed and influential books on science to be written during the Enlightenment. The scheme presented in *Epochs* was not tested against evidence and thus was a philosophical prehistory at best. In order to estimate the age of the Earth, one must first understand the order or relative sequence of geological events that make up Earth history. This is done by building a chronological sequence by examination of stratigraphic record using the Law of Superposition. The German mineralogists and mining engineers Johann Lehmann and Abraham Werner turned to this task in the last half of the 18th century.

At that time, the systematic study of stratigraphy as a tool for understanding the geological record was centered in the German mining districts of Thuringia and Saxony. German miners were interested in a reliable means of predicting the locations of ores. Johann Lehmann, surely influenced by the earlier work of Steno, formulated the Law of Superposition in its modern form. In any pile of depositional units in which the top and bottom of the pile can be identified, the order of succession from bottom to top is the order of deposition. More will be said of this law in the following chapter. The law does not produce ages in years for the layers, only the order of their deposition relative to one another. Using Lehmann's principle, the German geologist **Abraham Werner** (1749-1817) worked out the first practical geological time scale.

Werner's book was titled *Brief Classification and Description of the Different Types of Rock Formations*. In contrast to Buffon's seven epochs, Werner's system was based on first-hand, field observation. Werner classified the rock formations around Leipzig in southern Germany using lithology, (mineralogical form and composition), topographic position, and fossil content. He paid only passing attention to fossils as he felt the first two criteria were more reliable (Rudwick, 1985: 126).

In the Leipzig area Werner considered the unconsolidated materials on the Earth's surface to be the residue of Noah's flood. He, therefore, called this highest stratigraphic layer, the Alluvium. He also recognized a basement of crystalline rock which he thought was the original crust of the Earth. He referred to this formation as the Primary series. Above the Primary series he found sedimentary rock and referred to it as called the Flat-lying series (later, the Secondary series). Elsewhere in southern Germany, he found he thousands of feet of sedimentary rock of a different character between the Primary and Secondary series. He called that the Transitional series. Another series was found between the Secondary and the Alluvium—this formation he called Tertiary and renamed the Alluvium, the Quaternary.

Werner concluded that the major geological events in the history of the entire Earth were recorded in the order of their occurrence in the southern German stratigraphic series. He tried to cross-date the German and Italian geological records by matching the two sequences. Werner theorized that the bulk of the geological strata on Earth had been formed by deposition in water and considered the role of volcanic activity in such formation to have been decidedly minor. Consequently, the school of geological thought that emerged based on his work was dubbed **Neptunism** after the Roman god of the sea.

Werner correctly described the sequence of rock formations for southern Germany and in doing so demonstrated the importance of rock lithology and the Law of Superposition in stratigraphic studies. However, the relative order of geological events that he recorded was not a universal one as he supposed. The south German sequence could not be generalized to the entire world. Further, the Neptunist theory which was developed to account for this sequence was later proven to be in serious error.

Further, Werner's students, chiefly in Britain, complicated the application of his theory by attempting to pack his entire stratigraphic sequence into the *Old Testament* time frame of less than 6,000 years. Nonetheless, Werner made an important and lasting contribution to both geology and archaeology by systematically applying the Law of Superposition in sorting out the **order** of events in the geological record. Further, his attempts to correlate separate stratigraphic exposures by matching the lithology of the rocks was a critical methodological step in the development of the science of stratigraphy.

The Solution of the Fossil Enigma. Building on Werner's research, geologists working chiefly in England were able to expand and refine his techniques. These efforts begin with **William "Strata" Smith** (1769-1839), often called the father of English geology. Smith worked as a mining and agricultural engineer, canal builder, and ultimately, as a consulting geologist. His engineering and surveying activity took him all over the rapidly industrializing south of England. He became involved in mining at a time when the infant industry was seeking practical and economical ways of locating coal seams. Smith's studies of fossils were motivated in part by his professional interest in predicting the location of coal bearing rock strata.

By the late 18th century, experienced engineers and miners followed Werner's lead and used rock lithology to distinguish between different strata. They distinguished different layers on the basis of the form, composition, and character of their mineralogy. Once they identified a coal-bearing stratum, they used the Law of Superposition to find younger and older layers. They then tried to find similar strata in other exposures as a means of finding new coal deposits. However, predicting the order or position of coal-bearing formations by lithology alone was imprecise at best and often impossible.

Smith's realized that strata could correctly and reliably be distinguished on the basis of the distinctive fossils they contained, as well as by their lithology. He used the Law of Superposition to establish the relative order of events in a given geological

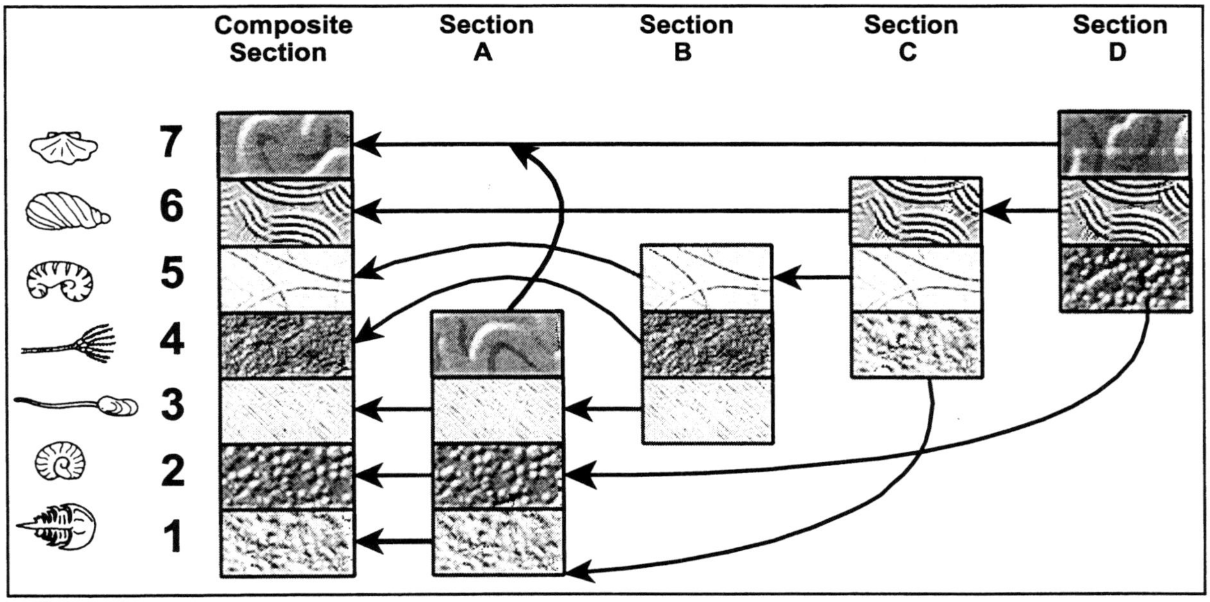

Figure 2-1: Smith's method of stratigraphic cross-correlation using organized fossils. In this schematic diagram, four distinct stratigraphic exposures are correlated or cross-dated by matching the distinctive fossil assemblages found in the various layers of each profile. Note none of the sections contains all of the strata represented in the composite stratigraphic column. Using this method, Smith was eventually able to order 32 distinct and successive rock formations and thereby compile the first composite stratigraphic column for England.

exposure. Then he distinguished layers on the basis of the different fossil assemblages that they contained. Finally, he cross-dated the exposure to other exposures by matching strata containing similar fossils.

In this way, he recognized 32 distinct, successive rock layers in England and formulated the **Law of Strata Identified by Fossils**. In 1799, Smith compiled a list of these strata together with their characteristic invertebrate fossils but did not get around to publishing either the list or the law upon which it was based until 1816 in *Strata Identified by Organized Fossils, Containing Prints on Colored Paper of the Most Characteristic Specimens in Each Stratum*, which presented both the law and an illustration of its application. In it Smith demonstrated that many of the geological strata of England contained distinct assemblages of fossil plants and animals. At a given outcrop or site, the strata can be distinguished by contrasts in the types of fossils and the frequency of their occurrence. He also showed that these fossil assemblages were successive, in a fixed, historical order.

Smith concluded that stratigraphic layers in different rock exposures that contained the same kinds and proportions of fossils were deposited at the same time. Fossil assemblages could thus be used to correlate or **cross-date** the layers from different stratigraphic exposures. By the 1830s, British researchers were actively applying method of "organized fossils." The delineation of the stratigraphic column on the basis of distinct fossil groups was one of the great conceptual breakthroughs of modern science.

Smith had used fossils primarily to establish the chronological sequence or order in which various strata had formed. His method worked however, only because the fossils were part of a particular sequence of organic evolution. As one moves up the geological column from the bottom to the top, the fossils differ due to the evolution of species over time. Smith was able to predict the order of strata using fossils because the fossil record contains a unique historical sequence of particular faunal assemblages. Thus, without realizing it, Smith laid the foundation for the **Law of Faunal Succession**. This is the principle "that fossil organisms succeed each other in the stratigraphic record in a definite and recognizable order, each geologic formation having a total aspect of life different from that in the formations above and below it." (Boggs 1987: 5, 15). Very shortly afterwards, Smith's column would be enlisted in the final battle over the fossil enigma, the struggle over organic evolution and natural selection.

Smith's 32 separate strata presented another problem. How could this vast amount of rock have accumulated in the 6,000 years or so since Creation? Smith does not appear to have doubted the Biblical chronology. Others in England were not so sure. If one concluded that the strata had accumulated by means of the normal processes of sedimentation, a 6,000 year period was too short for all 32 strata to have accumulated through natural processes.

If one did not reject the short chronology, it became necessary to assume that geological processes must have operated much faster in the past. Alternatively, there must have been immense catastrophic events unlike anything that can be observed in the modern world. Despite its failure to adequately explain all the new data that geologists were accumulating, there was no better explanation available.

Generally speaking, scientists, like most people, will not reject an explanation without first having another one to put in its place. The doctrine that emerged to replace Genesis was dubbed **Catastrophism**. Catastrophism as an explanation of the geological record is most closely associated with **George Cuvier** (1769-1832), a French comparative anatomist considered to be the father of paleontology.

Early in his research career, Cuvier described the fossil remains of mammals. His made detailed comparisons of these fossil remains with the skeletons of living mammals. As a result, he became convinced that the fossil record contained many species that had no living counterparts. Largely as a result of the work of Cuvier and his students, the succession of plant and animals forms in the fossil record became known in broad outline during the early decades of the 19th century. However, to Cuvier, the gaps in the record appeared to reflect abrupt transitions, rather than gradual changes (Sherwood 1969: 316).

He interpreted these gaps as evidence of separate catastrophes in the past that had resulted in massive extinctions. Cuvier concluded that the fossil record contained evidence for 27 catastrophic extinctions. These were caused by abrupt changes in climate or sea level. He rejected the Larmarckian view that species did not perish but developed gradually, one out of another.

Cuvier was not concerned with reconciling geological history with the Scriptures. A history of the Earth marked by 27 or more catastrophic extinctions was hardly an orthodox Christian view. He did not regard these abrupt revolutions to be supernatural in origin. To him they were merely unlike the normal processes of change observable on Earth at present. Therefore, he felt no obligation to compress the occurrence of these catastrophes into the 6,000 year Biblical chronology (Rudwick 1985: 131-132). But not all Catastrophists were as generous as Cuvier. A more moderate reading of the fossil record that emphasized Noah's Deluge as the major catastrophic event attracted a number of Christian geologists. Included among this group were the English geologists, Dean W.M. Buckland, who accepted evidence for only one catastrophe in the geological record, and the Rev. W.D. Conybeare, who made room for three of them.

In 1812, the Scottish geologist Robert Jameson translated Cuvier's four volume monograph, *Research on the Fossil Deposits of Quadrupeds Where One Reestablishes the Characteristics of Several Species of Animals That the Revolutions of the Globe Appear to Have Destroyed*. In the preface to this translation Jameson—not Cuvier—introduced and defended the view that the Deluge was a real event, albeit only the latest of numerous other abrupt catastrophes (Rudwick 1985: 134-135).

As a result, Cuvier's work came to be used in Britain by those seeking to defend the literal Biblical, short chronology. The Biblical interpretation of Catastrophism and the attempt to reconcile Genesis and geology seems to have been primarily an English phenomenon. Other English geologists proposed an alternative to Catastrophism that became known as **Uniformitarianism**. It is in the work of the uniformitarian geologists such as Hutton, Playfair, and Lyell that we see the emergence of the modern science of geology.

James Hutton (1726-1797), was a Scottish scientist who laid the theoretical foundations of modern geology, and to a lesser extent, modern archaeology. Extensive field observation in Scotland convinced Hutton that the ancient rock formations, mountains, and floodplains could best be understood by observing the processes of deposition and erosion currently operating on the Earth's surface.

Hutton considered Cuvier's catastrophes to be without cause or evidence and therefore useless as scientific explanation. In place of sudden, catastrophic events, he substituted the slow and persistent work of the forces of erosion and deposition. Hutton considered himself an "actualist." He would accept no explanation of geological formation if the process it postulated could not be found at work on the modern Earth. At the same time, Hutton saw these geological processes as timeless and cyclical. Erosion of the land leads to deposition in the sea, deposition in the sea generates pressure and heat, pressure and heat leads to volcanic activity and the uplift of new strata. To Hutton, the Earth was a vast self-renewing machine in which, in his famous phrase, "we find no vestige of a beginning, no prospect of an end" (Hutton 1788: 304). The emphasis which Hutton and his circle placed on heat and volcanic activity placed them in direct opposition to Werner and the Neptunists. They were given a contrasting label, **Vulcanists**.

In 1788, Hutton published his views as *Theory of the Earth; or an Investigation of the Laws Observable in the Composition, Dissolution, and Restoration of Land Upon the Globe*, in the Transactions of the Royal Society of Edinburgh. His heavy writing style made the book difficult going and it attracted little notice. Fortunately, his theory was ably interpreted by a friend and colleague at the University of Edinburgh, the astronomer and mathematician, John Playfair (1748-1819). In 1805 Playfair published a clear and pursuasive account of Hutton's ideas in *Illustrations of Huttonian Theory*.

Nearly three decades later, Hutton's approach to the geological record was even more successfully applied by **Charles Lyell** (1797-1875). Lyell was lawyer turned scientist whose fieldwork and precise exposition of Uniformitarianism largely defeated and replaced Catastrophism. Lyell's most important work is his *Principles of Geology: Being an Attempt to Explain the Former Changes of the Earth's Surface by Reference to Causes Now in Action* published in three volumes between 1830 and 1833. In it, Lyell followed Hutton and used only those processes currently shaping the Earth to interpret the geological record. Since the changes brought about by these forces are often almost imperceptible to us, implicit in this doctrine is the view that alterations of the face of the Earth were brought about slowly over vast stretches of time rather than abruptly by revolutions.

Like Hutton, Lyell saw the Earth as a balanced system in which forces of destruction were continuously countered by forces of construction. The assumption that the terrestrial system was in stable equilibrium, at first led Lyell to rule out the possibility of cumulative development in the plant and animal world. In the first edition of his *Principles* he attacked the "progressivism" of biologists like Lamarck and asserted that the paleontological record shows no evidence of fundamental or progressive change in fossil form (Cannon 1969: 78-79). Later, under the influence of Charles Darwin, Lyell modified his theory.

Uniformitarianism proved to be a powerful tool for understanding the Earth. So powerful, in fact, that after 1833, even Catastrophists accepted the doctrine in some form. Yet, its triumph was not total. Many scholars continued to attempt to reconcile it with Scriptural accounts. Lyell, himself, applied "uniformitarianism in a consistent manner only to inanimate nature, but left the door open for special creation in the living world" (Mayr 1972: 985). Nonetheless this general acceptance represented a crucial change of attitude which had some profoundly unexpected consequences.

Uniformitarianism purged geology of its mythic elements by limiting explanation to "causes now in action." As Bowler (1984: 7) notes, "For science to be possible, it was necessary to conceive the laws of nature so that they operated solely in a mechanical fashion, allowing the past (but not the future) to control the present by the normal rules of causality."

Associating Fossils and Stone Tools

Uniformitarian interpretations of fossils and the geological record were difficult to reconcile with Genesis. But the idea of people living on the Earth before 4004 B.C. was heresy. As Glyn Daniel notes "the moment you accepted the doctrine of uniformitarianism, in however liberal a form, you immediately drew attention to the fact that long before 4004 B.C. there existed animal life" (1960: 37). Cuvier flatly denied that "fossil men" were possible. When stone tools or human bones were found with extinct animals, the association was interpreted as accidental (Daniel 1960:34).

Yet, it was apparent to some that stone tools and fossils of extinct animals occurred together in cave and river terrace deposits. Little is known of **John Frere** (1740-1807) other than the fact that he discovered stone tools and the bones of extinct animlas together in a river terrace at Hoxne, England. Frere recognized the association of the bones and tools and described his observations in a letter to the Royal Society of Antiquarians in 1800. Speaking of the artifacts Frere asserted, "They are, I think, evidently weapons of war, fabricated and used by a people who had not the use of metals. The situation in which these weapons were formed may tempt us to refer them to a very remote period indeed, even beyond the present world" (quoted in Daniel 1962: 41).

But how remote a period? Frere's evidence was easily rejected by the scientific world at the turn of the 18th century. The short Biblical chronology for human creation still had a powerful hold on both the scientific and the lay communities. The battle between Catastrophism and Uniformitarianism had to be won before the evidence of the human antiquity could be recognized.

When **Jacques Boucher de Perthes** (1788-1868) asserted in late 1840s and 1850s that he had found an association of extinct animal bones and tools, he received a different reception than had Frere. Boucher de Perthes was a French customs officer with a long-standing interest in geology and prehistory. He examined strata exposed in gravel quarries on the river terraces of the Somme River in northern France.

In these exposures Boucher de Perthes found chipped stone tools (known now as handaxes) associated with fossil bones of extinct animal. He published his finds as evidence that human beings had existed far earlier than had been suspected.

In the scientific climate of the 1850s, the possibility of "fossil man" could no longer be ignored. Two English scientists, archaeologist John Evans and geologist Joseph Prestwich, visited Boucher de Perthes' sites. They had previously visited excavations at Brixham Cave in England where stone tools and the bones of extinct animals had been discovered in sealed deposits (Grayson 1983: 185-186). Charles Lyell visited shortly afterwards. On their return, they presented papers to the British scientific community agreeing that Boucher de Perthes had demonstrated that the handaxes of Abbeville were associated with the bones of extinct animals. Lyell wrote, "The disappearance of the elephant, rhinoceros, and other genera of quarupeds now foreighn to Europe, implies . . a vast lapse of ages, separating the era in which the fossil implements were framed and that of the invasion of Gaul by the Romans" (Lyell quoted in Grayson 1983: 187).

The solution of the fossil and the stone tool enigmas revealed the vastness of human prehistory. The solution of the fossil enigma had another profound consequence. The acceptance of the organic nature of fossils lead to the recognition that plant and animal life had changed dramatically over the eons. The scientific explanation of fossils set the stage for the next transformation in the concept of prehistory, the theory of organic evolution by natural selection.

The Enigma of Changing Species

By the middle of the 19th century two ideas about the past had been generally accepted by the scientific community: uniformitarian geology and the Three-Age chronology. As a result, great empirical progress had been made in both geology and archaeology. Geologists could construct composite stratigraphic sequences over great distances using the superposition, fossils, and lithology. While individual layers in this stratigraphic column could be placed in a relative order, no accurate estimate of their age was possible. When the processes that generated these rock formations were explained in Uniformitarian terms, it became obvious that a vast amount of time was needed. The Earth had to be much older than 6,000 years.

By the middle of the 19th century, most natural scientists had accepted the long chronology of the Earth. The stage was set for the scientific acceptance of a theory of evolution. Geology had provided convincing evidence that enough time had elapsed on Earth for such development to have occurred. However, geologists could not account for the paleontological data upon which the catastrophist argument was built. Geology recorded profound and extensive changes in species composition through time, but there was no recognition of "causes now in action" that might explain these changes.

Archaeologists and other students of the human past recognized that stone tools were human artifacts from a time before written records. Together with bronze and iron tools, stone implements were part of an ancient sequence of cumulative technological development. They had also recognized that stone tools were associated in the archaeological record with the remains of extinct animals. The age of these extinct forms could not be determined but was clearly great. Therefore, acceptance of the Three-Age System also forced scholars to reject the short chronology in human prehistory and led them to see that the human prehistoric record contained a sequence of progressive cultural development. Like their brethren in geology, archaeologists were unable to explain the process that had led to such development.

Then in 1859 Charles Darwin provided a unified solution to these problems and by adding third conceptual explanatory model to the study of the past, organic evolution by natural selection of chance variation.

Charles Darwin and the Theory of Natural Selection. Charles Darwin (1809-1882), the most celebrated natural scientist of the 19th century, originally intended to be a physician. His first exposure to surgery nauseated him, so in 1827 he enrolled at Cambridge to prepare for the ministry. His decision to select natural science as his life's work came to him only after his successful tenure as naturalist in the years between 1831 and 1835 on the exploratory ship, *HMS Beagle*. It was a decision that changed both his destiny and the direction of 19th-century biology.

Darwin is known most widely for developing the theory of evolution. His name is indelibly associated with the conflict between science and religion which emerged in the 19th-century debate over his work. History gives Darwin primacy in formulating the **Theory of Natural Selection of Chance Variation** to explain the evidence of changes in species over time.

Darwin's first comprehensive account of his theory was published in 1859 in *On the Origin of Species*. In that work Darwin mustered evidence of change in modern species. This evidence consisted in large measure of domesticated species of plants and animals that had been altered through time by human selection, the animals of the volcanic islands of the Galapagos, and the varied distributions of contemporary plant and animal species on Earth.

At the outset of his book, Darwin adopted the long chronology of Earth's natural history and the uniformitarian geology propounded by his friend and colleague, Sir Charles Lyell. His experiences in an earthquake on the coast of South America and his discoveries of marine fossils at high altitudes in the Andes had convinced Darwin that the Earth's crust is not stable but is dynamic. He recognized that a dynamic earth meant that the demands placed on species were ever changing.

From these and other strands of thought, Darwin wove together his theory of natural selection in order to account for such changes. Darwin's theory rested on three observations and two deductions (McKern and McKern 1974: 27):

Observation One: All organisms reproduce more than required to replace their own numbers.

Observation Two: Despite this tendency to multiply, the number of members of a given species remains relatively constant.

Observation Three: All living organisms vary. They resemble but do not exactly duplicate their parents.

Deduction One: There occurs a universal struggle for existence, both among and within species.

Deduction Two: Individuals with some advantage have the best chance for surviving and for reproducing their own kind.

According to Darwin, natural selection results from three characteristics that all life forms share. The first characteristic is **variability**, individuals of a particular species are different in their appearance and abilities. The second characteristic is **heritability**, children tend to look like their parents. The third characteristic is **differential survival**, all individuals do not survive to have children and some individuals have more children than others. If differential survival is not completely random, but is related to heritable variations, natural selection will lead to changes in the appearance of the species. More individual organisms are produced than can survive. In the struggle for resources, mates, territories, those with greater fitness will tend to leave more descendants in the population. That differential success leads to changes in the population over time. Individuals of a species live, breed, and die. They do not evolve. Only a population can evolve as a result of changes in the proportion of its members who have a particular trait.

Darwin envisioned natural selection, and the species change that resulted from it, to be a material process governed by chance. There was no inevitable direction, intention, or goal. He saw natural selection as a myopic process, meaning one that tends towards local maximization without regard to the long-term consequences. It does not store useful variations for future adaptation. Selection operates on the present generation; it can not be influenced by future conditions. In the Darwinian model, characteristics that do not contribute to current reproductive success tend to be eliminated.

At the same time evolution is an historical process. Once extinct, a form will not reappear in the fossil record at a later time even if the environmental conditions to which it was adapted return. As a consequence, as one moves up the geological column from the bottom to the top, the fossils in successive layers differ in unique and nonrandom ways. It is the existence of this particular and historical sequence that enabled William "Strata" Smith to correlate geological strata using the fossils they contain. However, Darwin did not conclude from this order that the historical sequence has been directed inevitably towards some goal or towards higher forms of life.

This point must be emphasized because in the 19th century the term "evolution" (derived largely from the writings of Herbert Spencer) took on the connotation of **inevitable progressive change** (Freeman 1974). This notion persists in popular thinking in our own time. However, inevitable progress is **not** what Darwin meant when he talked about "chance and necessity" governing the evolution of life.

Modern Genetics and the Emergence of the Synthetic Theory of Evolution. Since Darwin's time, examples of organic evolution that seem best explained with reference to his theory of natural selection have accumulated greatly. But where do these advantages or adaptive edges come from in the first place? Darwin simply did not and could not provide an adequate mechanism for the appearance and transmission of the countless variations that are preserved or eliminated by natural selection.

One model of heredity had been proposed in 1802 by **Jean Baptiste Lamarck** (1744-1829) in his book *Research on the Organization of Life* and more fully in 1809 in *Zoological Philosophy: An Exposition with Regard to the Natural History of Animals*. In these works Lamarck recognized that species had changed over time. He concluded that an internal force drove organisms to greater and greater levels of complexity. Lamarck explained this development by asserting that the traits acquired during the lifetime of an organism are transmitted to succeeding generations. Lamarck asserted that individuals pass characteristics to their offspring which they have developed during their lives in response to the needs created by their environment. This is known as **Theory of Acquired Characteristics**. Reduced to its logical extreme, the doctrine predicts that rats whose tails are cut off during their lives should produce stump-tailed offspring! Nonetheless, in the absence of a more compelling explanation of variation, even Darwin admitted a degree of Lamarckian acquired variation into his model.

The absence of a satisfactory account of the origin of variation in living populations remained the primary defect in Darwinism throughout the 19th century and was one of the reasons for the eclipse of Darwinism and its near rejection by the scientific community later in the 19th century (Bowler 1984: 234).

The foundation of modern genetics was laid down by the Augustinian Monk and botanist, **Gregor Mendel** (1822-1884). Between 1856 and 1863, Mendel carried out a series of breeding experiments with garden peas at his monastery at Brunn in Austria. He raised some 28,000 plants and studied crosses between variants differing in one, two and three hereditary traits (Oiby 1982: 417-418). Mendel's experiments demonstrated that hereditary traits were transmitted in accord with certain laws that governed what he referred to as "particles." Mendel's **Law of Particulate Inheritance** states that while each individual organism has a pair of particles that determine its specific hereditary trait, it transmits only **one** to its offspring. The **Law of Independent Assortment** states that the particles assort independently of one another in the process of inheritance.

Mendel reported his finding in 1866, but his work caused little notice at the time. Although Mendel's work was largely ignored during his lifetime, it was rediscovered independently by three different biologists in 1900. By this time biologists had independently arrived at results and conclusions similar to those reached by Mendel over thirty years earlier. After the turn of the century, Mendel was given belated credit for his formulation of the laws of heredity.

The combination of Darwin's Theory of Natural Selection with the Mendel's Laws of Inheritance in the early twentieth century lead to the development of the **Synthetic Theory of Evolution**. The Synthetic Theory combines Darwin's natural selection with modern genetics and the laws of inheritance (Dobzhansky 1937; Huxley 1963; Simpson 1944).

Subsequent work in genetics in our century identified Mendel's particles as genes. Genes are located on chromosomes, thread-like structures within the nuclei of all living cells. Mendel's Law of Independent Assortment is now known to apply to chromosomes, not to genes. The normal human cell contains 46 individual chromosomes which are arranged in 23 separate pairs. During **meiosis**, chromosome pairs segregate and assort independently to produce sex cells or gametes which contain only 23 chromosomes each. When the egg and sperm are united in fertilization, the two sex cells become one and thereby restore to 46 the number of chromosomes in the cells of the resulting embryo. In this way, the genetic characteristics of each parent are transmitted singly to offspring, one chromosome from each pair from each parent, in chance combinations. The random assortment of these chromosomes makes a large number of combinations possible.

The principal units of heredity are genes which are arranged or located at specific loci on chromosomes. Genes carry hereditary information and regulate the formation of cellular products (proteins). Genes can have alternate forms at a locus called **alleles**, that is, variants of a gene which occupy the same location in corresponding chromosomes. Individuals with different alleles generally do not show blended characteristics. For a given trait, one allele is usually dominant and is expressed while the other is recessive and is not expressed.

In recent years, biochemical research has revealed the chemical nature of the principle component of genes: deoxyribonucleic acid or DNA. DNA consists of a long coiled chain of paired nucleotides. Sequences of nucleotides encode the information needed to construct a new organism. The code is the same for all life forms on Earth.

Thus, it is now clear that the genes are not particles but segments of the DNA chain. Genes are generally stable, consequently the coded information they store in stays the same. Sometimes a change takes place in their sequence of nucleotides, however, and this change results in a **mutation**. Thus the chemical understanding of genetics has provided insights into the mechanisms for the appearance and transmission of the countless variations that are preserved or eliminated by natural selection in the classical Darwinian model.

We now largely understand the source of the variation between individuals which Darwin saw as being eliminated or gradually increased in a species over time by natural selection. As Stebbins and Ayala

(1984: 72) put it, "the science of genetics made it possible to identify the determinants of the traits on which natural selection acts as genes: heritable units of information governing structure, development and function."

Changes within a population come about largely as the result of genetic mutation that is, lasting alterations that arise as basically random errors in the sequence of nucleotides in molecules of DNA. Mutations represent the failure of a gene to duplicate itself. They are copying errors that sometimes create new alleles.

Evidence of such change comes from micro-evolutionary studies of fruit flies and other simple organisms as well as the analysis of family pedigrees, breeder's records and so forth. Such studies show that over time many mutations occur. However, most of these traits do not contribute to the survival of the individual organism. As a result, natural selection operates to eliminate them.

The classic Darwinian model of natural selection predicts that an equilibrium will eventually be reached in which mutation replaces genes at the same rate selection removes them. As a consequence of the operation of chance and necessity, "the population climbs a fitness gradient until it reaches a point from which all further movement can be downward only and here it ceases to change" (Elster 1983: 50).

Such equilibrium is upset however, when the selective pressures of the environment change. New environmental conditions generate new selective pressures that favor different arrays of traits. Mutations systematically eliminated in the old equilibrium state may now confer advantage on organisms. The geological record contains abundant evidence of environmental change in the past and this change was no doubt a primary stimulus to biological evolution.

However, natural selection of favorable mutations is not the only way that gene frequency can change within a breeding population. Other mechanisms include: **genetic drift** and **gene flow** or migration. Genetic drift refers to the tendency of the proportions of genes in each generation to vary by chance. Small, isolated populations are particularly likely to lose certain alleles simply due to sampling error over time. Gene flow refers to the introduction of new alleles into a breeding population due to interbreeding with related, but formerly separated, sub-species or varieties.

For example, the periodic creation of land bridges following the lowering of sea level during stages of continental glaciation in the late Cenozoic era facilitated gene flow between formerly isolated terrestrial fauna. Conversely, the periodic submergence of these bridges by the higher sea levels of subsequent interglacial stages divided terrestrial breeding populations and increased the likelihood that, over time, the genetic makeup of the now-independent populations would come to differ due to genetic drift. Connection and severance has the opposite effect on marine fauna. That is, "when a land connection forms, the newly united land faunas tend to converge and the newly divided marine faunas tend to diverge, whereas the breaking of a land connection gives rise to the opposite effect" (Hallam 1972: 62).

In human history, the expansion of Europe has contributed immensely to human gene flow and has resulted in the creation of whole new "races" including the "Neo-Hawaiian" (in which is combined Euro-American, Oriental, and Polynesian genetic material) and the American Negro (in which is combined African, European and American Indian genetic material) and others. The combination of genetics and natural selection into the synthetic theory of evolution has provided modern evolutionary biology with a unified and powerful theory that is a long way from the simple Darwinism of the 19th century.

Evolution and Scientific Prehistory. As noted above, by the middle of the 19th century, students of the human past had come to accept the view that human technology had passed through three stages: stone to bronze to iron. They also recognized that stone tools were associated with the remains of extinct animals of great antiquity. They had thus been forced to conclude that human prehistory represented an immensely long period, and was characterized by the development of increasingly complex culture. However, they were unable to explain the process that led to such cultural development.

From its very inception, Darwinism had a powerful influence on archaeology. In large part this is because Darwin's model of evolution by natural selection fit both the theory and data of prehistory available at mid-century. More importantly, however, Darwinism provided an explanation of human prehistory that made sense at an even higher level of generality.

Darwin did not exempt humankind from the sequence of evolution. In his later book, *The Descent of Man*, he asserted that humans and the apes were related by a common, as yet undiscovered, ancestral fossil form. This common ancestor came to be known by the unfortunate term of the "missing link." The scientific acceptance of Darwin's evolutionary model initiated a search for the fossil ancestors of humankind in the late 19th-century which, in short order, turned up such seemingly primitive and exotic forms of early humankind as Neanderthal in Germany, Pithecanthropus in Java and Sinanthropus in China. The quest for human fossil ancestry which began after 1859 continues unabated into our own time. The search for fossil hominids also uncovered the changes in stone tool technology. It became clear that increasingly complex forms of tool manufacture emerged in tandem with the evolving forms of fossil hominids. Hominid culture became more complex as the human capacity for culture increased.

In a related manner, the Three-Age system came to be seen as a **cultural evolutionary** sequence analogous to the sequences of evolutionary change witnessed in biology. Although the development of the doctrine of social and cultural evolutionism owes as much to Herbert Spencer and Lewis Henry Morgan as it does to Darwin, the doctrine of cultural evolution gained credibility and acceptance largely as a consequence of the success of Darwinism in biology. By the end of the century, the solutions to the fossil enigma and the stone tool enigma had been subsumed at a higher level by the doctrine of evolution. Prehistory had its first **unified scientific theory**.

Major Points Raised in Chapter Three

1. This chapter begins by making the distinction between the two categories of time used by students of the past: relative and absolute time. Their are two primary methods of organizing ancient material in proper relative order: stratigraphic analysis based on the **Law of Superposition** and **seriation**.

2. **Seriation** is the arrangement of archaeological materials in a presumed chronological order. In archaeological seriation, stylistic, functional or technological variation are used to date objects of human manufacture. Three kinds of seriation, sequence dating, item seriation, and frequency seriation, are distinguished.

3. Worsaae's **Law of Association** is defined and its utility in testing dates based on seriation is illustrated using the Three Age System.

4. The section closes with an examination of the cross-dating of relative sequences and the concepts of ***terminus post quem*** and ***terminus ante quem***.

5. An example of the strategy of relative dating using stratigraphy and seriation on a hypothetical site in a New England river valley is provided.

CHAPTER 3

ESTABLISHING RELATIVE SEQUENCES OF ANCIENT EVENTS

Introduction

The fundamental task of the archaeologist is the measurement of time and the ordering of events and objects into their proper sequence. Archaeologists generally distinguish between two kinds of time and time sequences: **relative** and **absolute** or **chronometric** time.

Absolute or chronometric dating assigns a date in years (or some other system of time units) to historic or prehistoric events, objects, or sites. For example, the assertion that Columbus discovered America in 1492 is a statement of absolute time. Be aware that the term "absolute" does not mean "absolutely correct." The phrase would have been a statement of absolute time even if the date were listed as 1493 or 1954. "Absolute" merely refers to the fact that the time has been measured in years. The measurement of absolute time is discussed in the next chapter.

In contrast, a relative chronology places events and objects in order from oldest to youngest. No estimate of age in years is made. A relative date is simply a statement that event A occurred before B and event B occurred before event C. For example, the assertion "American Indians arrived in the New World before the Europeans and the Europeans arrived here before the Chinese" is such an example. A relative date is merely an arrangement of events in their proper sequence of appearance in time, A before B before C. There is nothing in such statements to indicate that A occurred at or before 12,000 B.C., B after A.D. 1492 and C about A.D. 1850. The three methods of creating a relative chronology are **stratigraphy**, **seriation**, and **cross-dating**.

The concept of relative ordering of events or object in their proper sequence in time is a simple one. However, like so many things in archaeology it is not always easy to do in practice. Nonetheless, establishing a sequence of events is the most basic activity in the scientific practice of archaeology.

Simply to excavate or collect ancient objects without seeking to order them in their proper chronological sequence is merely treasure hunting. The passion for ancient objects, present in humankind throughout history, did not produce the science of archaeology. Only when that passion was combined with a desire to place ancient objects and events in proper chronological order did the scientific study of the past emerge. It is the geologists of the 17th, 18th and early 19th century, rather than the antiquarians and collectors of the same period, who are among the true founders of scientific archaeology. **Antiquarians** were satisfied to label an artifact and leave it at that; the coin was Roman, the ruin "a temple of the Druids," the stone axe left by the Danish Vikings (Daniels 1962: 23). In contrast, geologists developed and applied the fundamental techniques of relative dating, **stratigraphy** and the **Law of Superposition.** In a like manner, Christian Thomsen, seeking order in the apparent chaos of the Danish archaeological record, applied the logical principles of the **Three-Age System** to that record and thereby developed the second fundamental technique of relative dating, **seriation**.

Stratigraphy and the Law of Superposition

Stratigraphy may be defined in two ways. The word literally means "the study of layers." However, when geologists and archaeologists speak of "stratigraphy" at a given site or exposure, they are referring to the specific order of layers evident there. As we have already noted, a scientific understanding of stratigraphy was critical to the emergence of scientific prehistory and geology.

The **Law of Superposition** states that, for any pile of deposits, the order of succession from bottom to top is the order of deposition. That is, the position of a layer in **space** relative to other layers represents the order of its deposition in **time**. The law applies to when the layer was deposited, not necessarily to the age of individual artifacts in the deposit. The nineteenth century controversy over the association of stone tools and extinct animals could not be resolved until geologists and archaeologists could confirm that the extinct animal bones had not eroded from older sediments and been subsequently mixed with younger stone tools. Erosion and human activities such as excavation (for storage pits or graves) can mix materials that are not contemporaneous.

Before going further it is necessary to define the general term "layer." A **layer** is a general term for a single, discrete package of material which has been deposited in a given location. The material may include mixtures of rock, ice, sediment, soil, and cultural material. A layer is a depositional unit in a stratigraphic exposure that has a distinct upper and lower boundary. Layers are generally oriented parallel to the surface on or against which they were deposited (**Law of Original Horizontality**). A layer may be of any size and thickness.

Layers are formed by the action of wind, water, ice, and gravity. They are also formed through chemical precipitation. In addition, layers can be created by human activities, such as the discard of debris or the building of mounds, cities, or landfills. Whether of natural or human origin, all depositional units can be ordered by the Law of Superposition.

Primary and secondary deposition. The contents of a given layer were brought together during the formation of that layer. However, some of the contents may have been eroded and re-deposited many times before being incorporated into the layer in which they are found. Some of the contents of a layer may be described as being in **primary context** if, once deposited, the layer has remained essentially undisturbed. Conversely, some of the contents are said to be in **secondary context** if they have been disturbed and then re-deposited at least once. For example, the remains of an animal that died in a dry cave and whose bones were gradually incorporated into the layer accumulating there would be in primary context; those of an animal that died on a river bank and whose bones were washed downstream, disarticulated, and mixed together with the bones of other animals in an alluvial (water-lain) deposit, would be in secondary context. The terms primary and secondary context apply to specific contents within the deposit, not to the deposit as a whole.

An example of cultural stratigraphy. As an example of the use of stratigraphic analysis in the construction, and correction, of a cultural succession, let us examine the work done by Michael Waters in order to refine understanding of the prehistory of southeastern Arizona and the American Southwest. In the most widely accepted view of the North American cultural succession, the continent was first inhabited by people during the late Pleistocene epoch. The people were hunters and gatherers who adapted to the cooler, moister climate of the epoch and hunted the great game animals or "megafauna" like mammoth, bison, horse, and camel that were present in North America at that time. Their stone tools consisted of flaked stone artifacts including bifacially-worked projectile points. At the close of the Pleistocene around 10,000 years ago, much of the megafauna became extinct. In the millennia that followed, subsistence systems based on the collecting of a wide range of wild foodstuffs developed. Plant collecting occupied a more prominent position in the economies of the time and the stone industries came to include groundstone tools and milling stones for seed grinding. However, in 1926 Byron Cummings excavated a mammoth skull that appeared to be associated with milling stones at a site in the Whitewater Draw. Iin the 1930s, work by others seemed to confirm the discovery. The association of groundstone artifacts with extinct forms of late Pleistocene megafauna in the Whitewater Draw seemed to indicate that the prehistory of southeastern Arizona was quite different from the rest of the Southwest and North America at large.

On the basis of Cummings' and others work, the sites were used to define the Cochise Culture and divide it into three phases. The earliest of these, the Sulphur Spring phase, was described as a late Pleistocene megafaunal hunting culture with a stone tool technology that included milling stones, hammerstones and other groundstone artifacts but no projectile points. The date of the Sulphur Spring phase was estimated to be between about 12,500 and 11,000 B.P. (Before Present).

The association of milling stones and extinct fauna in the Sulphur Spring phase seemed to indicate that plant processing was present during the late Pleistocene. However, later work consistently failed to turn up associations between groundstone artifacts and extinct Pleistocene fauna. The Sulphur Spring phase began to appear increasingly anomalous.

In the early 1980s Michael Waters (1986a: 251-256) sought to resolve the problems with the Sulphur Springs phase. Waters reopened excavations at four of the original six site locations in the Whitewater Draw. His excavations consisted of long trenches cut across the sites in order to fully expose their stratigraphy in cross-sectional profiles.

Waters (1986b: 5) found that, although no single locality in the draw contained a complete section of the stratigraphy of the last 15,000 years, he was able to compile a composite section by cross-correlation of the partial exposures uncovered in his excavations. This composite profile (Figure 3-1) includes 16 major geological stratigraphic units labeled C through P from oldest to youngest. The internal subdivisions of these stratigraphic units where designated by numbers and lower case letters.

Waters' excavation of the four archaeological sites in the draw resulted in the recovery of Sulphur Spring artifacts, the first human burial reported for the phase, and the bones of extinct megafauna. Charcoal, used for absolute dating by radiocarbon, was also recovered. All of the Sulphur Spring artifacts recovered by Waters from his excavation trenches came from stratigraphic units Da, Db, and Dc (see the composite geological profile). These stratigraphic units also produced the remains of extinct megafauna, but all of the megafauna bones recovered were disarticulated and none showed evidence of butchering marks. Further, most of these bones were found scattered and isolated in their stratigraphic units.

At the single site where the bones of extinct megafauna were found in direct association with Sulphur Springs phase artifacts, both the bones and the artifacts site appeared to be in **secondary** (disturbed) context. That is, at sometime in the past both the bones and the tools had been eroded and moved by water action from their original locations to their current positions. Waters found no locations where megafauna bones and Sulphur Springs phase artifacts were associated with one another in a **primary context**. Waters' work also confirmed that the stratigraphic units found below the D units **do contain** articulated (more-or-less whole) remains of extinct megafauna. These articulated remains were apparently in primary archaeological context. The stratigraphic units below unit D, however, **do not appear to contain** Sulphur Spring artifacts; their deposition was therefore assumed to have predated the beginnings of the Sulphur Spring phase.

Since the co-occurrence of Sulphur Spring artifacts and megafauna bones in unit D was strictly in secondary context, Waters concluded that these materials had probably been brought together by the stream action that eroded the sites of the original, independent primary deposition sites of the bones and artifacts. On the basis of this stratigraphic evidence, therefore, Waters concluded that Sulphur Spring phase artifacts from the Whitewater Draw sites were **not contemporary** with the extinct megafauna recovered there.

Finally, based on the radiocarbon dates from his excavation, Waters concluded that the late Pleistocene dates for the Sulphur Spring phase were 2,000 to 2,500 years too early. His data indicated the phase actually appeared in the early millennia of the Holocene epoch sometime between ca. 10,000 and 8,000 B.P.

By showing that the anomalous Whitewater Draw sequence did not correctly represent the cultural sequence of the Southwest as a whole, Waters' stratigraphic analysis supports the theory that the emergence of a grinding technology in North America was an early Holocene event that followed the late Pleistocene extinction of the megafauna, and reflected the increasingly importance of plant gathering in the post-Pleistocene subsistence systems caused by those extinctions.

Seriation

The second major technique used by archaeologists to establish relative sequences of events is seriation. The word "seriation" is derived from the Latin word seriatim, meaning "in sequence." As defined by Rowe (1963), **seriation** is "the arrangement of archaeological materials in a presumed chronological order on the basis of some logical principle other than superposition."

Before objects can be arranged in a sequence, they must classified systematically into **types**. A type is a group of objects that share the same attributes or traits. A system or set of types is referred to as a **typology**. In general, archaeological types are defined using **stylistic**, **functional**, or **technological** attributes or some combination of all three. Stylistic classifications sort artifacts into types using such attributes as form or shape, size, decoration, and color; functional classifications sort objects into types on the basis of their use or purpose; and technological classification use criteria like raw material and attributes that reflect the technique of manufacture.

Types defined strictly on the basis of style are the most sensitive indicators of change through time; functional types are most useful in monitoring the changing nature of human activities; and technological types often mirror evolutionary or developmental change most clearly.

Individual objects can be sequence dated by arranging them in a presumed chronological order by means of some logical principle. According to Rowe (1963: 324-330) the logical principles used in arriving at such chronological arrangements fall into two categories: (1) those that rely on **similarity** or resemblance between objects as an indication of their proximity to or distance from one another in time and (2) those that assume **evolutionary** change within the development of a type of artifact. Perhaps the two most common evolutionary assumptions are that: (1) certain technical developments necessarily precede others, or (2) technical development proceeds from simple to more complex.

Seriation can be used to order individual artifacts or assemblages (collections of artifacts that are associated with one another). Seriation of individual artifacts is called **"sequence dating"** and was first developed by Sir Flinders Petrie to arrange Egyptian burial urns from earliest to latest based on the orderly changes in decorative styles through time. A common assumption about style is that decoration moves from functional to non-functional. Flinders Petrie's seriation of Egyptian burial urns was based on this assumption. He presumed that urns with actual working handles were the oldest, those with pictures of handles merely painted on them were next oldest, and urns with mere painted lines or symbols where the handles should be were the most recent (Petrie 1901).

Roman terracotta lamps can also be sequence dated using the principle of similarity. Throughout Antiquity, the oil lamp consisted of a chamber to hold the fuel (normally olive oil), and a nozzle to hold the wick. Most lamps consisted of a single wick but a few used multiple wicks. The technical design of terracotta oil lamps remained essentially the same throughout the 800 years of the Roman period (ca. 300 B.C. to A.D. 476) but the style of these lamps changed continuously. Thus, although the basic category of "oil lamp" is a functional class, distinctions between lamps must be made largely on stylistic grounds (after King 1983: 199-204).

Within Rome proper, the earliest recovered terracotta lamps date to the 3rd and to the 2nd centuries before Christ. These lamps were made on potter's wheels and then fired with a black glaze. They show little decoration or elaboration beyond the requirements of their purpose. Styles began to change in the late 2nd century B.C. until, by the Late Republican period of the 1st Century A.D., black glazed lamps had ceased to be popular; instead, lamps were finished with a red slip. The pointed nozzles of earlier times had given way to a blunt style and a variety of decorative features had been added, including sidelugs, applied handles, and raised point decorations. During the 1st Century A.D. a technological change in the manufacture of oil lamps occurred: they were made in molds rather than on the wheel. After this time, the differences between lamps from one period to the next were wholly stylistic in nature. Terracotta lamps continued to be made in molds during the early years of the Empire in the 2nd Century A.D., but the period witnessed the gradual development of a variety of new lamp shapes adapted from lamp styles of the Late Republican period. Most important was the "volute-lamp" characterized by two curved ornaments flanking the nozzle. The tops of the oil chambers on volute lamps commonly bore images of gods, animals, and mythological scenes shown in relief. During the 3rd century A.D., a new form, the "globular lamp" with rows of raised points replaced previous types. The globular lamp in turn gave way to a distinctive elongated lamp of Tunisian-style in the late 4th Century A.D.

The sequence of stylistic change in Roman lamps is very useful to Classical archaeologists. The lamps were objects of everyday use, were widely traded throughout the Roman Empire, and were sometimes buried in tombs and graves. Consequently, they occur in great abundance in the archaeological sites of the region. Since the sequence of stylistic change in the lamps through time is well-known, archaeologists are able to use the different styles of Roman lamps to relatively date the different occupations from which the lamps are recovered as well as the other objects associated with the strata.

Sequence dating is not applicable only to ancient artifacts. The principles that underlie it apply to material culture of all periods. Artifacts of all kinds from stone tools to pots, buildings, and other materials can be placed in a sequence by this means.

Another assumption about style that is commonly used in seriation is the view that decoration develops over time from directly representational to conventionalized or symbolic images. Deetz and Dethlefsen found that tombstone carvings in Colonial period cemeteries in the Boston area changed in this fashion. Tombstones erected during the 60-year period from 1710 to 1770 were generally decorated with grinning, blank-eyed skull with wings called the "winged death's head" motif. The earliest versions of this motif are complex and realistic; no one who sees them is in any doubt as to what they represent. However, as time passed, representations of the motif become less elaborate and more stylized. This change began with:

> the reduction of the lower part of the skull's face and the shrinking and final disappearance of the skull's teeth. A heart shaped design, which appears before the teeth finally vanish, becomes a kind of mouth. Following this change, the feather ends on every other row (of the wings) are reversed in direction producing a series of wavy lines crossed by arcs surrounding the head. When the arcs are omitted, the wavy lines soon take on the aspect of hair, and this hair goes through a sequence, marked by its becoming more curled and more extensive. The end result is final simplification of the elaborate hair style" (Deetz 1967:32-33).

Figure 3-2. Gradual stylistic change in grave stone motifs from the detailed Death's Head design to a more stylized version (after Deetz and Dethlefsen 1967).

The 1760s versions of the motif are so stylized and simplified, that only someone familiar with tombstone design conventions would recognize them as winged death's head at all (Figure 3-2). By the end of the decade the motif had disappeared from the graveyards altogether.

Technological change can also be used in sequence dating using the assumption that in technology there is a general movement from **simple** to the increasingly more **complex**. The icon of 20th century American culture, the automobile, provides an excellent illustration of the movement of technology from simple to more complex. The complexity of the automobile has increased profoundly during the last 90 years. Our familiarity with the sequence of those changes allows us to arrange cars of different periods relative to one another. Using this evolutionary assumption, the average American would have little trouble arranging say, the exhibits in the "classic" car museum assembled at Harold's Club in Reno, Nevada, into a chronological sequence which is generally correct.

The initial problem of such a seriation would be deciding on a suitable measure of technical complexity. After all, advanced automotive technical innovations generally do not appear at the same time throughout the auto industry. Further, radical change is more likely to appear first in sports or racing cars, next in pricey luxury models, and only much later (or not at all) in less expensive autos. Seriations based on the appearance of these innovations would consistently place "family cars" earlier in the chronological sequence than they in fact belong. Conversely, cars that were in some sense ahead of their time, such as the Cord of the 1930s, or stylistically anomalous, such as the legendary Edsel of the late 1950s, would probably be placed too late in our sequence.

This problem could be solved by developing a measure of complexity that is insensitive to difference in model function or price. One possibility would be weight per linear foot. All automobiles, regardless of make, model, or nation of manufacture, have become lighter through time due to the increasing use of new materials such as plastic, aluminum, and various lighter metal alloys. Use of these materials reflects the increasing technical complexity of the automobile. Measuring this complexity as weight per linear foot would be to use a purely technological attribute in classification that would over-ride differences between automobiles unrelated to their time of manufacture, such as different cost, function, style, or status associations.

A second form of seriation, **item seriation**, orders archaeological assemblages by the presence or absence of a variety of different artifact types. Since different artifacts are used for different lengths of time, it is often possible to arrange graves and other archaeological deposits into a sequence based on the assumption that each artifact type is in use for a limited time. It is also necessary to assume that an artifact type cannot return later in the chronological sequence. While a later artifact type may resemble an earlier one, it will be distinctly different in some way.

The Three-Age system described in the preceding chapter provides an example of item seriation based on technological development. In technology, certain technical developments necessarily precede others. The development of stone tools preceded that of bronze, bronze that of iron. Archaeological deposits containing stone tools are considered to be earlier in time than those that contain stone and bronze tools. Deposits containing stone, bronze, and iron tools are younger than those containing only stone and bronze.

The technical reasons for the precedence of stone tools over metal ones should be obvious: stone is common and it needs no special treatment to transform it into a raw material for tool manufacture. To make a rudimentary tool from a stone, all one need do is strike the stone with another stone until the desired shape is achieved. Not surprisingly, the earliest metal tools appear to have been made in a similarly simple fashion by "cold-hammering" native copper into a desired shape. Later native copper tools appear to have been "annealled" that is, heated in a low temperature fire and then hammered. However, to make effective tools of bronze or iron, one must convert ores into metals by "smelting" them at high temperatures. Mere campfires do not generate sufficient heat to reach the melting point of copper at 1085° Celsius. The techniques necessary to generate and control fire at sufficiently high temperatures for smelting do not appear to have been mastered until sometime after about 4000 B.C., when peoples in the region between Anatolia and the Caspian Sea first began reducing native copper or copper ore in the presence of charcoal (Hawkes 1973: 108).

Perhaps the first experiments with smelting were done in pottery kilns, a parallel technology that was well developed at this time. Only later were specialized smelting kilns equipped with stack and bellows built. The combination of tin ore with copper was probably accidental at first but experiment and observation would have quickly led to its intentional inclusion in the smelting process. Further trial and error would have led early metallurgists to the discovery that a ratio or 5 to 10 percent tin to 90 to 95 percent copper ore produces an ideal bronze alloy. But why must tools of bronze precede tools of iron? Again, the basic reason is smelting temperature. To create bronze one must smelt copper and tin (or some other ore) at temperatures around 1,085° to 1,100° Celsius. Although iron ore is smelted in a similar manner, the process requires temperatures above 1,537° Celsius.

It is clear that in order to smelt iron ore, one must first have developed the capacity, skill and knowledge needed to smelt the copper, tin and other ores used in bronze making. In order to smelt iron ore, the techniques needed to generate higher levels of heat in the smelting furnace had to be developed. The sequence of technical development of such furnaces seems to have begun with clay lined pits or ceramic vessels sunk in the ground. Burning coals and ore were probably mixed together in these receptacles. Such receptacles were probably eventually replaced by crucibles in which the ores were separated from coals. Sometime after about 2500 B.C., shaft furnace with draft supplied by blowing on reed or pumping bellows capable of reaching even higher temperatures were developed.

However, pure iron is inferior to bronze in a number of ways: it corrodes quickly, it was less easily cast in molds due to the higher temperatures involved, it was more brittle and softer than bronze. What Thomsen and other archaeologists called "iron" was actually a crude form of steel. Steel consists of iron into which carbon has diffused. Such carbon would have initially been introduced into the crude "bloomery" iron in the smelting process and during the "hammering" of the iron into tools. Overtime, metallurgists must have become aware of the superior properties which iron obtained through "steeling" and done it intentionally (Maddin et. al. 1977:122-131).

Thus, the development of iron tools conforms nicely to our evolutionary assumption that, in technology, certain technical developments must, of necessity, precede others.

A third kind of seriation, **frequency seriation**, uses difference in the proportions of artifact types to arrange archaeological assemblages relative to one another. The basic assumption of the technique is that new styles or technologies are introduced gradually. People do not discard all of their old possessions just because new styles or new, improved models have been introduced. Therefore, changes in the proportions of the old and the new types can be used to order assemblages.

Frequency seriation begins with the assumption that any given artifact style or decorative motif follows the same general pattern over time. Immediately after its invention, style "A" is but a minor element in the stylistic repertoire of the culture and its frequency of occurrence relative to contemporary styles B and C is low. However, if style A achieves acceptance, it grows in popularity and frequency of occurrence over time at the expense of B and C. Eventually, successive style D emerges, grows in popularity and begins to replace A in turn. Type A declines and, at the end of its popularity, once again represents a small percentage of the artifactual repertoire at the site. Eventually it disappears altogether. Archaeologists assume that the popularity of a particular style of artifact during a given period in the past was roughly proportional to the frequency of its occurrence in the archaeological sites that date to that period. Thus, the history of the appearance, growth in popularity, decline, and disappearance of the artifact is reflected in the changing frequency of occurrence of the type in the archaeological record. Archaeologists plot the frequency of a given artifact type against time and refer to the resulting line as a **battleship curve**.

Knowledge of the frequency of occurrence of the artifact types A, B, C, and D through time provides the archaeologist with a means of seriating the sequential occupations of undated archaeological sites. Archaeological sites contain the material remains of human activities that took place there at different periods of time. These discrete groupings of remains constitute separate "occupations" of the site. Since these occupations took place at different times, they contain different frequencies of the four artifact types. The occupations from the different archaeological sites can be arranged into a relative chronological sequence based on the frequencies of occurrence of the four types they contain. Although comparisons of artifact assemblages from different occupations are often done by simple inspection, a number of increasingly sophisticated mathematical techniques have been developed for manipulating frequency matrices and measuring the degrees of assemblage or dissimilarity in numerical terms.

Of course, cultural change is not always gradual. Stylistic change is affected by contact with other groups and trade with or conquest by foreign peoples can often radically alter the orderly change of indigenous styles. Since contact as well as isolation, chaos as well as order has characterized human prehistory and history, archaeologists cannot always assume that styles have followed smooth trajectories of development through time. For example, American Indian pottery, which appears to have changed in a generally gradual manner in Pre-Columbian times, was fundamentally transformed through exposure to the foreign forms and ideas introduced with the arrival of the Europeans.

Alternatively, older cultural forms are sometimes deliberately resurrected after long periods of neglect. For example, Italian artists and architects of the Renaissance consciously imitated the styles of the Classical Romans and Greeks. In a like manner, modern Hopi Indian potters have utilized the ancient pottery designs and forms of their ancestors recovered from archaeological excavations in the American Southwest. Archaeologists must be aware of such possibilities and attempt to validate the presumed sequence that results from the seriation against stratigraphic data where possible.

Do artifact types follow battleship curve trajectories throughout their histories and that the changing patterns of the battleship curves of many types of artifacts relative to one another may legitimately be used as a means of dating? To find out, two archaeologists, James Deetz and Edwin Dethlefsen mentioned earlier, undertook an imaginative study of tombstone decoration in Colonial New England grave yards. They used tombstones because these artifacts are both decorated and dated. Thus, the timing of changes in tombstone decoration style could be precisely determined. Deetz and Dethlefsen identified and plotted the frequency of appearance of three styles of gravestone decorations in eastern Massachusetts between 1720 and 1820: the "winged death's head" motif, the "cherub" motif, and the "urn and willow" motif. 2.

Deetz and Dethlefsen found that the the cherub and the urn-and-willow motifs, like the winged death's head discussed earlier, changed from more to less elaborate and from realistic to highly stylized as time passed. However, even more interesting was the changing frequency of occurrence of the three types of designs relative to one another. According to them, the winged death's head was the earliest motif in the cemeteries but:

> Sometime during the 18th century--the time varies according to location, the grim death's head designs are replaced—more-or-less quickly, by winged cherubs. This design also goes through a gradual simplification of form with time. By the late 1700s or early 1800s, again depending on where you are observing, the cherubs are replaced by stones decorated with a willow tree overhanging a pedestaled urn (Deetz and Dethlefsen 1967:29).

When the frequency of occurrence of these three motifs in the graveyards of eastern Massachusetts were plotted on a graph, their popularity careers formed three sequentially-overlaping battleship curves. The frequency of occurrence of the three motifs in the Stoneham, Massachusetts cemetery in north Boston illustrates these sequential curves nicely (See Figure 3-3). In the years between 1760 and 1770, fully 70 percent of the tombstones were decorated with death heads, 20 percent with cherubs, and a spare 10 percent with the newly appearing urn and willow. In

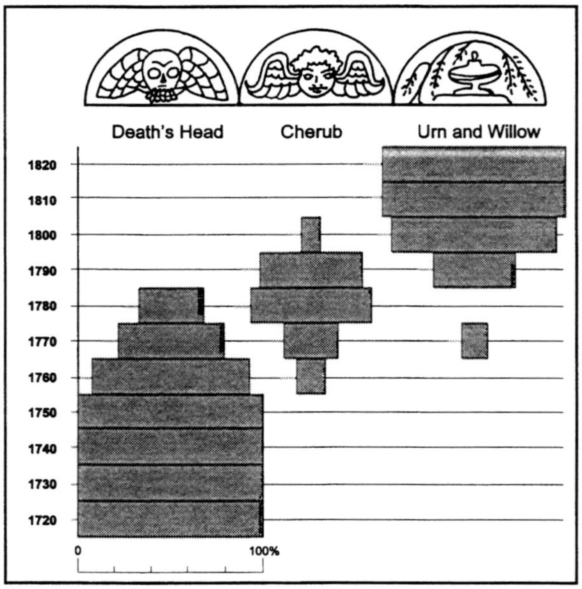

Figure 3-3. Changing popularity of gravestone styles in Stoneham, Massachusetts. The Cherub shows the full "battleship curve." After its introduction in 1760, its popularity increases steadily until 1780 after which its popularity declines (after Deetz and Dethlefsen 1967).

the decades that followed, however, the percentages changed until, by 1820, fully 100 percent of the tombstones displayed the urn and willow motif and the other two had completely fallen out of favor.

Thus, Deetz and Dethlefsen's controlled test revealed that (1) when the frequency of occurrence of Colonial tombstone design motifs are plotted over time, they do in fact form battleship curve-shaped distributions, and (2) in the Boston area, cemeteries of the same age showed essentially the same patterns of design motif appearance and disappearance from decade to decade. Thus, their work seemed to vindicate the essential assumption of similarity seriation that "like fits to like" because the extent of changes in style in artifacts is a function of time and to indicate that the dating of sites by reference to the frequency of artifact types is valid.

However, even though cemeteries in the Boston area showed similar design frequencies, we could not conclude that any New England Colonial cemetery in use between 1760 and 1770 would show a 70-20-10 percentage distribution of the three design motifs. Deetz and Dethlefsen's work in the Colonial-era cemeteries outside of Boston indicated that this was simply not the case.

The cemeteries beyond Boston indeed contained the three design motifs found within that city and exhibited changes in design frequencies which formed battleship curves over time. However, the pace of the appearance and disappearance of these design motifs in the tiny rural hamlets beyond Boston was slower by a decade or more than those occurring in the city itself. Geologists would say that the pace of design change was **time transgressive**, that is, it occurred at different times in different places. The authors concluded that, in addition to time, design change was a function of space (distance from Boston, the source of the innovation), socio-economic class (the willingness of wealthy urban Bostonians to change style and the reluctance of poorer, rural farmers to do so), and ideology (design changes relating to the changes in religious ideology that occurred in New England Protestantism in the 18th and 19th centuries).

Deetz and Dethlefsen offer archaeologists a cautionary tale. Although knowledge of the changing frequency of different artifact types through time can legitimately be used to order the occupations of undated sites relative to one another, the nature and pace of **stylistic change in artifacts can be due to variables other than time**.

For example, machine-made ceramic dishware is an important part of the archaeological record of the 18th and 19th centuries in North America. Changes in various types of transfer-printed pearlwares and whitewares make it possible to arrange archaeological sites from this period. According to Price (1979: 19), in "transfer printing," pottery makers used inked waxed paper to transfer colored designs from an engraved copper plate to the surface of a ceramic vessel. The method theoretically enabled manufacturers to decorate their wares with any design or color they chose, but in fact, transfer ware exhibits a gradual and sequential pattern of decorative change over time. As to color, the overwhelming majority of transfer ware manufactured before 1820 was blue, although black and sepia were sometimes used. During the latter part of the 1820s red, purple, lavender, green, brown, and brighter and lighter blues began to appear on transfer ware. Yet standard blue remained the dominant color well into the 1830s. Then, in the decade of the 1840s, multicolored printing came into use and rapidly gained popularity at the expense of standard blue.

The designs on transfer ware also changed gradually. Floral and geometric motifs and historical scenes were favored during all periods. However, the transfer designs of the early part of the century tended to fill the entire surface of the vessels, but over time, decorated and undecorated space came to be more equal. By the last half of the 19th century, undecorated predominated over decorated space on transfer ware vessels. Using these changes archaeologists can order historical sites by the proportion of the different kinds of transfer wares they contain.

Validating chronological arrangements. No

matter how precise the types definition nor logical the principles by which they are ordered, a relative chronological sequence remains hypothetical until it is tested against the actual order of occurrence of the objects in the archaeological record.

The best way to validate a sequence is through chronometric dating of the deposits in which the assemblages or artifacts are found. Chronometric dates provide an age estimate for the deposit. A series of chronometric dates can provide substantial support for the sequence. The various methods of chronometric dating are discussed in the next chapter.

The next best way to validate a sequence is through stratigraphic analysis. If the sequence is valid, later artifact styles will be found in strata that overly those containing earlier styles. While it would be rare to find a site with enough strata to verify the entire sequence, even two or three strata can provide support for the arrangement. Multiple stratigraphic sequences can be used to provide further support.

A third way to test a sequence is to use Worsaae's **Law of Association**. As you recall Worsaae argued that the objects found in a burial were usually all in use at time of burial. Further, these artifacts were likely made within 25-50 years prior to their interment in the grave. Of course, some exceptions to this law can be imagined. In some cultures people are re-buried many years after their deaths or the grave offerings are renewed. In other cases, the contents of graves have been disturbed by subsequent interments or looting. Despite these exceptions, archaeologists have been persuaded by field experience that Worsaae's law is generally sound.

For example, in the eastern Mediterranean, oil lamps are often recovered in Roman period graves and tombs. Let us suppose for a moment that the sequence of Roman oil lamp styles had been placed in chronological order strictly by means of seriation. The sequence would run from the earliest style A, the wheel-made, black glazed lamps, through B, mold-made, volute-lamp; to the latest style, C, the elongated, Tunisian-style lamp.

If the order is correct and the three lamp types are sequential in their appearance, then examples of the three types should either consistently be recovered in different graves; or, if sometimes found together, the associated types should be from adjacent periods, that is, type A only with type B or B only with C. We should never recover type A in an undisturbed grave containing type C lamps. If lamps of types A and C are in fact consistently found together in the same graves, all other things being equal, the arrangement of lamps in the presumed chronological order A-B-C is incorrect. In other words, greater credence is given to stratigraphic evidence than to the logical principle upon which the seriation of types A, B and C was based. "Theory is good," as Sigmund Freud was fond of saying, "but it doesn't prevent things from existing."

Cross-dating Relative Sequences

Prehistorians and historical geologists are not content with delineating the sequence of events at individual sites. While such delineation is the essential first step, they also seek to relate these specific events to human prehistory and geological history at large. To accomplish this aim, both kinds of scientists seek to construct **time-space grids**. A grid is a network arrangement of two sets of parallel lines that intersect at right angles. Grids are used to divide maps or charts into equivalent units or representing the various values of two or more variables. Time-space grids display the trajectory of geological history and human biological and cultural change through time and across space. The ultimate aim of prehistorians and historical geologists is the construction of a time-space grid for the entire Earth. Such a grid would allow the cross-dating correlation of events over the entire globe from the beginning of time to the present. Such a construct would constitute a worldwide geological, biological and cultural sequence and would allow the establishment of chronological ties between prehistoric events at different locations on the Earth. Of course, a complete and accurate time-space grid of this magnitude has yet to be constructed.

Basic to the construction of time-space grids in both archaeology and geology is the technique of **cross-dating**. Stratigraphic sequences at separate locations are said to be cross-dated when specific strata at each of them are determined to be of the same age. Cross-dating was first successfully used by the father of English geology, William "Strata" Smith. Recall that Smith sought to match the geological strata found in separate rock exposures in England by comparing the fossils contained within the layers. Smith developed what he referred to as the "Law of Strata Identified by Organized Fossils" in 1795.

In a given stratigraphic outcrop, Smith distinguished individual depositional units by comparing the types of fossils which they contained and by the frequencies of the various types. He then compared two depositional units in different outcrops and concluded that those strata which contain the same fossils in the same proportion were formed at the same time.

Cross-dating based on Smith's principle begins with the assumption that a correctly ordered geological stratigraphic succession is also a biological succession.

Cross-dating in archaeology is used to discover parallel or contemporary occupations at different sites. Smith's method is applied in archaeological cross-dating, but comparisons are made between the artifacts, rather than fossils, contained in separate strata. With Smith's method, it is assumed that artifacts which resemble one another were made and used at approximately the same time. Similar frequencies of artifact types recovered in the layers of different sites suggest that the layers were roughly contemporary.

Cross-dating often involves the use of artifacts that have known ages such as coins. When the artifact has a known age or age range, we can use the information to place a range on the age of the deposit. Archaeologists use two concepts for these age estimates. A ***terminus post quem*** ("end after which") is the earliest possible age for a deposit. For example, if a burial contains a Roman coin with a known age of A.D. 200 we know that the burial must have occurred **after** that date. A ***terminus ante quem*** ("end before which") provides an estimate of the latest possible age for a deposit. If the burial just mentioned is discovered below the foundation for a church constructed in A.D. 400, then we know that the burial must have occurred before the construction of the church.

Relative Dating: A Hypothetical Example

As an illustration of the utility of some of the foregoing techniques in constructing a relative sequences, let us examine Figure 3-4, a stratigraphic profile of a hypothetical site located on the floodplain of a river in New England. The site consists of four stratigraphic units numbered consecutively from the lowest, Stratum A, to the highest, Stratum E. Two features were found at this site. Feature 1 is a shallow pit dug into the surface of Stratum E. Feature 2 is a brick wall set on a stone footing in a shallow footing trench that had been dug down from the surface of Stratum II.

Six burials were excavated at the site. Burials 2, 3, 4, 5 and 6 were dug into Stratum B cutting through the weakly developed soil horizon at its surface. The trench of Burial 5 cut into the side of Burial 6 and the trench of Burial 2 cut into and disturbed Burial 3. The trench of Burial 1 was dug from the surface of Stratum C into Stratum B. In the process of this excavation, the trench for Burial 1 cut into Burial 2. All six of the burials contained skeletal remains of American Indian physical type and all contained pottery grave offerings:

(1) Burial 1 contained pottery of Types 3 and 4;
(2) Burial 2 contained pottery of Type 2;
(3) Burial 3 contained pottery of Type 5;
(4) Burial 4 contained pottery of Type 1;
(5) Burial 5 contained pottery of Type 5;
(6) Burial 6 contained pottery of Type 1.

For illustrative purposes, assume the archaeologist has two aims in excavating this site. He or she wants to reconstruct the relative sequence of events of the site and determine whether the sequence dating of the pottery in the vicinity is correct. Let us first turn to the relative sequence of events mirrored at the site as a whole.

Stratum A contains fossils of late Pliocene age animals. It contains no evidence of human activity. Stratum B is an **alluvial deposit**, that is, one laid down by water action. It consists of massive gravel and sand interbedded with silt and clay lenses. Although no cultural material was recovered from this unit itself, six grave pits were excavated into it. The interbedded sand and clay lenses suggest that the stream was a braided one that flowed in a number of dividing and reuniting channels. The stratum sits atop material of great geological age. Burials 2, 3, 4, 5, and 6 were excavated into Stratum B from the top of the layer.

Burial 2 cuts into Burial 3. Therefore, Burial 3 was excavated and then Burial 2. Similarly, Burial 5 cuts into Burial 6, indicating that the excavation and sealing of 6 preceded 5. The grave goods in Burials 3 and 5 are of the same type, as are those in Burials 4 and 6. Therefore, the two graves in each of these pairs were presumably contemporary. Taken together, the stratigraphic and artifactual evidence suggests that Burials 4 and 6 were dug into Stratum B at roughly the same time, then Burials 3 and 5, and last Burial 2.

Stratum C is also an alluvial deposit. A sharp boundary between the Strata B and C suggests a period with little deposition or a period of erosion before the deposition of C began. Stratum C consists entirely of fine-grained sand occasionally interbedded with silt and clay lenses. The fact that the Stratum IV is composed almost entirely of fine-grained sand and occasional clay lenses, suggests that different depositional environments characterized the formation of the two strata.

Sediments like those in Stratum C are commonly found to have been deposited in marsh environments where water travels sluggishly through heavily vegetated channels. Only fine-grained sediments are deposited in such settings because vegetation on the margins of the marsh filter the sediments derived from the surrounding slopes and remove the coarse fraction (cf., Waters 1986: 28). The unit contains a high

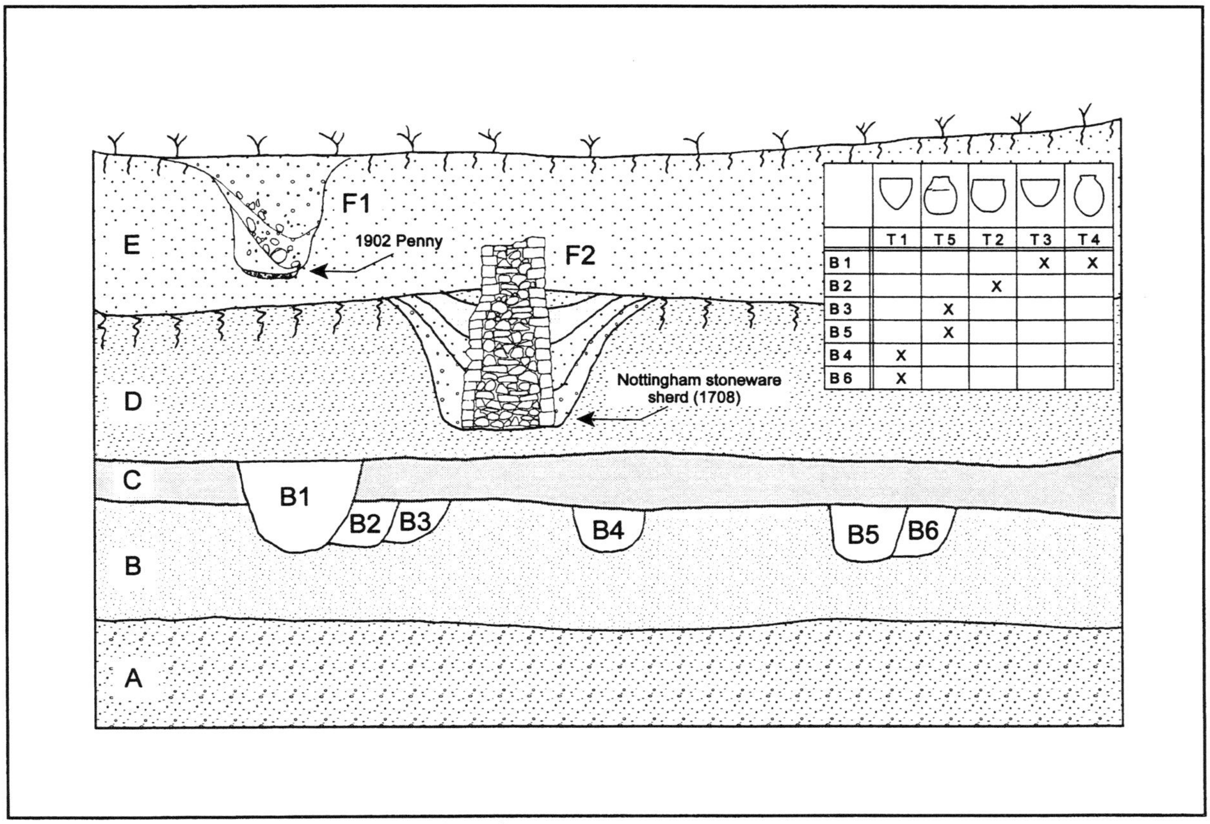

Figure 3-4. Hypothetical section illustrating relative dating.

percentage of organic materials. Stone tools, potsherds of Type 3 and 4, deer bones, freshwater mussel shells, and large concentrations of charcoal were recovered from Stratum C. The artifacts lack abrasion or other signs of water-rolling and this, together with concentrations of charcoal and the faunal remains, suggests that the cultural material was in primary, rather than secondary, context. A single grave, Burial 1, was excavated from the surface of Stratum C through the layer into Stratum B. Burial 1, which contained pottery of Types 4 and 3, cut into and disturbed Burial 2.

Stratum D is an alluvial deposit that consists of gravel and cross-bedded sand near its boundary with Stratum C and then exhibits a fining upward sequence of horizontally-laminated sand and silt. This fine sand and clay is in turn overlain by a darkly discolored clay deposit. Stratum D reflects a change in the course of the river. The lower gravels in Stratum D probably accumulated rapidly in the open river channel, while the fine sand and clays that cap the gravels probably represent a return of marshy, heavily vegetated backswamp conditions. The only cultural material recovered from this layer were isolated potsherds of Type 3 recovered from the lower gravel at the base of the unit. These sherds show evidence of abrasion and water-rolling and apparently were accumulated in this by water action. Presumably they were discarded on the surface of the floodplain somewhere upstream from their current location. Later, the changing course of the river's channel reworked and redistributed these floodplain deposits and incorporated the sherds and other materials they contained into Stratum D. These sherds are in secondary, rather than primary, context, and therefore cannot be taken as evidence that the site was occupied at this time.

At the top of Stratum D is a **paleosol** or buried soil horizon. The development of this soil horizon indicates that the alluvial deposition processes that formed Stratum D had ceased operating at this location and that the surface had stabilized and been exposed to weathering for some period of time. Presumably the river had changed course slightly at this time; the site location probably now formed the river's bank. The paleosol exhibits a well-developed soil structure which shows signs of having been formed under cool temperate environmental conditions much like those found in the region at present. Plant macrofossils and snail shells recovered in the humus suggest an open forest vegetation.

The top of Stratum D contains small quantities of

organic debris, flecks of charcoal, stone tools, potsherds of Type 3, and a fragment from the stem of a clay pipe. The cultural debris found in this layer suggests that the area was utilized for a variety of human activities. The mixture of American Indian and Euro-American cultural material indicates that the debris did not result from a single human occupation or activity set. Rather this materials appears to be part of a mixed deposit that contains the remnants of a number of cultural contexts. Feature 2, built on the surface of Stratum D, is a brick wall set on a stone footing in a shallow footing trench. Within the footing trench a lead musket ball and a Nottingham stoneware sherd were recovered. The Nottingham stoneware sherd had a potters mark with a date of 1708. Feature 2 constitutes what archaeologists call a "site-unit intrusion," meaning it marks the appearance of an altogether new culture complex hitherto unrepresented at the site. Specifically, it is evidence of the arrival of Euro-Americans in the vicinity. The potter's mark on the Nottingham stoneware sherd provides us with a starting point or *terminus post quem* date: we know that the footing trench must have been filled after the Nottingham stoneware vessel was fired in 1708.

Stratum E is a massive alluvial deposit that consists of sand and silt. It contains chunks of mortar, brick fragments, broken glass, and organic matter. This final deposit seems to have formed rapidly, perhaps during one or more large floods. The top of Stratum I is the modern ground surface of the site. A weakly developed modern soil horizon is visible at the top of this stratum but it is insufficiently pronounced to justify designating it as a separate stratum. Visible on this surface is Feature 1, a shallow pit dug into the surface of Stratum I. The pit contained a penny with a minting date of 1902 and a ballpoint pen.

The penny provides us with a *terminus post quem* for Feature 1. The penny could not have been buried before 1902, so the pit had been filled sometime after that date. The presence of the ball point pen, a contemporary artifact that did not become common in American culture until the 1950s, suggests this fill was in fact quite recent.

In sum, the site reflects the following events:

1. The earliest traces of human activity at the site occur at the top of Stratum B when burials 4 and 6, then 3 and 6, then 2 were interred there. The exposure of Stratum B ended with the establishment of marshy conditions, perhaps indicating the presence of a backswamp of the river. A human grave, Burial 1, was interred through Stratum C at this time.
2. The backswamp or marsh disappeared during the subsequent episode of river channeling that followed in Stratum D. The immediate locality of the site does not seem to have been occupied at this time.
3. Then, deposition slowed at the site and an open temperate forest vegetation grew. Cultural debris found scattered in the Stratum D paleosol suggests human presence.
4. The first appearance of cultural material of Euro-American origin is indicated by the construction of the brick wall into Stratum D and could not have occurred before 1708. The wall was subsequently buried after 1708 by the flood deposits of Stratum E. The last event represented at the site is the excavation of the shallow pit at the modern land surface and its subsequent filling with modern debris sometime after 1902.

Let us now turn to the second problem confronted at the site: the determination as to whether the sequence dating of the pottery in the vicinity was correct. Let us say that the presumed chronological order of pottery forms had been established by similarity seriation as running from the earliest form, Type 1, through Types 2 to 3 to 4, and ending with the latest pottery form, Type 5. How well does this sequence accord with the stratigraphic evidence from the site? First, following the Law of Association, we can assume that the pottery in the individual graves was in use at the time of the burial. Therefore, the stratigraphic superpositions of the graves provides us with an initial pottery sequence as follows:

1. Since Burial 5 cuts into Burial 6, the excavation and sealing of 6 must have preceded 5. Therefore, the pottery of Type 1, which Burial 6 contained, is probably older than the pottery of Type 5 recovered in Burial 5.
2. Burial 1 (pottery Types 4 and 3) cuts into Burial 2 (Type 2) which in turn cuts into Burial 3 (Type 5). Therefore, these graves suggests that pottery of Type 5 is older than Type 2 which in turn is older than Types 4 and 3.
3. Pottery Types 4 and 3 further co-occur in Stratum C and this co-occurrence, together with their association in Burial 1, indicates that the two types were in use at the same time. However, only Type 3 sherds were recovered in Stratum D suggesting that the type continued to be made and used after the disappearance of Type 4. Taken together, the stratigraphic evidence at the site suggests

a ceramic sequence that begins with Type 1 followed by Type 5, then Type 2, then to 4 and 3 at the same time, then to 3 alone.

4. Thus, the relative sequence of pottery types obtained at the site through stratigraphic analysis and the Law of Association is different than the sequence obtained by seriation. Does this mean that the seriation was in error? Perhaps, but data from a single site is rarely sufficient to cause the abandonment of a well-established pottery sequence. Such evidence cannot be ignored, however. If the findings from this site are confirmed in later excavations, a revision of the pottery sequence will surely follow.

Summary

At the beginning of the foregoing section, we distinguished between two kinds of time: relative time and absolute or chronometric time. Statements of relative time merely assign a position in time to events or objects in terms of the occurrence of other events or objects; A precedes B, B precedes C and so forth. We then discussed the major methods and concepts used by archaeologists to establish **relative chronology**. These included **stratigraphy**, three types of **seriation**, and **cross-dating**. A seriation must be tested using other methods of age estimation such as chronometric dates, stratigraphy, and Worsaae's Law.

Separate stratigraphic sequences can also be cross-dated by determining the absolute dates of the layers in each sequence, it is to the construction of absolute chronologies in ancient studies that we turn in the next chapter.

Major Points Raised in Chapter Four

1. The second category of time is absolute time. Beginning with a discussion of the use of **ancient calendars** to date prehistoric events.

2. **Varve counting** involves the use of distinct, sequential stratigraphic patterns or "signatures" to cross-date separate deposits, a technique later applied to cross date tree rings and deep sea cores.

3. **Dendrochronology** is the most accurate geochronological technique available. The tree ring growth sequences are also useful sources of paleoclimatic data.

4. **Carbon-14 or radiocarbon** dating was the first dating technique to make use of the phenomenon of isotopic decay. It is the most widely used absolute dating technique in modern archaeology.

5. We discuss **Potassium/Argon** dating in conjunction with the development of chronology of the Earth's magnetic history.

6. For events that occur between the range of potassium/argon and radiocarbon archaeologists are turning increasingly to **thermoluminescence** and **electron spin resonance** to estimate ages up to about 500,000 years ago.

7. The study of **geomagnetism** in the sediments recovered from deep sea cores can be cross-dated with one another and with terrestrial sediments. They contain important information about Quaternary climate change.

CHAPTER 4

ABSOLUTE OR CHRONOMETRIC DATING

Introduction

Absolute or chronometric time is time measured in years, for example, A.D. 1492, 25,000±1,000 years ago, 5,024 B.C., or 4.5 billion years ago. Although **relative dating** past events and artifacts by stratigraphy, seriation, association, and cross-dating remains the heart of the logical structure of archaeology, recent developments in geology, physics, astronomy, biology, and chemistry have provided prehistorians an array of techniques for assigning dates in years to ancient objects and events. In this section we study major techniques for obtaining absolute or chronometric dates.

Absolute Time Measured by Ancient Calendars

The clock and the calendar have governed the lives of peoples in the West for centuries. However, the Western or Gregorian calendar was not the first system used for reckoning and calculating time. Archaeological evidence suggests that in the late Pleistocene epoch, some 25,000 or more years ago, hunting and gathering peoples living in Western Europe kept track of time by counting moon phases. Such a lunar calendar would have conferred important adaptive advantages on hunting peoples exploiting seasonally migratory animal species like reindeer and salmon. Agricultural people have an even greater need for keeping track of the seasons as a guide to planting.

For example, Stonehenge in England appears to have been used for measuring time by tracking the positions of the sun and other key stars. A person standing at the center of the concentric circles of upright stone pillars at the site can observe the position of key astral bodies relative to lines of sight fixed between the upright pillars. Some of the stones appear to align with the position of the sun on the horizon during the summer and winter solstices. Astronomers and archaeologists have suggested that these alignments were used to calculate the date, and perhaps, to predict eclipses. The purpose of the other stone alignments at Stonehenge remains unknown. **Megalithic monuments** that are similar in some respects to Stonehenge are found elsewhere in England as well as in Scotland, Ireland, Spain and France. Some of these monuments may have been calendars also. Unfortunately, the absence of written records makes their original purpose difficult to interpret.

Archaeologists have been more fortunate in some of the other areas of the world where calendars and dating systems were developed. Early civilizations in the Near East, Central America, Peru, and China not only used accurate calendars, they inscribed buildings, tombs, temples, and artifacts with dates calculated from these calendars, thereby earning the gratitude of future archaeologists. Prior to the development of independent dating methods in the second half of the 20th century, inscribed dates were the only means of assigning absolute dates to past events.

The Egyptian calendar proved a particularly useful system for establishing the chronology of events in the ancient Near East after about 3,000 B.C. Calendric records were kept and some dates inscribed on buildings. The Egyptians also recorded the dates of the appearance of the star Sirius. Astronomers have been able to calculate these dates and use them in calibrating the Egyptian calendar against our own. The Sumerian King List from Mesopotamia has been tied to the Egyptian calendar and this list has sometimes proven to be a helpful chronological tool for prehistorians.

The complex calendar developed by the ancient Maya of Guatemala, Belize, and Mexico has also been extremely useful. The Maya carved dates on a wide range of monuments, buildings, and objects. In Peru, the Inca king lists has been used in the Andes area as a chronological aid but correlation with chronologies based on ceramics and other archaeological materials has often proven to be difficult.

The use of ancient calendars as a means of absolutely dating archaeological objects and events is complicated by a number of problems. If these calendars were not used in association with a system of writing or permanent record keeping they are largely useless as an archaeological dating tool. Even ancient calendar-using peoples who kept written records generally reserved dated inscriptions for elite artifacts or buildings. Seldom if ever were the everyday objects, such as pottery, tools, or habitation structures, dated in this manner and it is precisely these everyday objects that form the bulk of the archaeological record.

Given the general limits of calendric dates found in antiquity, archaeologists have had to develop other dating techniques. The remainder of this chapter will be devoted to the the discussion of the most commonly used techniques, including: varve counting, dendrochronology, carbon 14 or radiocarbon dating, geomagnetic dating, potassium-argon dating, and the use of deep-sea cores.

Varve Counting

Shortly before the beginning of World War I, Baron Gerard de Geer of Sweden published a paper demonstrating that the thin clay "varve" sediments found exposed in banks over much of northwestern Europe originally formed in meltwater basins created by the retreating Pleistocene glacial ice. In addition, each of these layers is an annual deposit so it can be counted in order to determine the date of formation. De Geer reahced this conclusion by studying the nature of sediment formation in modern Scandinavian glacier-fed lakes located at the margins of the remnant Pleistocene ice sheet, and by invoking the principle of uniformitarianism in interpreting identical deposits found well south of the glacier in modern times.

The movement of glacial ice across the landscape results in the scraping, grinding and dislodging of rock and earth beneath the glacier's path. Some of this debris is picked up and incorporated into the glacial ice and some is dislodged and pushed to the front or to the side of the ice mass as it moves. When the glacier begins to melt, water from the glacier carries some of this sediment into basins or lakes at the edge of the glacier. This sediment-laden meltwater gives the glacial lakes their characteristic azure-blue color, called *Gletschermilch* (literally, "glacier milk"). When the fast-moving streams of melt water reach the lake, this debris accumulates as sediment layers on the lake bottom. Baron de Geer's analysis of these sediment beds demonstrated that each spring there is an influx of material into the lakes. First the coarser silty and sandy materials settle and then, as spring wears on into summer, the finer clay materials are deposited.

The finest particles commonly remain in suspension in the lake waters and during the winter, gradually settle to the bottom of the lakes. When the returning spring thaw again brings large quantities of coarse-grained sediments into the lake, a distinctive contact zone is formed between the new material and the fine grained sediments deposited during the winter. The annual temperature controls the rate at which the source glacier melts. Therefore, variations in temperature from year to year result in annual varve deposits of varying thickness. Thick varves are produced in warm years, thin varves in cold ones.

In the course of his studies, de Geer recognized that similarities in varve sequences located within a kilometer of each other were due to the similarity of local climates. He was able to use variations in varve thicknesses to cross-date the sequences by identifying patterns of thick and thin sediments in each. (Butzer 1971: 188; West 1977: 191).

Cross dating the separate varve deposits was crucial, since the continental glacier had retreated northward and thus had left no single deposit containing all the varves formed since the end of the Pleistocene. De Geer sought to match unique patterns of thick and thin rings, working from varve sequences of known age to nearby sequences of unknown age farther south. Actually, varve thickness alone is too crude a measure; in order to achieve full precision in cross-dating one must also consider varve color and grain coarseness. Eventually de Geer was able to erect a chronology of the retreat of the ice sheet for the last 17,000 years (West 1977: 191-192). Later de Geer's student and collaborator, Liden, linked the chronology to historic time (Wilson 1974: 47).

Using this master varve chronology, de Geer was able to obtain the first chronometric date for an event of the late Pleistocene epoch. He dated the end of the Pleistocene at 7,912 B.C. with the draining of the Baltic ice lake, which agrees well with later C14 dates (Butzer 1971: 188). The European varve sequence is also the longest known in the world. Its validity has been independently confirmed through the radiocarbon dating of varve layers (Tauber 1970). A comparable varve chronology for North America has not yet been accomplished. However, recent attempts to radiocarbon date North American varve sequences suggest that the problems caused by the lack of a starting point for the chronology may eventually be overcome.

Unfortunately, de Geer's method is of little direct

use in archaeology. Archaeological sites are not found in direct association with varve banks (although human artifactual materials occasionally are found in them). Varve counting has been used to date archaeological sites by indirect means. Sharer and Ashmore (1993: 317) cite the example of Ertebølle sites of the European Mesolithic period in Finland that were dated with reference to the varve chronology.

Varve analysis has also proven to be an excellent means of dating and reconstructing the paleo-environmental changes that took place during the Late Pleistocene and Holocene periods in northwestern Europe, North America, and even South America and east Africa. The pollen of different plant species is deposited along with sediment in each varve. When the percentages of various pollen taxa change, we infer that the paleoenvironment changed also.

Whatever its current limitations, varve analysis remains significant in the history of archaeology and geology because it was the first chronometric dating technique developed. By linking a natural process to the passage of time, de Geer alerted us to the existence of "natural clocks." The search for other such clocks continues to this day.

Further, de Geer's study methods served as a prototype for other dating techniques. At the heart of de Geer's achievement was his assumption that varve thickness was a function of local environmental conditions and that local variations in annual temperature created distinctive sequences of thick and thin varve layers which could be used as signatures or fingerprints to cross-date adjacent varve beds. De Geer's assumptions and methodology were applied with similar success to the dating technique known as dendrochronology that was developed in the American Southwest in the first decades of the 20th century.

Dendrochronology

Dendrochronology, or tree-ring dating, was the second absolute dating technique to be developed, but even today, it remains the most accurate yet devised. Trees grow outward from their centers by forming concentric layers or "rings" of wood fiber during the spring and summer. Certain tree species, including Douglas fir, yellow pine, oak, spruce and sequoia, form tissue during each growing season in a generally continuous layer throughout their entire stem and root systems. It is this growth pattern that creates the ring pattern visible in a cross-section of a log or on tree stumps. The size of these growth rings is a function of the tree's heredity, age, and the environmental conditions experienced throughout its life.

These growth rings make it possible to determine the age of the tree by counting the number of rings. The first ring visible under the bark is the ring that was forming when the tree was cut down. The age of the tree at the time of its cutting can be determined simply by counting the number of rings between the outer ring and the first ring at the pith or center of the log.

Variation in ring size is due largely to climatic variation from year to year. Such climatic variation creates specific ring patterns in all trees growing in the same area. Tree rings which deviate from average size, that is, are notably narrow or broad, form distinct and recognizable arrangements of rings referred to as "signatures." Signatures are unique since the precise sequence of wide and narrow rings climatic variations are never exactly the same from year to year. Year-to-year fluctuations in temperature and moisture that are similar throughout a region produce variations in ring structure. Dendrochronologists attempt to identify the specific signatures found in trees of known age in logs of unknown age from the same region. When such patterns are identified, it is possible to cross-date the undated specimens and to determine their ages. The age of wood recovered in archaeological sites can be determined in this manner and, by inference, the age of the archaeological remains associated with such wood.

Of course, botanists have known for centuries that certain tree species add a new concentric layer or ring of wood for every year of growth and that the relative thickness or thinness of the ring reflects the climate of the growing season during that year. It was not until the first decade of the 20th century however, that a serious attempt was made to at reconstruct climatic history using the sequential variations in tree ring thickness. Serendipity played a large role in the application of this knowledge as a dating technique. Neither a geologist nor an archaeologist conceived of the method of dendrochronology; rather it emerged as a spin-off of the research of an astronomer, Andrew Elliot Douglas.

Douglas began his professional career at Harvard under the tutelage of the celebrated astronomer, Percival Lowell. Around the turn of the century, Lowell commissioned Douglas to establish an astronomical observatory in Flagstaff, Arizona. Douglas accomplished this work and in 1906 was appointed Professor of Physics and Astronomy at the University of Arizona in Tucson. He also vigorously pursued his own research into the relationship between sunspots and the Earth's climate. Astronomical observation of the sun indicate that sunspot activity follows an 11.5 year cycle. If sunspots and climate were correlated, Douglas reasoned, similar periodic variation should be evident in the Earth's rainfall and temperature record. He tried to compare meteorological records with the records of sunspot activity to determine whether significant correlations existed between them (Hitch 1982: 300). Douglas soon found that the historical records did not cover enough time for his purpose.

Douglas hit upon the notion of using the growth patterns of tree rings as his climate records and began by examining freshly cut yellow pine stumps in the Flagstaff and Prescott regions. Yellow or ponderosa pine trees commonly live for more than one hundred years and add rings annually. In addition, their growth is quite sensitive to variations in rainfall and thus the thick and thin patterns of their annual rings closely reflect the specific patterns of light to heavy rainfall that they experienced during their lives. Douglas extended his weather records back as far as possible using freshly cut yellow pines. In the course of this work he recognized that tree rings formed distinct sequences of narrow and wide rings, and that matching sequences could be found on all the yellow pine trees growing in a region.

In attempting to extend the tree-ring record farther and father back in time, Douglas found that he could match the signatures he had identified in the living trees with those found in cross sections of beams and other old wood used in older buildings of the region. In this way he was able to cross-date wood of unknown age with that dated to the present. Douglas's method of cross-dating wood using the temperature dependent signature patterns is directly analogous to de Geer's method of varve correlation.

Douglas extended his master chronology back some 500 years before enlisting the collaboration of the archaeological community and applying his method to wood specimens drawn from prehistoric Indian sites throughout the Colorado Plateau. By 1929, he had established a master sequence of tree-ring growth pattern from the present back to A.D. 1280. In addition, Douglas has constructed a "floating chronology" of sequential tree-ring growth of nearly 600 years that was internally consistent but could not be tied to the present had been constructed. Douglas and his archaeological colleagues concentrated their efforts on dating wood from the Hopi Indian towns of Arizona. On August 22, 1929, a specimen from an ancestral Hopi site near the town of Showlow in central Arizona produced a ring sequence that linked the two sequences. In a single night, the problematical floating chronology was joined to the master chronology and a sequence of tree-ring dates extending back to A.D. 700 was established. At the same time, the absolute ages of 40 major southwestern Pueblo Indian ruins were determined (Haury 1962).

In the two decades that followed, the Southwestern archaeological chronology was carried back into the first millennium B.C. After World War II, Edmund Schulman, carried the tree ring chronology back still further using the giant California sequoias (*Sequoiadendron giganteum*), some of which were more than 3, 000 years old.

In 1954 and 1955, Schulman discovered that under certain conditions, the bristlecone pine (*Pinus aristata* Englemann) grew to be even older than the sequoias. At elevations of over 10,000 feet in the White Mountains of California, Schulman found individual trees of the Great Basin variety of the bristlecone pine (now classed as *Pinus longaeva*) that had lived to be more than 4,000 years old. Further, in the cold, dry climate of the White Mountains the hard and resinous bristlecone pine decays extremely slowly. Using dead wood collected from bristlecone pine groves, the master chronology was extended further back (Ferguson, 1968: 839).

The long chronology of the Bristlecone pine. Sadly, Schulman died shortly after beginning research with the bristlecone pine. His work was continued at the Laboratory of Tree Ring Research by C. W. Ferguson. By 1968, Ferguson and his co-workers had constructed a master chronology for the western United States back to 5132 B.C., some 7,100 years before the present (Suess 1980). Both the length and the accuracy of this master chronology made it an ideal standard to test the validity of radiocarbon dating and to calibrate radiocarbon years to calendar years.

Since then, the master chronology for the western United States has been pushed back to 8000 years before the present. Dendrochronologies using German and Irish oak extend back even further.

Methods of dendrochronology. In addition to ponderosa pine (*Pinus ponderosa*), oak (*Quercus*), sequoia, and bristlecone pine, a host of other species of trees and shrubs can be dated by means of dendrochronology. Some of the major dateable plants include: beech (*Fagus* spp), hemlock, juniper, pinyon pine (*Pinus edulis*), limber pine (*Pinus flexilis*), Jeffrey pine (*Pinus jeffreys*), bigcone spruce (*Pseudtsuga macrocarpa*), Douglas fir (*Pseudotsuga menziesii*), white fir (*Abies concolor*), balsam (*Abies* spp), and artemsia.

Wood specimens for dating purposes are collected in a variety of ways. It is common to collect entire pieces of wood uncovered in archaeological sites. On the other hand, samples are removed from living trees or from large wood specimens that must remain *in situ* by means of a specially-designed drill with hollow drill bit. Using this tool dendrochronologists extract "radial growth ring samples" or circular cores about four millimeters in diameter.

Once in the laboratory, the surfaces of the wood samples must be prepared by cutting or sanding to expose the cellular structure and ring pattern. Samples prepared in this manner can then be cross-dated with a master chronology using a **skeleton plot** (Bannister 1969: 195). The widths of individual rings are converted to a relative scale which can be compared with relative scales of other specimens. As the diameter of a tree increases with age, the width of its outer growth rings tends to diminish. It is therefore necessary to standardize tree ring thicknesses in order to separate the changes associated with the tree's age from those changes due to climatic variation (Fritts 1971: 423-4). Each measurement of ring width for a particular year is divided by the expected growth for that tree and age determined for other trees in the area (Fritts, *et al.*, 1981: 196).

Construction of a master chronology for an area begins with the collection of radial core samples from living trees and works backward from these modern samples of known date to successively older and older specimens by cross-dating these specimens with archaeologically-recovered ones. In North America, more than 1000 regional tree-ring chronologies have been developed (Fritts, Lofgren and Gordon 1981: 196). Where chronologies have been established, specimens are compared to the master chronology so that the age of the outermost ring can be estimated.

Dendrochronology can be used to date archaeological sites whenever three conditions are met. First, tree growth must generally be limited by a single environmental factor, either temperature or moisture. Second, the prehistoric and historic population of the region must have made extensive use of wood. Third, the wood must be sufficiently well-preserved so that it still retains its cellular structure. Practically, this means either that wood is found in settings that inhibit decay, dry caves, bogs, and permafrost, or that it was used in the constructing roofed, stone buildings that insulated it from decay.

Not all trees produce clearly defined annual rings. Some trees more than one ring per year. Other produce irregular or indistinct rings that cannot be counted. Other trees are relatively unaffected by variation in environment. The clearest signatures are produced in environments where one factor limits tree growth. In arid and semi-arid regions ring width varies directly with intensity and durations of rainfall. In subarctic regions ring growth varies with temperature. In temperate and subtropical climates, ring width and climate are less closely correlated.

Obviously, if dendrochronology is to be applied to the archaeology of a region, the former inhabitants of that region had to have made extensive use of wood. However, the ancient use of wood is a necessary but not sufficient precondition for the application of tree ring dating. For example, the native peoples of the forested regions of both Germany and the North American Midwest made extensive use of wood. However, the Indians of the Midwest did not incorporate wood in the construction of well-preserved, stone buildings. Insulated from the elements, the wooden beams of the German castles and cathedrals survived. Consequently, a master tree-ring chronology has been constructed for Germany that extends back before the time of Christ while the technique remains unusable in Midwestern prehistoric archaeology.

The utility of dendrochronology. Dendrochronology is used both for archaeological dating and for paleoenvironmental reconstruction. The most useful specimens are those that still retain vestiges of their outer bark. When the bark is present, the outer ring can be identified with certainty and the final cutting date of the specimen ascertained with complete confidence. Next in utility are specimens that lack bark but still exhibit the beetle galleries excavated directly beneath the bark. The cutting dates of specimens that lack both bark and beetle galleries can sometimes be determined if their surfaces exhibit certain characteristics associated with outer rings. Least reliable are those specimens whose age is based on the assumption that the last observable ring was near the outer ring.

Even with a good ring sample, the age estimate of the archaeological site can be incorrect. The most common type of error is due to the ancient re-use of wood. Rather than cut new timber, Southwestern Indians commonly removed beams from abandoned structures and either used them to build new structures or burned them as fire wood. In instances of re-use, the cutting date of the wood is not the construction date of the structure from which it has been recovered by the archaeologist. Southwestern Indians commonly remodeled or repaired aging structures by removing rotting wood and replacing it with fresh cut timber. In this case the dendrochronological estimate is younger than the construction of the structure.

Many of these errors can be avoided simply through careful archaeological excavation, observation, and recording in the field. Others can be eliminated by collecting and dating all of the wood in a given archaeological site. Such a strategy was used by Jeffrey Dean at the large 13th Indian pueblo village sites of Betatakin and Kiet Siel in the Tsegi Canyon of northern Arizona. By dating virtually every piece of wood and plotting their locations in these huge pueblos, Dean was clearly able to distinguish between new construction, repair, and reuse. On the basis of this detailed chronology, Dean was able to generate a

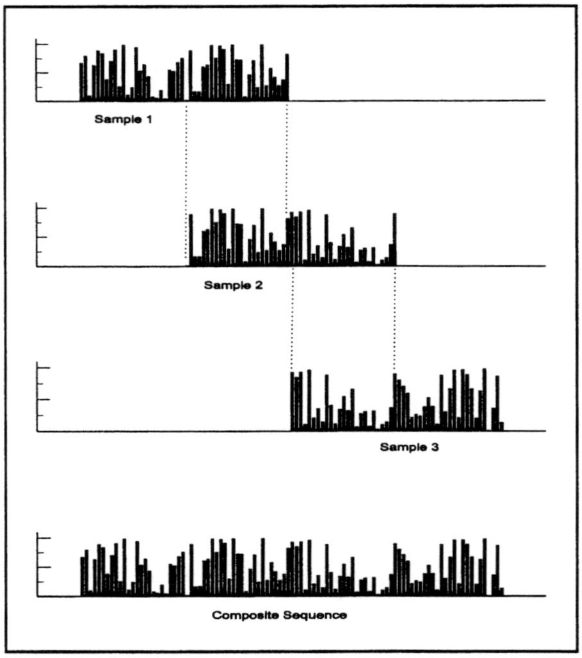

Figure 1. Tree ring thicknesses are standardized and converted to a simple bar chart showing changes in the thickness of the rings over time. Separate samples are linked by matching identical patterns (in this case twenty-five year patterns). The composite sequence shows changes in tree ring thickness over more time than any one sample.

series of highly plausible hypotheses about the history and social organization of the people who had inhabited them (Dean 1970).

As with all archaeological dating techniques, the dates obtained by means of dendrochronology should be cross-checked for consistency with dates obtained by other means, including artifact seriation, archaeomagnetism, and radiocarbon. When these methods are followed, dendrochronology provides the most reliable absolute dates of any technique available to the archaeologist or geologist. Consequently, regions where it can be used have the most securely dated prehistoric and historic archaeological sequences. In the southwestern U.S. dendrochronology has been used in more than 1,000 archaeological sites that range in age from the historic period back to the earliest permanent American Indian structures of the first millennium B.C. Unfortunately, dendrochronology can only be used in a limited number of regions.

Dendrochronology has also been used by archaeologists in Germany, Switzerland, England, and Ireland since before World War II. Oak preserved in bogs or stone cathedrals, castles, farmhouses, and boats have proven to be particularly useful. Even oak panels used by Dutch artists have been dated by dendrochronology as a means of authenticating the paintings themselves (Bannister and Robinson 1975: 212).

Using dendrochronology, Russian archaeologist were able to date the early settlement of the medieval city of Novgorod to about A.D. 884. Vast numbers of pine and spruce logs were used to pave the streets of the medieval settlement. As each level of logs gradually sank into the mud beneath the road surface they were replaced by new logs. Between the 10th and the 15th centuries, 28 successive street levels were laid down and preserved in the boggy soil (Bannister and Robinson 1975: 213).

Tree-rings also have proven to be useful in paleoenvironmental and paleoclimatological reconstruction. Because tree ring growth is a complex and multivariate phenomenon, the use of the tree rings in climatic reconstruction is more difficult than using them simply for dating purposes. The simplest way to determine what tree-ring width means in terms of climate is to compare recent meteorological data with tree-rings from the same period. Using weather records and tree-rings for the period between 1901 and 1962, Fritz, Lofgren and Gordon (1981: 197), developed a calibration equation that allowed them to covert tree-ring widths into estimates of temperature, rainfall and other meteorological parameters. Using these data, Fritz and his collaborators generated a series of "weather maps" for 19th century North America.

In recent years, the ratios of various isotopes of hydrogen and oxygen are measured. The different isotopes of hydrogen and oxygen respond to temperature differently. Different proportions of these isotopes in the tree's rings reflect changes in the environment during its life. Trees are thus subtle thermometers whose isotopic ratios can provide us with evidence of climatic change (Libby 1983; Wilson 1981: 215-217). Isotope ratios from tree-rings also provide data on anthropogenic environmental destruction in history. For example, isotope ratios in the bristlecone pine rings, post-dating 1850, reflect unprecedented increases in the amount of carbon dioxide in the atmosphere. This change has been interpreted as reflecting the massive expansion of agriculture, deforestation and industrialization that occurred on Earth beginning in the late 19th century (Wilson 1981: 227-228).

Carbon-14 or Radiocarbon Dating

Until 1950, the dating methods available to the prehistorian consisted of stratigraphy, seriation, cross-dating, the use of ancient calendars, varve counting, and dendrochronology. Although ancient calendars, varves, and tree rings provided extremely accurate absolute dates, they were available for use only in a few areas of the world. Truly universal chronometric dating systems, systems that could be used throughout the world, were needed. Fortunately, the scientific foundation for such systems was already being developed in chemistry and nuclear physics. In the second half of this century, the radiometric dating systems that were devised would transform the measurement of time and the understandin of the past. The first of these radiometric techniques to be developed was **radiocarbon dating**.

Radiocarbon dating, the most widely used absolute-dating technique in archaeology and Quaternary geology, was developed by the chemist W.F. Libby. In 1947, Libby and Grosse demonstrated that the isotope, carbon-14, was found in all living things. Libby and his students then began seeking ways to measure the level of natural carbon-14 in the Earth's atmosphere.

By 1951, this search had led Libby to develop a dating method which used the decay of carbon-14 in dead organic matter. Realizing the archaeological potential of this method, Libby sought the collaboration of the American archaeological community. Archaeologists provided Libby with organic archaeological specimens of known or presumed age in order to test his new method. The success of these tests stimulated the widespread adoption of radiocarbon dating in archaeology and geology and ultimately brought Libby the Nobel Prize for Chemistry.

The radiocarbon technique provided archaeologists with a way of assigning absolute ages to organic material without any of the geographic limitations that characterized either varve counting or dendrochronology. Not surprisingly, the general applicability of the radiocarbon technique combined with the prestige of modern physics led the archaeological profession to embrace the technique with wholehearted enthusiasm. Despite some nagging discrepancies in the radiocarbon ages assigned to Egyptian archaeological material dated by reference to the Egyptian calendar. At first, most professionals simply assumed that the ancient Egyptian calendar was inaccurate. Eventually the profession was to come to learn that the Egyptian calendar was accurate and that the radiocarbon ages were too young for reasons we will explain shortly.

Radiocarbon dating is possible because of interconnections between the physical nature of the universe and what Libby called "the unique chemistries of life." The elements of the periodic table consist of atoms with different arrays of neutrons, protons, and electrons. The atomic number of an element refers to the number of protons it contains; the atomic mass refers to the number of neutrons. All forms of a given element have the same chemical properties. However, some elements can exist in two or more forms characterized by different numbers of neutrons and therefore by different atomic masses. Alternate forms of the same element are called **isotopes**.

There are three isotopes of carbon: ^{12}C, ^{13}C, and ^{14}C. All forms of carbon have six protons, but ^{12}C has six neutrons, ^{13}C has seven neutrons, and ^{14}C has eight neutrons. The designations 12, 13 and 14 are the **atomic mass**, that is, the sum of the protons and neutrons present in the nucleus of an atom of each of these isotopes. Like those of other elements, carbon isotopes vary in their relative stabilities. Carbon-12 is the most stable isotope and is therefore the most common form of carbon found on the Earth. Approximately 99 percent of all carbon atoms are ^{12}C.

The least stable carbon isotope, that is, the one subject to constant disintegration, is carbon-14. Only about one in one trillion atoms is ^{14}C. This still means that there are approximately 50 billion ^{14}C atoms in one gram of carbon. Carbon-14 would have disappeared from the Earth long ago save for the fact that new carbon-14 is continually being produced. The Earth is continually bombarded by high energy radiation from outer space called cosmic rays. Primary cosmic rays consist chiefly of the nuclei of hydrogen, helium, carbon, nitrogen, oxygen and various heavier atoms. These particles collide with nuclei in the upper atmosphere to create secondary cosmic radiation of free protons, neutrons and other elementary particles.

Natural carbon-14 is formed in the upper atmosphere by this secondary radiation when free neutrons react with molecules of nitrogen-14. Nitrogen molecules (^{14}N) and neutrons (n) produce carbon-14 and hydrogen (^{1}H) (Libby 1952: 2).

$$^{14}N + n = {}^{14}C + {}^{1}H \qquad (1)$$

Nitrogen is abundant in the atmosphere so the production of carbon-14 appears to be dependent on the supply of neutrons. Libby made two important assumptions about the nature of this process. First, he assumed that the newly formed carbon-14 rapidly oxidized to form carbon dioxide ($^{14}CO_2$), mixed quickly with lower atmosphere, and became evenly distributed throughout the earth. Second, he assumed that the production of carbon-14 had remained constant through time. He therefore concluded that, despite the constant decay of carbon-14, the proportion of carbon-14 to carbon-12 remains relatively constant because of the constant production of new carbon-14 in the upper atmosphere and the rapid combination of this new carbon-14 into carbon dioxide.

Although Libby's first assumption appears to be essentially correct, the second was not. Since the neutrons are produced by the primary cosmic ray bombardment, the quantity of neutrons is dependent upon the intensity of the cosmic radiation striking the atmosphere. Current research indicates that the magnetic field around the Earth screens the atmosphere from much of this cosmic bombardment. It has also been determined that the magnetic field has fluctuated in intensity through time. It seems likely, therefore, that when the field is stronger, fewer cosmic rays reach the upper atmosphere, fewer neutrons are released, and less carbon-14 is produced. We will return to this point later.

As noted above, the carbon-14 isotope is "heavier" than that of carbon-12, that is, it has a higher atomic mass due to the presence of eight rather than six neutrons in its nucleus. It is these extra neutrons that render carbon-14 unstable and subject to radioactive disintegration or "decay" over time. This radioactive decay involves the release of a neutron from the nucleus of a carbon-14 molecule. Whenever a neutron leaves, an electron termed a Beta particle (B-) is emitted and a molecule of nitrogen-14 is produced (Libby 1952: 4).

$$^{14}C = B- + {}^{14}N \qquad (2)$$

Once produced, carbon-14 molecules do in fact seem to mix rapidly in the atmosphere and combine chemically like molecules of ordinary carbon-12. Because plants use carbon dioxide in the photosynthesis, they incorporate carbon-12 and carbon-14 into their systems in proportion to the amount present in the atmosphere. Despite the constant decay of carbon-14, the ratio remains constant in living creatures because they are continually taking in fresh quantities of the isotope.

However, when the organism dies it ceases to acquire new carbon-14 to replace that which is decaying. Thus, the organism's death starts the "radioactive clock." As the carbon-14 in the dead organism decays over time, less and less of the total carbon in the organism is carbon-14.

The disintegration of carbon-14 atoms takes place at a constant rate. Every 83 years 1% of the carbon-14 atoms decay. That leaves 99%. In 83 more years 1% of the 99% remaining decays (1% times 99% or .99%) leaving 98.01%. In 83 more years 1% of the 98.01% remaining decays (1% times 98.01 or .98%) leaving 97.03%. After 5,730 years one-half of the available carbon-14 has decayed. In another 5730 years half of what was left has decayed. For carbon-14 then, 5730 years is the **halflife** of the isotope. The rate of decay is not affected by environmental conditions such as temperature or moisture.

Since this decay occurs with the emission of beta particles, the amount of carbon-14 left in a substance can be determined by counting the number of beta emissions per gram of carbon; the fewer the beta emissions, the less carbon-14 and the older the specimen. In this way Libby found that, by measuring the number of beta particles emitted over a fixed period of time (generally 1,000 minutes), he had a simple means of calculating the number of years since the organism had died.

After about seven halflives (40,000 years), conventional instruments can no longer measure the remaining radioactivity. Specimens greater than roughly 40,000 years in age must be artificially "enriched" if they are to be dated. Enrichment proceeds by concentrating the gas or liquid used in counting. The use enrichment process extends the range of the radiocarbon method to materials 75,000 years old (Stuiver, M., C.J. Heusser, and I.C. Yang 1978).

The laboratory methods for estimating radiocarbon dates. A specimen to be dated is examined and sources of contamination, such as plant rootlets, are removed. The sample is then soaked in hydrochloric acid to remove inorganic carbon compounds or carbonates which tend to concentrate near the surface of the organic specimen which they have contaminated.

The specimen is then converted to a gas (either carbon dioxide, methane, or acetylene) by burning or by acidification. This gas is passed through a series of traps in a high-vacuum system which is designed both to remove any electronegative and radioactive impurities and to concentrate the original carbon. The concentrated carbon gas is then piped into a Geiger counter where the number of Beta particles emitted over a 1,000 minute interval (about 16.7 hours) are counted. The more Beta particle emissions, the younger the specimen. Of course, the Geiger counter must be carefully shielded to prevent it from counting "background radiation," that is, natural radiation and mesons of cosmic radiation in the atmosphere. Alternatively, the original specimen may be converted into an organic liquid like benzene for analysis by means of the liquid scintillation method.

For very small samples, accelerator mass spectrometry (AMS) is generally used. This method uses a subatomic particle accelerator to directly count the individual C-14 atoms in the sample. Spectrometry eliminates the problem of background radiation, involves a much shorter counting period, and requires far smaller quantities of organic material

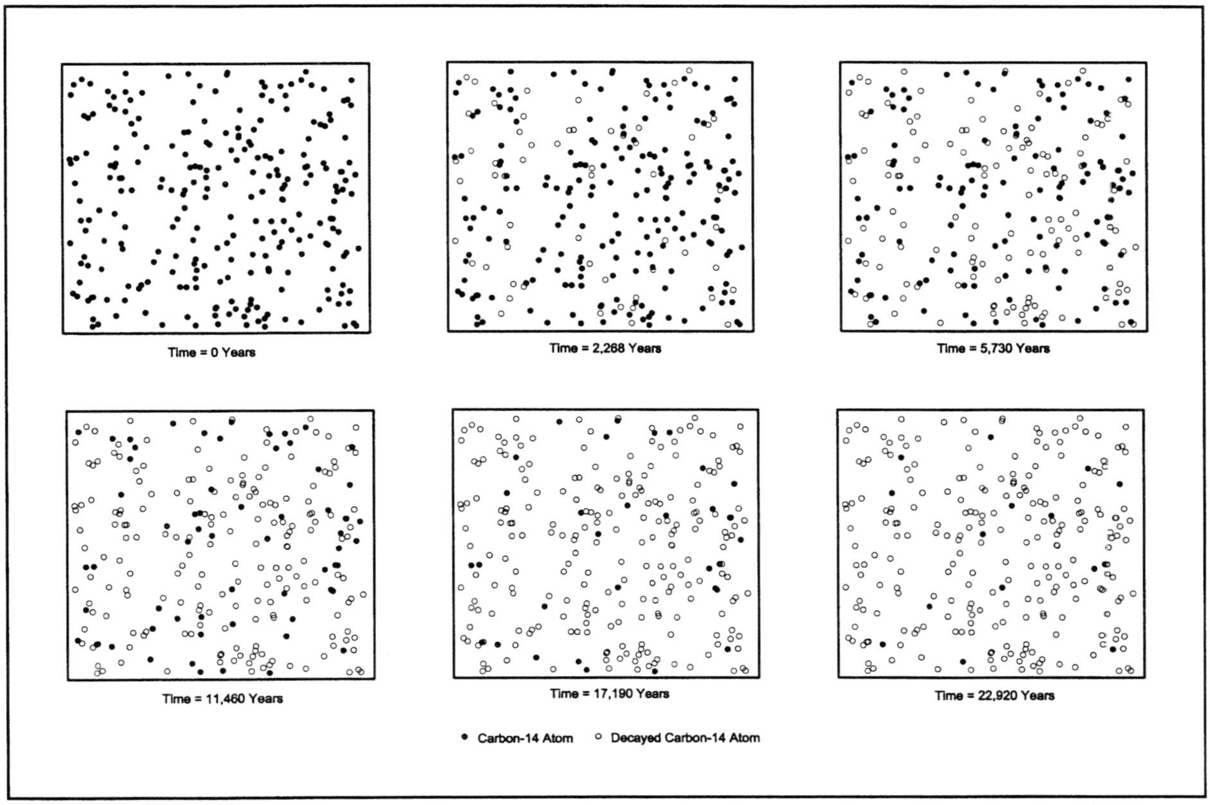

Figure 2. When an organism dies it stops absorbing new carbon-14 atoms. Over time more and more of the carbon-14 originally present in the sample decays. After four halflives (22,920 years), only 1/16th of the original carbon-14 atoms remain in the sample.

(Pavlish and Banning 1980). Although it was initially expected that AMS dating would extend the maximum age of radiocarbon dating to 100,000, this has not happened. Nonetheless, the method remains important because it has sharply the minimum sample size needed for radiocarbon dates. Whereas formerly at least a gram or more of material was needed for radiocarbon dating, using the AMS method, on needs only provide a single seed or small fleck of charcoal for direct dating (Bowman 1990: 34).

In any case, it must be recalled that the decay of carbon-14 is a random process; when a particular carbon-14 molecule will decay cannot be predicted with precision. All that can be said is that, after X period of time, a quantity of molecules will, on the average, have disappeared. Thus, our 1,000 minute counting run provides only a **sample** of the actual number of beta emissions from the true population present in the specimen. Therefore sample variability and error must be taken into account and more than one sample is needed to produce a statistical "sampling distribution." Consequently, carbon-14 laboratories always run samples at least twice and then use a chi-square test to determine if runs are in statistical agreement. If they are, no more sampling runs are necessary. The Beta counts of these two runs are averaged.

The standard deviation is a measure of the degree of variability between the sample runs. The larger the standard deviation, the more variability was observed between the various counting runs and, therefore, the less reliable the data.

In sum, a radiocarbon age estimate provides the prehistorian with a range of values and a statement of the probability that the true age of the dated material falls within that range. For example, a C-14 date of 10,000 ± 200 B.P. indicates a 68% probability that the actual date of the specimen falls between one standard deviation on either side of the mean (that is, between 9,800 and 10,200 B.P.) and a 95% probability that the date falls between two standard deviations of the mean (between 9,600 and 10,400 B.P.). Radiocarbon dates are computed "before the present." That is, before 1950, the date established by the U.S. Bureau of Standards as the $^{12}C/^{14}C$ equilibrium level. Before present dates are listed as B.P.

The range of the radiocarbon method. The standard radiocarbon method is useful for dating events that fall in the time range beginning roughly 100 years ago and extending back some seven to ten halflives (40,110 to 57,300 years ago). The usefulness of radiocarbon dating on material of the late historic period is limited by the fact that such dates are presented as ranges. Such a range is of little utility to the historical archaeologist who probably can date the object with more precision by other means. Further, the large-scale burning of fossil fuels which began in Europe and American during the 19th century has complicated the use of radiocarbon dating in historic archaeology, as such burning dumped large quantities of fossil carbon into the atmosphere. Since such "old" carbon was composed almost entirely of C-12, this burning altered the natural ratio of C-12 to C-14. The widespread testing of atomic bombs at high altitude that took place during the 1950s and early 1960s had just the opposite effect: it "enriched" the upper atmosphere with carbon-14 (Bennett 1979: 456).

A different problem occurs when archaeologists seek to date objects greater than ten halflives (57,300 years) of age. Radioactive decay of C-14 from a gram of modern carbon produces about 15 Beta particles **per minute**; a gram of carbon from a specimen 57,300 years old will produce only about 2.2 Beta particles **per day** (Bradley 1985: 52). Thus, after ten halflives, the amount of carbon-14 in the sample has become so small that precise dating by conventional measurement of Beta particle decay is extremely difficult and the standard deviation is generally quite large.

Problems of Calibrating Radiocarbon Years and Calendar Years

Recall that when Libby first introduced the radiocarbon method, he dated a series of organic archaeological specimens of known age. Prominent among which were artifacts dated by reference to the Egyptian calendar and, although discrepancies between the C-14 dates for these artifacts and their Egyptian calendar dates were observed, this variation was largely discounted. However, given the significance of the Egyptian calendar in the cultural sequences developed for the eastern Mediterranean region, the problem of its calibration with the radiocarbon method did not disappear. Instead, the inconsistency between these two chronological sources became more obvious and more troublesome to the profession until, in the late 1960s, chronologists determined to resolve the problem.

The solution came from an unexpected source: the radiocarbon dating method was cross-checked against another, earlier dating technique, dendrochronology. By 1969 the master tree-ring chronology for the western United States had been extended back nearly 7,500 years to 5450 B.C. The entire master chronology was dated by radiocarbon method in ten-ring (e.g., ten year) increments. Blocks of dated wood from sequoia and bristlecone pine trees were used in this study.

The C-14 dates obtained for each of these ten-ring blocks were then plotted against their actual calendar ages obtained from the master chronology. This plot revealed a number of important discrepancies. First of all, a definite oscillation appeared between the dendrodates and C-14 dates for the years between A.D. 1700-1500. However, there was a close correspondence between the two sets of dates from A.D. 1500 back to 500 B.C. Then, beginning between about 500 and 1000 B.C. and extending back to 5450 B.C., the dates returned by the carbon-14 method were as much as about ten percent **too young**. In the crucial millennium between 3000 and 4000 B.C. of Predynastic Egypt, carbon-14 dates were as much as 800 years too recent (Fergueson 1968: 845). It rapidly became apparent that the Egyptians, had been right!

The variation observed between carbon-14 dates of the sequoia and bristlecone pine samples and the exact ages of these samples as determined by dendrochronology led to the conclusion that "radiocarbon years," (calculated by the rate of decay of carbon-14), are simply not the same as "calendar years," (measured by the rotation of the Earth around the sun). Further, the reality of these oscillations between C-14 years and calendar years appears to be supported by deep-sea cores, varve chronologies, deep-ice cores and other methods.

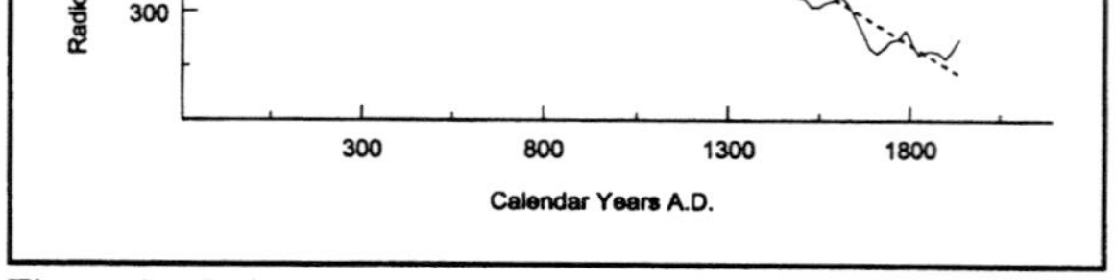

Figure 3. Calibration of radiocarbon dates to calendar dates for the last 2,000 years. The dashed line represents equivalence between radiocarbon and calendar dates. When the curve is **above** the dashed line, the radiocarbon ages are **older** than the calendar date. When the curve is **below** the dashed line, the radiocarbon ages are **younger** than the calendar date.

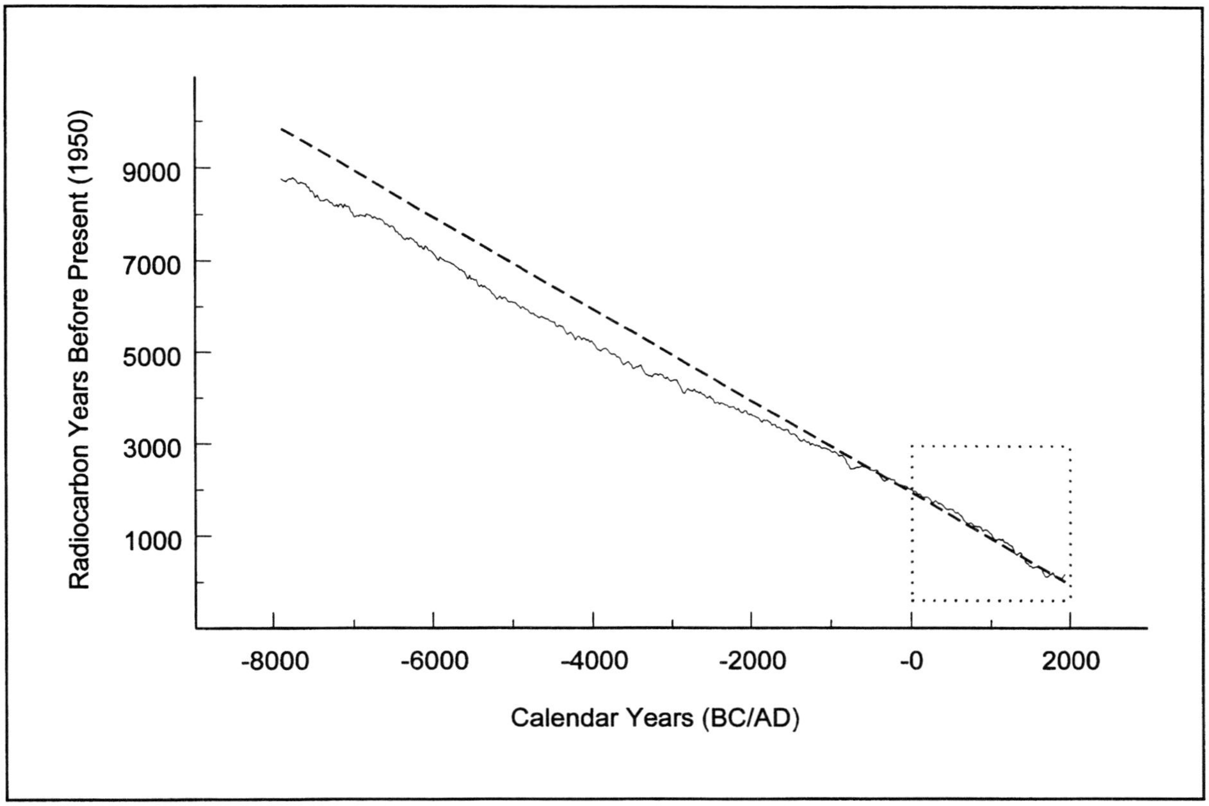

Figure 4. Calibration of radiocarbon dates over the last 9,840 years. The boxed area is illustrated in Figure 3. The dashed line represents equivalence between radiocarbon and calendar dates. When the curve is **above** the dashed line, the radiocarbon ages are **older** than the calendar date. When the curve is **below** the dashed line, the radiocarbon ages are **younger** than the calendar date.

The calibration work also indicated that one of Libby's initial assumptions was in error: the carbon-14 content of carbon dioxide in the atmosphere had not remained constant through time. We will return to the implications of this in the discussion of paleomagnetism and potassium/argon dating below.

In any event, it is now possible to make adjustments or "calibrations" in conventional carbon-14 dates using dendrochronological sequences based on Bristlecone pine and German and Irish oaks to compensate for the known oscillations over the last 9,840 years (Stuiver and Reimer 1993). The process involves the use of one of several calibration curves derived from the comparisons of carbon-14 dated dendro-specimens and allows scientists to change the "radiocarbon years" which they measured into "calendar years."

In Chapter 3, we discussed the Glyn Daniel's seriation of megalithic monuments in the Near East and Europe. You will recall that Daniel's presumed chronological sequence postulated that the civilizations of the Near East were **older** than those of the western Mediterranean and Europe. Specifically, he assumed that the megalithic tombs found in Spain, Brittany, and the British Isles were modeled on tombs and pyramids of Egypt and were built under the inspiration of ideas that had diffused to Europe from the Near East via Malta, Greece, Italy, and southern Europe.

The calibration of radiocarbon dates has resulted in a profound reappraisal of this traditional relative sequence of events. It is now apparent that the megalithic monuments of western Europe are older than both the pyramids of Egypt and the stone tombs and monuments of Greece and Malta. What Colin Renfrew has called a chronological "fault line" has become apparent and it looks as if Europe, far from being a cultural back water in prehistory, may actually claim preeminence over the Near East in certain areas of cultural development during the 5th and 4th millennia B.C. (c.f., Renfrew 1973).

This example reveals how the various archaeological dating techniques can serve as a cross-check on one another. Discrepancies between the Egyptian calendar lead to the calibration of C-14 by means of dendrochronology; discrepancies between the newly calibrated C-14 dates and the Old World relative sequence lead to a revision of our understanding of the prehistory of that region which, over the last one hundred years, had been built on the basis of meticulous seriations and cross-dating of architecture and artifacts.

Potassium/Argon Dating and the Geomagnetic Reversal Scale

Potassium/argon dating is another method of absolute age determination based on the radioactive decay of isotopes. The technique is most widely used in geology and geophysics to date volcanic rocks. Additionally, certain sedimentary rocks, soils, and sediments can be dated in this manner (Dalrymple and Lanphere 1969).

Potassium-40/argon-40 (and the related argon-40/argon-39 technique) can date geological events back to the origin of the earth some 4,500,000 years ago. Geophysicists have used the technique to date the changing polarity of the Earth's magnetic field over the last five million years. Primary igneous rock formations related stratigraphically to archaeological sites in Africa and elsewhere also have been dated by this means.

The element potassium, one of the most common constituents of the Earth's crust, is found in three isotopic forms: potassium-39, potassium-40, and potassium-41. Potassium-40 is unstable and decays with a halflife of approximately 1.31 billion years. Potassium/argon dating is based on the fact that the decay of potassium-40 produces either calcium-40 or argon-40. Argon-40 is released from a rock when it is heated sufficiently; therefore, volcanic lavas and molten magmas contain no argon. The high temperatures associated with volcanic activity sets the clock to zero. When the rock crystallizes, most of the argon-40 generated by the decay of their potassium-40 is retained in its crystalline structure (some escapes, and this is a source of minor dating error).

The amount of argon-40 contained in a volcanic rock is thus proportionate to the number of years that have passed since the rock cooled. To date a specimen, the argon-40 in a volcanic rock sample is driven off with heat and measured. The age of the rock is calculated as a function of the ratio of potassium-40 to argon-40 in the sample (Dalrymple and Lanphere 1969).

The potassium-40/argon-40 and the argon-40/argon-39 systems are the only radiometric techniques that can be used to date igneous rock ranging in time from the formation of the Earth to, theoretically at least, as recent as 10,000 years ago (Gentner and Lippolt 1969:88, 93). Recently improvements in potassium/argon dating was used to date the eruption of Mount Vesuvius that buried Pompeii in A.D. 79 (less than 2,000 years ago), but usually the technique is used to date older materials (Renne, Sharp, Deino, Orsi and Civetta 1997).

Another practical limitation of the technique is the fact that material of archaeological interest is rarely associated with igneous rocks that are especially rich in potassium. Consequently, Quaternary studies have a "time window" during the crucial years between about 500,000 years ago and 40,000 to 50,000 B.P.

Nevertheless, potassium/argon dates have been obtained for areas of particular archaeological interest and importance in East Africa, such as Olduvai Gorge, the Omo Valley, and Hadar. This chronometric framework is making possible a clearer interpretation of the crucial evolutionary events of the Late Pliocene and Early Pleistocene epochs.

Dating shifts in the Earth's magnetic field. The use of potassium/argon dating also has proven to have important implications for studying human cultural and biological evolution during the Pleistocene epoch. Success in dating changes in the earth's **geomagnetic polarity** gives us a worldwide master chronology of geomagnetic change.

The Earth's magnetic field is aligned roughly along its axis of rotation. This magnetic field is believed to be generated by electric currents with the Earth's molten core. Recently, geophysicists have discovered that the magnetic properties of the Earth have not remained constant. Through time the intensity of the magnetic field has fluctuated, Earth's magnetic poles have "wandered," and the polarity of the field has periodically reversed.

The changes in the polarity and intensity of Earth's magnetic field are reflected in the **remnant magnetization** of soils, sediments, and igneous sedimentary rocks. Iron oxides are sensitive to magnetization. When these compounds are heated above the Curie point (500 degrees Centigrade), they lose their magnetism. Then, when the iron oxides cool a few degrees below the Curie point, they become susceptible to re-magnetization. Generally, The Earth is the magnetic principal field determining the nature of their re-magnetization. Rocks, soils, and sediments containing iron oxides that have been heated above the Curie point provide a record of the Earth's magnetic field at the time the material cooled. Over the last thirty years, geophysicists have plotted the history of fluctuations in the intensity of the magnetic field and reconstructed the sequence of reversals in the field's polarity. Studies of stratigraphic sequences of volcanic rocks have made it possible to identify and date changes in the Earth's magnetic field (Matthews 1974:116).

Based on these studies, the time between polar reversals has varied considerably. The longer geomagnetic polarity intervals are called **epochs** and are named after scientists; shorter ones are called **events** and are named after the location where they were first recognized (Dalrymple 1972: 111). Mankinen and Dalrymple (1979) have recently determined the dates of the major epochs by means of potassium/argon method as follows:

Magnetic Epoch	Years Before Present
Brunhes (normal)	present to 730,000
Matuyama (reversed)	730,000 to 2,470,000
Gauss (normal)	2,470,000 to 3,400,000
Gilbert (reversed)	3,400,000 to 5,260,000
Epoch 5 (normal)	5,260,000 to ?

The reversal of the earth's magnetic field occurs on a worldwide scale and takes place rather quickly in geological terms. Therefore, reversal boundaries can provide an unambiguous way to distinguish the late Cenozoic geological time periods from one another (Butzer and Isaac 1975:901-903; Bowen 1978:195). Using magnetic epoch boundaries, the time scale of the Pleistocene and Pliocene is divisible as follows:

Geologic Epoch	Years Before Present
Middle Pleistocene	730,000
Lower Pleistocene	1,720,000 to 730,000
Pliocene	5,260,000 to 1,720,000

The lack of resolution in the potassium/argon technique makes it difficult to precisely determine the age and duration of the short magnetic reversal events. Dates for the beginning and ending of such events have been more precisely established by extrapolating from the potassium/argon determinations of the age of basalts in sea-floor spreading anomalies and by reference to deep-ocean sediment cores. The two independent methods have produced results which are in rather close agreement (Berggren *et al.* 1980: 281-283). Because of their great implications for terrestrial archaeology, deep sea cores will be discussed in the final section of this chapter.

Thermoluminescence and Electron Spin Resonance

Beyond the age range of radiocarbon dating and for sites, such as caves, that lack rocks suitable for potassium-argon dating, archaeologists are increasingly depending on **thermoluminescence** and **electron spin resonance** for age estimates (Aitken 1990). Thermoluminescence was originally developed to estimate the age of fired clay objects like pottery or ceramic figurines but can now provide ages of burned rock, stalagmitic calcite, and wind or water-borne sediments. Electron spin resonance has been applied to tooth enamel, mollusc shells, and coral.

In thermoluminescence dating a sample is ground up, rapidly heated to 500° C, and the amount of light emitted is measured. The sample is reheated and measured a second time. The extra light emitted during the first heating is a result of electrons trapped

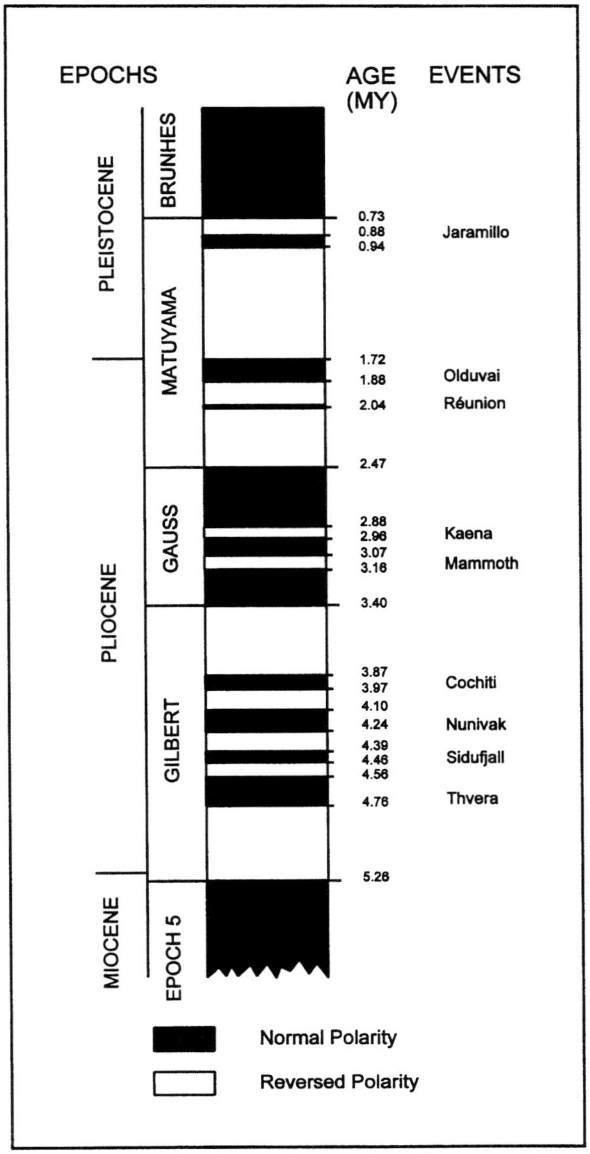

Figure 5. Worldwide absolute time-stratigraphic sequence of magnetic reversal epochs and events recorded in igneous rock and dated by the potassium/argon technique (after Maniken and Dalrymple 1979).

in the crystalline structure of the sample. The number of trapped electrons is a function of the amount of background nuclear radiation that the sample has been exposed to since it was last heated (or since it last crystallized). This background radiation comes from minerals in the sample and from very low levels of radioisotopes in sediments containing the sample. Heating the sample releases the stored energy and sets the "clock" to zero. The age of the sample can be estimated if we know the annual radiation dose for the site and the sensitivity of the material in holding the energy obtained from the radiation. The first is estimated by taking measurements at the site. The second is estimated by subjecting pieces of the sample to measured doses of radiation (Aitken 1990: 142-143).

Thermoluminescence is capable of dating burned flint and chert (commonly used in making stone tools) as old as 100,000 to 500,000 years. Calcite from cave stalagmites as old as 500,000 years can be dated as long as the material has not recrystallized (which resets the clock). The technique has also been applied to sediments since light can reset the clock. Artifacts in an ancient floodplain in Siberia have been dated to about 250,000 years old by this means (Waters, Forman and Pierson 1997).

Electron spin resonance provides an alternate way of evaluating the dosage of nuclear radiation that does not require heating the sample. While thermoluminescence measures the trapped electrons by allowing them to escape, electron spin resonance measures the electrons by their response to high-frequency electromagnetic radiation in the presence of a strong, steady magnetic field which is gradually changed. When the magnetic field reaches a certain strength, the trapped electrons resonate and absorb some of the electromagnetic power. The more electrons trapped, the greater the age, and the greater the amount of absorption (Aitken 1990: 188; Ikeya 1986). For stalagmitic calcite and travertine ages of up to one million years are possible. Ages of several million years are possible for tooth enamel.

Archaeomagnetic Dating: The Direct Use of Paleomagnetism to Archaeology

Under certain circumstances, changes in the Earth's magnetism may be used as a means of directly dating archaeological remains. This technique is referred to as **archaeomagnetic dating.** This technique is based on the fact that the fired clays encountered in archaeological sites have a very stable remnant magnetism which records with fair accuracy the magnetic field of the earth at the time of magnetization, that is, at the time of their final heating.

Archaeological sites commonly contain burned earth structures, pottery kilns, or fire pits in which clay was heated beyond the Curie point, cooled, and then left undisturbed in its original location. Provided the clay has remained in same position it occupied when it was last heated and cooled, the declination, inclination and intensity of Earth's magnetic field at time of its final firing may be determined. With the information, the location of the poles at the time of the firing may be calculated. Knowledge of the location of the poles at the time of the clay's firing provides the archaeologist with the date for that event. The analysis of the fired clay sample thus provides us with a date for the last use of the hearth and, by implication, the material associated with it.

There are some limitations to this method. First of all, the fired clay generally must be *in situ* if an absolute date is to be obtained by means of palaeomagnetism. Second, the sample for archaeomagnetic dating must be collected with great precision in recording the exact orientation of the fired clay. Finally, the quality of the preservation of fired clay in archaeological features such as hearths and structures varies from region to region and some types are simply inadequate for this purpose. Nonetheless, the advantages of the archaeomagnetic dating far offset its drawbacks and it promises to be as accurate as dendrochronology over a far longer time period and without the latter technique's geographic limits.

The geomagnetic stratigraphy of ocean sediments. The changing magnetic properties of the Earth have been recorded over time in sediments that have accumulated on the floors of the world's oceans. Stratigraphic sections of these sediments, called **deep sea cores**, are removed in 10 to 30 meter-long units by means of a "piston corer" dropped over the side of a research vessel. Sediments retrieved in this manner consist of unconsolidated organic and inorganic material that settled to the ocean bottom to form deep deposits.

The organic or biogenic fraction consists of the remains or byproducts of organisms that live in the sea. Classed by chemical composition, this fraction consists principally of carbonates (e.g., **foraminifera**) and silica (e.g., **radiolaria**) (Shackleton 1975: 3). The inorganic fraction of deep sea sediments consists of material that precipitated from the sea water, erupted from underwater volcanoes, or was transported into the oceans from the continents by wind, by rivers, or by floating ice. Commonly the inorganic fraction of deep sea sediments contain magnetic minerals. Once deposited on the sea floor, the magnetic minerals tend to take on the orientation of the earth's magnetic field. Allowing for the disturbance of the position of some of these minerals by burrowing animals, it is possible to use them to determine the polarity and intensity of the earth's magnetic field at the time that the sediments of which they are a part are laid down (Matthews 1974: 116-118).

Relatively recent deep sea sediments can be dated directly by the radiocarbon method but for older deposits the ratio of the isotope thorium-230 to protactinium-231, two isotopes which decay from from uranium-235 and uranium-238 respectively is used to date the sediment. Unfortunately, the range of the technique is not much longer than radiocarbon: protactinium-231 decays to equilibrium after about 150,000 years. Deep sea cores older than 150,000 B.P. can only be indirectly dated with reference to the geomagnetic reversal scale established by potassium/argon dating igneous rock. This is done by determining the sequence of polarity reversals recorded in a deep sea core. The boundaries of the geomagnetic reversal epochs represented in the core are dated by reference to the established terrestrial potassium/argon chronology. If the rate of sedimentation in the beds from which the core was drawn appears to have been constant over time, the age and duration of the magnetic reversal event represented in the core "are determined by extrapolating rates of sedimentation from known reversal transitions or by interpolating between magnetic transitions" (Berggren *et al.* 1980: 281-283). This method has led to a more precise determination of the age and duration of the various geomagnetic reversal events than could be achieved through potassium/argon dating alone. Figure 3 is the composite geomagnetic reversal time scale based on a both on deep sea core analysis and igneous rock dated by potassium/argon (Berggren *et al.* 1980: 283).

When a change occurred in the Earth's magnetic field, it was recorded more-or-less uniformly in contemporary sediments accumulating throughout the world's ocean basins. Therefore, a newly extracted deep sea core can be cross-correlated with previously extracted cores by comparing the magnetic events it records with those of the master core sequence. Of course this is not always so easy in practice as the rate of sediment accumulation in a given location is subject to variation, burrowing animals can disturb and mix the materials, and ocean currents can sweep some of the accumulations away and deposit them in other areas. All of these factors complicate the cross-correlation of individual core sections.

Deep sea cores and oxygen isotope stratigraphy. Refinement of the geomagnetic reversal scale is not the only contribution made by the study of deep ocean sediments. The organic fraction of in deep sea cores contains an unparalleled record of the Earth's climatic history. This climatic record is derived from four kinds of data:

1. the changing arrays of microfossil assemblages including foraminifera, nanoplankton, diatoms, silicoflagellates, and radiolaria;
2. temperature sensitive changes in the direction of the "coiling" of the shells of certain species of foraminifera.
3. the changing ratios of oxygen isotopes which reflects changing ocean chemistry between the present day to the Jarmillo magnetic reversal event at around 0.9 million years ago.
4. the changing calcium carbonate-content in the equatorial Pacific Ocean back as far as the Jarmillo event (Berggren *et al.* 1980: 279).

The study of ocean microfossils, especially foraminifera, has made particularly valuable contributions to the understanding of the climatic history of the Earth. Foraminifera are marine protozoa which form shells by extracting calcium carbonate, silicon dioxide, and calcium phosphate from ocean water (Bowen 1984: 60). These organisms are ubiquitous in the world's oceans and, when they die, their shells or "tests" fall to the bottom and accumulate in the sediment. The temperature requirements of different species of protozoa varies and the direction of shell coiling within certain species is temperature dependent. Therefore, a record of the temperature of the ocean during the formation of the sediment can be obtained by extracting foraminifera shells and other microfauna along the length of a deep sea core and noting (1) the direction of shell coiling in certain species, (2) the first appearance or the final disappearance of various species, and (3) the ratio of key species to one another through time. Interestingly,

> biostratigraphic events, such as beginnings and terminations of ranges, appear in some cases to be closely associated with magnetic events. This fact offers the stratigrapher the possibility that a single event in the history of the earth may be expressed in both physical (magnetic properties) and biological (termination or beginning of a range) data. The event therefore becomes a concept that allows correlation among various types of data (Matthews 1974: 119).

Further, independent chemical record of temperature change is obtained from the foraminifera shells. As we have seen in the case of carbon, different isotopes of the same element do not behave identically in all chemical processes. For example, the lighter oxygen-16 isotope evaporates from water faster than the heavier isotope, oxygen-18. Since evaporation is a temperature dependent process, (1) the ratio of oxygen-16 to oxygen-18 in sea water reflects the ocean temperature at any given time, and (2) oxygen-bearing substances that precipitate from water register the isotopic ratios of their water source. By plotting the oxygen-16/oxygen-18 ratios in the micro-plankton shells extracted along the length of a deep sea core one obtains an independent, and more reliable, measure of temperature that serves as a cross check on the parallel plots of changes foraminifera species ratios. Further, unique sequences of $^{16}O/^{18}O$ isotope ratios form "isotopic signatures" that are useful in cross-dating deep sea cores and tying them into the master Pleistocene ocean core chronology in the same way that specific climatic signatures are used to tie varve banks or tree-rings from an individual tree into the master chronology.

Stable isotopes from foraminifera shells, in combination with the study of microfossil assemblages and calcium carbonate content, has provided the most precise means of estimating: (1) fluctuations that have occurred over time in the global continental ice volume, (2) the paleotemperatures of ancient oceans, and (3) the timing of the worldwide continental glacial advances and retreats. Further, over the last thirty years, a chronometric framework or scale for the Earth's oceans has been constructed based on oxygen isotope stratigraphy.

The oxygen isotope scale consists of a succession of absolutely-dated stages of isotopic variation shown by deep sea cores. It has become the standard method for globally correlating oceanic events. At present, this stratigraphic succession consists of 23 separate stages that extend from the present back to the Jarmillo magnetic reversal event about 900,000 B.P. but recent work has tentatively added 40 more stages and extended this scale backward to the base of the Pleistocene at 1.88 mya. (Williams, Thunell, Tappa, Rio and Raffi 1988: 222).

Oxygen isotope stratigraphy has proven important to the understanding of the terrestrial events of the Quaternary as well. The oceans contain a more complete stratigraphic record of the Pleistocene epoch than do the continents because, in the northern hemisphere at least, each successive ice advance obliterated much of the evidence of previous glacial and interglacial stages. Deep sea cores also contain a far more complicated sequence of Quaternary temperature and climate change than has been discerned in the terrestrial sedimentary record of the epoch. A fully satisfactory cross-correlation of deep sea cores and terrestrial Pleistocene stratigraphy has yet to be achieved (cf. Kukla 1977) and the nature of the relationship between biological and chemical change in the ocean on one hand and glaciation and climatic change on land is not fully documented. Nonetheless, the succession of absolutely-dated stages of the oxygen isotope scale is the best stratigraphic framework presently available for Quaternary events on land or sea. Of course, the late Pliocene is currently still beyond the earliest oxygen isotope stage. Therefore, the best framework for arranging events before about 1.88 mya is still the geomagnetic reversal scale. These two chronological scales are used in the remainder of this book for calibrating the complex cultural and evolutionary events of the Paleolithic period in a worldwide time-space grid.

Using the boundaries between oxygen isotope stages, we may finish assigning absolute dates to the Pleistocene time scale begun earlier in this section:

Epoch	Years Before Present
Terminal Pleistocene	13,000
Upper Pleistocene	128,000
Middle Pleistocene	730,000
Lower Pleistocene	1,720,000 to 730,000
Pliocene	5,260,000 to 1,720,000

Conclusions

The development of geochronology over the last forty years has done more than refine our ability to date the past; it has also demonstrated the essential unity of science and the self-correcting force of its methods (c.f., Damon 1970). When the calibration of C-14 against the bristlecone pine master sequence revealed discrepancies between radiocarbon years and calendar years, it called into question Libby's assumption that carbon-14 had been produced in the upper atmosphere at a constant rate through time.

Carbon-14 is produced when the free neutrons that form the secondary cosmic ray bombardment strike nitrogen-14 atoms. Since these free neutrons are produced by the primary cosmic ray bombardment, their quantity is dependent upon the intensity of primary cosmic radiation striking the atmosphere. Research in recent years indicates that the magnetic field around the earth screens the atmosphere from much of this cosmic bombardment. The construction of the paleomagnetic chronology by means of potassium/argon dating of volcanic rock has demonstrated that the Earth's magnetic field has fluctuated in intensity through time. It seems likely therefore, that when the field is stronger, fewer cosmic rays reach the upper atmosphere, fewer neutrons are released, and less carbon-14 is produced.

The recovery of deep sea cores has permitted the construction of a chronology of the Pleistocene that emphasizes world-wide climatic shifts. By combining a variety of chronometric dating techniques with geochronology and an analysis of the composition of the microfossils in the cores, we know much more about the timing and magnitude of the periods of intense cold known as the Ice Age than we did just 50 years ago.

Major Points Raised in Chapter Five

1. Students of the past have developed a systematic body of method for collecting, recording and analyzing the material remains of past human activity.

2. The first section deals with the archaeological site formation process that have created the archaeological record in the first place. Archaeological sites consist of six categories of material evidence: occupational surfaces, features artifacts, ecofacts, human remains, and datable samples.

3. The nature of organic decomposition and decay is discussed and special circumstances are noted were, due to an absence of water, oxygen, or temperatures above freezing, organic material does not decay.

4. Archaeological context and excavation methods are discussed. The five kinds of relationships that make up the archaeological context are discussed together with the conventional scientific field methods used to establish them.

5. The aims of archaeological field methodology and the relationship between the ethnographic descriptions of culture done by anthropologists and the "paleo-ethnographic" descriptions done by archaeologists.

6. The section ends with an analysis of the excavation and interpretation of the Coxcatlan Cave in the Tehuacan Valley of Mexico to provide a concrete example of the application of these methods.

CHAPTER 5

BASIC FIELD METHODS IN ANCIENT STUDIES

Introduction

Transcending mythic prehistory and transforming philosophical accounts into scientific ones demands nothing less of archaeologists than **their continued and unflinching application of the brute force of the scientific method to questions about the past**. The logic of this method demands that they test their theories and models about the past against the actual patterns of ancient material remains.

To accomplish this, archaeologists have developed a systematic set of methods adapted from engineering surveying and geology to recover material patterns at archaeological sites. These basic methods are discussed in this chapter and their application is illustrated using the Coxcatlan Rockshelter site in the Tehuacan Valley in Mexico. In describing the field and laboratory methods, we turn to the relationship between **what** we know scientifically about the past and **how** we come to know it.

Archaeological Site Formation Processes

Shortly after mid-morning in the late summer of A.D. 79, the long dormant Italian volcano Vesuvius erupted spewing forth an immense black cloud that spread westward towards the Bay of Naples, blotting out the sun. In the hours that followed, toxic fumes filled the air and volcanic ejecta of all kinds rained down on the countryside. By the time the eruption ended three days later, virtually all life along the coast had been snuffed out and Roman towns as far north as Herculaneum and as far south as Stabiae had been buried. Yet as terrible as this disaster was, it would have long been forgotten by all but a few historians and Latinists had not chance excavations in the 18th century revealed the existence of buried ruins and statuary beneath the volcanic debris. These ruins, eventually recognized as the remnants of the doomed city of Pompeii, produced a treasure trove of ancient relics.

In the century that followed, treasure hunting at the site gave way to archaeological excavation. Systematic work revealed that the blanket of volcanic ash that had destroyed Pompeii had also preserved its final moments. Excavators found buildings, artifacts, art treasures, privies, brothels, and shops frozen in time much as their occupants had abandoned them. Pots were recovered in the very rooms, sometimes on the very stoves, where their owners had been in using them on that day in A.D. 79. Rooms still contained their furniture, stores their merchandise. Even the remains of Pompeiians who had been unable to flee the site were recovered. Although bodies buried in the ash had decayed, the space that they had occupied remained. By carefully pouring plaster of Paris into these spaces, excavators were able to recover casts of these bodies. As a consequence of this preservation, the excavation of Pompeii has given the world a remarkably complete picture of life and culture in Roman Italy in the first century after Christ. Justifiably, the site is regarded as one of the most significant finds in the history of archaeology.

In another sense, however, Pompeii's legacy to modern archaeology has not been entirely beneficial. According to Lewis Binford, the very completeness of the remains of that site has created a persistent mental impediment to the scientific understanding of the archaeological record in general. That impediment, which Binford has christened the "Pompeii principle," is the conscious or unconsciously held notion that the archaeological record contains the remains of the past frozen in time. It is the assumption that the archaeologist excavates buildings, artifacts, pots, and tools more-or-less in the location of their former use and that these material remains directly mirror the way of life that produced them. In fact, it is seldom so simple. More commonly, the material remains recovered at an archaeological sites were brought together by a series of cultural and environmental transformation processes peripheral, even entirely unrelated, to the manner in which they were used in the socio-cultural system that produced them. It is these processes which create the entities that archaeologists refer to as "sites."

An archaeological **site** is a location that contains material evidence of past human activity in a natural matrix. The term may be applied to locations ranging in size from small clusters of a few artifacts to the remains of whole cities, from scatterings of stone tools on the surface of a plowed field to deep cave or river valley deposits measured in tens of meters.

The material evidence recovered from such archaeological sites can be arranged into six categories: occupational surfaces, features, artifacts, ecofacts, human remains, and datable samples.

Occupational surfaces are the areas such as room floors, streets, courtyards, and pathways upon which people carried out activities. Occupational surfaces form the boundaries between depositional strata. An archaeologist must distinguish between the items that were deposited during the occupation of a surface and those deposited after it was abandoned (Rathje and Schiffer 1980: 189).

Features are nonportable units, facilities, or artifacts that cannot be removed from their site matrix without destroying their integrity. Distinctions must be made between constructed and cumulative features (Sharer and Ashmore 1993: 404-405). **Constructed** features are in some sense the result of intentional human effort and activity. Common examples include: fire hearths, graves and burial mounds, storage or refuse pits, caches, ovens and kilns, post holes and post molds, roads, irrigation works, and the remains of buildings including houses, temples, palaces, and earthworks. **Cumulative** features are those formed by incidental accretion or removal, rather than design or intentional effort. Examples include: garbage middens, and concentrations of debris resulting from activities like skinning or stone tool manufacture.

Artifacts are objects that have been manufactured, modified, and used by human beings. From the archaeological perspective, artifacts are distinguished from features very simply: artifacts are portable, features are permanent. According to Oswalt (1976: 24, 18), artifacts must also be distinguished from **naturefacts**. While naturefacts are "natural forms, used in place or withdrawn from a habitat, that are used without prior modification by creatures," artifacts are "the end products resulting from the modification of a physical mass in order to fulfill a useful purpose" (Oswalt 1976: 18, 24). Artifacts are a diverse class. Objects as varied as roughly-shaped and pointed sticks used for prying open mussel shells, stone or metal projectile points, pottery vessels, glassware, and computer hardware all may be considered artifacts.

Ecofacts are natural objects and materials that have not been manufactured by human beings and yet have bearing on the interpretation of an archaeological site (c.f., Binford 1972: 147-148). Bower (1986: 15) distinguishes between environmental and cultural ecofacts. **Environmental** ecofacts are natural objects and materials whose presence on an archaeological site is primarily due to the operation of natural forces rather than human activity. The pollen and spores, soils, and the sediments recovered from archaeological sites generally constitute environmental ecofacts. Their analysis is the chief means of reconstructing the site's ancient environment. **Cultural** ecofacts are natural objects whose presence or abundance in the site **is** the result of human action. For example, human excreta and food debris like animal bones, seeds and husks are cultural ecofacts. Their presence at an archaeological site is the result of specific human food preferences and hunting, storage, and food processing techniques and they provide insights into the operation of the cultural system that produced them.

Human remains are commonly recovered on archaeological sites. Skeletal remains are by far the most common human material recovered, however, under especially favorable conditions, soft parts of the body are sometimes preserved. These remains are excellent sources of information on ancient disease, growth patterns, mortality rates, population structure, life expectancy, and diet. They also reflect cultural practices like cranial deformation, surgical techniques, sacrifice, warfare, and mortuary beliefs and customs.

Datable samples are the materials used to obtain relative or absolute dates. Such samples include wood specimens for dendrochronology, varves, organic material for radiocarbon dating, unworked volcanic deposits for potassium/argon determinations and so forth.

In the process of excavating a given site, the archaeologist groups the above six classes of evidence into separate occupations. An **occupation** may be defined as

> a spatial cluster of discrete objects which can reasonably be assumed to be the product of a single group of people at a particular locality and deposited there over a period of continuous residence (Dunnell 1971: 202).

The specific configuration taken by six classes of

evidence in the occupations at a given site is the product of the **cultural formation processes** (C-transforms) and **environmental formation process** (E-transforms) that operated there. According to Rathje and Schiffer (1982: 106), archaeologists and prehistorians are interested in ancient objects and material in two states: "**systemic context**, where the artifacts are a part of activities within an adaptive system, and **archaeological context**, where artifacts have been deposited and are no longer involved in activities." The cultural formation process consists of human activities which result in objects becoming part of the archaeological context. Rathje and Schiffer (1982: 107) categorize cultural formation processes as: reuse, deposition, reclamation, and disturbance.

Reuse is "the retention of items in systemic context that, after use, might otherwise have been discarded" or "a change in the user, use or form of an artifact, following its initial use" (Schiffer 1987: 28). **Secondary use**, such as proping open a door with a broken automobile starter motor, or the **recycling** of old materials into new raw material, such as is commonly done today with aluminium cans, or **lateral cycling**, when an object is passed from person to person as an heirloom, are examples of variations of the reuse process.

The **deposition** process involves the discharging of material to the environment, whether through accidental loss or intentional discard. Activities as diverse as the caching of food, the disposal of trash, the interment of the dead, and the abandonment of a structure are all forms of the cultural deposition. It is the opposite of reuse (Schiffer 1987: 47).

The **reclamation** process involves the transformation of materials "from archaeological context back into systemic context." It involves the reuse of previously deposited artifacts and structures. Reclamation occurs, for example, when scavengers root through trash deposits or grave robbers remove the contents of a tomb (Schiffer 1987: 99).

The **disturbance** process refers to human activities that unintentionally modify the archaeological record. Road building, plowing, land clearance, flooding are examples of disturbance processes. Reclamation and disturbance processes both alter the archaeological record but in the case of disturbance, archaeological material does not re-enter systemic context (Schiffer 1987: 121).

The interaction of these cultural formation processes has a profound impact on the archaeological record. This interaction determines what will and will not leave the systemic context and enter the archaeological context. The environmental formation process refers to the interaction of archaeological remains with the natural environment of which they are a part. In this process, natural forces transform the material products of human cultural activity into archaeological remains. These natural forces or agents of deterioration are chemical, physical and biological in character. They include:

1. sediment deposition and soil formation;
2. erosion, mass wastage, and earthquakes;
3. organic decay brought about by such organisms as aerobic and anaerobic bacteria (especially gram positive bacteria or actinomycetes), protozoa, arthropods (insects, spiders, crabs, centipedes, etc.), annelids (segmented worms), and so forth.
4. disturbance by plants;
5. chemical changes such as solution, oxidation, hydration, and dehydration;
6. periodic freezing and thawing; and
7. periodic wetting and drying.

The fate of most archaeological remains is decay and dissolution. All organic material is subject to decomposition and the process takes place continuously. However, organic decomposition involves the interaction of oxygen, water, living organisms, and requires temperatures above freezing. In the absence of any of these key elements, organic dissolution is held at bay and preservation is fostered.

The absence of water. Permanent dryness is an environmental condition that characterizes many cave and rockshelter locations in arid and semi-arid regions of the Earth. The interiors of these landforms are shielded from direct rainfall and are characterized by low humidity. Consequently, the sediments of dry caves commonly contain well-preserved plant and animal remains, fecal material, human remains, and organic artifacts. The Mexican site of Coxcatlan rockshelter discussed below, is a dry site of this type. In a similar vein, the interiors of sealed tombs and structures in arid and semi-arid regions are also ideal preservative environments. The excellent state of preservation characteristic of Egyptian mummies owes as much to the dehydrating effect of the Nile Valley climate as to the vaunted skills of ancient embalmers.

The shortage of oxygen. Permanently wet conditions found in bogs and swamps also promote organic preservation. At the bottom of these bogs, the operations of microorganisms gradually reduces the oxygen in the water and creates a nearly oxygen-free or anaerobic environment. The shortage of oxygen in these waterlogged environments in turn:

> slows down the activity, and multiplication, of the bacteria and fungi that decompose organic matter. Litter builds up and becomes

> compressed under its own weight into peat. As the density increases, it further restricts the access of both oxygen and bacteria into the energy-rich resource which lies just beneath the skin of vegetation. The constant evaporation of water from the surface of the mire keeps the whole system cold and further restricts the growth of microorganisms (Moore 1987: 46).

Human remains and artifacts incorporated in peat bogs, marshes, or lake bottoms often exhibit remarkable preservation. Star Carr, a Mesolithic period site found during bog draining in the north of England, is a classic example. The anaerobic environment of the bog preserved large quantities of wooden and antler artifacts and bones of elk, deer, auroch, and other animals over 9,000 years old. This material provided a detailed picture of the subsistence base of early Holocene peoples in northwestern Europe. Human remains, some dating to the Iron Age and before, have also been recovered from the peat bogs of Denmark and Great Britain. Although stained a dark red-brown from the tannin in the bog waters, the excellent state of preservation of these "bog people" has sometimes deceived the discoverers of these 2,500-year-old bodies into concluding that they had recovered a recent murder victim.

Temperatures consistently at or below freezing. The permafrost conditions of the Arctic regions of the world both contribute to remarkable preservation and impede archaeological excavation. Under special circumstances, natural refrigeration has contributed to the preservation of organic remains outside the Arctic. A unique example of such preservation is found in the burial mounds of the Scythian peoples of the Altai mountains of Soviet Siberia and Outer Mongolia. According to Artamonov,

> Through an accident of their construction these graves have been frozen virtually since they were made more than 2,000 years ago. In each of the mounds the burial chambers were covered with a layer of massive boulders. This layer was not airtight, and the burial chambers were soon filled with freezing air during the cold Altai winter. Because cold air is heavier than warm, the frigid air below the boulders was not penetrated by warm air during the short Altai summers; only still colder, denser air could settle into the chambers. thus the graves became natural refrigerators. Here, in these chambers of eternal frost, the bodies of ancient chieftains, with their horses, clothing and varied possessions, have been preserved from decay (Artamonov 1965: 100).

Unfortunately for the archaeologist, such conditions of preservation are exceptional. Generally, organic, and even some inorganic, materials enter archaeological context only to disarticulate, decay, and disappear.

Archaeological site resolution. Objects begin in systemic context and, subsequent to the operations of the cultural and environmental formation processes, become part of the archaeological record. During this transformation, the order and arrangements of objects that existed in the systemic context becomes increasingly disordered. In the course of these transformations, the material remains of cultural systems move from a more-ordered state to a less-ordered state.

Physicists use the term **entropy** to refer to the quantity of disorder or randomness in a system bearing energy or information. In information systems, entropy represents the "noise" or random errors that occur in the transmission of signals or messages. As time passes, the environmental formation process introduces increasing disorder or "noise" into the archaeological context, reduces the site's "resolution" and complicates any "reading" or interpretation of the ancient systemic context based on the information that the archaeological context contains. It is the archaeologist's task to **reconstruct the former systemic context of objects based on their archaeological context**.

Success in this task is dependent in large measure on the degree of **resolution** in the sites under interpretation. Resolution refers to the degree to which a site's archaeological context reflects the systemic context that originally produced it and the clarity with which this reflection can be seen. A site's resolution is determined by the extent and nature of the C-transforms and E-transforms that have contributed to its formation. Specifically resolution is affected by (1) whether the original deposition was the result of a single cultural activity or environmental event or of multiple events; (2) whether or not multiple events involved the same or different kinds of activities (Binford 1981: 276); (3) whether or not the site represents a primary or secondary deposition; (4) whether or not the site has been affected by reclamation or disturbance processes; and (5) the nature, duration, and intensity of the chemical, physical, and biological forces of deterioration that have operated at the site. It is possible to plot sites on a scale in terms of the degree and clarity of their resolution. Sites like Pompeii, which clearly reflect the ancient systemic context that produced them, are at the high end; sites of low resolution, such as modern landfills or a "blowout" site, are at the opposite end (Table 1).

Pompeii entered the archaeological record as the result of the operation of a single E-transform: the rapid deposition of a large blanket of volcanic ash from Vesuvius. After that deposition, the operation of biological, chemical, and physical decay altered the remains. Otherwise, little further disturbance appears to have occurred until the 18th century. Perhaps no higher resolution is possible in a archaeological site than what is met at Pompeii.

The Great Death Pit at Ur This site is located within the Royal Cemetery of the ancient Sumerian city of Ur on the floodplain of the Euphrates River in southern Mesopotamia (modern Iraq). The site consists of a deep shaft or pit containing two vaulted tombs, the skeletal remains of more than 70 individuals, numerous animals, rich grave goods, and personal artifacts in great profusion. The pit was constructed, filled, and sealed sometime around 2,500 B.C. during the Early Dynastic period in the cultural sequence of Sumer. The two tombs were built at different times, one atop and connected to the other. The lower tomb was supposed by the excavator, Sir Leonard Woolley, to have contained the remains of a king. The disorder within this tomb suggested to Woolley that the body had been removed when the tomb was opened and plundered, perhaps during preparations for the construction of the next tomb built above it. This later vault contained the remains of an apparent queen, Shub-ad, which lay in state on a raised platform surrounded by the skeleton of a number of females, thought to be her ladies-in-waiting. Outside of this tomb, a large, open courtyard had been excavated in the pit. This courtyard was connected to the surface by a steep earthen ramp down which two wagons, drawn by oxen and attended by male servants, had evidently been backed. In the courtyard, excavators recovered the skeletons of fifty-nine individuals. The majority of these skeletons were of females wearing lavish jewelry. A few males, thought by their dress and helmets to have been soldiers, were also recovered. Woolley concluded that the main internments in the Death Pit occurred at the time of the burial of Queen Shub-ad and that they represented the human sacrifice of her attendants as part of the funeral rites. He further contended that the ordered distribution of the bodies indicated that the attendants, grooms, and soldiers descended willingly into the pit and died there, if not voluntarily, at least passively.

Like Pompeii, the Ur Death Pit entered the archaeological record as the result of a single event. Unlike the former site, however, intentional C-transforms, rather than an unintentional E-transform, created the archaeological context at Ur. Initially, the tomb of the king was constructed and filled. This tomb was then reclaimed and looted, apparently during the process of preparing the next tomb. At least some of its contents were then re-used. The second tomb and the death pit were then constructed, filled and closed. Once the site was abandoned, E-transforms, including organic and inorganic decay, became significant. Later disturbance seems nonexistent until the site was excavated early in this century.

In contrast to the site of Pompeii, the Death Pit represents only one aspect of socio-cultural life at the Ur: elite death rites. Nonetheless, the archaeological context recovered in the Death Pit reflects the ancient Sumerian systemic context in a number of clear and dramatic ways. The materials in the Death Pit constitute a "grave lot," that is, they were all buried together. We therefore we can refer to Worsaae's Law of Association. The Death Pit grave lot associates the jewelry, weapons, musical instruments, costumes, and other elaborate artifacts that jointly played a part in the elite component of the ancient Ur systemic context around 2,500 B.C. The Death Pit contains tangible evidence that by 2,500 B.C., social formations in Sumer were sufficiently powerful and authoritarian enough to legitimately demand that members sacrifice themselves on behalf of the collective. Further, the Pit seems to indicate that Ur was socially-stratified, occupationally specialized, and yet socially-unified at this early period.

However, although resolution is high, it is not perfect. We must also recall that this context does not provide us with a "snapshot" view of society at Ur as do the remains of Pompeii. The Pit contains the material remains of a political drama, albeit a particularly horrible one. It therefore displays only those aspects of society favorable to the powers that orchestrated the drama. The tensions, ambiguities, and social conflicts that must surely have existed in the city are not apparent.

The **Olsen-Chubbeck site**. Located in eastern Colorado on the Great Plains of North America, the Olsen-Chubbeck has produced evidence of a bison hunt by Paleo-Indian peoples of the Plano complex around 6,500 B.P. The site consists of "a continuous bed of bones lying within the confines of a small arroyo, or dry gulch (Wheat 1967: 45)." A 200-foot long section of this arroyo was excavated. Within it were found the skeletal remains of more than 200 individual *Bison bison occidentalis*, an extinct species ancestral to the modern American buffalo (*Bison bison*). The arroyo had apparently served as a trap into which the animals had been stampeded.

Excavation revealed that only the bison at the top of the arroyo had been fully butchered; lower in the deposit skeletons were partially or nearly complete, at the bottom of the arroyo fully articulated skeletons were recovered. The bone segments from the butchered animals provided clues to the hunters' butchering techniques, the meat yield of the kill, and the size of the hunting band. The position of the herd in the arroyo led to inferences about the hunting strategy.

Like Pompeii and the Ur Death Pit, the Olsen-Chubbeck site entered the archaeological record as the result of a single event. Like the Ur Death Pit, C-transforms were more significant in creating the archaeological context at Olsen-Chubbeck than E-transforms, but unlike Ur, the deposits that resulted from the operation of those C-transforms were unintentional. The deposit was created initially by the Paleo-Indian stampede and the butchering techniques of the hunters. Once the site was abandoned, E-transforms, including scavenging animals, organic decay, and the filling of the arroyo, became significant. Later disturbance seems insignificant until our own century when modern overgrazing and subsequent erosion exposed the site and amateur archaeologists began its reclamation. The current archaeological context reflects only one aspect of the Paleo-Indian systemic context: their hunting and butchering technology. Thus, although the site's resolution is high, the picture it presents of the former the systemic context is limited.

The **Great Temple at Tenochtitlan**. The ancient city of Tenochtitlan is located in the Valley of Mexico. It was the capital of the Aztec empire until 1519 when it was conquered and sacked by Spanish conquistadors under Cortez. Once the city was under their heels, the Spanish razed the temple and incorporated what remained of it into a huge public plaza upon which they built a large church. The siege of the city had been a bitter one and the destruction of the temple was undertaken partly as retribution against its Aztec defenders, and partly as an act designed to symbolize the new order in the Valley. The construction of a Roman Catholic church on the ruins of the temple reflected the intent of the Spanish to obtain ideological as well as military domination in the New World. In the centuries that followed, the streets and buildings of the Spanish-American metropolis of Mexico City covered the remainder of the temple compound. The temple was rediscovered in 1978 by workers excavating below street level in the modern city.

Like Pompeii, the great temple of Tenochtitlan entered the archaeological record as the result of a single catastrophic event. However, the event that created the archaeological context at the Great Temple was a C-Transform of a type totally absent at Pompeii. Initially, the temple furnishings were stripped and the statuary smashed or defaced. Next the superstructure was thrown down. Rubble fill from the temple and other building was packed around the base and sides of the structure and a pavement was built above it. E-transforms immediately began operating on the organic and inorganic remains interred at the site. There, subject only to the relatively minor disturbance of later building construction, the temple remained until its rediscovery 459 years later. Despite the generally good state of preservation, the extent to which current archaeological context reflects the 16th century Aztec systemic context is severely limited by the destruction that formed the site and by the existence of the modern city atop much of the compound. We can interpret these remains correctly only through use of the contemporary historical accounts of them. Site resolution is moderate.

A tell site in southwestern Asia. A **tell** is a mound of cultural and natural debris built up as a consequence of long-term human occupation and abandonment. **Tells** are particularly prominent and important cultural landforms in southwestern Asia and the eastern Mediterranean. In contrast to the site of Pompeii, mound sites are the products of multiple events over hundreds of years. Consequently, the C- and E-transforms that operate in their formation are varied and complex. According to Butzer and Freeman:

> The rate and type of buildup differ on living floors, in streets and alleys, or in and around community structures, such as civic buildings, walls, terraces, and drainage systems. Furthermore, during times of demographic expansion, construction in a village mound outweighs decay, so that garbage or collapse rubbles due to accident or selective razing show little net accumulation because they are cleared away and dumped elsewhere. But during times of progressive demographic decline or catastrophic destruction, through natural or human disasters, garbage and collapse rubble build up and then are affected by running water, by biogenic agents, and by wind.... Once abandoned, the mound is slowly lowered by compaction and weathering, as well as by erosion through gravity movements, surface water, and wind; such sites may also be used as quarries for rock, mud, pottery temper, and fertilizing compounds; finally, shifting streams can undermine and erode a mound. As the

> topographic form is gradually flattened, potsherds move down-slope and initially concentrate around the lower periphery of a mound, where trampling reduces them in size; eventually the sherd debris is increasingly buried within water-lain sediments or turned down into the soil by plowing (Butzer and Washburn 1986: xii).

The complexity of the C-transforms that operate to form the archaeological context of a **tell** make a direct reconstruction of its systemic context difficult. Trash disposal and settlement maintenance activities that took place during periods of occupation make it unlikely that the archaeologist will recover artifacts in the structures or activity areas where they were actually used, such as was the case at Pompeii. In an active village, buildings that are abandoned either come to be filled with trash and other debris unassociated with their original use or are repaired and reused. Of course, the **tell** settlements were sometimes destroyed suddenly in accidental fires, military attacks, or by earthquakes or other natural disasters. Such destruction levels in **tells** are more Pompeii-like in character, that is, their archaeological context more closely reflects the former systemic context. However **tells** were commonly re-occupied at some period after their abandonment and later construction often destroys the evidence of earlier occupation. In general, resolution of individual events in **tells** is poor to moderate. However, because **tells** contain long occupation sequences, we can say that their cumulative resolution is quite good.

A modern sanitary landfill When modern objects cease to be of any value, they are often placed in a trash container outside the location of their last use, picked up, carried off, and deposited in a sanitary landfill. Scavengers commonly pick through the most recently deposited refuse in a landfill, but earlier material is generally inaccessible to them as it has been bulldozed into a relatively horizontal layer and covered, first with clean earth, later with more refuse. Material deposited in the landfill at the same time is subjected to roughly similar C-transforms of reclamation and deposition. This material comes to be located together in the same cultural layer. Within this layer, objects are subjected to similar environmental transformations but the effect of the environment on specific objects varies. In any case, objects which had no functional association in their systemic context have become associated in the archaeological record. All that can definitely be concluded about materials in the same layer at a landfill is that they were **disposed of** at approximately the same time. We might also infer from this association in the archaeological context that the objects were used contemporaneously in their systemic contexts. However, we cannot be sure of this on the basis of stratigraphic position alone. Thus, because objects in a landfill were used elsewhere and their association with one another is largely spurious, the resolution in a modern sanitary landfill (or an ancient dump, for that matter) is low.

The Olduvai Gorge hominid site FLK 22. Olduvai Gorge is a deep canyon located in Tanzania in East Africa. The gorge cuts through stratified formations that began to be deposited in the early Pleistocene epoch some 1.72 million years ago. The volcanic tuffs in these formations have been dated by means of the potassium/argon and fission track methods providing tight chronometric age for the archaeological sites recovered within them. Additionally, the gorge has produced some of the best archaeological evidence of hominid biological and cultural evolution yet recovered. Particularly significant is the archaeological association of early fossil hominid skeletal remains with crude stone tools and animal bones. The site of FLK 22 contains such an association. Fossil remains classified by the Leakeys as **Zinjanthropus** were found at FLK 22. They are now classed as *Australopithecus bosei*. The abundance of bird and amphibian bones, together with some fish remains and crocodile teeth and other evidence, suggests the site was formed in a wooded, lake side environment.

Louis and Mary Leakey, the discoverers of the site, interpret these associations as evidence that early hominids had established a base camp to which they carried their game animals. Some scholars have gone so far as to infer from these associations that the site is evidence that these early hominids shared food and practiced a division of labor wherein males left the base camp to hunt while females remained to tend their offspring and to gather vegetal food in the camp's vicinity (Isaac and Isaac 1975). However, as Binford (1981) points out, some of the animal bones recovered there show evidence of having been gnawed by scavengers and the bone assemblage from the site contains a very high percentage of fragmentary bones. After comparing the patterns of bone breakage found in the Olduvai sites with patterns observed to be the result of predatory and scavenging animal activity, Binford concluded that destruction of bone by non-hominid predators and scavengers was clearly one of the major processes contributing to the formation of the FLK 22 assemblage (Binford 1981: 261). He also ran a series of statistical comparative analyses of the bone assemblages from Olduvai hominid sites in general and concluded that far from containing the remains of a single hominid occupation or activity set, the FLK 22 "Zinj" site is a mixed deposit that contains the remnants of a number of contexts including the material remnants of a lake margin habitat; the remains of animals who probably died a natural death and whose remains were scavenged; the remains of animals killed elsewhere, scavenged, and transported to the site by either hominids or non-hominids; the remains of animal kills that actually took place on the site (Binford 1981: 276)

Far from being the material remains of a single kind of hominid base camp activity, the site is really a "palimpsest of many different event episodes" in which, at different times, both hominids and non-hominids played a part (Binford 1981: 251). Because of the number and diversity of the events that contributed to the formation of the site, its resolution is low.

A **west Texas blowout site** Like many semi-arid regions of the world, parts of West Texas have been extensively overgrazed at various times in the recent past. Overgrazing results in the removal of grasses and shrubs that protect the land from wind and water erosion. As erosion occurs, successive levels of the topsoil are removed exposing heavy material on the new land surface. Archaeological "blowout" sites are created when wind erosion removes the soil and sediment matrix at a open-air site exposing the heavier cultural material on the surface. Archaeological materials exposed in this way no longer have stratigraphic contexts. Blow out sites are characterized by mixed surface assemblages of artifacts and ecofacts dating to many periods. Once exposed, they are subject to environmental destruction on the site's surface; decay, trampling by animals, or transportation from their original location by water erosion. Reclamation by relic collectors also often takes place. Like FLK 22 Zinj, a blowout site is a "palimpsest of many different event episodes." However, unlike that site, erosion of the matrix, rather than its deposition, is the primary E-transform active in blowout formation. Archaeologists can date some of the artifacts by means of seriation but little

Table 1. Scale of Archaeological Site Resolution

Resolution	Site	Site Resolution Criteria 1	2	3	4
High	Pompeii	S	P	RY,DN	S,W
	Death Pit at Ur	S	P	RY,DN	S,W
	Olsen-Chubbock	S	P	RY,DY	L,I
	Great Temple, Tenochtitlan	S	P	RY,DY	S,I
	Southwest Asian tell	M	P/S	RY,DY	S,I
	Modern landfill	M	P	RY,DN	S,W
	Olduvai Gorge homind site	M	P/S	RY,DY	L,I
Low	West Texas "blowout" site	M	S	RY,DY	S,I

Criteria for Placement on the Site Resolution Scale:

1. Whether the original deposition that created the site was the result of a single cultural activity or environmental event or multiple events (**S=single, M=multiple events**);

2. Whether the site represents a **primary (P) or secondary (S) deposition**;

3. Whether or not the site has been affected by reclamation or disturbance processes (**RY=reclamation has occurred, RN=no reclamation; DY=disturbance has occurred, or DN=no disturbance**);

4. The duration (**L=long, S=short**), and intensity (**W=weak, S=strong**) of the chemical, physical, and biological forces of deterioration that have operated at the site;

information can be obtained other than knowledge that people of a particular period once were present in the area. Thus, blowout sites are characterized by extremely low resolution.

Archaeological Setting and Archaeological Excavation Methods

The nature of ancient material and its place in an archaeological site are the results of direct human activity and of the natural processes that foster accumulation, preservation and decay. Understanding the relationships between materials and their natural matrices in archaeological settings is critical for understanding the nature of the human activity that took place. These relationships may be classed as **locational**, **positional (provenance)**, **stratigraphic**, **contextual**, and **associational.** Knowledge and understanding of them is obtained through systematic site reconnaissance and excavation. Systematic excavation of archaeological sites constitutes the "experiments" of the sciences of archaeology and prehistory. However, the systematic excavation of an archaeological site results in that site's **destruction**. Unlike the experiments of the physical sciences, archaeological excavations can be preformed only **once**. The archaeologist has but a single opportunity to achieve scientific understanding of a site's unique contents. It is therefore critical that relationships within the data be recovered as completely and accurately as possible in order that the site may be "reconstructed" in the laboratory and in the final site report. To achieve this goal, a basic set of engineering surveying methods and conventions have been adapted for use in the excavation and recording of archaeological sites. These methods provide simple and elegant solution to the problem of describing archaeological spatial relations. The application of these methods to the excavation of archaeological contexts permits archaeologists to locate sites of past human activity; recover and record positional, stratigraphic, contextual, and associational relationships in the archaeological context; to reconstruct the site's archaeological context in the laboratory; and to use the understanding of these relationships to reconstruct the ancient systemic context in the final site report.

Locational relations Willey and Philips (1958: 18-21) suggest archaeologists use the terms **site**, **locality**, **region**, and **area** to refer to geographic units of increasing scale. The position of an archaeological site in the landscape of a geographic locality, region, or area is referred to as its location. The location of an archaeological site is established either through accidental discovery or reconnaissance. Archaeological site **reconnaissance** refers to the systematic search for and identification of archaeological sites. Once a site is located, it is assigned a number and it's position is plotted on a topographic map and its size and nature are assessed. There are two basic methods of site reconnaissance: the direct examination of the ground surface by pedestrian survey or the indirect examination of the surface or subsurface using remote sensing equipment. Remote sensing involves either the use of aerial or satellite imagery or radar to examine the ground surface for archaeological remains, or a subsurface sensing technique carried out at ground level using proton magnetometers or resistivity detectors, pulse detectors or other means (Sharer and Ashmore 1993: 196-220). Direct surface survey takes many forms but it always involves the location and assessment of archaeological site by directly examining the ground on foot or by vehicle. Once a site has been located, plotted and assigned an inventory number, its surface remains are mapped, and a collection is made of artifacts and ecofacts found on its surface. Material collected from the site's surface is processed in the laboratory, dated by means of seriation, and used to estimate the time and duration of the site's occupation. Surface data is critical in deciding which sites demand subsurface testing and full scale excavation.

Positional (provenance) relations Positional or provenance relations refer to the place or location in three-dimensional space occupied by archaeological objects. Archaeologists use basically four surveying conventions in describing the position of archaeological material in vertical and horizontal space. These conventions include the use of control grids, triangulation, datum points, and plan and section drawings.

Control Grid. An archaeologist uses a "control grid" on a site in order to record the location of objects or points on the **horizontal** plane. The construction of a grid begins with the establishment of a point of origin on the site. The orientation of the grid may be based on geographic north or magnetic north or it may be arbitrary. The origin is labeled with grid coordinates that may be arbitrary (e.g. N 500 E 500) or based on an established grid system (e.g. Universal Transverse Mercator system). The horizontal surface of the site is then divided into a grid by laying out a (potentially infinite) number of equally spaced lines parallel to each of these axis. In this fashion, the grid divides the site into a network of adjacent and identical cells or "grid squares." Each square is assigned a different number. Within the area enclosed by the grid, any point can be located simply by determining the number of it's grid square. This is done by counting the number of grid squares along each axis to the point. Each of these distances is measured parallel to the other axis (see Figure 1).

Triangulation. When the horizontal location of an object or point must be specified with greater precision, archaeologists use triangulation. Triangulation is a method of determining a position by taking bearings to two fixed points known distances apart and computing the resultant triangle. Since the dimensions of a grid square are known, the distances from any two of the square's corners to a object within its boundary unambiguously describe the location of that object. The angles of the triangle need not be computed. Note that in Figure 1, that by specifying the distances between the northeast and the northwest corners of the square and points X and Y within them, we have described the unique position of each point.

Datum An archaeologist establishes a datum point on an archaeological site in order to record the elevation of objects or points, that is, their location on the vertical plane. The datum point represents a fixed point in **vertical** space against which all other elevations at the site are referred. Although it is sometimes desirable to determine the elevation of the datum point relative to sea level, this is not always necessary. More often, the archaeologists specifies that the primary, "alpha" datum at the site is equal to 100 meters and computes all other site elevations from that value. For example, an object 10 centimeters below datum has an elevation of 99.90 meters. Ideally, the point of origin for the control grid is also the site datum and is located in the southwest corner of the site.

Plan and section drawing. Archaeologists record three-dimensional positional information recovered at a site on carefully drawn site maps. Plan view maps show the horizontal position of excavation units, artifacts and features. Archaeologists commonly construct plan-view maps of each occupation floor, feature, or structure they uncover. Vertical position or elevation is recorded on sectional maps, often referred to as profiles. Stratigraphic profiles are discussed in the next section. The production of plan and sectional views of the archaeological context has been markedly simplified in recent years through the use of computer plotting programs. Provided accurate measurements have been recorded, it is now literally possible to reconstruct archaeological sites in three-dimension on the computer screen.

Stratigraphic relations. Vertical and horizontal space are only two of the dimensions of interest to the archaeologist; time is the third one. Time is commonly recorded in the accumulated layers found in archaeological sites and geological formations. In Chapter 3, a layer was defined as a single, discrete unit of material deposited, formed, or constructed in a given location. Stratigraphy may be defined in two ways. The word literally means "the study of layers." However, when geologists and archaeologists speak of the "stratigraphy" at a given archaeological site or geological exposure, they are referring to the specific order of layers evident there. Strata of interest to the archaeologist include soils, sediments, and cultural layers.

Layer segregation. The recognition and segregation of individual layers is the initial problem confronting a stratigrapher investigating a geological or archaeological site. Segregation is generally made on the basis of differences in color, texture, composition, permeability, and contents among the layers (Michels 1973: 25). Once their boundaries have been recognized, the individual layers are minimally described in terms of six attributes (modified after Farrand 1985: 22):

1. The nature of the layer's **contact relations**, that is, the extent to which the boundary of the layer is sharp and clear or gradational, and the presence or absence of unconformities.
2. The **sedimentary structure**, that is, the degree of lamination and the presence of crossbedding, ripples, mud cracks and other features within the layer.
3. The **specific color** of the layer together with an assessment of the degree of uniformity or variety of color throughout it.
4. The presence of **internal facies changes**, that is, lateral or vertical changes in the layer's mineralogy and sedimentary structure, or in the concentration, dispersal, and nature of the artifacts and ecofacts that it contains.
5. The layer's **lateral relations**, especially its proximity to other layers or to constructed and cumulative features like hearths, pits, graves, and middens.
6. The presence and position of **datable samples** in the layer.

In addition to this field description, the samples of the layers' contents should be examined in the laboratory. This because the texture, grain size, mineralogy, and chemical composition of the material that makes up that layer can tell a great deal about the processes involved in its deposition, as well as, about rates of formation. Such examination also provides information about the manner in which the layer has been modified following deposition by weathering, erosion, and the actions of humans and other organisms. Following description, layers are interpreted. The order of the formation of the layers is determined using the Law of Superposition. Correctly reading the stratigraphy exposed at an archaeological site can sometimes be difficult. Strata can be complicated, distorted or mixed by erosion, burrowing animals, the subsequent digging of pits, foundation, and graves by site occupants, pothunting and a host of other cultural and environmental formation processes. These formation processes must be recognized if the strata are to be correctly interpreted.

If for no other reason than this, the careful recording of stratigraphic data would be important in archaeology. There is a more important reason, however. As Farrand notes,

> Sedimentary strata are real units, each with a definable top and bottom (either sharp or gradational), that encompass all the physical space within the site. Moreover, sediments are always present in a site, whereas artifacts or fossils may be lacking in any given stratum. Therefore, stratigraphy based on the sediments (i.e., lithostratigraphy of the geologist) *is the most logical means of organizing all the other data gathered from the site* (1985: 22; italics added).

The conventions of stratigraphic excavation in archaeology As we saw in Chapter 2, the development of the study of layers made a scientific understanding of the past possible. Stratigraphic position is extremely important to our understanding of a complex archaeological site. Excavation of the site systematically removes this crucial stratigraphic information unless steps are taken to preserve it. Archaeologists often use **balks** to preserve stratigraphic profiles within a site.

A **balk** is a portion of unexcavated earth left between grid squares during the excavation of an archaeological site. When a grid square is excavated, its contents are removed and the relations that existed between the strata it contained are destroyed. However, the stratigraphy is exposed on the vertical face of the pit. When the adjacent grid squares are excavated, this vertical face is commonly retained as a balk. In this way the pit's layers may be studied and referred to as the excavation proceeds.

A **stratigraphic profile** is a scale drawing of layers exposed on the vertical plane at an archaeological site or in a geological formation. An adequate profile accurately represents the size, structure, color, facies, and lateral relations evident in the original stratigraphic exposure at the site.

Contextual relations. Layers constitute the context in which the materials they contain were recovered. The term **context** means more than merely the place or location in space occupied by archaeological objects. The layer or feature in which an object was found contributes to the artifact's context. Both the order and the process of the layer or feature's formation influences our understanding of the objects that it contains. The major convention for recovering contextual relations at an archaeological site is the method of removing earth by natural layers. Excavation by natural layers follows logically from the exposure of stratigraphic profiles described above. Once segregated and described, individual layers are removed as **units.** The artifacts, ecofacts, human remains, and datable samples they contain are processed separately from the contents of all other layers. This material is catalogued and assigned a number that refers to the specific layer from which it was taken. This number remains with the material throughout its processing and allows researchers to place it in proper stratigraphic context in future.

Associational relations. The notion of context in archaeology is closely tied up with the concept of association. The stratigraphic context of an archaeological object consists of its location and surroundings. Artifacts from the same context are associated if they were deposited by the same process at the same time.

Associations can be either **singular** or **recurrent**. That is, they can either unique co-occurrences of particular kinds archaeological material or co-occurrences which are found to in more than one instance (Schiffer 1987: 19-21). Both singular and recurrent associations are important archaeological data. While it is common archaeological practice to assume that the association of objects in the archaeological record results from their association in the systemic context that produced them, Schiffer asserts that such an assumption is unwarranted. In fact, many processes, both cultural and non-cultural, can create such associations. One has only to return to our example of the modern landfill to realize that objects found together in the archaeological record were not necessarily used together in the past. Since different formation processes can produce similar patterns in the archaeological record, it is up to the archaeologist to demonstrate that the patterns he uses to test his hypotheses were not created by processes other than those he has inferred. In any event, the ability to demonstrate that certain classes of archaeological material are singularly or generally associated at a given archaeological site depends on the quality of the field methods and records employed at the site. The ability to demonstrate that such associations exist through time and space at other sites depends additionally on the cross dating of individual sites with other such sites in the region.

Site Relationships and Archaeological Interpretation: an Example

To illustrate the utility of recording stratigraphic, contextual, and associational relationships at archaeological sites, let us examine a brief example. Suppose that we have excavated a test trench at the mouth of a rockshelter at the foot of a limestone cliff. In our test trench we exposed and segregated two sediment layers above the limestone floor of the shelter. We numbered the layers in the order of their excavation: the top layer as Stratum I and the bottom as Stratum II. **Stratum I** ranges between four and seven centimeters thick and consists chiefly of bird droppings, organic debris, and fine silt apparently deposited by wind action. This layer also contains an array of small and mostly fragmentary mammal bones. Ninety percent of these bones are those of rodents and other small mammals; the remaining ten percent those of white-tail deer. Two projectile points or "arrow heads" were the only cultural material recovered in this layer.

Stratum II ranges between 12 and 15 centimeters thick and is composed partly of *eboulis* or rock spalls of limestone that were broken from the walls and ceiling of the cave by frost weathering. The remainder of the matrix of the stratum consists of fine particles of sands, silts and clays derived from the chemical and physical weathering of the shelter's limestone or introduced into the cave by water and wind action. The stratum contains fairly large quantities of chipped stone debris, cores from which this debris appears to have been struck, and a number finished stone tools. The bones of mammals have also been recovered from the stratum. Ninety percent of these bones were those of bison, white-tail deer, and elk. Small rodents make up the remaining ten percent. In addition, the stratum contains wood ash, plant seeds and macrofossils, and various other forms of organic debris. Near the back of the cave, a small quantity of dry **coprolites** (fossil fecal remains) were recovered and identified as of human origin.

The most striking contrast between the two layers is seen in the species of animal bones recovered in them. The bone remains recovered in Stratum I of our hypothetical site reveals a very different array of animal species than those in Stratum II. The ratio of rodent species to larger species becomes completely reversed when we move from Stratum I into Stratum II. Different species of "microfauna" like mice and voles tend to have fairly specific moisture and temperature requirements. Their limited tolerances therefore make them highly sensitive to variation in rainfall and temperature. Because of this sensitivity, changes in their frequency of occurrence in an area is often be used to measure alterations in local climate. At our site, almost 90 percent of the animals bones recovered from Layer 1 are those of species of small rodents while the remaining 10 percent of the faunal array consists mainly of white tail deer. In contrast, remains of such rodents constitute less than 10 percent of the animal bones recovered in Stratum II; the bulk of the species represented in that faunal array are from larger species including white tailed deer, elk, and bison. **Does this dramatic reversal indicate that the natural environment around the cave changed between the formation of the two strata?** To answer that question, we must consider the site relationships within and between the two strata.

The material recovered from these two layers have very different contexts. First, they are successive; the deposition of Stratum I followed that of Stratum II in time. Second, the nature of their formation was different. The shape and texture of the fine sediment particles of Stratum I suggest that the layer was deposited chiefly by wind action. The wind-blown or aeolian materials of Stratum I are mixed with organic fecal matter deposited by birds nesting in the cave. This layer is best classed as a natural sediment as it contains only limited material of a human origin.

Much of Stratum II consists of material of non-human origin such as the colluvial material or rock fragments that broke off the cave walls and roof and fell to the floor, and the various silts, sands and clays. However, the remaining contents of the layer suggest human agency. Most obviously, the stone tools and chipping debris it contains were presumably dropped or discarded there by the people who the cave at various times. In addition, the bones of large animals like deer, bison, and elk would not be expected in such a setting if they had not been brought there after their deaths. In the absence of evidence of predator or scavenger gnawing or breakage on these bones, or indications that they had been deposited there by water or gravity, we are probably safe in attributing them to human agency as well. Besides the human coprolites, we can be less certain of the origin of the other organic debris and the wood ash. Since the ash was scattered throughout the deposit rather than being concentrated in hearths or firepits, we cannot rule out the possibility that it represents the natural burning of the layer. On balance, however, it is apparent that Stratum II may be considered a culturally-modified sediment as much of its contents reflects either human action or human presence during its formation.

In terms of the preceding example, the large mammal bones in Stratum II are associated with wood ash and large quantities of stone tools. The small animal bones are associated with bird feces and low quantities of stone artifacts. The different contexts and associations of the materials recovered in the two layers has some immediate consequences for their interpretation. Any reconstruction of the environment that characterized the area around the cave site during the formation of the two layers must first have to account for the association of human artifacts with the bones in Stratum II and of bird feces with the bones in Stratum I. It seems clear from these associations that the Strata I and II were formed by very different processes. The picture of the environment presented by examining the faunal remains in the two layers must be tempered by the recognition that most of the small bones recovered in the first layer probably past through the stomachs of owls nesting in the cave and were derived from their fecal matter. Therefore, the specific array of animal species found in the layer is chiefly due to the selective action of predatory birds. Stratum II, on the other hand, appears to consist largely of the residue of the human occupation of the cave. The animal bones contained in this layer are more likely to represent species favored and selectively hunted by human hunters.

When the different **associations** of the bones in the two strata are considered, it becomes clear that the different bone arrays were generated by different formation processes. Although the forest may be filled with mice, human hunters tend to disregard them and concentrate instead on the larger animals. It is thus more parsimonious to explain the difference in the ratios of rodent bones to larger animals in the two layers as due to the difference in the way in which humans and predatory birds exploit the same environment.

Since the different percentages of species may be attributable to the processes of layer formation, we cannot use such differences as evidence that the climate changed around the cave between the formation of Strata I and II. Thus the context in which the bones were recovered and the material associated with them is as important as the bone frequencies of climatically-sensitive species of microfauna. We may safely disregard the increased frequency of mice bones in Stratum I as evidence of climatic change and instead take the simple **presence** of the same species of mice in both layers as sufficient evidence that the the climate outside the cave remained the same during the formation of the two layers.

On the other hand, what about the *eboulis* in Stratum II and its absence in Stratum I? Since *eboulis* or rock spalls are produced by the alternate freezing and thawing of moisture in the fissures of the rockshelter's limestone, doesn't this indicate a harsher, colder climatic regime outside the cave during the formation of Stratum I? This may be so but the answer to this question is equivocal. It is possible that human presence in the rockshelter at may account for this difference between the two layers as well. That is, by increasing the humidity and then heating the walls and ceilings with their camp fires, the human occupants of the rockshelter may have created micro-climatic conditions that produced the *eboulis*.

Therefore, in the absence of any other evidence, the differing contents of Stratum I and Stratum II **cannot** be taken as indicating local climatic change around the cave between the formation of the two layers. As this hypothetical example illustrates, stratigraphic, contextual, and associational relationships must be considered if the archaeologist is to address the question of climate change at the site.

Ethnography and Paleoethnography

In the foregoing sections of this chapter we have focused on the formation process that have created the archaeological record and on the field methods archaeologists have developed to recover data from this record. In this section, we focus on the problems of interpreting and synthesizing data about the past collected in this manner. Anthropologists describe the lifeways of contemporary peoples by writing **ethnographies**, that is, holistic descriptions of their customary behaviors, beliefs, and attitudes. An ethnography generally includes accounts of natural and social environment; technology; settlement and community patterns; economy, including the subsistence system, seasonal round, craft and manufacturing specializations, and trade patterns; demography; diet and nutrition; status of health; social organization, including kinship and marriage systems; political organization; religion and ideology; and language.

Since they are able to observe these customary practices themselves, or can learn of them through historical records, anthropologists can **directly** test their hypotheses about the nature of a given socio-cultural system. Prehistorians seek similarly to describe the lifeways of ancient peoples, but since they can only test their hypothesis **indirectly** against the material remains of vanished socio-cultural systems, much is irretrievably beyond their reach. In the terms we have been using thus far, prehistorians seek to reconstruct the vanished, ancient systemic context by testing their ideas against the evidence found in the contemporary, but remnant, archaeological context. To relate, and yet to distinguish, the aims of anthropologist and prehistorian, it might be said that the latter writes "ethnographies of the dead" or simply **paleoethnographies**. The paleoethnography stands in the same relation to standard ethnography as the shadow does to the body. Yet the prehistorian has a distinct advantage over the anthropologist: the four million years of prehistory provide a vast population of socio-cultural systems to draw upon in comparative research. Further, the prehistorian often can recover the entire history of an ancient socio-cultural system and learn "how it all turned out." The prehistorian has the entire movie, and need not be content with the mere snapshot view available to the anthropologist.

The three-part structure of the paleoethnography. In constructing a paleoethnography, the prehistorian constructs a model of the socio-cultural system and then attempts to understand how and why that system changed through time. This modeling is done in three interrelated and sequential steps: the construction of the culture's chronological sequence; the description of the culture's history, and the explication of the culture's processes.

The term **culture sequence** refers to the chronology of a given socio-cultural system. As we have emphasized throughout this book, the establishment of the order of past events, whether at a single site or in an entire region, is the first and most fundamental responsibility of the prehistorian and its successful accomplishment is what distinguishes him from the mere pothunter. The culture sequence is established by means of the relative dating techniques of stratigraphy, seriation and cross-dating aided by the absolute or chronometric techniques like radiocarbon dating, dendrochronology, and varve counting. The first step in building a culture sequence for a given region is the establishment of a "master chronology." To this end, the archaeologist generally seeks a large, deeply-stratified site in order to recover as long and complete a series of sequential occupations as possible. Surrounding sites are then excavated and cross-dated with the sequence from this initial deep site. In this way, an increasing complete sequence of events can be assembled for the region. At the same time, and in much the same way, an environmental sequence for the region must be reconstructed.

The next step in this process is the description and synthesis of the **culture history** of the region, that is, of the forms taken by its material culture, the distribution of those forms over space, and the changes that occurred in both form and distribution through time. The archaeological site reconnaissance is a crucial part of the strategy of culture history. One site is no longer enough; the prehistorian is concerned with the entire range of variation in a culture over space and time and for this he needs sites of all types in his study. Finally, culture change must be monitored through detailed studies of artifact form and distribution. The chief aim in this research stage is the construction of a time-space grid for the region under study. To this end, the various individual site strata and occupations correlated in the first stage are grouped into a series of sequential cultural **phases**. A phase may be defined as a group of chronologically-related occupations. Occupations that are close to one another in time and exhibit similar or related cultural assemblages are considered to belong to the same phase. Phases, in turn are grouped into **periods**. Periods refer to longer blocks of time than do phases and apply to larger geographical regions. Like phases, they are arranged sequentially and may be defined in either relative or absolute time.

The explication of **culture process** is the final step in the construction of the paleoethnography. In this research step, the prehistorian attempts to explain the functional interrelationships within the socio-cultural systems and of the forces generating change over time. Understanding of the culture process in a specific region is best undertaken as part of the effort at understanding human culture at large. Therefore, in at this stage of research, the prehistorian attempts to integrate his regional data into the **stages** of cultural and biological evolution of human prehistory in general.

In sum, the goal of the anthropologist, the archaeologist and the prehistorian is essentially the same, the understanding of human systems of society and culture. However, the anthropologist approaches his subject directly, the archaeologist and prehistorian must use proxy data drawn from the material remains of ancient systems and their environments.

A Case Study: Culture Sequence And Culture History at Coxcatlan Cave

The general logic and the specific utility of archaeological field methods in the construction of paleoethnographies can best be understood by examining an actual site example. We will use the excavation of the **Coxcatlan Rockshelter** located in the Tehuacan Valley in the south central Mexican state of Puebla. The excavation of this large, dry rockshelter was undertaken by Melvin L. Fowler and Richard "Scotty" MacNeish as part of the Tehuacan Archaeological-Botanical Project, a monumental interdisciplinary search for the origins of corn agriculture in Mesoamerica. Their work at Coxcatlan is described by them in the fifth volume of the project final report (Fowler and MacNeish 1972). Altogether, full scale excavations were undertaken at twelve major sites in the valley in the course of this project. The

Tehuacan Valley is a long, narrow landform surrounded by high mountains known as the Sierra Madre de Oaxaca. The climate is hot and dry. Coxcatlan Rockshelter is located at it's southern end in the side of a tall bluff on the slopes of the valley's side. It is above an intermittent stream channel in a semi-arid thorn forest. The shelter, which formed as the bluff was under cut, measures some 30 meters long and nine meters wide. The stratigraphic deposits in the cave are over four meters deep. As the interior of the shelter is shielded from rain and it contains no natural springs or seeps, the deposits are entirely dry. Consequently, desiccated remains of plants, animals, human remains, feces, and perishable cultural debris were recovered there in great abundance and in an excellent state of preservation.

MacNeish found the rockshelter with the help of local informants. He determined upon an extended excavation of the site after the initial test excavations of its deposits produced both ancient cultural material and tiny, well-preserved cobs of primitive corn. Following this discovery, MacNeish states that the excavators

> made a contour map and set up a one-meter grid system; each square of the grid system was given a number based on the north-south grid coordinates. Next, we dug an initial square or trench into the center of the site by trowel from above with a conscious effort to remove and record each artifact as to its position in each square and stratum or substratum. Once this initial sounding was completed to the bottom of the archaeological deposit, we then shifted our technique, for we felt that we could have good control at defining the cave stratigraphy if we could excavate it by trowel, spoon, or paintbrush from a vertical profile that clearly showed all the nuances of the strata. The first step in this direction was a thorough study of the walls of our sounding to determine the strata, which were given a number. Then drawings and photographs were made of these profiles along the grid lines. We continued this process until we had long scrolls of profiles for every grid line in every direction. When the stratigraphy was complex or special features were encountered, we made other profiles. These good records of the strata were supplemented by soil and pollen samples from each one (MacNeish 1978: 98-99).

The physical stratigraphy in the cave deposits was very clear, so segregating them did not prove to be difficult. Individual depositional units or layers, called "zones" by Fowler and MacNeish, were assigned Roman numerals in the order of their discovery, that is, from the top of the trench downward. Twenty-eight such depositional units were identified in the cave stratigraphy. The zones had generally been formed through the build-up of plant remains, broken artifacts, feces, animal bones, and other human living debris. Fowler and MacNeish (1972: 225) noted that in general, a layer of relatively undisturbed organic debris was visible in the profile at the point where the accumulation of a specific zone ceased. These undisturbed layers, capping and terminating the growth of each zone, were labeled "floors" by Fowler and MacNeish. Such layers are referred to as **occupation surfaces** in the usage adopted in this chapter. The floors at the site were numbered from the bottom of the trench upwards; twenty-eight floors were recognized in the rockshelter.

Recall that an **occupation** is a spatial cluster of discrete objects which can reasonably be assumed to be the product of a single group of people and deposited over a period of continuous residence. Within the zones, the excavators discerned 42 separate occupations. Like the floors, the occupations in Coxcatlan Rockshelter were numbered from the bottom of the deposits to the top, that is from earliest to latest.

The excavators also found a number of **features** in the cave. Roasting pits, fire pits and charcoal areas, cache (storage) pits, and burials found in the cave were examples of constructed features because they had been deliberately built to provide a setting for an activity. The cave deposits also contained simple accumulation features including refuse areas and middens. The roasting pits, cache pits, middens and other features often contained food remains and other cultural ecofacts. These proved to be critical to the understanding of the manner of the diet and subsistence system of the rockshelter's occupants and the ways this system changed through time. Particularly noteworthy were the macrofossils of the primitive ancestors of corn, beans, gourds, squash and other economic plants recovered in these features. Using these, and specimens recovered in other dry rockshelters in the valley, MacNeish and his collaborators were able to reconstruct the sequence of botanical change that led eventually to the domestication of these plants in Mesoamerica.

Environmental ecofacts were recovered both in the stratigraphic zones and in the features. The major environmental ecofact was plant pollen extracted directly from the sediments and from human coprolites. The excellent preservation in the shelter favored the recovery of large quantities of plant macrofossils and animal remains. All of these data were used in reconstructing the sequence of environmental change that occurred through time in the vicinity of the rockshelter. That sequence revealed natural changes in the environment that followed the close of the Pleistocene as well as anthropogenic environmental change and destruction following on the human population increase and subsistence change.

The strata in the Coxcatlan Rockshelter contained large quantities of **artifacts** including chipped and ground stone, bone, ceramics, and textiles. Chipped stone tools and projectile points occurred throughout the deposit. Since such implements are particularly prone to stylistic change over time, they proved to be excellent time markers useful in cross-dating the Coxcatlan occupations with those of surrounding sites. Grinding stones, ancestral to the milling stones of historic time, also change form through time and these formal changes were generally correlated with the increasingly quantities of domesticated plants in the food remains recovered in the rockshelter's zones.

Given the abundance of organic material and the relatively recent date of the deposits, radiocarbon dating was the chief means of establishing absolute dates in the rockshelter. Multiple radiocarbon dates were run on **datable samples** recovered in the Coxcatlan zones and features from the top of the deposits to the bottom. The earliest of 28 occupations at the site was dated by C14 to greater than 10,000 B.C. Absolute dating also revealed that, with the exception of interruption between 2,300 B.C. and 900 B.C. deposition in the rockshelter was more-or-less continuous from that time until the present (MacNeish 1964:174).

The construction of the culture sequence at Coxcatlan Rockshelter. The composite stratigraphic profile that resulted from Fowler and MacNeish's careful excavation and recording of the stratigraphy at Coxcatlan cave is shown in Figure 2. In their final report, Fowler and MacNeish described the stratigraphic zones in the cave from 28 through 1 in terms of color, composition, and the occupations, artifacts, and ecofacts they contained. They described the features recovered at the cave in the same sequential fashion.

In the process of these descriptions, the authors draw conclusions about the way of life or the cultural practices and environmental conditions that resulted in the formation of each zone. Assuming that the 28 separate zones or layers in the Coxcatlan stratigraphic profile were correctly segregated and that the preconditions of the Law of Superposition regarding the identity of the top and bottom layers were met, then the succession of the layers from zone 28 to 1 is the order of their deposition. To determine whether or not the stratigraphic sequence developed at Coxcatlan was complete and could be generalized to the Tehuacan Valley at large, it was necessary to conduct a large scale surface site reconnaissance of the entire region and to cross-date the Coxcatlan zones and occupations with those of other sites in the region.

Cross-dating the Coxcatlan stratigraphic sequence. Recall that cross-dating as a technique of relative dating was developed earliest by William "Strata" Smith for use in reconstructing the geological sequence of England. The principles Smith developed, however, apply equally well to archaeology with sites like Coxcatlan. In order to cross-date two independent stratigraphic sequences, Smith compared them layer by layer. In a similar way, cross-dating was undertaken between the Coxcatlan sequence and those of 11 other major sites excavated in the Tehuacan Valley as part of the project's investigations. Especially useful in this regard were the stratified sites of Purron Cave, El Riego Cave, Abejas Cave and San Marcos Cave. However, similarities between layers in two site sequences could seldom be established based on similarities observed in the plant and animal remains they contained. Instead, layers were assumed to be contemporary when it could be established that their cultural inventories were similar. The occupations of open-air sites were later correlated with the master cultural sequence in a similar manner. MacNeish was also able to cross-date site sequences by comparing the absolute dates obtained by radiocarbon dating organic material from the various layers. According to MacNeish:

> In combination with our finding in the Purron cave, which contains 25 floors that date from about 7,000 B.C. to 500 B.C., we have an almost continuous record (the longest interruption is less than 500 years) of nearly 12,000 years of prehistory. This is by far the longest record for any New World area (MacNeish 1964: 174).

Thus, a master cultural sequence was established for the entire Tehuacan Valley based on the logic of Smith's method.

The construction of the culture history of the Tehuacan Valley. The cross-correlation of stratigraphic sequences from the sites in the Tehuacan Valley resulted in the establishment of a detailed master sequence and relative chronology. The next step was the grouping of the successive zones, occupations, and features into cultural **phases**. Recall that a phase is a unit into which are grouped site occupations that are close to one another in time and exhibit similar or related cultural assemblages. A sequence of seven cultural phases were identified, defined, assigned date ranges. This sequence of phases applied to the entire Tehuacan Valley and extended in time from the end of the Pleistocene to the Spanish Conquest and provided the framework for the culture history of the valley. Each phase was characterized by a different way of life than that of the phases preceding and following it. In describing each of these lifeways, the authors constructed paleoethnographies of the socio-cultural systems that successively occupied the Tehuacan Valley (Table 11 in Fowler and MacNeish 1972: 231).

Ajuereado phase. (12,000 B.C.? to 6,500 B.C.) During this phase, the environment of the valley was somewhat cooler and moister than it is at present. At the beginning of the phase, the environment supported a number of late Pleistocene mammalian species, including the horse and a form of antelope, that became extinct by the end of the phase. The human population of the valley was low. Simple flaked stone projectile points and other implements formed the centerpieces of the technology. Subsistence was based on the hunting and trapping of mammals and, to a less extent, the gathering of wild plants. Such activities were probably closely synchronized with seasonal changes in the availability of key wild species. In addition to hunting and gathering, some time after about 8,000 B.C., it is probable that the Ajuereado peoples began the experimental manipulation of the wild ancestors of corn. Most of the hunting and gathering peoples studied by anthropologists have been organized into simple socio-political formations known as egalitarian bands. Based on analogies with these peoples, MacNeish assumed that the Ajuereado phase peoples were similarly organized. Little could be determined of their religion, nothing of their language.

El Riego phase. (6,500 B.C. to 4,800 B.C.) During this phase, the climate of the valley came to resemble that of modern times. The beginning of this phase are marked by the appearance of a subsistence change: the collecting of plant foods replaces hunting and trapping as the predominant subsistence activity. The plants relied upon during this phase were wild. The technology included grinding stones of various types. Sites seem to have been occupied for somewhat longer periods of time than during the Ajuereado phase and population in the valley may have been slightly higher. A number of burials were recovered from the period hint at the existence of part time religious practitioners.

Coxcatlan phase. (4,800 B.C. to 3,500 B.C.) By onset of this phase, fully 10 percent of the plant remains are those of domesticated plants and the number of domesticated species present has lengthened considerably. "Coxcatlan projectile points tend to be smaller than their predecessors; scrapers and choppers, however, remain much the same. The polished stone implements include forerunners of the classic New World roller-and-stone device for grinding grain: the mano and metate." (MacNeish 1964: 176). Occupation sites are fewer in number but larger in size. This suggests that either the population was growing or the bands were remaining in one location longer or perhaps both.

Abejas phase. (3,500 B.C. to 2,300 B.C.) A fundamental shift in subsistence marks the onset of the Abejas phase in the valley. By this time, domesticated plants contribute to the diet. Long blades made of obsidian are added to the stone tool repertoire as are stone bowls and oval metates. Small pithouse villages, apparently the first fixed settlements in the valley, were built along river terraces. These house clusters may have been occupied throughout the year. Population apparently continued to rise although a band level of socio-political organization is still assumed (MacNeish 1964: 176).

Purron phase. (2,300 B.C. to 1,500 B.C.) This phase, the least well-known in the valley sequence, was characterized by a key technological change: the development of pottery and the beginning of the replacement of stone vessels with ceramic ones. The agricultural subsistence system was characterized by the cultivation of more hybridized types of maize during the Purron phase.

Ajalpan phase. (1,500 B.C. to 900 B.C.) This phase is characterized continued population growth and resulting increase in the scale and complexity of village life. The crude pottery of the Purron gives way to markedly more refined wares in the Ajalpan phases and ceramic anthropomorphic figurines, apparently part of the religious system, appear.

Santa Maria phase. (900 B.C. to A.D. 50) The subsistence system during this phase comes to be based on irrigation. Population rises sharply at this time. Temple mounds appear and the artifactual assemblages of the phase include artifacts of foreign origin. The influence of the early Gulf Coast civilization of the Olmecs appears to have been important in the valley at this time. The simple egalitarian band socio-political organization of the valley probably had given way to more complex tribal forms of organization by this time.

Palo Blanco phase. (A.D. 50 to A.D. 700) By this phase, the environment appears to have been altered substantially by deforestation and erosion. The valley was characterized by large-scale irrigation works and large ceremonial centers built on its hilltops. The is evidence that occupational specialization, especially in the production of salt, is an important aspect of the valley economy. The scale of the archaeological remains suggest that the population was quite large and growing. The socio-political system of the valley was probably characterized by a stratified class and centralized authority structures. The foreign cultural influence of the Olmecs appear to have waned in face of new influences from the civilization of Monte Alban to the south and west.

Venta Salada phase. (A.D. 700 to A.D. 1,500) By the onset of this phase, the population of the valley was very large and the bulk of its foodstuffs were derived from large-scale irrigation agriculture. Trade and commerce with surrounding regions continued to be important but the impact of Monte Alban declined in favor of that of the Mixtecs. The political authority in the valley was clearly centralized, backed by a standing army and supported by a complex religious ceremonialism. At the end of the phase shortly before the Spanish conquest, the Tehuacan Valley came under the control of the Aztecs of central Mexico.

Once the Tehuacan sequence was complete, it was possible to integrate it into the master chronological sequence for western Mesoamerica at large. This was done by cross-dating it with other major sequences in the area including the Valley of Mexico, the Oaxaca region, the Olmec region on the eastern coast of the Gulf of Mexico, and a number of less well-known regions (Tolestoy 1978).

From the foregoing culture sequence and culture history, MacNeish constructed the final part of his paleoethnography. He explained the functional interrelationships within the socio-cultural systems of the Tehuacan valley. Then he the delineated the forces generating change in them over time. Finally, he placed these specific, local systems in the stages of cultural and biological evolution that characterize human prehistory in general from the hunting and gathering life of the late Pleistocene and early Holocene, to the rise of agriculture in the middle of the latter epoch, through the emergence of civilization and the beginning of history.

Conclusions

Deeply stratified dry depositional sequences of the scale found at Coxcatlan Rockshelter and in the other shelters in the Tehuacan Valley are unfortunately not always available to the archaeologists. Nonetheless, occupational surfaces, features artifacts, ecofacts, human remains, and datable samples are recovered at all kinds of sites and the methods and logical procedures used in constructing culture sequence and culture history are universally applicable. These methods are simple and the relationships they uncover in the data, locational, positional (provenance), stratigraphic, contextual, and associational, are logical and obvious. But what Clausewitz said of war, that it is **simple**, but never **easy**, applies equally well to archaeological methodology. And, if they are to work, these methods must be applied systematically and continuously no matter how difficult the field situation.

Yet systematic application is only the first part of what, at the beginning of this chapter, was called the "unflinching application of the brute force of the scientific method to questions about the past." The second part involves the testing of theories and models about the past against the patterns of ancient material remains recovered in this first part. Only through this two-fold process of systematic recovery and continuous testing are students of the past able to transcend mythic prehistory and transform philosophical accounts of the past into scientific, "paleoethnographic" ones.

PART II:

PREHISTORY: HUMANKIND IN DEEP TIME

Major Points Raised in Chapter Six

1. "Deep time," is the immense age assigned by modern science to the Universe and, closer to home, to the Earth. Deep time begins with the catastrophic mutual annihilation of matter and antimatter called the Big Bang. In the first section of this chapter, consideration is given to earliest deep time and to the development of the planet during the Hadean, the first of the four geological eons.

2. An outline of the evolution of life from the simplest Precambrian fossils to the mammals of the Cenozoic era is provided in the second section. This outline is intended to place the very recent appearance of humankind in the Earth's evolutionary history in proper temporal perspective.

3. The third section deals with the evolution of the primate order. The section begins by considering the archaic primates and strepsirrhine primates or prosimians and proceeds to a discussion of the general characteristics shared by all primates, including modern humans.

4. The fossil and molecular biological evidence used in reconstructing the primate evolutionary career is treated in the fourth section of the chapter.

5. The chapter ends with a brief consideration of how the study of living, non-human higher primates contributes to an understanding of the evolution of the behavior of humankind.

CHAPTER 6

DEEP TIME AND EVOLUTION

Introduction

"Deep time" is the name often given to the vast span of years that constitute the age of the Universe and of our Earth within it. At least since the time of James Hutton and Sir Charles Lyell, enlightened students of the past have recognized this age to be of an immense duration. However, accurate assessment of deep time is a decidedly contemporary scientific achievement.

Deep Space, Deep Time. In 1929, the American astronomer, Edwin Hubble, made a dramatic and unexpected discovery: the galaxies that surround the Earth in every direction are not fixed in space as astronomers had long assumed. Rather, they are moving apart. This observation, and his many subsequent confirmations of it, lead Hubble to the inescapable conclusion that the Universe is expanding. Despite its unmeasurable size at present, it had once been small. Infinitesimally small, Hubble deduced, and infinitesimally dense. The Universe that we know originated when, back beyond time, that infinitesimally small and infinitesimally dense entity flew apart in a furious explosion known as the **"Big Bang"** (Hawking 1988: 8-9). Hubble's extraordinary insight has come to shape the view of space, time and the Universe held by most contemporary astronomers and cosmologists.

The Beginning of Deep Time. But when did the catastrophic mutual annihilation of matter and antimatter of the Big Bang occur? To measure the amount of time that has past since that grand explosion is to ascertain the age of the Universe. Two primary astrophysical methods have been used to effect this measurement: (1) measuring the rates at which galaxies that surround the Earth are receding from our Milky Way Galaxy and (2) calculating the relative ages of the various stars in our galaxy. The results achieved by these measurements are largely dependent upon the assumptions made in formulating them. However, by requiring that the independent methods produce a consistent result, astrophysicists have generally been able to agree that the universe is some 15 ± 1.5 billion years old (Dalrymple 1991: 402-403).

Fortunately, the age of the Earth can be ascertained in a less subjective manner than that of the Universe as a whole. This is because geological strata near its surface can be directly dated using nuclear methods based on isotopic decay. As noted in the preceding chapter, the decay rates of the radioactive elements have been established in the laboratory and these rates have been shown to be constant regardless of temperature and pressure variation. In addition, a number of different isotopic techniques exist and can be used to cross check one another. Using potassium/argon, uranium/lead, lead/lead and samarium/neodymium isotopic methods, dates in excess of 3.8 **bya** (billion years ago) have been obtained from geological strata in various locations in Greenland, western Australia, southern Africa, and North America. However, 3.8 bya must be regarded as a minimum age for the Earth. Rocks recovered from the Moon by NASA and dated using hafnium/tungsten have returned ages that range between 4.5 and 4.52 bya (Lee, Halliday, Snyder, and Taylor 1997). As the Moon most likely formed as an aggregate of collision debris from the early Earth, its parent planet must be as old or older. Finally, the most ancient meteorites recovered on Earth date to between 4.5 and 4.6 billion years in age. These meteorites are probably debris cast off during the formation of our solar system and of the Earth within it. Taken together, these three sets of dates suggest that the Earth formed sometime around 4.5 to 4.6 bya. Significantly, the calculated ages of the Universe in general and of the Milky Way Galaxy in particular are consistent with the age range of the Earth (Dalrymple 1991: 398-403).

Evolution of Earth and Life

The formation of our Solar System from an interstellar cloud of dust and gas is thought to have begun more than 10 billion years after the Big Bang. Condensation of the solar nebula formed the Sun, the planets, and most of the other objects that rotate in the orbital plane of the solar system. Matter seems to have condensed outward from the center of this nebula, eventually leading to the formation of the Earth (Cloud 1988: 3, 38; Weinberg 1977).

The history of the Earth during the first 600 to 800 million years of its existence as a planet is referred to as the **Hadean** eon after Hades, the Greek word for hell or the underworld. The Hadean eon lasts from about 4,600 to 3,800 **mya** (million years ago). This is a huge time span to be sure but, unfortunately, the nature of the Earth during the Hadeon can only be indirectly reconstructed. Nonetheless, from what we know of its lifeless surface, the eon appears well named.

> Earth's primordial surface was probably too hot, too exposed to lethal radiation, too dry, and too battered by large planetesimals or meteoroids for life to emerge or survive during much or all of that long interval (Cloud 1988: 32).

It is during the Hadean eon that the Earth's magnetic field and atmosphere formed. By the close of eon, the solid crust of the Earth took shape and its first ocean appeared.

The subsequent eon is called the **Archean,** a word derived from the Greek that means literally "primitive." The Archean eon began about 3,800 mya and lasted until approximately 2,500 mya. The earliest microbial life appeared on Earth at the beginning of the eon. The Archean is followed by the **Proterozoic** eon, a name derived the Greek that means literally "before life," a label now known to be inaccurate. The Proterozoic eon lasts from ca. 2,500 to 570 mya. During this time the individual segments or **plates** that form the Earth's continents emerged on the Earth's **mantle**, the constitution of atmosphere changed from **anoxic** to **oxic,** and the Earth experienced its first continental-scale glaciation (Cloud 1988: 248).

Precambrian Origins of Life. Geologists refer to rocks from the vast expanse of time represented by the Hadean, the Archean, and the Proterozoic eons collectively as **Precambrian** in age. Thus, the interval from ca. 4,600 to 570 million years (nearly 88 percent of geological time) is designated as a prelude to the remaining 12 percent represented by the post-Cambrian from 570 mya until the present. Geologists call this latest eon the **Phanerozoic**, a term derived from the Greek and meaning literally, "visible life."

The ancient rocks of Precambrian age imperfectly reflect the processes that formed the surface of the Earth. This is due to the limited extent of Precambrian-age exposures and to the degree of deformation and erosion to which they have been subjected over time. Further, the fossil evidence of life found in Precambrian rock is sparse indeed while that of the subsequent eon is rich almost beyond measure.

The limited nature of the Precambrian fossil evidence makes reconstruction of the origins of life difficult. Nonetheless, it is probable that all forms of life on Earth have evolved from unicellular organisms. The manner in which these early single-celled creatures were first formed is unknown However, as Bertrand Russell put it,

> their origin is no more mysterious than those of helium atoms. There is no reason to suppose living matter subject to any laws other than those to which inanimate matter is subject, and considerable reason to think that everything in the behavior of living matter is theoretically explicable in terms of physics and chemistry (Russell 1948: 36).

If life originated as a self-organizing process of chemical evolution, then theoretically it should be possible to replicate that process in the laboratory. In fact, experiments in this direction appear to be laying the groundwork for such replication. Peptides are organic compounds composed of two or more amino acids. Such compounds constitute key components of the DNA chain. Recent experiments have demonstrated that, under certain conditions, two otherwise competitive, self-replicating peptides will enter into a **symbiotic** or mutually beneficial relationship. That-is-to-say, separate peptides have been found to work together and thereby enhance one another's ability to survive in the face of natural selection. The discovery that symbioses can occur naturally in organic chemical systems is remarkable. The appearance of such symbioses is a plausible first step in the transition from inanimate chemistry to animate organisms (Lee, Severin, Yokobayashi and Ghadirii 1997). Of course, this does not mean that chemists can now cook up animate organisms in their laboratory retorts. Rather, it suggests that, if experiments along these lines continue to prove fruitful, science may soon be able to explain the origins of life, as Russell said we should, strictly in terms of the laws of physics and chemistry. If that proves to be the case, we must further be prepared to discover that the operation of those laws elsewhere in the Universe has, under similar conditions, similarly produced life. In fact, evidence of such parallel evolution may already be in our hands. A meteorite recovered recently in Antarctica was determined to have originated on Mars. On the surface of this meteorite were found small traces of carbonate, a key organic compound, together with tiny, tube-like structures teasingly and provocatively reminiscent of fossils (McKay *et al.*1996; McKay, Gibson, Thomas-Keprta, Vali 1997. By no means all scientists are convinced that the meteorite in fact contains traces of Martian life (Bradley, Harvey and McSween 1997) and it will probably take an actual visit to Mars to settle the question.

But whatever was–or was not–occurring on Mars or elsewhere in the Universe, it seems plausible that life emerged on the ancient Earth out of the ordinary constituents of the atmosphere, crust, and water in some reducing environment. Limited geological evidence indicates that, by the end of Hadeon eon, the Earth possessed a biosphere with extensive oceans and a thick anoxic or neutral atmosphere that lacked free oxygen (0) and consisted chiefly of carbon dioxide (CO_2) and nitrogen (N_2). Despite an apparent decrease in solar luminosity, the Earth was probably quite warm at this time. This is because the high percentage of CO_2 it the atmosphere would have created a pronounced "**greenhouse effect**" in the biosphere (Maher and Stevenson 1988: 612; also cf., Kasting and Ackerman 1986).

A great deal of confusion surrounds discussions of the greenhouse effect in the contemporary press. Actually, the meaning of the phrase is quite simple: by allowing light energy radiated from the sun to pass through it and warm the Earth and then by impeding the loss or re-radiation of heat energy into space, the atmosphere of the Earth behaves much as a gardener's greenhouse. This occurs because visible and near-infra-red sunlight can pass through the atmosphere unimpeded while middle-infra-red heat re-radiated from the Earth's surface is adsorbed by atmospheric gases, most notably water vapor and CO_2. Once adsorbed, this thermal energy is re-radiated both outward into space and back downward to the Earth's surface. The CO_2 rich atmosphere of the Archean Earth must have been warm indeed. However, as the primitive anoxic atmosphere of the planet changed over time to an oxic one, its energy balance changed as well and temperatures within its biosphere must have dropped accordingly. Such would have been the conditions in this early Archean biosphere wherein the posited earliest organisms first developed.

Autotrophy. An autotrophic organism manufactures the organic constituents it needs from inorganic materials. Most plants containing chlorophyll are autotrophic: they manufacture organic materials from water, carbon dioxide, nitrates, and other salts using sunlight as the source of their energy. A **chemoautotrophic** organism obtains energy from a simple inorganic reaction, the nature of which vary according to species. For example, sulphur **archaeobacteria** is a chemoautotroph that oxidizes hydrogen sulphide to sulphur while hydrogen archaeobacteria oxidizes hydrogen sulphide to hydrogen. Archaeobacteria of the foregoing kinds, or species very like them, are strong candidates for the role of earliest and most primitive living organisms on our planet (Maher and Stevenson 1988: 612; Lake 1988).

The strategy for determining what the earliest organisms must have been like is two-fold: molecular biologists attempt to reconstruct their evolutionary tree by comparing the DNA sequences of living organisms and working backward (Lake 1988; Penny 1988), paleontologists seek to recover actual fossil remains of these first creatures.

On this latter front, rocks from western Greenland dating to the very beginnings of the Archean eon shortly before 3800 mya have recently been found to contain minute quantities of elemental carbon trapped in apatite deposits. Provocatively, the isotopic compositions of this carbon falls within the range found in living microorganism. Since chemical processes other than those of living organisms might have produced the carbon, these samples cannot be taken as conclusive evidence that life was present on Earth at the end of the Hadeon or at the onset of the Archean eon. Nonetheless, the carbon traces from Greenland, like the faint patches of carbonate on the Martian meteorite, are suggestive indeed (Holland 1997).

More certain evidence of life comes from western Australia. There, carbon-rich chert deposits dating to the middle Archean eon between 3,500 and 3,300 million years ago have produced fossils tentatively identified as **cyanobacteria** or **blue-green algae**. Such organisms produce oxygen through photosynthesis (Schopf and Packer 1987: 70). These fossils suggest that, by the early Archean eon, a varied and possibly extensive microbial community was contributing oxygen to the $C0_2$ and N_2 of the primitive anoxic atmosphere of the Earth.

Somewhat younger microfossils of algae and fungi have been recovered from Precambrian rocks in North America suggesting that by the early Proterozoic eon (ca. 2500 to 1600 mya), the most complex life form on Earth was still blue-green algae, now growing in large, dome-shaped mats in shallow seas. In a process that has continued into the present, blue-green algae extract fine sedimentary particles from the water and binds them in layers within the mats. These mats have been preserved in the fossil record as the celebrated fossils called **stromatolites**.

By about 1200 mya during the middle Proterozoic (ca. 1600 to 900 mya), blue green algae appear to have colonized the land. The recent discovery that cyanobacteria was growing on land at this early date was unexpected and raises the possibility that terrestrial microbial communities were sufficiently extensive to have affected weathering, erosion, sedimentation and geochemical processes during the Precambrian (Horodyski and Knauth 1994). Whether growing in the sea or on the land, blue-green algae was no doubt adding significant quantities of oxygen to the primitive atmosphere.

Evolution of the Eucaryotes. Bacteria and cyanobacteria or blue-green algae are exceedingly simple **procaryotic** organisms. Procaryotes are made of cells containing the transparent gel or viscous fluid known as **cytoplasm** that forms the stuff of life in all living organism. However, the chromosomes in the cells of procaryotic organisms are not paired with one another nor separated from the cytoplasm by a nuclear membrane.

Eucaryotic organisms may have evolved as early as 2000 mya during the middle of the Proterozoic eon (Knoll 1992: 623). Eucaryotic cells are the basic structures found in all organisms excepting, of course, bacteria, blue-green algae, and viruses. As such, the eucaryotes constitute one of the three principle domains of life; procaryotic bacteria and archaebacteria constitute the other two.

The chief characteristic of eukaryotic cells is that their nuclei are separated from their cytoplasm. Secondly, in addition to having membrane-bounded nuclei, eucaryotes are able to surround external particles and bring them into their cytoplasm, an important defense against invading bacteria. Third, most eucaryotes have membrane-bound **organelles** within their cytoplasm. These organelles fulfill various specialized functions. For example, respiration is performed by organelles known as mitochondria and organelles called chloroplasts carry out photosynthesis.

A limited number of living eukaryotic species that lack organelles have been discovered and these primitive organism have been used to model what the earliest eucaryotes must have been like (Vogel 1997: 1604). Since the energy producing mitochondria have their own genes, evolutionary biologists hypothesize that the organelles originated when a bacterium entered a primitive eucaryotic cell and the symbiotic relationship resulting from this intrusion provided both organic life forms with greater selective advantage than either had acting alone. According to Knoll (1992: 622), molecular data indicates that such "endosymbiotic" events must have occurred independently many times in prehistory. Finally, most eucaryotes are capable of rapid multicellular growth and differentiation. The evolutionary success of the eucaryotes are largely due to these attributes (Knoll 1992: 622).

By the late Proterozoic, relatively complex, multi-cellular eucaryotes appear to have been well established in the oceans. Recent studies of phosphate deposits in South China dating to 580 mya have revealed fossil remains of primitive sponges together with thousands of multicellular embryos. These fossils are clear proof of the presence of metazoan or multi-cellular animal life on Earth before the close of the Proterozoic eon (Li, Chen and Hua 1998)

The Great Unconformity. A break or gap in the geological record is called an **unconformity**. An unconformity is said to be present in a geological exposure when it is evident that the normal succession in the rocks has been interrupted, for example, when they show a change from an environment of deposition to an environment of erosion and then back to an environment of deposition. Such interruptions can be brief (in which case they are referred to as **diastems**) or of lengthy duration. The greatest unconformity in the geological record occurs between the middle of the Proterozoic eon and the Cambrian period of the subsequent Paleozoic era. The record of late portions of the Proterozoic eon, an interval perhaps as long as 100 million years, is missing. During this time the vast seas of the early and middle Proterozoic eon, apparently contracted. The lengthy period of erosion attested to by the Great Unconformity must have been initiated by this oceanic contraction. Sea level is the general or ultimate **base level** towards which erosion on the land progresses. A drop in sea level in the late Proterozoic would have steepened the **gradient of flow** or degree of slope of the rivers and streams draining the bare continents. Water must surely have rushed over the as-yet-unvegetated land surfaces of the young Earth in pursuit of this new ocean base level,scouring it, and carrying its rock and sediment to the sea. Deposition resumed in the subsequent Paleozoic era but no record has yet been discovered for that key interval of time represented by the Great Unconformity. It appears that we have irretrievably lost the evidence of a critical episode in the early evolution of life on Earth (Burrus and Spiegel 1976: 253-254).

The Paleozoic Era. The next great era in geological time is called the **Paleozoic**. The name, derived from the Greek, means literally "old life."

The Paleozoic era begins around 570 mya with the **Cambrian** period and closes some 325 million years later at the end of the **Permian** period about 245 mya. In the geological record, the late Precambrian erosional surfaces of the Great Unconformity are generally covered by sedimentary rock formations formed during the slow **transgression** of the Cambrian seas, that is, the intrusion and spread of these ancient oceans onto the land. The Paleozoic era contains the first clear evidence of widespread expansion of the sea into the interiors of the Earth's continents. This pattern of marine transgression and recession continued throughout the Paleozoic era with major cycles during the Ordovician, the Silurian, and the Devonian periods.

Paleozoic Plate Tectonics. Although we can be certain that the shapes and arrangements of the Precambrian continents were very different from the present, it is difficult to reconstruct their patterns with any precision. Still, our understanding of the nature of these ancient landmasses has been profoundly expanded in recent years by the development of one of the most revolutionary ideas in the history of science: the **Theory of Plate or Global Tectonics** mentioned in Chapter Two. According to this theory, the thin shell that constitutes the Earth's crust is not a solid mass. Rather, it is divided into between ten and 25 independent segments or plates. Although the plates themselves are more-or-less rigid, the zone directly below them, known as the mantle, consists of more viscous material. The individual plates that make up the Earth's continents float upon the mantle and move independently across it. Plate movement is generally thought to be due to the convection or transfer of heat from the Earth's core through the mantle to its crust. The upward diffusion of this heat creates two thermal boundaries: a hot lower one in the mantle and a cold upper one at its surface. Convection between these two boundaries slowly deforms the mantle like a fluid. It is this deformation that breaks the cold, brittle rocks on the Earth's crust into separate plates and slides them across the mantle (Davies 1992: 493).

The moving plates often collide with one another at their margins. In such encounters, one plate may descend beneath the other or both may slide past each other. In either case, collisions create zones of deformation that result in mountain building, volcanism and earthquake activity. When plates move apart, new material wells up from the mantle between them to create mid-oceanic ridges and, eventually, new oceanic plates (Moores 1995: 134).

Paleomagnetic and stratigraphic studies indicate

Geologic Time Scale--Deep Time on Earth

<table>
<tr><th>EONS</th><th>ERAS</th><th colspan="2">PERIODS</th><th>EPOCH</th></tr>
<tr><td rowspan="5">PHANEROZOIC</td><td rowspan="3">CENOZOIC</td><td colspan="2">Quaternary</td><td>Holocene
Pleistocene</td></tr>
<tr><td rowspan="2">Tertiary</td><td>Neogene</td><td>Pliocene
Miocene</td></tr>
<tr><td>Paleogene</td><td>Oligocene
Eocene
Paleocene</td></tr>
<tr><td>MESOZOIC</td><td colspan="2">Cretaceous
Jurassic
Triassic</td><td></td></tr>
<tr><td>PALEOZOIC</td><td colspan="2">Permian
Pennsylvanian (Upper Carboniferous)
Mississippian (Lower Carboniferous)
Devonian
Silurian
Ordovician
Cambrian</td><td></td></tr>
<tr><td>PROTEROZOIC</td><td></td><td colspan="2"></td><td></td></tr>
<tr><td>ACHEAN</td><td></td><td colspan="2"></td><td></td></tr>
<tr><td>HADEAN</td><td></td><td colspan="2"></td><td></td></tr>
</table>

that the clustering and dispersal of the continents has been occurring for at least the last 2 bya of geological history (Tronpette 1994). Details of the continental movements in the Proterozoic are sketchy because the paucity of fossils in the deposits from that eon makes stratigraphic cross-dating difficult. Evidence improves by the Paleozoic era, however, and geologists are certain that, in the 150 million year interval between the Silurian and the Permian, the separate continental plates had, one by one, begun to converge. By the end of Paleozoic era, successive collisions between the individual continents resulted in the assembling of a single super continent geologists have named **Pangaea**. Needless-to-say, the uniting of Pangaea was accompanied by extensive volcanic and tectonic activity (Eicher, McAlester, and Rottman 1984: 76). Once together, the Pangaean super continent was to remain in more-or-less one piece for 100 million years But life on Pangaea was not easy. Computer simulations indicate that the continent's great size must have had a profound effect on temperature and wind circulations and would have made the climate there one of harsh extremes (Waters 1992: 42).

Explosion of Life. In contrast to the nearly lifeless rocks of the Precambrian era, formations dating to the Cambrian period (ca. 570 to 505 mya) contain abundant and complex fossil assemblages. The luxuriant, even riotous, nature of these assemblages has led paleontologists to refer to the evolution of life during the period as the "**Cambrian Explosion.**" The fossil record is dominated by simple algae-like sea plants together with a highly varied array of marine invertebrates. Many of these early Cambrian fossils bear scant resemblance to anything living today; some in fact seem more like creatures from a mad man's dreams than real organisms that once lived and breathed on Earth. For example, a species of "armored lobopod" has been aptly named *Hallucigenia* and characterized as a cross between a centipede and the Michelin Man (Bengtson 1991: 184). Yet, in addition to including a host of organism which have no living descendants–or even analogues, Cambrian fossil beds have also been found to contain the ancestors of modern invertebrates like sponges, bivalves, sea slugs, lobsters, shrimp, jellyfish, salt coral, sea pens, and segmented worms. Perhaps the most celebrated marine invertebrate taxa to appear during the Cambrian period is the **trilobite,** a distant ancestor of the modern crab.

Although the Cambrian (along with the subsequent Ordivician) is known as the **Age of Marine Invertebrates**, the earliest **chordates** (phylum Chordata) appear during the period about 540 mya. The simple, early chordates have rod-like cords of cells that forms the chief axial support of their bodies. Although the ancestral organism from which the earliest chordates sprung has yet to be identified in the fossil record, paleontologists presume it consisted of little more than a bilaterally-symmetrical tube of cells with a mouth and perhaps tiny thread-like oars or cilia to move it through the water. The have christened this hypothetical ancestor Urbilateria, literally "earliest bilaterally-symmetrical organism." (Pennnisi and Wade 1997: 36). The first chordates differentiated from Urbilateria presumably by evolving hollow nerve cords along their dorsal surfaces.

True **vertebrates** (Subphylum Vertebrata) evolved from these primitive chordates sometime in the Cambrian period. In vertebrates, the axis supporting the body has become a true vertebral column or spine. Additionally, vertebrates have skulls that contains their brain and skeletons made of cartilage. It is to this subphylum that we refer all higher animals classes including fish, amphibians, reptiles, birds, and mammals. Despite, or perhaps because, of all the spectacular evolutionary innovations that took place during the Cambrian, the close of the period is marked by a series of mass extinctions.

Rock formations dating to the subsequent **Ordovician** period (ca. 505 to 438 mya) are particularly rich in clam-like **brachiopods,** a phylum of marine animals with two-valved shells, in **graptolites**, a class of marine animals that formed tiny cup or tube shaped exoskeletons and lived in large branching colonies, and in **crinoids,** echinoderms that clutched the floors of the ancient seas with dainty stems and, flower-like, stretched lilly-shaped heads upwards towards the light. Ordovician formations have also produced fossils of the first fish-like vertebrates called **ostraderms**. These curious creatures were jawless and had bodies enclosed in bony armor plate.

Terrestrial plant life appears minimal at best during the Cambrian and Ordovician and the landscapes of these periods must have been barren, empty places. Nonetheless, late in the Ordovician, animal life emerged on dry land. Traces of what appear to be millipede burrows from Late Ordovician period paleosols found in Pennsylvania have been interpreted as the first evidence of fully terrestrial animal life (Ratallack and Feakes 1987).

The close of the Ordovician is marked by a short period of continental scale glaciation and a decline in worldwide temperature. These events are correlated with a large-scale, global wave of extinctions. Fully one-third of the families of brachiopods disappear along with many species of trilobites, bryozoans and much of the reef-building fauna. Graptolite taxa were also markedly impoverished. The exact cause of these extinctions will probably never be known but their geographic and ecological pattern suggests glaciation and global cooling may have been important factors (Raup and Seploski 1982; Stanley 1987: 71-27).

Paleozoic Plant Life. Throughout the Paleozoic era, plant life in the seas continue to evolve from the early multi-celled blue-green algae. Primitive plants began the colonization of the land in earnest during the **Silurian** period (ca. 438 to 408 mya). The selective pressures operating on the dry land were harsher than those in the oceans. First, to avoid drying out, the cells of land plants have to be encased in some manner and must possess some means of internally conducting water to their tissues. Second, absent the buoyant powers of the sea, land plants had to evolve some structural means of supporting themselves on dry land. Finally, reproduction on land had to occur largely without dependence on water as a medium in the fertilization process.

The first of these selective barriers was over come with the appearance of the earliest trachecophytes or **vascular plants.** Vascular plants have internal vessels that conduct water, mineral salts and synthesized food materials through their systems. Inferential evidence suggests that such plants evolved in the late Silurian period. Their appearance is one of the most significant events in the history of the plant kingdom. The Silurian is also marked by the appearance of the primitive "psilophyton" flora that lacked true leaves and consisted largely of either horizontal rhizomes (root-like underground stems) or of creepers with spiny stems (Clark and Stearn 1960: 316).

The problem of reproduction outside the sea was first solved by the evolution of plants able to generate single or multi-celled **spores** that detach from the parent and produce a new individual. Still later evolution resulted in plants that reproduced by means of **seeds.** In contrast to a spore, a seed is a fertilized ovule enclosed in a protective outer coat. A seed contains both an embryo plant and a food source to nourish it. The evolution of the seed was a major adaptive break through from the standpoint of pollination, protection and dispersal of the embryo (Rothwell and Scheckler 1988). Collectively, these attributes represent the first successful evolutionary solutions to the structural problem facing plants seeking to adapt to the land.

Unequivocal fossil evidence for the presence of vascular tissue in plants appears in the fossil record during the subsequent **Devonian** period (ca. 408 to 360 mya). At this time, terrestrial plants began a period of explosive evolution that produced two major divisions or **phyla** in the plant kingdom: the **Pteridophyta**, which reproduce by means of spores, and **Spermatophyta**, the seed-bearing plants. Devonian Pteridophyta include three kinds of plants still present on Earth: evergreen club moss, horsetail, and the true ferns. The earliest seed-bearing plants include seed ferns and the **Cordaitales**, a primitive order of trees that bore naked seeds and had strap-like leaves (Clark and Stearn 1960: 318). These early seed-bearers, which disappear in the Permian period, are the first known representatives of that great subdivision of the Spermatophyta known as the **gymnosperms** (Gymnospermae). Cordaitales are distantly related to modern conifers (order Coniferopsida) that includes plants like pine, spruce, cedar, yew, larch and spruce. Conifers constitute the most important gymnosperm taxon on the modern Earth.

The seeds borne by the gymnosperms are naked or exposed in cones or other organs. It is this attribute that distinguishes the gymnosperms from the other, later subdivision of seed-bearing plants known as the **angiosperms** (Angiospermae). The angiosperms, discussed below, are known as the flowering plants and bear seeds that are encased in fruit.

Evolution during the Devonian produced a terrestrial flora that grew in swampy lowland groves and, through time, assumed considerable size and density. Fossil evidence suggests that these ancient plants grew in thick clusters with deep, well-developed root systems able to support trees that may have reached heights of 30 meters or more. Over time, Devonian vascular plants began to move out of the moist lowlands to colonize the drier upland interior.

A paleosol recently unearthed in Antarctica and dated to middle Devonian times contains the fossilized remains of the most ancient, well-drained forest ecosystem soil known to science. This well-differentiated forest soil, coupled with the fossilized remains of root systems of an impressive size and depth, indicate that by the Devonian period, plants had assumed the key role they have played ever after in the ecology of the Earth: the break up and dissolution of bedrock through their growth.

For a variety of chemical reasons, soil scientist conclude that bedrock weathering by plants at this time would have reduced the levels of carbon dioxide in the ancient atmosphere of the Earth (Retallalck 1997). In fact, a recently developed method of estimating ancient carbon dioxide levels through Earth history has produced data in essential agreement with this prediction. Using this method, carbon dioxide levels appear high in the early Devonian and then decline precipitously to reach levels approximately those of the present day during the late Carboniferous and early Permian periods. Carbon dioxide levels rise only moderately during the Jurassic period and then generally decrease again from the Cretaceous to the present day.

The Devonian decline in atmospheric carbon dioxide was followed in the subsequent Carboniferous and Permian periods by the longest and most extensive episode of continental glaciation known for the Paleozoic era. The correlation between declining levels of carbon dioxide in the atmosphere and the onset of the Permo-Carboniferous glaciation is not surprising. As noted earlier, carbon dioxide is a key "greenhouse gas." Because of its strong absorption bands in the middle-infra red, CO_2 absorbs thermal emission radiated from the Earth's surface and re-radiates much of it back into the biosphere. Thus, any decline in atmospheric carbon dioxide would tend to allow more of the earth's heat energy to escape into space and result in a cooling in the lower atmosphere. The decline in atmospheric carbon dioxide followed by the onset of glaciation during the late Paleozoic era is interpretable in terms of the Greenhouse Theory of climate change (Berner 1997: 545).

In any case, the Devonian woodlands must have been strange, silent places. Although they supported a dense and varied assemblage of primitive plants, they were home to but a modest array of animal species. This was to dramatically change, of course. The early Devonian witnessed the appearance of the **spring-tails**, the first known insect-like creatures. Spring-tails very likely fed on decaying vegetation and thus evolved in concert with terrestrial plant life. By the middle of the period, true **insects** (class Insecta) had appeared. Members of this class are characterized by three part bodies (head, thorax and abdomen), thoraxes lined by three pairs of legs and paired antennae on their heads. True insects mastered flight during the Devonian, no doubt to take advantage of the richness of the evolving tree canopy (Cleal 1992: 541). Diversifying since the Devonian into a million or more species, insects are arguably the most successful life forms on Earth.

Sea and Land. Life in the Devonian seas also continued to evolve and diversify. True fish (class Pisces), **ectothermic** or cold-blooded vertebrates, fully aquatic, gilled and finned, appeared in the late Devonian (ca. 374 to 360 mya). They have flourished ever after. So abundant are these creatures in the fossil record of the Devonian and subsequent Mississippian, the two periods are referred to collectively as the **Age of Fish**. Most notable in this finy array are the **cartilaginous** fish (Chondrichthyes), the marine vertebrate taxon that includes modern sharks, rays and skates. Although the skeletons of these groups lack true bone and consist entirely of cartilage, this absence has not prevented them from being among the most persistent and successful vertebrate taxons in the fossil record.

At this point it is worth recalling that it was his recognition that the teeth of living sharks were morphologically similar to the mysterious "glossopetrae" that had been used as charms and amulets in Mediterranean Europe for generations that lead the 17th Century scientist, Nicolas Steno, to conclude that fossils were the petrified remains of ancient creatures. Steno was right, of course. We now know that the glossopetrae are in fact the fossilized teeth of Paleozoic era sharks.

Perhaps the most significant developmental step taken during the Devonian was the evolution of the **amphibians** (class Amphibia) from true fish. Class Amphibia today includes frogs, toads, newts, salamanders and variety of burrowing worm-like tropical animals. Amphibians were the first vertebrates to inhabit the land. However, their break with the sea was not complete. Most members of this class fertilize their eggs in water rather than by direct coition and must therefore return to aquatic or marine habitats to breed. Yet despite the richness of Devonian evolution, the period ends with a catastrophic cycle of marine animal extinctions (Raup and Sepkoski 1982).

The subsequent **Carboniferous** period is subdivided into the earlier **Mississippian** (ca. 360 to 320 mya) and the later **Pennsylvanian** (ca. 320 to 286 mya) periods. During this time, vast, damp forests appear to have covered the Earth. Carboniferous swamps accumulated immense deposits of dead plants that were compressed by their own weight. Over time these deposits became the great coal beds of the modern world. Needless-to-say, they were also ideal fossil-forming environments

Sometime during the Mississippian period the first four-legged vertebrates appear in the fossil record. The earliest known quadrupeds are the **microsaurs**. These were amphibian-like creatures no more than 10 to 15 centimeters in length. The earliest microsaurs date to 312 mya but paleontologists hypothesize that this line began as early 360 mya at the very dawn of the Carboniferous period (Moffat 1997).By the end of the Carboniferous period, climatic conditions on land had changed dramatically. Glacial ice sheets sat astride the poles and the great lowland tropical forests had catastrophically receded.

The Permian period (ca. 286 to 245) which follows the Carboniferous is commonly referred to as the **Age of Amphibians** as such creatures appear to dominate its fossil record. Worldwide temperatures apparently continued to decline during the Permian. Geological formations dating to period are characterized by extensive evaporites, that is, by deposits of salts that had precipitated from water and then been concentrated by evaporation, and by spectacular "red beds," sedimentary rock formations reddened by iron oxide. Both types of formations reflect increasingly arid conditions.

The Great Paleozoic Crisis. The era that had began so hopefully with the dramatic explosion of life on Earth in the Cambrian was to end in turmoil and crisis. A stunning array of worldwide catastrophes overtook the Paleozoic at the Permo-Triassic boundary separating it from the era that succeeded it. We have already mention the continental glaciation that had begun in the Carboniferous boundary. This was followed by rapid sea level change around the now-united continent of Pangea. Shifts in the carbon isotope ratio in the atmosphere took place at this time and a successions of massive eruptions of flood basalts spread lavas over large regions of the Earth's surface (Erwin 1993). Recent dating of the Siberian Traps, the largest of these massive outpouring of flood basalt, indicate that they formed at the Permo-Triassic boundary. Such sustained and long term eruptions must have pumped large quantities of sulfur dioxide into the upper atmosphere. Unlike carbon dioxide, sulfate aerosols are not greenhouse gases. Rather, dispersed as fine-grained solid particles in smoke or haze, they do not absorb heat but reflect it. Once in the atmosphere, the reflective properties of these aerosols can affect the magnitude of the Earth's **albedo**. The Earth albedo is the total of the Sun's radiation reflected from the Earth divided by the total incidence of radiation striking our planet. A massive infusion of sulfate aerosols into the stratosphere conceivably could have reflected sufficient solar radiation back into space to increase the Earth's albedo, reduce the greenhouse warming of the globe and cool the Earth enough to cause the Antarctic ice cap to melt (Campbell *et al.* 1992).

But if these atmospheric changes were not enough, late Permian-age chert formations from Japan and British Columbia have recently been found to contain evidence that a long term, worldwide "**superanoxia**" or depletion of oxygen took place in the deep and boundless ocean that swirled around the great continent of Pangea. At the same time the adaptive problems faced by marine creatures were further complicated as the Permian world ocean began to "stratify," that is, to separate into layers of water of differing densities one atop the other (Isozaki 1997).

Finally, at the close of the Permian period ca. 245 mya, two biotic exigencies of uncertain cause succeeded one another in rapid fashion. These exigencies resulted in the most severe mass extinctions in the history of animal life on Earth. It is estimated that following the close of the Permian, fully 80 percent of marine invertebrates that had evolved throughout the immeasurable ages of the Paleozoic era had been extinguished (Stanley and Yang 1994). Perhaps even more significant is that fully eight of 27 known orders of Permian insects became extinct at the close of the period and three more orders disappeared shortly after the onset of the Triassic. In the 390 million year history of the insects, this is the only mass extinction that they have been known to suffer (Kerr 1993: 1370).

While the specific cause of this mass extinction remains unclear, ocean superanoxia and stratification of course are strongly implicated in the catastrophic impoverishment of the marine life that brought the curtain down on the era (Isozaki 1997). Extinction on land, meanwhile may have been caused by rapid global cooling in combination with acid rain both triggered by the massive Siberian flood volcanism.

Whatever the cause or causes, from the perspective of the present, this great dying at the end of the Permian

> marked the end of the 300-million-year reign of the "old life" of the Paleozoic era–typified by the last of the trilobites–and made way for more diverse and predatory life, including the dinosaurs (Kerr 1996: 1080).

The Mesozoic Era. The Paleozoic era is succeeded by the Mesozoic, a name derived from the Greek meaning literally "middle life." The era dates to between 245 and 66.4 mya and is subdivided into three periods: the **Triassic** (ca. 245 to 208 mya), the **Jurassic** (ca. 208 to 144 mya), and the **Cretaceous** (144 mya to 66.4 mya).

Mesozoic Plate Tectonics. Great changes in the Earth's crust took place during the nearly 177 million years of the Mesozoic era. The huge Paleozoic super continent of Pangaea finished its consolidation and began to break up and spread apart near the end of the Triassic and the beginning of the Jurassic periods. This separation appears to have begun with the opening of a vast and continuous marine seaway called the **Tethys Sea** below the equator. This waterway created two new continents: **Laurasia** on the north and **Gondwanaland** on the south. By the Early Cretaceous, Gondwanaland itself had begun to separate into the proto-continents of South America, Africa, Antarctica, and India.

By the end of the Cretaceous, North America had begun to pull away from Laurasia and South America had separated entirely from Gondwanaland. As the Americas drifted west from Pangaea, the Atlantic ocean trough opened along an extensive rift zone that later became the underwater mountain range called the mid-Atlantic ridge. Concurrently the developing Indian subcontinent broke from the east coast of

Gondwanaland and moved northward creating and widening the Indian Ocean in its wake. The uplift of the Himalaya mountains began when the Indian continental mass first collided with the southern margin of Laurasia in the following Cenozoic era. During the early Mesozoic,

> the amount of land on the Earth's surface was nearly equally distributed between the northern and southern hemispheres. Today, after prolonged separation of the fragments of Pangaea, two-thirds of all land lies north of the equator (Eicher, McAlester, and Rottman 1984: 122-123).

Mesozoic Plant Life. The terrestrial vegetation of the early Mesozoic era was much like that of the preceding Permian period. However, during the latter era, the **cycads** or sago plants, and the **cycadeoids**, a now-extinct taxa, appeared and flourished. Both of these taxa contained palm-like plants that reproduced by means of naked seeds. The most important botanical development of the Mesozoic period was the appearance of the angiosperms or true flowering plants. The angiosperms apparently evolved from, and then replaced, the cycadeoids. Unlike the naked seeds of those more primitive plants, the seeds of angiosperms are encased in a ripened ovary or fruit. This evolutionary development ignited a rapid radiation of the flowering plants across the Mesozoic Earth. So successful were the angiosperms that today they make up the bulk of the visible flora on Earth (Clark and Stearn 1960: 360-361). Not surprisingly, the evolution of the flowering plants appears to be related to the extraordinary diversification and radiation of insect taxa. Modern insects are intimately linked to flowering plant pollination and reproduction. Cretaceous-age flower fossils suggest they had begun to forge these ties during the Mesozoic era (Cleal 1992: 541; but cf., Labandeira and Sepkoski 1993).

Mesozoic Age of Reptiles. By the middle of the Triassic period, animal life on Earth appears to have recovered from the catastrophic Permian mass extinctions. Perhaps the most notable evolutionary trend in this recovery is the evolution of the **reptiles** (class Reptilia) out of the amphibians. The generalized structure of the terminal Paleozoic era **cotylosaurs** (order Cotylosauria) makes them likely candidates for the first member of this taxon. Reptiles are distinguished from their amphibian forebearers primarily by their horny skins and ability to reproduce entirely on land by enclosing the embryos of their offspring, together with nutrients to sustain them, within protective eggshells. The domination of the entire Mesozoic fossil record by this single class of vertebrates has lead paleontologist to refer to the era as the "**Age of Reptiles.**"

The most significant class of Triassic reptiles, the **theocodonts**, were the common ancestors of the dinosaurs and crocodilians. True dinosaurs first appear on Earth in the late Triassic period. From this small beginning sprang a truly noble line of land-dwelling animals that evolved rapidly and successfully for nearly 135 mya. Dinosaur taxa spread widely across the Mesozoic world during that enormous time span: some as bipeds, others as quadrupeds, some as ponderous giants, others as lurking, scurrying pygmies. They filled niches as carnivores and as herbivores and inhabited the full range of climate zones and geographies on the Mesozoic Earth, even taking to the air and returning to the oceans. Many paleontologists have consider it likely that at least some dinosaur taxa were **endotherms**, that is warmed blooded creatures able to generate their own heat, rather than cold blooded ectothermic animals dependent upon external sources of body heat like modern reptiles (but see Ruben, Jones, Geist and Hillenius 1997). Still others credit certain dinosaur taxa with complex social behavior and parenting skills.

Although the questions of whether or not the dinosaurs were cold or warm blooded remains in dispute, the **mammals** (class Mammalia), which appeared on Earth at about the same time, were clearly in the endothermic category. The roots of the mammalian line can be traced to an extinct order of mammal-like reptiles known as the **therapsids** (order Therapsida) that appeared in the Permian period. The first true mammals probably evolve from the therapsids during the Triassic period. Although they lived as contemporaries of the dinosaurs, primitive mammals remained a distinctly minor element in the ecology of the Mesozoic Earth. Their time, however, would come.

At the end of the Triassic period around 208 mya, the fossil record contains evidence of another mass extinction event. Over a period of less than a million years (rather abrupt in geologic terms), a large number of dominant marine and terrestrial animal taxa disappeared and few, if any, new species originated (Olsen, Shubin, and Anders 1987; Raup and Sepkoski 1982).

A marked recovery in animal life took place during the subsequent **Jurassic** period, however. A major evolutionary event characterizing this rebound was the appearance of the true birds some time after 150 mya. The earliest undisputed member of this new taxon is the celebrated fossil form known as ***Archaeopteryx***. While the feathered wings evident on *Archaeopteryx* fossils clearly indicate their avian nature, they also exhibit a number of skeletal features that reflect derivation from early species of flying reptiles. But which species? In recent years many paleontologists have concluded that the birds evolved from the carnivorous, bipedal suborder of dinosaurs known as the **theropods** (Cracraft 1988: 389). This identity was strengthened recently with the recovery in northern Spain of a the well-preserved fossils of a early Cretaceous nestling bird ten million years younger than *Archaeopteryx* yet still exhibiting a skull of a distinctly dinosaurian aspect (Sanz *et al.* 1997; Morell 1997). The issue is far from resolved, however. A strong case has recently been made linking the birds directly to theocodonts, bypassing the dinosaurs altogether (Burke and Feduccia 1997).

Cretaceous Extinctions. As was the case at the end of the Triassic period, the end of the Cretaceous is marked by profound changes in the nature of life. At the close of period around 65 mya, the great dinosaurs and many other life forms of the Mesozoic era disappeared from the fossil record. In geological terms, the Cretaceous period mass extinctions took place suddenly and their very abruptness makes them difficult to explain. Until quite recently, these extinctions were accounted for by most paleontologists as due to worldwide changes in climate that had begun in the late Cretaceous. According to this scenario, the complete extinctions of the dinosaurs is foreshadowed five to 10 million years before the end of the period by a decline in their numbers and distribution in the fossil record.

A more extraordinary and imaginative explanation has been advanced in recent years by Luis Alvarez and his colleagues (Alvarez *et al.* 1980). Alvarez, a Nobel prize-wining physicist, was surprised to learn that, throughout much of the world, unbroken stratigraphic sections through the Cretaceous-Tertiary boundary commonly encounter a one to 10 centimeter clay layer. This layer contains **iridium** (a rare and precious metal resembling platinum) and **spherules** or tektites, spherical bits of glassy rock two to three millimeters across, that appeared to have formed from molten silicate droplets.

Alvarez and his collaborators astounded the scientific community when they identified the now famous "Iridium Layer" as evidence that one or more mighty asteroids or comets struck the earth around 65 mya. They hypothesize that such an impact generated a giant vapor cloud out of which the spherules condensed. They suggest further that the impact would have sent shock debris across the globe, created an immense fireball and then cast up a huge plume of dust into the atmosphere. As this dust plume ascended, atmospheric winds would have spread it into a globe-embracing cloud blocking sunlight from the Earth's surface for a year or more. In their scenario, the mass extinctions of terminal Cretaceous biota resulted both from the initial impact and from the disruption of the food chain that followed it on the darkened Earth. Publication of the Alvarez hypothesis ignited a scientific debate that continues to the present. It also sent geologist scrambling into the field to test his prediction that one or more giant impact craters dating to the end Cretaceous awaited discovery. As a result of these efforts, an enormous impact crater of the proper age and aspect, the Chicxulub structure, has in fact been identified in the subsurface of the Yucatan Peninsula of Mexico (Alvarez, Claeys, and Kieffer 1995; Jo Morgan *et al* 1997).

A catastrophic event of this magnitude has rarely been postulated to explain changes in the fossil record since the time of George Cuvier. Moreover, such extra-terrestrial causes are now being proposed for other mass extinction events. For example, evidence of an large scale impacts by extraterrestrial bodies in the late Devonian (Claeys, Casier and Margolis 1992), the Permio-Triassic boundary (Kerr 1996) and the late Triassic (Olsen, Shubin, and Anders 1987: 1027-1028) have all been interpreted as indicating that meteorites colliding with the Earth were responsible for mass extinctions at the end of those periods as well as the Cretaceous.

Geologists generally consider the abrupt changes in the fossil record evident at the close of the Ordovician, Devonian, Permian, Triassic and Cretaceous periods to represent the five most significant mass extinctions in Earth history. Surely a genuine intellectual sea change has occurred in the discipline when mainstream geologists who cut their teeth on Lyell's brand of Uniformitarianism are willing to explain four of the "big five" mass extinctions as due wholly or in part to catastrophic collisions with extra-terrestrial bodies.

Of course, meteoric impacts meet Sir Charles Lyell's Uniformitarian criteria. That is, cosmic collisions can now be considered observable "causes now in action." Not only is the crust of the Earth dotted with ancient impact craters, in 1908 a giant fireball struck northern Siberia near the remote village of Tunguska, flattening 2,000 square kilometers of forest and generating atmospheric shock waves that circled the globe twice (Zebrowski 1997: 216). Further, during the summer of 1994 astronomers on Earth were astonished to observe 21 fragments of the comet Shoemaker-Levy 9 bombard the planet Jupiter and kick up a huge cloud of debris from the planet's surface (Levy 1995). The remarkable photographs of the colossal impacts of these fragments taken by the Hubble Telescope and the Galileo Spacecraft demonstrate irrefutably that such catastrophic encounters occur in our Solar System. Observed events of this kind have transformed Alvarez's catastrophist explanation into what Berggren and Van Couvering (1984) call the "New Uniformitarianism," a contemporary blend of the insights of Lyell and the Catastrophists.

Yet, even if extraterrestrial bodies did collide with the Earth in the past, did they always leave mass extinctions in their wake? Apparently not. Dates for the Manson structure, four large impact craters discovered in South Dakota, indicate that a large extraterrestrial body or bodies struck Earth about 74 mya (Izet, Cobban, Obradovich and Kunk 1993). Similarly, seismic profiles, drill core samples and other data suggest that an impact crater 90 kilometers in diameter dating to 35.5 mya lies buried beneath Chesapeake Bay, Virginia. While both of these collisions were surely massive, it appears they struck the Earth without triggering catastrophic extinctions in the biosphere (Koebert, Poag, Reimold and Brandt 1996). But does "not always" mean "not at all"? The debate continues between "Neo-Catastrophist" or "New Uniformitarian" geologists who accept the link between collisions and the extinctions and their more conventional brethren who prefer to look closer to Earth for the cause (for example, cf., Hallam 1987).

Whatever the cause (or causes) of the end Cretaceous mass extinction, the changes that occurred in the fossil record then were profound:

> Before the Cretaceous extinction, very few vertebrate species were smaller than a cat. After the extinction there was virtually no vertebrate larger than a cat. The living world had changed dramatically, and the Age of Reptiles had given way to the Age of Mammals (Lewin, 1984a: 3).

The Cenozoic Era. The most recent era of geological time is called the Cenozoic. The name is derived from the Greek and means literally "recent life." The Cenozoic era dates from the end of the Cretaceous period around 66.4 mya to the present time. Lasting slightly less one-third as long as the Mesozoic, but one-fifth that Paleozoic, and a mere *one-sixtieth* the duration of the Precambrian, the Cenozoic is the shortest of the four great eras of geological time.

The Cenozoic era is divided into two periods, the first of which is called the **Tertiary** (ca. 66.4 to 1.72).The Tertiary period is in turn divided into five epochs: the **Paleocene**, the **Eocene**, the **Oligocene**, the **Miocene** and the **Pliocene**. Geologists commonly group the first three of these epochs togther as the **Paleogene** (ca. 66.4 to 23.7 mya) and refer collectively to the last two as the **Neogene** (ca. 23.7 to 1.72 mya). The second period of the Cenozoic era is called the **Quaternary** The Quaternary period dates to between ca. 1.72 mya and the present. It is subdivided into the **Pleistocene** epoch (1.72 mya to 10,000 BP) and the **Holocene** or **Recent** epoch (10,000 BP to the present).

Cenozoic Plate Tectonics. The Cenozoic saw the dramatic continuation of the global realignment of the Earth's continental plates and oceans. Although the continents had begun to achieve their modern shapes by the beginning of the Cenozoic, they were still a long way from occupying their current positions on the globe. During the era, the uplift of Alps and the convergence and joiing of the Eurasian and the African continental plates resulted in the destruction of the equatorial Tethys sea and joined Africa and Europe for the first time in geological history. Conversely, Australia, Antarctica, and South America, continued to move farther apart and Greenland was severed from Europe (Pomeral 1982: 147) These movements enlarged the oceans between the continents and, by the middle of the Eocene, rendered the break between American and Eurasia complete. As a result

> mammals, which during the Palaeocene and early Eocene had migrated from North America towards Europe across the northern continent, could now only cross by way of the Bering isthmus, impermanent because of its low elevation (Pomeral 1982: 16).

Meanwhile, the collision of the Indian segment into Asia that began 50 mya resulted in the uplift of the Himalayas and the Tibetan Plateau (Harrison, Copeland, Kidd and Yin 1992). In the North America, mountain building resulted in the contraction of the Caribbean and the uplift of the Sierra Nevada and the Coastal Range mountains. .Meanwhile, the Andes, which had begun to rise during the Mesolithic era, continued to be elevated (Pomerol 1982: 18). Late in the Cenozoic, South America was joined to North America along the mountainous spine of Central America (Marshall, Webb, Sepkoski and Raup 1982).

In the Southern Hemisphere, the isolation of Antarctica created the cold circum-Antarctic current. In the Northern Hemisphere, the separation of Greenland and North America from Europe opened a passage that channeled cold waters from the Arctic into the Atlantic. The changes in ocean current circulation patterns brought about by these separations appear to have had profound climatic and biogeographic effects as the cycles of glaciation that followed in the late Pliocene and Pleistocene epochs may, in part, have been triggered by them (Eicher, McAlester, and Rottman 1984: 153). Plate realignment also resulted in the growth of mountains and in heightened volcanism that further contributed to the cooling of the Earth's climate during the era (Cloud 1988: 418, 435).

Finally, the convergence of the Eurasian and the African continental plates at the straits of Gibraltar and Arabia resulted in the creation of the Mediterranean basin. This basin was filled with run-off from the land and water from the Atlantic until around 6.5 mya. At that time, the growth of the Antarctic ice cap lowered worldwide sea level and cut off the flow of water from the Atlantic into the the basin. Over the next 1.2 million years of the Late Miocene, the Mediterranean basin was desiccated and refilled numerous times at roughly 1,000 year intervals. It re-filled for the last time around 5.3 mya, an event that marks the beginning of the Pliocene epoch (Cloud 1988: 440; Hsu 1983).

Cenozoic Climates. Not surprisingly, deep-sea cores dating to the last 66 million years record a profound climatic transition. Shortly before the start of the Cenozoic era, Cretaceous temperatures globally are estimated to have been 6 to 12° Centigrade higher than present day values. It seems likely that the Earth had no permanent ice during that period. In contrast, the Quaternary, the most recent period of the Cenozoic era, has witnessed widespread and repeated continental glaciation and the establishment of permanent ice caps at both poles of the Earth. The decline of mean annual temperature on Earth during the Cenozoic was not a continuous process, however. Paleoclimatic records indicate that the Earth became cooler in a fluctuating but step-wise fashion as the Cenozoic era wore on.

Data from the deep sea cores suggest that the Earth's climate cooled markedly at the boundary between the Cretaceous and the Paleocene (ca. 66.4 mya). The cool conditions of the Paleocene period gave way to global warming during the Eocene period (57.8 to 36.6 mya). Near the end of the Middle Eocene, an ice cap formed on Antarctica. World climate remained cool between 38 and 30 mya during the Late Eocene and the Early Oligocene. After about 30 mya, world climate began to move erratically towards relatively warm conditions once again during the late Oligocene.

A particularly detailed history of the climate change that took place during the Miocene, Pliocene and Pleistocene has been reconstructed in recent decade. This detailed climate curve is based largely on the analysis of data from glacial ice cores extracted from the Greenland ice sheet and from deep sea sediment cores. These data reveal that the Earth began to cool again during the early Miocene period and that, as a consequence,

> the Miocene marked the transition from an ice-free world with well-mixed oceans to one that was glaciated and had highly temperature-stratified oceans like those of today (Schoell *et al* 1994:1122).

A further result of this cooling was the re-establishment of a permanent glacial ice cap on Antarctica. Beginning some time after 10 mya in the Late Miocene, continental glaciation began to occur in cycles in the Northern Hemisphere. The cooling trend continued in a step wise decline throughout that epoch and the subsequent Pliocene. Continental scale glaciation begins in the late Pliocene after ca. to 2.3 mya. Glacial and interglacial cycles become especially frequent from 800,000 BP. until the end of the Pleistocene about 10,000 years ago. Pliocene and Pleistocene climatic fluctuations will be discussed in greater detail in the next chapter.

Cenozoic Age of Mammals. The fossil record reveals that, during the Cenozoic era, the class Mammalia expanded from the 3 or 4 orders and 20 families known from the Tertiary period to the 18 living orders, 120 families, and 4,500 species known on Earth today (Cloud 1988: 436). As noted earlier, true mammals evolved from mammal-like reptiles during the Jurassic period more than 200 million years ago. These early taxa appear to have occupied marginal niches as insectivores in the environments of the Mesozoic era. Following the end-Cretaceous extinction of the dinosaurs, however, the descendants of these formerly insignificant insect-eaters became thrive, spread and fill numerous empty ecological niches once occupied by dinosaurs and other reptiles.

But how was the early mammalian line able to effect this opportunistic adaptive radiation into the ecospace left vacant by the failure and extinction of so many other taxa? This success may be due in large part to their unique physical characteristics. Perhaps most importantly, mammals are warm-blooded and capable of maintaining a constant body temperature through internal, rather than external, heat. Mammalian hair or fur aids in the maintenance of proper body temperature by serving as insulation. These two traits in particular allowed these organisms to adapt to a wide range of temperature environments. Compared to other animal taxa, mammals have a greater brain to body weight ratio and a correspondingly larger and more complexly-organized cerebral cortex and central nervous system. The mammalian heart provides more efficient respiration as its four-chambers separate body and lung circulation. Finally, mammalian reproductive success is enhanced by the internal gestation of the offspring and general tendency of mammalian parents to provide extended care for their young after birth (after Burrus and Spiegel 1976: 263).

Class Mammalia consists of three sub-classes. Listed in the order of their appearance in time and the degree of their relative evolutionary development, the three sub-classes are the monotremes, the marsupials, and the placentals. The **monotremes** are the most primitive surviving forms of mammals. They have body hair and secrete milk, but retain repitilian traits such as egg-laying. In this, they are no doubt similar of the mammal-like reptiles of the Mesozoic era. Today this sub-class is limited to Australia and includes only the duckbill platypus and two genera of spiny anteaters. The **marsupials** are the next most primitive surviving form. The marsupials and the placentals appear to have diverged from a common monotreme ancestor in the early Cretaceous period (ca. 144 and 97.5 mya). Marsupial embryos are not connected by placentas to their mother's tissue. Instead, marsupial offspring are born in an underdeveloped state and crawl to their mother's nipple and pouch where they continue their development. Today marsupials are native only to Australia, Tasmania, New Guinea and its adjacent western Pacific island, and North and South America. The opossum and kangaroo are two of the best know members of this sub-class.

The great majority of mammals living today are **placentals** characterized by reproduction in which the embryo develops in the maternal uterus attached to the maternal tissues by a highly organized placenta. Fully 16 of the 18 living orders of mammals are members of the sub-class Placentalia. The earliest placental mammal fossils were recovered from early Cretaceous deposits in Mongolia, Eurasia and the interior of North America. This distribution has generally been taken to indicate that the distribution of the placentals was limited to the super continent of Laurasia in the Northern Hemisphere. It has long been concluded that placental and marsupial mammals remained confined to Laurasia until the emergence of a landbridge or island route about 65 mya allowed both kinds of mammals to enter South America. Marsupial continued the journey and arrived in Australia during the Paleocene period a short time later. Since no Australian placental mammal fossils other than those of certain bat species were known to be older than five mya, it has been generally concluded that terrestrial placental mammals did not enter Australia until the early Pliocene epoch. Recent recovery of teeth and jaw of what was apparently a primitive placental mammal in a southeastern Australian dating to the early Cretaceous age ca. 122 mya has complicated this picture (Rich, Vickers-Rich, Constantine, Flannery, Kool, van Klaveren 1997). If this find is indeed an early placental mammal, its presence in Australia at this early date may indicate that the higher mammals diverged from the monotremes far earlier than had been expected, perhaps to a time before the break up of Pangea. If that is true, the vast and continuous expanse of that as-yet-unbroken super continent would have rendered their dispersal across the Earth easier–and earlier–than has hither to been imagined (Wuethrich 1997).

In any case, the Paleocene-Eocene boundary is marked by an increasing frequency of extinction of species in the archaic mammalian sub-classes and their fairly rapid replacement by placental mammals. The faunal transition that takes place at the beginning of the Eocene epoch (ca. 57.8 mya) marks the beginning of the placental domination of the Earth's environmental systems (Russell 1975: 29).

Surely the most singularly remarkable order of placental mammals to appear in Cenozoic era were the **proboscideans.** With their spectacular tusks and their prehensile fifth limb or trunk adapted to grasping and probing their surroundings, the proboscideans include mammoths, mastodons, a host of less familiar taxa and, of course, the modern Indian and African elephants.

The earliest members of the Proboscidea order diverged from the other mammals in Africa during the early Paleocene epoch sometime between 65 and 60 mya. The incisors of the earliest proboscideans have begun to specialize into small tusks but these forms lack many of the attributes of their later descendants. For example, the Eocene age proboscidean, Moeritherium, had a very small trunk and was no larger than a wild boar Pomerol 1982: 38).

The early evolution and diversification of the proboscidians was confined to Africa by the Tethys Sea. However, when the Miocene destruction of the Tethys Sea removed this barrier, members of this order were able to spread to Europe and Asia. By the late Miocene, they crossed the Bering Straits landbridge that connected northeastern Siberia with Alaska and spread south into North America. When the Panamanian landbridge emerged in the late Pliocene, they migrated across it as well and spread into South America. During the Tertiary period, proboscideans occupied a wide range of habitats from the tropics to the tundra, in process diversifying into some 40 genera. At the close of the Pleistocene epoch, however, mammoths and mastodons became extinct in the Northern Hemisphere and the remaining probocideans, including ancestors of the modern Indian and African elephants, became limited to the subtropical and tropical regions of the Old World (Jeheskel and Tassy 1996).

An even more successful evolutionary line was that of the hoofed, grazing, herd animals known collectively as the **ungulates.** Ungulates fall into two orders: the **Artiodactyla,** animals such as pigs, hippopotami, camels, deer, antelope and goats classed together because their hooves consist of either two or four toes, and the **Perissodactyla,** such as horses, rhinoceros and tapir, whose hooves consist of an odd number of toes. The explosive diversification and radiation of the ungulates is correlated with the evolution and development of the family of plants known as the **grasses** (*Gramineae*) during the late Cenozoic. The emergence of the grasses was certainly the most important land plant evolutionary event of this era. As the late Cenozoic wore on, treeless **savannah grasslands** came to occupy one-third or more of the Earth's available land surfaces. In the process, grasses diversified into more than 8,000 species thereby providing a host of new ecological niches for grazing animals.

Fierce and fleet-footed predators evolved in tandem with the ungulates on the great grasslands of the late Cenozoic era. An impressive example of these hunters is ***Smilodon***, an early member of the cat family (Felidae) and the largest species of the so-called "stabbing cats" or sabertooth tigers. Smilodon measured more than three meters in length and stood more than 1.2 meters high at the shoulders. Although Smilodon was limited to North America, related stabbing cats were widely distributed across the Europe, Asia and Africa during the Pliocene and Pleistocene epochs. The powerful jaws and enormous, projecting, dagger-like upper canine teeth of Smiledon and the other stabbing cats suggests they were efficient predators. ***Homotherium,*** another of the stabbing cats, had a short back that sloped toward the tail like that of the modern hyena. A den site of this species excavated in Texas produced the remains of over 70 baby mammoths, a number that suggests cooperative group hunting on the part of this species (Adams 1997: 909).

In the long run, however, an even more significant evolutionary event was the Miocene appearance of the ancestors of the modern cats, carnivorous species including lions, tigers, leopards, jaguars and numerous smaller felines not forgetting our own domestic tabby.

Emergence of the Primates

The evolutionary roots of the **primates,** the order of placental mammals to which humans belong, are to be found in the late Cretaceous or early Paleocene period. The first archaic primates seem to have evolved sometime between 70 to 60 mya and were present on Earth by the Early Paleocene period (ca. 57.8 to 52 mya). The earliest of these archaic primates resembled rats or squirrels in appearance and probably subsisted on insects and seeds collected on the forest floor. Over time selective pressures forced these creatures to expand their search for food into the trees. Modern south Asian tree shrews trace their ancestry to these archaic Paleocene primates and closely resemble them. Modern tree shrews

> have large eyes and good vision and can use their hands effectively for holding food. Their fingers have sharp claws, rather than the flat nails characteristic of other primates. Tree shrews are territorial, omnivorous, and extremely active, they dart about constantly in the trees screaming and fighting with one another ...Tree shrews bear some anatomical resemblance to the true shrew and were formerly classified along with shrews as insectivores. However, they are anatomically more like the primate Lemur, and they are now usually classified as primates; they actually constitute a transitional form between the two groups (Columbia University Press 1975:2781).

Other close living descendants of the archaic primates include voles, moles, and hedgehogs.

Pre-Monkeys. While modern tree shrews differ only slightly from their Paleocene period fossil ancestors, the primate line that began with them changed radically over the next 55 million years. The first primates of modern aspect seem to have evolved from these archaic, insect-eating primates sometime in the Early Eocene epoch (Martin 1988: 660). The earliest true members of the primate order belong to the sub-order of **Strepsirrhini** which today includes lemurs, tarsiers, and lorises (Voger 1997). This suborder is often referred to by its older term, Prosimii, a name that translates literally as "pre-monkeys."

During the Eocene period, temperature of the Earth appears to have been warmer than at present. Consequently tropical and subtropical environments were located in the northern hemisphere. Although modern non-human primates are largely restricted to the Southern Hemisphere, the fossil record indicates that their ancestors were widely distributed throughout the vast jungles once located in what is today Europe and North America (Martin 1988: 660).

Strepsirrhine primate species diversity and geographic distribution reached a peak in the Eocene.

At the close of the period, however, the sub-order became regionally extinct in Europe and North America. From the Oligocene onwards, Strepsirrhines have been confined to the Southern Hemisphere.

Today Strepsirrhines are limited in distribution to parts of Africa and Asia and modern lemurs and tarsiers survive in abundance only on the island of Madagascar. The separation of Madagascar from the mainland of Africa took place before the emergence of monkeys and apes. Consequently, Strepsirrhines continued to fill arboreal niches there without competition from these later primate forms. Tragically, their habitat is now threatened by a still later primate: *Homo sapiens sapiens*, who, in the last twenty years, has deforested the island at such a frightening rate that the final extinction of many of these prosimian forms is likely to occur before the end of the century.

Primate Characteristics. Despite their ancient lineage, Strepsirrhines possess the basic characteristics shared by primates generally. These include (after Clark 1959: 43; Harrison and Weiner 1964: 27-38).

1. Largely herbivorous diet. Aside from the more primitive insectivores, most primates are either frugivores, omnivores, or herbivores. Accommodation to this diet has entailed some changes in the primate digestive system.

2. Distinctive dental pattern. Primate dentition lacks certain elements of the primitive mammalian dentition and preserves a simple cusp pattern of the molar teeth. The chief determinant of this dentition appears to be the herbivorous nature of the primate diet.

3. Prehensile hands and feet and specialized forelimbs. Primates have prehensile hands and feet capable of grasping and holding objects. This is because their digits, especially the thumb and big toe, exhibit great mobility and highly efficient organs for searching and probing. Further, these digits have flattened nails and highly sensitive tactile pads on them instead of claws. These organs allow primates to climb with ease, to manipulate objects both with strength and with delicacy, and to explore their environments with precision and sensitivity by tactile means. The grasping ability of primates is further enhanced by their specialized forelimb. Over time the forelimbs develop into an exploratory organ that can be rotated, flexed, and extended with facility and accuracy. The forelimb is thus an ideal instrument for probing and examining the animal's immediate surroundings.

4. Visual acuity. Arboreal life must have created powerful selective pressures favoring the ability to judge distance and to respond to visual cues. The elaboration and perfection of primate visual apparatus have resulted from this selection. In the higher primates at least, this selective process resulted in color vision, large eyes relative to the over all size of the skull, and the positioning of the eye orbits towards the front of the skull creating stereoscopy (three-dimensional vision).

5. Reduced reliance upon the olfactory sense compared to other mammals. Over time the apparatus of smell became progressively less developed in primates. This attenuation was presumably related to the greater importance of vision in the primate adaptation. These changes have been accompanied by the progressive shortening of the primate snout or muzzle.

6. Enlargement of the brain. Most living primates have a high ratio of brain weight to body weight. The fossil record indicates that a progressive expansion of primate cranial capacity has taken place over time. Structurally, this enlargement and elaboration has been especially evident in the region of the cerebral cortex and its dependencies.

7. Reduced number of offspring per birth Most primates have one to three offspring per birth. This appears largely to be the consequence of their increasingly efficient gestational processes for nourishing the foetus before birth and prolonged nurturing of offspring after birth. Harrison and Weiner suggest that the difficulty of caring for the new-born in the trees would create selective pressures favoring the birth of a small number of offspring at any one time. They assert that reduced birth numbers have further evolutionary consequence by reducing competition between offspring and thereby prolonging gestation, and prolonging infancy. According to them:

> If only one or a few fetuses are carried, selection for rapid development will be reduced. The same situation holds after birth when offsprings rely on maternal care for survival. Thus an extension of development time is favored by a reduction in litter size. It is characteristic of primates that the pre-maturation period of their extended and dependence upon maternal care is longer. These factors offer the ideal prerequisites for a plasticity in behavior based on learning (Harrison and Weiner 1964: 38).

8. Complex social behavior. Complex social relationships are characteristic of all primates and is commonly credited to the great sophistication and development of the primate brain. Based on field observations of nonhuman primate populations, however, conclude that primate cognitive capacities are relatively less well-developed outside the social realm. They conclude that primate social behavior is not simply made possible by primate intelligence, greater intelligence is a result of such behavior in the first place. According to them, "at least some aspects of primate intelligence evolved to solve the challenges of interacting with conspecifics" (Cheney, Seyfarth, and Smut (1986; also cf., Goodall 1987).

These eight characteristics admirably suit primates for life in the trees and it is not surprsing that the so-called the **Arboreal Theory** dominated thinking about the natural selective forces operating in primate evolution. However, many other mammal species have adapted to life in the trees without evolving all of these characteristics. For example, such efficient arboreal animals as tree squirrels, opossums, and palm civets, for example, have laterally facing eyes, no stereoscopic vision, no significant reduction in the olfactory sense, claws on all digits except the thumb, and no opposability, of either the thumb or big toe. For this reason, Cartmill (1974) proposes the **Visual Predation Theory** of primate origins. In his view,

> the close-set eyes and grasping extremities typical of extant primates are adaptations to some activity other than simply running about in the trees; arboreal life *per se* cannot be expected to transform a primitive tree shrew-like primate into a lemur.

Instead he suggests that the primate visual characteristics were

> predatory adaptations comparable to the similar specializations seen in cats and owls...allowing the predator in each case to gauge its victim's distance accurately without having to move its head (Cartmill 1974: 439).

Further, notes Cartmill, the grasping, clawless hands and feet enabled these early primates to slip up upon their prey and grasp it with both hands without losing hold on the branch that supported them. Cartmill claims that the primate complex of characteristics is thus a product of the predation by Strepsirrhines on insects and other prey in the canopy and lower strata of tropical forests (Cartmill 1974: 441). We are descendants not simply of arboreal animals but **arboreal carnivores**.

In any event, Strepsirrhines did not remain the only form of primate for long. Sometime in the Eocene epoch, the primate order began to differentiate into two suborders: **Strepsirrhini** (or Prosimii) and **Anthropoidea**. Anthropoidea, the **anthropoids**, is a taxon that groups monkeys, the Great Apes, and humankind together. LeGros Clark provided the classic definition of the anthropoids:

> the members of the group are distinguished by their man-like appearance, which catches the attention of even the most unsophisticated eye. The human resemblance of the apes and monkeys on analysis resolves itself into a few outstanding characters such as the relatively voluminous and rounded brain-case; the flatness of the face; the position of the eyes, which look directly forwards and so appear rather close-set; the shrunken and fixed appearance of the external ear; the alertness and versatility of the facial expression, conditioned partly by the differentiation and complexity of the facial muscles; the mobility of the lips, and especially the upper lip which is not cleft and bound down to the gums as in most lower mammals; the absence of a naked moist surface (rhinarium) round the nostrils; the possession of a real 'hand' employed for grasping purposes, combined with the functional independence of the thumb; and the presence of flattened nails on the digits of the hand and foot (Clark 1959: 48-49).

The earliest fossil evidence of the appearance of the anthropoids is from Late Eocene period deposits in the Pondaung Hills of central Burma. These deposits date to about 40 mya and have produced fossil teeth that reflect the distinctive Anthropoidea molar pattern. Although the direct fossil evidence is lacking, it has been concluded largely on anatomical grounds, that the suborder Anthropoidea had separated into two superfamilies during the Late Eocene: the **Old World monkeys** (Cercopithecoidea) and the **New World monkeys** (Ceboidea). The New World monkeys became isolated from the other Anthropoidea breeding populations and thus play no further part in the evolutionary stream that runs to the hominid line. Thus, nothing more will be said of them here.

For the Old World monkeys, the fossil evidence improves greatly in the early Oligocene epoch as the remarkable fossil beds in the Jebel Qatrani formations of the Fayum Depression in Egypt were formed during this time. The Early Oligocene period dates to between 36.6 and 30 mya. The basalt layer that caps the Jebel Qatrani formation has been dated to 31 mya by means of the potassium/argon method. The fossil beds below it, therefore, must be older (Fleagle, Bowen, Obradovich, Simons 1986). Simon (1985: 102) averages the rate of sediment accumulation and estimates that the upper fossil beds date to at least 33 mya.

The ancient Oligocene environment represented in the Jebel Qatrani deposits seems to have been a heavily vegetated delta formation characterized by meandering streams that joined the greatly expanded Mediterranean Sea in saline estuaries. Inland from the ocean the environment was characterized by expanses of freshwater with dense floating vegetation. Reconstructions of the ancient environment of the Oligocene Fayum suggest that the best modern environmental analog is found today in the swampy region bordered by forest and grassland along Lake Victoria in Uganda (Olson and Rasmussen 1986). Such environments create ideal fossil forming conditions. Primates, living in the trees that lined the banks of the ancient streams and ponds, must at death have regularly been incorporated into their accumulating alluvial and lacustrine sediments. Nonetheless, it is jaws and teeth, rather than full skeletons, that form the most abundant primate fossil.

The most significant primate fossils recovered from the Jebel Qatrani formation are the earliest known examples of the primate subfamily Dryopithecine. The **dryopithecines** were a highly varied taxa that through time included fossil species no larger than the modern marmoset to species larger than the modern gorilla. Dryopithecines flourished in Africa and Eurasia from the Oligocene to the Pleistocene (Simons, Andrews, and Pilbeam 1978: 127-128).

The earliest and smallest of the dryopithecines from the Fayum is *Oligopithecus savagei*. Simon (1985: 102) estimates its age to 35 mya or earlier. Despite its primitive characteristics, it is clearly neither a prosimian nor a proto-monkey. Rather, it is evidently the oldest Old World anthropoid yet recovered. As such, it may be regarded as ancestral to all later monkeys, apes and humans (Simons, Andrews, and Pilbeam 1978: 128, 143).

Stratigraphically higher deposits in the Jebel Qatrani deposits produced fossils of ***Aegyptopithecus zeuxis***, a larger and more robust Late Oligocene descendant of *Oligopithecus*. *Aegyptopithecus* was a fruit-eating, arboreal quadruped whose skeletal remains are markedly sexually dimorphic (Simon (1985: 103-104). Although the skeletal features of *Aegyptopithecus* are still very generalized, Simon considers it to: (1) represent the earliest member of the **hominoid** superfaimly and thus to mark the point of divergence between the Old World monkeys on one hand, and the hominoids (which includes the gibbons, orangutans, gorillas, chimpanzees/bonobos, as well as humankind or the **hominids** (family Hominidae) on the other and (2) indicate that the higher primates or hominoids originated in Africa (Simon 1985: 102).

The close of the Oligocene ushers in the Miocene epoch (ca. 23.7 to 5.3 mya). Dryopithecine fossils of the genus *Proconsul* recovered from early and middle Miocene deposits in western Kenya are thought to be a key to understanding the differentiation of the Old World monkeys from the hominoids. These Miocene dryopithecines show important developments in skull, teeth, and wrist that allow them to be classed among the earliest definite hominoids. An absence of fossil evidence makes the dates of the divergence of either the gorillas or the chimpanzees impossible to ascertain at present.

But when did the hominoids differentiate into separate lines leading respectively to the Great Apes or **pongids** (family **Pongidae**) and to the hominids? Until quite recently, human palaeontologists assumed on anatomical grounds that a great evolutionary gap existed between hominids and other non-human primates. Correspondingly, they concluded that, although the pongids and ourselves were descended from a common ancestor, the split between the two lines probably occurred sometime in the Middle Miocene (ca. 16.6 to 11.2 mya).

Therefore, when a primate form, referred to as ***Ramapithecus***, was found to have evolved from the dryopithecine line sometime between about 14 and 10 million years ago, palaeontologists seeking an ancestral Miocene hominid scrutinized it closely. In contrast with the forest habitat of the dryopithecines, the ramapithecines appear to have adapted to open savannah woodland environments. Further, ramapithecine fossils from India and Pakistan exhibited features of jaws and teeth resembling the hominid pattern. This evidence lead Elwyn Simons to announce that the Miocene Ramapithecus was the long-sought earliest member of the human family, Hominidae. He reached this conclusion based on his analysis of *Ramapithecus brevirostris*, which he found had small canine teeth, a high-arched palate, and upper molar rows that diverged toward the rear of the mouth, all features distinguishing the fossil from living apes and allying it to human beings and their recognized ancestors. (Cartmill, Pilbeam, and Isaac 1986: 413). This identification was both logical and attractive: the ramapithecine adaptation to a new, less circumscribed habitat and combined with their hominid-like dental pattern suggested that they had adopted a more omnivorous diet and were probably a major step in the evolutionary processes which led to the emergence of humankind. No less a figure than Charles Darwin first proposed that the canine teeth of our early ancestors became reduced when the first sticks and stones supplanted teeth as weapons. Even so, while accepting this formulation in principle, most human paleontologists found it difficult to believe that dependence on tools began as early as the Miocene.

In any case, by the early 1970s, *Ramapithecus* became widely regarded as the first hominid and the notion that the hominids and pongids had diverged by the Late Miocene was generally accepted. This consensus did not stand for long, however. It was rejected just a decade later, not on the grounds of fossil morphology but on evidence drawn from a rather different science altogether: molecular biology (Cartmill, Pilbeam, and Isaac 1986: 413).

Molecular Clocks

Evolutionary biology is founded on two propositions. The first is that all living organisms are related to one another by descent. The second is that a succession of changes in the genetic makeup of populations of organisms has occurred between generations through time. The history of life is the history of evolution. Evolutionary biologists attempt to reconstruct evolutionary history by building **phylogenies** or evolutionary family trees of the organisms they study.

Until quite recently, the principal means of constructing such phylogenies was to compare living and fossil organisms and determine the degree of relationship or separation between them based on their morphological similarities and differences. In recent years, however, it has become commonplace to construct evolutionary phylogenies by comparison and analysis done at the molecular, rather than the morphological, level. This kind of analysis has been made possible by the automation of DNA sequencing and by the general availability of efficient computer programs able to perform complex phylogenetic analyses on such sequences (Huelsenbeck and Rannala 1997). Molecular phylogenies have been used both as a check on, and a replacement for, the traditional methods of morphological comparison. When used together, molecular and morphological analysis make it possible to test evolutionary hypotheses with an efficiency that is beginning to revolutionize our understanding of the fossil record.

Time and Molecular Change In early 1960s, Emile Zukerkandl and Linus Pauling (1965), two biochemists at the California Institute of Technology, advanced a revolutionary hypothesis: that the differences between the DNA sequences of two species was proportional to the time that had passed since the they had diverged from a common ancestor. This idea received wide scientific attention at the time. It was quickly recognized that, if true, their hypothesis that the pace of protein evolution was regular meant that DNA sequences constituted "molecular clocks" timing the pace of the evolution of life.

However, when first introduced, the whole notion appeared counter-intuitive. Taxonomists had long recognized that the evolution of the morphology of organisms through time was not a regular or continuous process. In fact, they knew that morphological change had occurred in fits and starts. From their perspective, there was no reason to expect that the evolution of **macromolecules**, that is, of the proteins, nucleic acids, and carbohydrates that make up the DNA chain, would have been a smooth and continuous, time-dependent process either. Nonetheless, it was soon demonstrated that

> molecular and morphological changes were not linked in lockstep; a single base substitution in DNA might produce small morphological change, large morphological change, or no change at all, depending on where [in the DNA chain] the substitution occurred. And the notion of neutral evolution began to emerge too, a useful, but not essential, support for the clock notion. A complex set of factors combines to make the chance and consequence of change at one nucleotide position different from that at another. As a result macromolecules offer a quasi-clock based on stochastic changes, not a perfect metronomic clock (Lewin 1986a:561).

The overall rate of mutation rate change in DNA sequences during evolution can be estimated from interspecies DNA sequence differences by examining "phenotype neutral mutations," that is, changes that have little or no effect on the external characteristics of an organism (Britten 1986: 1393).

The mutations that have occurred in **serum albumin** through time are examples of phenotypically neutral DNA changes. Serum is the fluid left when blood clots; it differs from plasma in that it has lost the proteins that cause clotting. Serum albumins are the major protein components of the serum of vertebrates. Vincent Sarich and V.M. Wilson (1967) proposed that the immunological characteristics of serum albumin had evolved at a constant rate through time. From their analysis they concluded that the amount of difference between serum albumin of the various primate species were proportionate to the date of the differentiation of the various branches of the primate line. In their view, these differences could be read as a record of the primate family tree or phylogeny Using this method, Sarich and Wilson inferred that a three-way split between the gorilla, chimpanzee/bonobo, and human lineages took place a mere five million years ago.

Sarich and Wilson's work touched off a fiery debate between palaeontologists and molecular biologists that continues to this day. Additional research along these lines forced taxonomists to gradually concede that the pongid-hominid split occurred sometime between five and 10 mya. Yet, even this broad range of time unsettled those convinced on morphological grounds of the hominid status of *Ramapithecus*. After all, such a range implied that the ramapithecines, whose remains dated from 14 to 17 mya, could not be ancestral exclusively to the hominids. In fact, more complete fossil remains have recently been recovered and these indicate that the ramapithecines were probably not ancestral to the hominids at all but to the orangutans (Cartmill, Pilbeam, and Isaac 1986; 413-414).

Hominoid trichotomy? The molecular data presented other challenges to the morphological reconstructions of hominid phylogeny. Recall that in the primate family tree based on anatomical characteristics, chimpanzees/bonobos and gorillas are classed together as great or anthropoid apes in the family Pongidae. Morphologically they are considered quite distinct from our family, Hominidae. One of the major anatomical criteria of this split is the fact that both gorillas and chimpanzee/bonobos share the anatomical complex that underlies **knuckle-walking**, their mutual means of locomotion. In contrast, hominids are upright, bipedal locomotors.

Yet much of the molecular evidence suggests a different grouping. Despite the fact that both use their knuckles in walking, the closest anthropoid molecular relationship is not between gorillas and chimpanzees/bonobos, but rather between chimpanzees/bonobos and humans! The molecular biologist Morris Goodman and his colleagues (Goodman *et al* 1994), for example, reject the separation of apes and humans at the family level and asserts that gorillas, chimpanzees, and humans should be classed in the same **sub-family**. This would place the classification of the differences between gorillas, chimpanzees/bonobos, and humans on the level of the **genus**. At present, few taxonomists are prepared to accept such a radical re-ordering of higher primate phylogeny. After all, in order to do so, they would have to conclude that the key morphological trait of knuckle-walking either evolved independently among the chimpanzee/bonobos and gorillas after they had diverged from one another or that ancestral hominids were knuckle-walkers or that, in some sense, both might be true. This has lead to the suggestion that a three-way split occurred in the ancestral hominoid population resulting in three lineages separated from one another by roughly three same evolutionary distance. Despite increasing numbers of molecular studies, the question of whether or not the hominoid family tree ends in a trichotomy remains unresolved (Bailey 1993: 107).

What ever the eventual outcome of this taxonomic dispute, the current reading of the timing of the major differentiations in the phylogeny of the superfamily Hominoidea during the Neogene period based on molecular data appears tentatively as follows:

1. The gibbons split off between 22 and 18 mya in the early Miocene period.
2. The orangutans between 16 and 13 mya in the middle Miocene.
3. The gorillas split off in the Late Miocene sometime between 10 and 8 mya.
4. This last split leaves humans and chimpanzees briefly united, until, at the very end of the Late Miocene between about 7.7 and 5.5 mya, they differentiated into two separate lines (these dates after Lewin 1987: 274; 1988: 239).

Based on fossil remains, the Neogene period phylogeny of the Hominoidea now appears as follows:

1 Hominoids first diverged from the Anthropoids in the late Oligocene between 26 and 23 mya in Africa.
2. The early Miocene fossil record from East Africa indicates that the hominoids were a highly diversified taxon between ca. 23 and 17 mya.
3. Hominoids appear in the fossil record of Europe and Asia at the very end of the early Miocene period. They apparently radiated northward when the Tethys Sea barrier that had confined them to Africa was removed by the convergence of the European and Asia continental plates.
4. Ramapithecine fossils date to between 14 and 10 mya and appear to represent the divergence of the orangutans in the middle Miocene.
5. The date of the divergence of gorillas and chimpanzee/bonobos can not be ascertained on morphological grounds at present due to the near total absence of their ancestors from the fossil record.
6. If the fragmentary remains recovered at Lothogam, Kenya are those of a hominid, they indicate that the taxon had diverged between 6 and 5 mya.
7. Australopithecine fossil remains of indisputable hominid affiliation recovered

from Ethiopia indicate that the hominids had definitely diverged from the other anthropoids by 4.4 mya.

On comparing these two data sets it becomes apparent that, although great gaps in our knowledge remain, the hypothesis that the evolutionary divergence between Great Apes and humans drawn from the molecular data is not inconsistent with the fossil evidence.

In fact, chronological problems aside, the primary insight provided by hominoid DNA analysis–the unexpected genetic closeness of chimpanzees and humans–is beginning to receive limited confirmation in the fossil record. For example, it has been observed that early Australopithecines and modern chimpanzees both have curved finger and toe bones. This feature, absent in modern humans, facilitates tree-climbing and suggest important behavioral as well as anatomical continuities between the early hominids and the Great Apes (Stern and Susman 1983). The closeness of the relationship may also be reflected in later, more evolved hominid forms as well. When arm and leg bones of the first member of genus *Homo* were recovered at **Olduvai Gorge** in Tanzania, East Africa, physical anthropologists were surprised by the primitive morphology of the bones and by the parallels between them and those of chimpanzees. We will turn to this question in the following sections.

Studying Contemporary Non-Human Primates

The measurement of the rates of DNA sequence evolution at the molecular level together with recent human fossil finds have forced a re-evaluation of the phylogenic relationships between the Great Apes and humans. In narrowing the evolutionary distance between us, molecular studies have also underscored the fact that understanding non-human primates, particularly the chimpanzees, bonobos and gorillas, is critical in interpreting human evolution. In fact, much has been learned in this regard from the study of non-human primates. Particularly notable is the work done among free-ranging primates in the tropics over the last 50 years by physical anthropologists and zoologists. Key studies of primate ethology include that of Sherwood Washburn and Irven DeVore (1961a, b) on the olive baboons of Nairobi National Park in Kenya, of George Schaller (1963) and Dian Fossey on the mountain gorillas of central Africa, of Jane Goodall (1986) on the chimpanzees of Gombe Stream National Park in Tanzania, fo Tayayoshi Kano (1992) on the bonobos or pygmy chimpanzees of the Congo of Birute Galdikas on the Orangutans of Borneo, and of James Loy (1970) on the Rhesus monkeys on Cayo Santiago island off Puerto Rico. The work of these and other scholars has sharpened scientific understanding of primate behavior and revealed its range and complexity. In doing so, they transformed primate **ethology**, the study of primate social behavior, into a credible scientific discipline. Their work has also provided models and hypotheses about the evolution of the behavior of our early hominid ancestors (for example cf., Washburn 1961, Loy and Peters 1991).

However, in studies of free-ranging animals, the forest undergrowth and the size of group territories makes it difficult to monitor the behavior of primates for long periods. While such long-term observations can easily be made in zoos, the behavior of animals as intelligent as higher primates like the chimpanzee is generally distorted by the conditions of captivity. To overcome both of these difficulties, a chimpanzee colony was established on a two-acre island in the Burger's Zoo in Arnhem, Holland. This colony has become a viable breeding population numbering some 25 individuals. An observation blind was built adjacent to the island and the colony has been under systematic observation since 1971 by the primate **ethologist**, Frans de Waal. Using this blind for uncounted hours, Waal has been able to observe chimpanzee behavior in detail. He reported his findings in a book provocatively titled, *Chimpanzee Politics* (1982). Waal's title sounds like a promotional ad for *Planet of the Apes*, that odd science fiction-adventure movie of the 1960s that starred Charlton Heston as a bewildered human lost among the intelligent great ape rulers of the post-Apocalypse Earth. Yet Waal did not name his book idly. In fact, his observations revealed that the Arnhem chimpanzees struggled constantly among themselves for power and sexual dominance. In the course of their struggles, Waal found the chimpanzees employed strategies, subterfuges, threats and brides in a manner so thoroughgoing that they might even have impressed that most cynical Renaissance student of (human) political manipulation, Niccolò Machiavelli. In addition, studies of non-human primates have revealed that sexual dissembling and covert sexual liaisons are also not limited to humankind. Limited DNA studies done on chimpanzees in a reserve on the Ivory Coast indicate that fully 54 percent of the infants were fathered by males other than those in the local social group (Wrangham 1997).

Finally, primate ethologists have recognized one other set of behaviors shared by chimpanzees and humans: a propensity for intra-species violence and aggression. Both chimpanzees and people kill their own kind and do it frequently. In their recent book, with the overheated title of *Demonic Male: Apes and the Origins of Human Violence*, Wrangham and

Peterson present the accumulating evidence of lethal behavior among chimpanzees and ask

> Does this mean chimpanzees are naturally violent? Ten years ago it wasn't clear...In this cultural species, it may turn out that one of the least variable of all chimpanzee behaviors is the intense competition between males, the violent aggression they use against strangers and their willingness to maim and kill those that frustrate their goals...As the picture of chimpanzee society settles into focus, it now includes infanticide, rape and regular battering of females by males (Wrangham and Peterson 1997: 108-109).

Further, the authors argue, chimpanzees and humans share violent natures because we both inherited it from our last common Pliocene hominoid ancestor. Wrangham and Peterson view violence as a chimpanzee-human **homology**, that is, a similarity due to common descent. Further, they consider the chimpanzee form *Pan troglodytes* to be evolutionarily more conservative than humans and therefore a good model for reconstructions of the behavior of the last common ancestor that we both share.

But whatever the source of these traits, why do apes and people continually pursue these lethal struggles and this sordid dissembling? For our species, this is a complex and elusive question and one that has occupied philosophers, theologians and psychologists for millennia. For the chimpanzees, however, the answer has a simple Darwinian clarity: **reproductive success**. To the winners goes the spoils of copulation. Further, high status appears, among chimpanzee females at least, to be correlated with greater levels of infant survival (Pusey, Williams and Goodall 1997)..

What the Apes Tell Us. Studies both of free-ranging and captive primates populations, have demonstrated that the apes have complex modes of social interaction and acquire much of their behavior by learning. They have also shown that, like ourselves, non-human primates forge complex ties linking their social organization to the demands of their specific ecological settings. Finally, such studies have made it clear that relationships of power, aggression, dominance, sex and kinship structure the lives of non-human primates no less than they do those of their human primate relatives.

However, this work also reveals that, close as we are to them, a great gulf, both behavioral and physical, separates our species from the Great Apes. We walk upright on two legs and are freed by this distinctive gait to use our hands to manipulate tools and, alas, weapons of all kinds. We speak, more importantly, we conceptualize and symbolize, with language. Our brains, teetering precariously atop our vertebral columns, are proportionately much larger than those of the chimpanzees and any of the Great Apes–too large for our own good, some would say–but much larger none-the-less. How, in the course of the evolution of the primate line, did nature conjure up such a prodigy? Or, for that matter, such a monster? It is to this question of hominid evolution, and to the Pliocene and Pleistocene context of that evolution, that we turn in the next chapter.

Major Points Raised in Chapter Seven

1. The first section of chapter carries the discussion of Deep Time into the Pliocene and Pleistocene geological epochs and provides a context for understanding the evolution of the hominids. The Pliocene epoch is described in terms of its tectonic history, climate and faunal evolution. Particular attention is paid to the Villafranchian land mammal age of the late Pliocene and to the evidence of human evolution contained in Pliocene-age strata in the Great Rift Valley of East Africa.

2. The tectonic history of the succeeding Pleistocene epoch is addressed in the second section of this chapter. Pleistocene-age climate change and the continental-scale glacial advances and retreats occasioned by it are examined. Sea level fluctuations, changes in river gradient and flow and the marked alterations in the distributions of plants and animals correlated with the cycle of Pleistocene continental glaciation also discussed.

3. The third section deals with the evolution of the hominids during the Pliocene and Pleistocene epochs. The section begins with a description of the first definite hominid form, the Australopithecines, and continues through the earliest members of the genus *Homo* and the many variants within and derived from the *Homo erectus* clade or grade.

4. In the fourth section, the cultural period known as the Paleolithic or Old Stone Age is introduced and distinguished from the units of geological time during which the Paleolithic took place. The earliest Stone Age begins with the first appearance of stone artifacts in Africa. The Oldowan industrial tradition is introduced and its replacement by the Acheulean tradition is discussed. The dispersal of African hominids bearing this stone tool tradition to Europe and Eurasia is then outlined. The section closes with a consideration of the technological "roll over" that marked the onset of the Middle Paleolithic period.

5. In the fifth section, the Pleistocene epoch and Paleolithic period in Asia are discussed. The Chopper-Chopping Tool Complex is introduced and its nature and geographic distribution contrasted with that of the Acheulean tradition.

6. In the final section of this chapter, an overall evaluation of the culture of the Early Paleolithic period is undertaken. It is concluded that hominid behavior at this time is paleocultural, deeply conservative, slow to change and yet, in the long run, possessed of profoundly revolutionary consequences for later hominid cognition and culture.

CHAPTER 7

HOMINID EVOLUTION IN THE PLIOCENE AND PLEISTOCENE EPOCHS

Introduction

The evidence of human evolution is contained in geological and archaeological contexts dating to the Pliocene and the Pleistocene epochs. Like the earlier time units of Earth history discussed in Chapter 6, the Pliocene and the Pleistocene epochs are defined chiefly through stratigraphic analysis, chronometric dating of sediment series, and changes in the fossil record within these sediments. However, because they are more recent, these periods are better preserved and more precisely understood than those from earlier times. The Pliocene dates to between 5.3 and 1.8 mya, the Pleistocene between 1.8 million and about 10,000 years ago. The last 10,000 years of Earth history, termed the Holocene or Recent epoch, will be discussed in Chapter 8.

The Pliocene Epoch

The term "Pliocene" is derived from the Greek and means literally "most recent time." It was coined by Sir Charles Lyell in the first edition of his *Principles of Geology* published between 1830 and 1833. Lyell used the term in reference to fossil-bearing strata in the Mediterranean Basin which lay unconformably above rocks of the late Miocene epoch. He considered strata to be of Pliocene age if about 45 to 50 percent of the shallow water marine mollusk and echinoderm fossils they contained were species still living in the present. This definition works very well in the Mediterranean stratigraphic column largely because a catastrophic event on gigantic proportions marks the Miocene-Pliocene boundary there. Elsewhere in the world, the beginning of the Pliocene is seldom so easily discerned.

Recall that tectonic shifts in the continental plates destroyed the Tethys Sea during the Miocene epoch. These shifts both created the Mediterranean basin and largely blocked most connections between it and the world ocean. Late Miocene deep-sea cores record an expansion in the percentages of *Globigerina*, a cold-adapted Foraminifera species in the late Miocene between 11 and 10 mya and again between 7 and 5 mya. These expanded percentages are as pronounced as those correlated with the stages of continental glaciation in the Pleistocene and have been taken to indicate that the polar ice expanded during these intervals. The particularly sharp decline in worldwide temperature during this last interval lead to the growth in the Antarctic ice sheet and a corresponding decline in sea level worldwide. The drop in sea level severed ties between the Mediterranean basin and the surrounding world ocean.

In the absence of inflow from the ocean, evaporation lowered the level of the Mediterranean and progressively increased the salinity of its waters. The Mediterranean continued to shrink and the floor of its basin was eventually transformed into a vast, semi-arid salt flat. During this period, evaporite deposits accumulated to the depths of upwards of five kilometers. These events, dated to between 6.5 and 5.3 mya, constitute the Messinian stage in the late Miocene stratigraphic column.

Near the end of the Miocene, worldwide temperatures rose and, as a consequence, the Antarctic ice sheet again retreated. Rising sea level occasioned by this warming trend eventually breached the land barrier between the Atlantic ocean and the semi-arid Mediterranean basin and carved a channel at the Straits of Gibraltar. When the land barrier burst, it must have released a titanic rush of water that filled the desert basin to a depth of 200 meters in a matter of days. It is this spectacular event that marks the beginning of the Pliocene. It is recorded in the stratigraphy of the classic outcrop at Capo Rosello, Sicily. In that famous Sicilian outcrop

> grey and black evaporite beds laid down on the floor of the desiccated Mediterranean are exposed a the foot of the beach cliff. These are directly overlain, in a knife-sharp contact, by chalk-white microfossil oozes that formed at the bottom of an ocean several kilometers in depth (Tattersall, Delson and Van Couvering 1988: 467).

In southern Europe, at least, the Pliocene epoch begins with this grand theatrical flourish. Since its dramatic terminal Miocene refilling, the Mediterranean has remained an ocean. In fact, during much of the Pliocene, its margin of extended farther than it does at present.

Pliocene Plate Tectonics Major changes continued to occur in the structure of the Earth's crust during the Pliocene epoch. Especially significant were the active and extensive volcanism and the continued mountain building of that time. The uplift of the Himalayas began in the Middle Eocene or Early Oligocene with the collision between the Indian continental plate and the southern margin of Laurasia. However, the maximum deformation of these mountains occurred during the Pliocene and Pleistocene epochs. At that time

> the northern edge of the Indian continental mass was actually pushed beneath the southern margin of the Asian continent. The result was an extraordinarily thick continental crust that deformed into the highest mountain range on earth (Eicher, McAlester, and Rothman 1984: 153).

The Alps also continued to experience uplift and over thrusting during the Pliocene.

In the New World, the uplift and erosion of the Andean Mountains of South America and the Sierra Nevadas, the Cascades, the Rocky Mountains, and the Appalachians in North America continued throughout the Pliocene. In tandem with these events, the continued up warping of the Colorado Plateau resulted in deepening erosion of the Grand Canyon (Kurten 1972: 130).

Pliocene Bridges and Barriers. At the very end of the Early Pliocene about 3.4 mya, the Isthmus of Panama was uplifted above sea level. This uplift connected the North and South American continents and blocked the ancient sea passage between the Pacific and Atlantic oceans. The impact of this event in the Late Pliocene was twofold. First, the newly-emergent Isthmus of Panama created a land bridge that resulted in an interchange of terrestrial mammals between North and South America. Most notably, this bridge allowed North American placental mammals to invade a continent where marsupials, ungulates and edentates had evolved in virtual insolation since the Mesozoic era (Marshall, Webb, Sepkoski and Raup 1982). At the same time, the emergent Isthmus created a **land barrier** that divided and isolated the marine fauna of the region into separate breeding populations in the now-separated Carribean Sea and Pacific Ocean.

Thus, the emergence of the Isthmus changed the selective pressures exerted on animal life in the region by fostering gene flow, species competition, and extinction on land and isolation and genetic drift in the sea. Second, the powerful ocean current that had flowed westward from the Atlantic around the northern rim of South America into the Pacific basin was blocked by the Isthmus. After the early Pliocene epoch, this current was diverted northward into the Gulf Stream. Thus empowered, the warm waters of the Gulf Stream pushed farther north and brought increased precipitation, and presumably greater accumulations of glacial ice, to the cold land around the northern rim of the Atlantic (Eicher, McAlester, and Rottman 1984: 157). In this way, the closing of the Isthmus during the Pliocene indirectly contributed to the onset of continental glaciation during the Pleistocene epoch.

Conversely, throughout most of the Cenozoic era, Alaska was connected to Siberia by a broad isthmus across the Bering Straits that enabled terrestrial fauna to move freely between Asia and North America. Throughout its existence, this isthmus impeded the interchange of marine fauna between the Arctic ocean and the North Pacific and prevented cold Arctic currents from flowing south around Alaska. At the end of the Miocene this situation was reversed. At that time, the land bridge across Beringia was submerged. Then, at the end of the Pliocene, the bridge re-emerged again. After the close of that epoch, it appeared and disappeared intermittently throughout the Pleistocene. The Beringian land bridge was the route that brought humankind into North America.

The uplifting of mountain ranges played a similar role in isolating plants and animals. For example, the fossil record suggests that, prior to the uplift of the Andes, the Amazon and Magdalena river valleys were interconnected and supported similar arrays of fish species. The uplifting of the Andes in the Late Miocene and Pliocene periods, raised an impenetrable barrier between these rivers. Consequently, the array of fish species that each river now contains is markedly different (Lundberg, Machado-Allison, and Kay 1986).

Pliocene Climate. During the early Neogene, a lush, tropical rainforest vegetation characterized the North Atlantic region and other parts of the Earth. Britain, for example, supported palm forests and mangrove swamps at that time. All of had disappeared by the late and terminal Miocene when temperatures on Earth become cooler and drier.. By beginning of the Pliocene epoch, tropical forests had been replaced by expanded grasslands and more cold-tolerant forest vegetation including *Sequoia*, *Tsuga*, and other conifers and the Antarctic ice cap appears to have attained its modern proportions (Goudie 1983: 23).

Then, during the early Pliocene, worldwide temperatures once again began to rise (Kennett 1996:5: 54). There is some uncertainty among paleoclimatologists concerning the impact of this global warming. For example, specialists remain divided as to whether the Antarctic ice sheet collapsed at this time or remained stable. At the present time, compelling physical evidence can be mustered on behalf of both hypotheses. The answer to this question is important to the interpretation of Pliocene as such a collapse would have raised global sea level by 60 meters or more and severed key landbridges between the continents (Burckle, 1995: 4-5). In any case, the warm temperatures of the early Pliocene have not been seen on Earth since.

Sometime after ca. 4 mya, the Greenland ice cap formed (Crowley and North 1991: 202) and by 3.2 mya worldwide temperatures had sharply declined (Kennett 1995: 57). Near the close of the Pliocene at around 2.36 mya, the world experienced a severe cooling event that signaled the onset of continental-scale glaciation in the Northern Hemisphere.

Recent evidence indicates that a large asteroid or comet some 0.5 kilometer in diameter struck the southeast Pacific ocean ca. 2.3 mya. The near correlation between global cooling and this impact begs the question "Did this late Pliocene extraterrestrial impact push the cooling Earth towards continental scale glaciation?" Kyte, Zhou and Wasson (1988: 64) calculate that this impact would have injected a quantity of water vapor into the stratosphere equal to the total amount water vapor that the atmosphere layer currently contains. They hypothesize that the injection of such a huge volume of water to altitudes of 30 kilometers or more may have saturated the upper stratosphere and caused global-scale clouds to form at high-altitudes. A cloud blanket of sufficient size to engulf the globe might have reduced the amount of solar radiation reaching the Earth. Such reduction would have increased the Earth's albedo. An increase in albedo in turn might have been sufficient to decrease global temperatures and, perhaps within the space of a single season, significantly increase the percentage of the Earth's surface under snow (Kyte, Zhou and Wasson 1988: 64).

An extraterrestrial cause for the onset of late Pliocene glaciation is by no means the only one that has been advanced to explain it. At present, there are three other major competing theories accounting for the cycles of worldwide climate change that lead to glaciation: the **Greenhouse theory**, the **Astronomical (or Milankovitch) theory** that posits orbital forcing as the source of climate change, and the **Chaos theory** that internal instability or "chaos" inherent in the Earth's climatic system is its cause. As noted earlier, Greenhouse theorists consider the expansion and contraction of worldwide glaciation to be due to changes in atmospheric temperature that follow decreases and increases in atmospheric carbon dioxide (CO_2) and other greenhouse gases. The Astronomical or Milankovitch theory posits that glaciation is caused by periodic, long term variations in the rotation of the Earth on its axis and in the angle and shape of the Earth's orbit around the Sun. This theory holds that such long term alterations affect the position of the Earth's surface relative to the Sun and therefore influence **insolation**, that is, the total amount of solar radiation, received by the Earth. According to this theory, variations in the geometry of the Earth's orbit and rotation affect the Earth's albedo, its atmospheric temperature and hence, the extent of glaciation or deglaciation that occurs on its surface (Muller and MacDonald 1997).

The notion that internal instability in the Earth's climatic system is the cause of continental glaciation is based on the mathematics of Chaos theory. Central to Chaos theory is the observation that dynamic systems whose change through time is governed by fixed laws move in cycles and yet also behave randomly. According to this theory, long term climate change on the Earth is both cyclic and random and can be accounted for–but never wholly predicted–by reference to the dynamic interaction between key variables constituting the system such as temperature, precipitation, glacial and sea ice, ocean circulation patterns, atmospheric CO_2 levels, cloud cover, humidity, snow cover and so on. According to this theory, the causes of glaciation are largely internal within the Earth's atmosphere rather than external to it.

Whatever the casue of the sharp temperature decline at 2.5 mya, climate on Earth began to moderate once again between 2.2 and 1.6 mya. By this time, however, much of the warm-loving early and middle Pliocene vegetation in the North Atlantic was locally eradicated, first by these late Pliocene temperature declines and eventually by the still more intense cold of the following Pleistocene epoch (Tatersall, Delson and Van Couvering 1988: 468; Goudie 1983: 24) It is during the Pliocene epoch that the zonal climate of the Earth took on the modern character that it shows today (Cloud 1988: 145).

Neogene Climax of the Age of Mammals. The diversity of land mammals, especially large herbivores, reached its peak during the middle of the Miocene epoch some 15 mya and retained a broad species continuity for another 10 million years or more. The rich array of ungulates in North America during this time was particularly striking containing as it did "camels as tall as giraffes; horses as diverse as African antelopes; rhinos proportioned like hippos and living a similar semiaquatic existence" (Webb, Hulbert and Lambert 1995: 91). This ungulate divesity suggests that Middle Miocene climate tended to foster complex habitat mosaics that included riparian forests and woodland savannahs and thereby supplied optimum feeding conditions for a wide range of grazers, browsers and mixed feeders. However, beginning in the Late Miocene about 5 mya, faunal assemblages in Asia, Africa, North and South America experienced a devastating series of extinctions. These faunal transformation follow a worldwide change in climate that took place between ca. 8 and 6 mya and stimulated an expansion in the world's grassland ecosystems at the expense of its forests and woodlands (Cerling *et al* 1997). For instance, the semi-arid steppe vegetation of the North American Great Plains was first established between 7 and 5 mya (Axelrod 1985: 131).

Grassland and savannah biomes can support far greater mammalian biomass than corresponding areas of forest or woodland but these taxa tend to be more narrowly specialized grazers. The **Astian** land mammal age of the Early and Middle Pliocene was characterized by continued specialization and diversification of the mammalian taxa evolving in adaptation to the rich and expanding steppe habitat. .

The early Miocene in North America witnessed the evolution of ***Merychippus,*** a small, three-toed ancestor of the horse. Expansion of the grasslands, in tandem with the emergence of key land bridges, permitted ***Hipparion***, a, small Middle Miocene descendant of *Merychippus*, to spread from North America to much of the Old World during the remainder of the Neogene (Kurten 1972: 133-134). ***Pliohippus,*** a small animal about the size of a pony with a single functional digit on each foot, evolves from this root during the Pliocene (Pomerol 1982: 143). Severance of the Bering Straits land bridge during the Pliocene then ended this exchange for a long time and contributed to the differentiation of Old and New World equine taxa.

One of the most dramatic and successful adaptation was made by the **bovids** *(Bovidae)*, the family of ruminants characterized by unbranched, hollow horns. The family includes cattle, goats, sheep, antelope, gazelle and other taxa. In particular, the Old World fossil record indicates that antelope, gazelles, and related taxa experienced a geometric increase during the Pliocene (Kurten 1972: 138-139)

Another result of continued Pliocene mammalian specialization and diversification was the appearance of species of truly memorable proportions: the mastodonts, rhinoceroses, hippopotamus, and giant ground sloths of the period are the largest land mammals known on Earth since the Cretaceous extinctions. Pliocene carnivores were equally impressive: giant bears and stabbing or saber-toothed cats continued to found in both the Old and New Worlds. The distribution of the lion-sized hyena, ***Percrocuta,*** was limited to the Old World, however (Kurten 1972: 143-144). These carnivores apparently preyed upon such mammals as deer, antelope, giraffes, various proboscideans, horses, and camels. The late Pliocene epoch in the Old World saw the emergence of the **Villafranchian** land mammal age after about 3.5 mya. This age has much in common with the late Astian land mammal age which it follows but is distinguished from the former by the appearance of *Leptobos* or true cattle, bison, and buffalo, *Elephas*, true elephant, and *Equus*, true or one-toed horse. The earliest Pleistocene elephant in Europe was the *Elephas* (*Archidiskodon*) *meridionalis*, a curved-tusked species ultimately of South Asian origin that was adapted to forested environments. *Meridionalis* stock is ancestral to a host of later Pleistocene proboscidean species, including *Mammuthus meridionalis*, the earliest Old World form to immigrate into the North America across the Bering Straits land bridge (Agenbroad 1984: 94). Mastodonts, primarily distinguished from the true elephants on the basis of their dentition, and loxondonts, primarily distinguished by their straight-tusks, were both present in Europe and Eurasia in the Villafranchian. Mastodonts became extinct in Europe early in the Pleistocene (Butzer 1971: 261). The rhinocerotids of the genus *Dicerorhinus*, present on Earth since the Oligocene, continued to thrive in the Villafranchian. The genus survives into the present with the Sumatra rhinoceros and is thus one of the oldest mammalian genera in existence (Kurten 1971: 91). Unfortunately, slaughter by modern hunters places its continued existence in doubt.

Pliocene Hominids in Africa Fossil evidence recovered from strata in southern and eastern Africa indicates that the hominid line–of which modern humans are the only surviving member--had begun to differentiate from the Old World primates in the latter part of the Pliocene epoch. As noted earlier, the first definite hominids are collectively known as **Australopithecines**, although some investigators prefer to place the very earliest known hominid form into a separate genus, ***Ardipithecus***. Currently the earliest evidence of hominid differentiation has been recovered in Ethiopia from contexts dating to around 4.4 mya. Beginning with this late Pliocene date, Africa has produced a nearly continuous record of hominid evolution into the present.

The first, but not the oldest, Australopithecine fossil discoveries were made in South Africa in the celebrated cave sites at Taungs, Sterkfontein, Swartkrans and Makapansgat. Unfortunately, for a variety of geological reasons, these cave deposits cannot be directly dated. At present, the date ranges of the South African Australopithecine fossils can only be be estimated using the presumed ages of the animal species found together with them in various cave sites. In order to obtain a full picture of hominid evolution in the Pliocene archaeologist have had to turn their attention to East Africa and to a truly remarkable geographic feature known as the **Great Rift Valley**.

The Great Rift Valley is a giant, elongated depression in the Earth's crust that stretches southwards some 6,400 kilometers from the Dead Sea in Israel to Zimbabwe in southern Africa. From the stand point of human evolution, the most significant portion of the Rift is the East African segment which is occupied today by the republics of Kenya, Tanzania, Uganda and the southern portions of Sudan and Ethiopia.

Bounded on each side by sharp faults or fracture zones, the structure of the Rift Valley is what geologists call a *graben* (German for "ditch"). Rift *grabens* are found in many places on the Earth. Geologists consider most such structures to be locations where continents are beginning to separate and create new oceans. Some, however, may represent remnants of failed continental break ups (Morgan 1991: 188). The Great Rift Valley appears to be of the former type. Given the role rifts appear to play in global plate tectonics, it is not surprisingly that extensive earthquake and volcanic activity is generally associated with them..

Rifting began in Africa during the Miocene epoch some 23 mya (Green, Achauer and Meyer 1991: 199). From the Miocene epoch until the present, faulting and other tectonic processes have created undrained interior basins that collect runoff from the surrounding uplands. Interior basins in the East African segment currently contain the great lakes of Victoria, Turkana, Malawi, and Tanganyika and a host of lesser ones. These lakes are the largest bodies of freshwater in Africa. They are also among the deepest lakes there, or for that matter, anywhere in the world. The structural basins in the Rift that contain these lakes have few exits. Typically, water does not flow readily out of them to the sea. Much of it either evaporates directly from the lake surfaces or remains impounded on the valley floor. As a consequence of this, lake levels in the Great Rift Valley are very sensitive to fluctuations in rainfall and temperature in their watersheds. Ancient or "fossil" shorelines surround these lakes at elevations above the current level testify to the fact that, at various times in prehistory, rainfall was higher or temperature was lower or both. A proxy measure of prehistoric climate change in East Africa has been obtained by dating and then plotting the elevations of the ancient shorelines around the Rift Valley lakes (Scholz and Rosendahl 1988: 1647).

Run off from the uplands coupled with wind erosion over time has brought vast quantities of sediments into the interior basins of the Rift. Frequent volcanic eruptions in this tectonically active zone have contributed thick layers of **tufts**, that is, sediments formed of compacted volcanic rock fragments, to the basin sedimentary deposits. Most recently, extensive erosion during the late Pleistocene and Holocene has dramatically exposed these deposits in many parts of the Great Rift making it possible to study their stratigraphy in detail. The faunal and pollen assemblages contained in these sediments and the chemical properties of the paleosols that formed on them after they were deposited provides additional data for the reconstruction of ancient Rift Valley climate (Cerling and Hay 1986).

It is the volcanic tufts in these eroded exposures that makes the stratigraphic column of the Rift Valley so valuable to world prehistory, however. First, the tufts contain a paleomagnetic record that tracks the changes that have occurred in the Earth magnetic field during Pliocene and Pleistocene epochs. Second, these volcanic deposits can be directly dated by potassium/argon, fission track, and other methods. Absolute ages have thus been assigned to the geological deposits that make up the stratigraphic column there. This has allowed development of a detailed and refined chronology of the magnetic history of the Rift Valley stratigraphic column as well as the faunal, flora and climatic records that the column contains.

Yet, the Rift Valley paleomagnetic and paleoecological sequence would be of chiefly geological interest were it not for the record of hominid evolution that is also incorporated it. Fossil evidence of this evolution is especially well represented in deposits found in three large regions of the East African segment of the Great Rift: Olduvai Gorge in Tanzania, the Lake Turkana Basin in northern Kenya and southern Ethiopia and the Awash Valley in Ethiopia (Aguirre 1997: 129-133). These three regions have been subjected to intense scientific scrutiny over the last thirty years ago. This effort has been amply rewarded. Numerous fossil hominids have been recovered from these strata in associations with animal fossils and, more rarely, with crude stone tools.

The unparallel richness of this record appears in large part due to the interior drainage conditions common in the Great Rift. The shallow saline lakes of the undrained basins forms water holes that are irresistible to animals inhabiting this harsh, semi-arid land. Throughout the Plio-Pleistocene, many hominids and other animals drawn to the edges of these lakes never left them. Some met death instead and were rapidly submerged in the sediment-rich waters.Others died elsewhere in the watershed and their remains were washed down into the lake basins. In the fullness of time their remains were fossilized.

These fossils would have remained buried in the sediments unmolested had not been for the subsequent uplift of some of the Rift Valley formations beginning in late Pleistocene times. The on-going erosion of these uplifted former lake sediments continually exposes the fossils they contain. In order to find them, archaeologists systematically comb the thinly vegetated surfaces of the deposits and closely examine the banks of the erosional channels that cut through them. Excavations are undertaken in promising locations but, in the Great Rift Valley, nature remains the chief archaeologist. And what an excavator she has proven to be!

Late Pliocene deposits dating to ca. 4.4 mya eroding in the Awash Valley of Ethiopia have produced fossil remains of the earliest definite hominid yet recovered, a gracile form of Australopithecine, *Australopithecus* (or *Ardipithecus) ramidus*. Fossils of a particularly robust line of Australopithecines have been recovered in late Pliocene deposits dated to around ca. 2.5 mya at Koobi Fora in the Lake Turkana basin. Fossilized remains of the gracile forms of the Australopithecines are absent from Pliocene deposits younger than about 2.2 mya. but robust forms continue to be found in Plio-Pleistocene deposit as recent as 1.2 mya in the Rift Valley. Finally, Pliocene deposits in the Lake Turkana Basin dating to around 2.2 mya have produced fossils of the earliest member of our genus, the large-brained hominid form known as *Homo rudolfensis*.

In sum, the Pliocene sedimentary deposits in the East African section of the Great Rift Valley contain the earliest, the longest and the most complete sequence of human biological and cultural evolution in the world. These sedimentary rock formations have also produced the earliest direct association between fossil hominids, faunal remains, and stone tools. Moreover, all of this can be cross-dated with worldwide magnetiostratigraphy, biostratigraphy, deep sea sediment cores and with higher-latitude glacial-interglacial cycles. These late Pliocene strata are overlain in many parts of the Rift Valley by fossiliferous deposits belonging to the subsequent epoch, the Pleistocene. It is to this subsequent epoch and to its record of hominid evolution that we turn to in the next section.

The Pleistocene Epoch

Sir Charles Lyell coined the name for the Pleistocene epoch. He classified strata as Pleistocene in age if it contained fossils of marine mollusks species 90 to 95 percent of which still exist in the modern world. In accord with Lyell's original definition, the 1948 International Geological Congress placed the boundary between the Pliocene and the Pleistocene at the base of the Calabrian marine stage. This stage was thought to mark the earliest appearance of cool-water molluscan and foraminiferal species in southern Europe. However, in the years that followed the IG Congress, at least four other criteria were also commonly used to mark the beginning of the Pleistocene epoch: (1) the first appearance of the Villafranchian mammal age characterized by the evolution of the true horse, true cattle, and true elephant, (2) the onset of the earliest glaciation, (3) the evolution of the earliest hominids and (4) and appearance of the earliest stone tools.

Unfortunately, subsequent research has revealed that Villafranchian fauna is mostly of late Pliocene age, that glaciation is a late Neogene process that is by no means limited to the Pleistocene, that cool-water marine fauna appear in strata earlier than the Calabrian (Berggren and Van Couvering 1974: 54) and that both the first hominids and the first hominid stone tools appear earliest in the late Pliocene. Needless-to-say, these inconsistencies have led to a good deal of disagreement concerning the placement of the Pliocene/Pleistocene boundary.

The problematic nature of the boundary lead the International Union of Geological Sciences (IUGS) to name a commission charged with establishing the beginning of the Pleistocene once and for all. After extensive study, the commission proposed that the

Plio-Pleistocene sequence of deep-water marine beds at Vrica, Calabria in southern Italy be recognized as the "stratotype," that is, as the standard, definitive example of this boundary for the world. This decision places the beginning of the Pleistocene once again in the stratigraphic position where Lyell first intended it. The absolute age for the beginning of the Calabrian in the Vrica type section has been calculated as 1.8 mya (Pasini and Colalongo 1997: 40). This date is now universally recognized as the worldwide chronostratigraphic boundary between the Pliocene and Pleistocene epochs.

Within the Pleistocene epoch itself, the geomagnetic polarity reversal marking the end of the Bruhes epoch and the onset Matayama epoch at 780,000 BP. has been adopted as boundary between the early and middle Pleistocene. The boundary between the middle and late Pleistocene has been set at beginning of the final or Eem interglacial (at oxygen isotope stage 5e). A date of 128,000 BP. for the onset of this last interglacial before the end of the Pleistocene has been extrapolated from radiocarbon-dated stratigraphy and is therefore "softer" than the magneto-stratigraphic determinations for the Pliocene-Pleistocene boundary and the middle Pleistocene. The Late Pleistocene thus dates from between 128,000 BP. to 10,000 BP. and encompasses the final interglacial and the final glacial stages of the epoch.

Pleistocene Plate Tectonics By the onset of the Pleistocene ca. 1.8 mya, the Earth's continents and oceans had largely reached their modern positions. This does not mean that the great tectonic plates were quiescent during the epoch. Rifting continued in the Great Valley and the uplift of the Tibetan Plateau reached its maximum elevations during the Pleistocene. This continued upthrust appears to have deeply affected wind and moisture circulation patterns in South Asia during the epoch. The continued frequency of volcanic eruptions and earthquakes along the boundaries of the great tectonic plates further attests to the fact that the Earth was not at rest during this epoch any more than it has been in times earlier. In fact, the Pleistocene epoch seems to have witnessed a global increase in explosive volcanic activity and other igneous events (Kennett and Thunell 1975).

Pleistocene Climates and Glacial Cycles. In the 19th century, geologists first recognized that glacial ice once covered a wide areas of now unglaciated regions of the Northern Hemisphere during the Pleistocene. Following this recognition, they began referring to the epoch as the "Ice Age." While that phrase is still common in everyday speech, Earth scientists generally avoid the label because it carries the implication that continental glaciation was unique to the Pleistocene epoch. We follow this convention and do not refer to the Pleistocene as the Ice Age. Yet, while acknowledging that the Pleistocene was by no means the only "ice age" in Earth history, we must note that there is no more important feature of the Pleistocene epoch than the fact that, throughout the its duration, glacial ice repeatedly covered vast tracks of land in the Northern Hemisphere, altered the nature and distribution of its plant and animal life and changed the levels of the Earth's oceans. Further, the frequency, severity, and duration of continental glaciation during the Pleistocene renders the epoch unusual. In the higher, northern latitudes at least, continental glaciation was the major creative force ssculpting the Pleistocene landscape.

The first major cycle of continental glaciation in the Pleistocene began about 800,000 BP. After that time, the northern latitudes of the Earth together with the higher elevations in its southern latitudes were subjected to multiple100,000 year cycles of glacial expansion and retreat. Near the end of the final interglacial/glacial cycle of the Late Pleistocene (ca. 128,000 to 10,000 BP), an ice sheet ranging from 1.9 to 3.3 kilometers thick stretched from Scandinavia to central Germany and from Ireland to western Siberia (Grosswald 1980; but cf., Peltier 1994). A separate glacial complex formed in the Alps encased all but the highest peaks of that mighty range in ice. Mountain glaciers were also present in the valleys of the Pyrenees and the Himalayas. In North America, two ice sheets merged in the late Pleistocene to form an ice mass that extended from the Arctic as far south as the modern states of New York, Ohio, Illinois, Montana and Washington. At the same time mountain glaciers formed and expanded in the Southern Hemisphere and the Antarctic ice sheet calved great numbers of ice bergs into the South Atlantic.

From the 19th century until quite recently, geologists counted only four or five major Pleistocene glacial cycles. Although the precise number of Pleistocene ice cycles is still to be settled, these low numbers are now rejected as far too conservative. Deep-sea cores and stratified deposits of loess contain evidence of anywhere from 10 to 20 cycles of glaciation and interglaciation from the late Pliocene to the late Pleistocene.

During stages of glaciation, mean annual temperatures in northern latitudes ranged between 15 to 18 degrees Centigrade colder than they are today and 1 to 6 degrees Centigrade cooler in the middle latitudes and tropics than they are there at present. In many regions of the world, the climate during periods of glacial maxima was more arid than it was during interglacial or is in modern times. This greater aridity was due to reduced ocean evaporation, greater quantities of permanent ice and greater tropical atmospheric stability (Butzer 1976:40). Finally, each successive interglacial period appears on average to have been slightly cooler than the one that preceded it

Glacial cycles, succeeded one another at intervals of 100,000 years or less. These cycles were punctuated by substages called **interstadia** (singular, **interstade**). Interstadia are short periods of a thousand years or less during which time the ice masses retreat (but do not disappear) and the climate warms. The end of each successive glacial stage ushered in a period of ice remission called an **interglacial.** Although it takes tens of thousands of years for a continental ice sheet to build up, such bodies appear to decay rapidly in only a few thousand years. The interglacial periods that follow the rapid contraction and disappearance of continental ice average 10,000 years or more in duration.

For a long time, geologists regarded glacials and interglacials as more-or-les stable climatic periods. However, the climatic records recovered from the Greenland and Antarctic ice cores and deep sea sediment cores recovered from the North Atlantic have challenged this interpretation. These data indicate that both glacial and interglacial periods were punctuated by significant climate variability. During the glacial period of the Pleistocene at least, temperature shifted violently from extreme cold to relative warmth back extreme cold over the space of a few thousand years. These climate changes were marked by temperature shifts of up to 10° Celsius and occurred on a regular, cyclical schedule of between 1200 and 6000 years. These cyclical climate shifts occurred during interglacial intervals as well but the amplitude of the temperature changes that accompanied them were much less extreme. (Oppo, McManus and Cullen 1998).

Data from both deep sea corers and those extracted from glacial ice indicate that climatic oscillations were most extreme when continental ice sheets were either growing or disintegrating That the end or the beginning of a glacial cycle would be characterized by pronounced and violent climate variability is perhaps not surprising. After all, the transitions from glacial to interglacial climates and back again represents a major reorganization of the ocean and atmosphere circulation system of the Earth (Barry 1992: 523). What is most remarkable about these transitions, is the great rapidity with which they appear to have taken place. Finally, the correlation of the record from ice and deep sea cores with terrestrial climatic indicators from other regions indicate that climate instability was a global phenomenon during the Pleistocene (Thunell and Mortyn 1995).

Sea Level Changes. Marked changes in worldwide sea levels occurred during the Pleistocene in concert with the cycles of glaciation and interglaciation. These changes were due both to **eustasy**, changes in the volume of water in the oceans, and **isostasy**, adjustments in elevation of the topography in response to change in the weight of the ice, sediment or water bearing down upon it.

During periods of glaciation, great quantities of water were withdrawn from the oceans and converted into ice. This withdrawal resulted in the repeated lowering of worldwide sea levels by as much as 100 to 120 meters beneath modern levels. Lower sea levels exposed large portions of the continental shelves as dry land and created land bridges that connected previously isolated land masses. In this manner England was periodically connected to the continent of Europe; Java, Borneo, and Sumatra to the Malay Peninsula; Japan to Korea; Australia to Tasmania and perhaps to New Guinea; Ceylon to India; and of course, Alaska to Siberia.

Conversely, sea levels rose as much as 20 meters or more above modern levels due to the melting of glacial ice during interglacial periods. Raising interglacial sea levels reduced the amount of land exposed on the continental shelves and once again isolated land masses from one another. The cyclical isolation and reconnection of land areas during the Pleistocene had a powerful impact on worldwide patterns of plant and animal distribution and evolution as well as upon human movement and interaction.

However, due to isostasy, the actual level of the oceans relative to the land did not change uniformly across the globe. During period of glacial advance, the weight of ice masses on the continents of the Northern Hemisphere compressed and lowered their coast lines. The removal of the ice during deglaciation had the reverse effect and the coasts of the northern continent rose relative to the sea due to **isostatic rebound** while the increased weight of rising water depressed sea floor by **hydro-static loading**. The responses of the lithosphere and the ocean surface to such loading and unloading were immensely varied and complex. In fact, it therefore now seems unlikely that geologists will ever be able to reach their goal of establishing a Neogene eustatic sea-level curve for the entire world (Farrand 1992).

River Gradient and Flow. Changes in sea level also affected the gradients of rivers flowing into the oceans. In general, lower sea levels steepen river gradients and cause rivers to create benches or terraces in their valleys as they down cut and deepen their channels. Conversely, higher sea levels tend to flatten river gradients. Thus, interglacial episodes tend to be periods of river aggradation when rates of water flow slows down causing rivers to deposit their sediments and gradually fill their valleys upstream from their deltas. This process is also affected by a number of other factors including the isostatic rebound of land following the disappearance of its ice burden, the depression of the sea bed by hydro-static loading and increases in the volume of river discharge as the glacial ice melted.

Biogeographic Change. The Pleistocene epoch was characterized by extensive changes in the distribution of plants and animals. These changes were largely in response to cyclical glacial advances and retreats and to the climatic oscillations which accompanied these cycles. Periods of glacial advance in the Northern Hemisphere lead to direct species change and to the southward displacement of vegetation zones towards the equator. In more southerly regions of the globe, biogeographic changes were caused as changing rainfall patterns and declining temperatures which tended to push mountain vegetation to lower elevations. Extensive changes in plant and animal biogeography were facilitated by land bridges during periods of glacial advance and impeded by rising sea levels during periods of interglaciation.

Rapid climate change also heightened the pressure of natural section on breeding populations, many of which were isolated and then reunited by the changing sea levels. As a result, the pace of mammalian evolution and extinction accelerated (Lundelius 1976: 56).

Pleistocene Land Mammal Ages The associations of species through time that constitute the Pleistocene land mammal ages are well-defined in Europe, North America, and South America but remain incomplete for Africa and Asia. Provisional cross-dating and correlation of the land mammal ages on these three continents has been done based on the absolute dating of Pliocene and Pleistocene stratigraphy in each area (Anderson 1984: 41).

As noted earlier, the Villafranchian land mammal age begins in the late Pliocene sometime after ca. 3.5 mya. and is characterized by the evolution of true elephants from the mastodons, true horse from the one-toed *Phiohippus*, and true bovines including cattle, bison and buffalo from cattle-like antelopes. The Villafranchian ends in the early Pleistocene about one mya and is followed by the **Cromerian** land mammal age. The Cromerian age begins with the onset of the first major glacial cycle of the Pleistocene after ca. 800,000 BP. It is characterized by the extinction of many of the Pliocene Villafranchian genera and the appearance in the Northern Hemisphere of **European brown bear** (*Ursus actos*), **cave bear** (*Ursus spelaeus* Rosenmuller and Heinroth), **reindeer** (*Rangifer tarandus* Linné) and **musk ox** (*Ovibos moschatus* Linné) and a great many other species adapted to conditions of extreme cold (Butzer 1971: 262; Dorozynski 1994; Kurten 1968: 253; Lundelius 1976: 50).

Apparently in response to falling mean annual temperatures, contracting forests and expanding steppe and grasslands, the Villafranchian forest elephant, *Elephas meridionalis* became extinct and was replaced during the Cromerian by two more specialized forms derived from *meridionalis* stock: *Elephas (Archidiskodon) trogontherii*, whose teeth suggest it was a grassland grazer, and *Elephas (Palaeoloxodon) antiquus*, a specialized temperate-woodland browser (Butzer 1971: 261). Changes also occur in the rhinocerotid family during the Cromerian. *Dicerorhinus etrucus* becomes extinct and its place was taken by rhinoceros, *Dicerorhinus merki*. The Cromerian also witnessed the general replacement of the stenonid horses (whose modern representatives include zebras and donkeys) by the caballoid horses (the ancestor of the modern horse) over much of Europe and Eurasia. This replacement also appears related to the increasingly cooler climate (Forsten 1988: 23). The Cromerian age closes in the middle Pleistocene epoch around 350,000 BP.

The **Steinheimian** age begins about 350,000 BP. and comes to an end at the close of the Pleistocene epoch around 10,000 BP. The cycle of continental glaciation and interglaciation continued to affect the character of the animal life during the Steinheimian stage. The Steinheimian faunal assemblages of the interglacial stages and interstadials between the various glacial advances included warm-loving species such as **cave hyena** (*Hyaena perrieri*), **hippopotamus** (*Hippopotamus antiquus*), **straight-tusked elephant** (*Paleoloxodon antiquus*), **giant deer** (*Megaloceros giganteus* Blumenbach), and **cave lion** (*Felix leo* Linné) (Kurten 1968: 13, 29-30, 134-135). The successively cooler temperatures of each interglacial period is reflected in the changing character of the Cromerian and Steinheimian interglacial bestiaries as one approaches the present.

By the end of the Steinheimian land mammal age, Europe and Eurasia supported a cold steppe-tundra vegetational assemblage with large mammals like the cave bear, **polar bear** (*Ursus maritimus*), **woolly mammoth** (*Mammuthus primigenius* Blumenbach), **woolly rhinoceros** (*Coelodonta antiquitatis* Blumenbach), **Saiga antelope** (*Saiga tatarica*), wisent or **woodland bison** (*Bison bonasus*), auroch or **wild cattle** (*Bos primigenius*), horse (*Equus* sp.), reindeer and numerous other cold-adapted forms (Lundelius

1976: 50-51). Woodland forms that survived there were probably relegated to river valleys and other protected refugia on the northern margins of the Mediterranean basin.

The close of the Pleistocene is marked by a worldwide wave of extinctions of animal taxa on a scale that is without precedence since the Cretaceous. The Steinheimian age fauna of Europe and Eurasia is dramatically altered and markedly impoverished. As Butzer (1971: 483) notes, "several genera or families were eradicated for the first time since the Villafranchian, and elephants and rhinos are now absent from temperate Eurasia for the first time since the Tertiary." More will be said of this extinction in the next chapter.

Pleistocene in Subsaharan Africa The fluctuations and changes in rainfall and temperature that took place from the Plio-Pleistocene to recent times had a deep impact on the evolution, distribution, dynamic interaction and extinction of animal populations in Subsaharan Africa. Key trends evident in the African fossil record include: (1) a remarkable increase in the number and distribution of bovids, (2) the increasing prominence of large predatory cats of the genus *Panthera* and the corresponding decrease and extinction of stabbing or saber-tooth cats, (3) and extinction of *Hipparion* as w ell as numerous species of suids, proboscideans, and carnivores. The cycles of aridity and the expansion of the grasslands that follow from them apparently influenced the success of certain migrate species as well. For example, the true horse (*Equus*) from the New World arrives and adapts to the widening belt of savannah grassland south of the Sahara. Of course, no discussion of the evolution of African fauna during the Pleistocene would be complete without noting that the epoch is the setting for the later evolution of the hominid line. The genus *Homo* appears near its beginning and the late Pleistocene witnesses the appearance of anatomically modern *Homo sapiens sapiens*. These two evolutionary events are important in this context because some, perhaps many, of the animal taxa that disappeared from the fossil record during the Pleistocene may have been extinguished by hominids (Klein 1984).

The strata of Africa, particularly that of the Great Rift Valley of East Africa, continues to provide us with the clearest fossil evidence of our early Pleistocene evolutionary career. Fossil remains of robust Australopithecines and of early *Homo* that first appeared in late Pliocene strata in the Rift Valley continue to be found there in rock formations of early Pleistocene age. Stratigraphically, these two early taxa briefly overlap the ***Homo erectus*** clade or grade, a more advanced hominid form which appears in the East Africa stratigraphy in contexts dating to between ca. 1.9 and 1.6 mya. *Homo erectus* crania were also excavated earlier in this century on the island of Java and presumed to be younger than those found in Africa. However, when these crania were radiometrically dated recently they returned ages of between 1.8 and 1.6 mya. If these dates are correct, they indicate the great dispersal of ancestral humankind out of Africa (1) began shortly after the evolution of this more advanced taxon and (2) occurred earliest in the tropical and subtropical regions of the Old World beyond Africa. It also indicates that the stratigraphy of the Great Rift Valley is no longer the chief repository of the story of human evolution. We must look throughout the Old World for it.

Remarkable combinations of formation processes like those of the Rift Valley were largely absent in Europe and Eurasia. Hominid remains there have been recovered chiefly from Pleistocene deposits accumulated in rockshelters, caves and sink holes. Unfortunately, the formation processes that create such deposits are complex and only partly understood. Further, they lack volcanic tuff layers that would facilitate the dating and cross-dating of their stratigraphy. Nonetheless, these sites and the hominid fossils they contain provide a rich and tantalizing--if fragmentary--picture of Pleistocene human evolution of *Homo erectus* and its many descendants. It is to a more detailed discussion of the sequence of hominid evolution from the late Pliocene to the middle Pleistocene epoch that we now turn.

Hominid Evolution in the Plio-Pleistocene

As we have seen, research along the multiple paths of molecular biology, paleoecology and human paleontology over the last 30 years has revolutionized our understanding of hominid evolution. While the particular events of the evolutionary career of our line can never be reconstructed with precision, its broad outlines have come into focus. We know with certainty that our taxonomic family Hominidae, which includes all living and fossil forms of humankind is divisible into two genuses: *Australopithecus*, all of whose members are now extinct, and *Homo*, which contains both fossil and living forms. From the standpoint of our species at least, we now know that three most significant events of the last six million years have been (1) the emergence of the hominid line within the Primate order, (2) the evolution of the genus *Homo* within the hominid line, and (3) the emergence of anatomically modern ***Homo sapiens sapiens*** from this early hominid substrate late in the

Pleistocene epoch. .

Genus ***Australopithecus*** In 1924, the paleontologist Raymond Dart was given the fossilized skull of a juvenile primate that had been collected during mining operations at the Taungs rock quarry in South Africa. Recognizing it as unique, Dart nonetheless confused its identity by christening the hitherto unknown species to which it belonged *Australopithecus,* literally, "southern hominoid" Subsequent analysis coupled with the recovery of numerous additional specimens in southern and eastern Africa have clarified this picture. The Australopithecines are hominoids surely enough, but they are clearly *not* apes. These creatures stood upright and walked on two legs in a most un-apelike manner. This trait alone qualifies them as the first definite hominids known to science. Later fossil recoveries have revealed the genus *Australopithecus* to have been a varied one. Six, and perhaps seven, separate forms have been recognized by scholars.

On the basis of DNA analyses, the differentiation of the hominid line from the hominoids is calculated to have occurred sometime between 5 and 7 mya. Currently, however, the fossil evidence of this differentiation is more ambiguous. Although a 5 to 5.6 mya fossil jaw recovered at the Lothagam site near Lake Baringo in Kenya is generally considered to be from a hominid, the earliest unequivocal hominid fossils recovered so far are from late Pliocene-age strata dated by potassium/argon to around 4.4 mya in the Awash region of the Rift in southern Ethiopia. These fragmentary remains were those of a small-brained creature that stood upright and walked on two legs like Australopithecine specimens found earlier in Africa. However, differences in skeletal morphology lead the excavators to conclude that their discovery not only represented the earliest definite hominid but a new genus, ***Ardipithecus ramidus***, that predates the Australopithecines (White, Suwa and Asfaw 1994). Their conclusions, and the reality of the genus *Ardipithecus*, has yet to be widely accepted by paleoanthropologists. Pending the discovery of further specimens, most prefer to class the form as ***Australopithecus ramidus***.

Another team of excavators, working in the Lake Turkana region of the Rift Valley in northern Kenya, have recently recovered slightly younger fossils of what they take to be still another form of Australopithecine. Fossils of this form, provisionally classed as ***Australopithecus anamensis***, have been recovered from strata in the Koobi Fora region that have been dated by means of the argon/argon method to between 4.05 and 4.12 mya. (Leakey, Feibel, McDougall, Ward and Walker 1998). A number of examples of the slightly later form, ***Australopithecus afarensis***, the celebrated "Lucy," have been recovered in the Awash region of Ethiopia and dated to between 3.8 and 3.5 mya. Later still is ***Australopithecus africanus***, a form which first appears in southern Africa between ca. 3.2 to 2.2 mya.

The most pronounced distinction made among the various species of the genus Australopithecus is between the so-called "**gracile**" and "**robust**" forms. As the names suggests, this distinction is one of based on inferences about relative body size and ruggedness of the different species as well as contrasts in their cranial and dental morphology. Although overlap occurs, in general, gracile Australopithecines species (ncluding *A.ramidus, A anamensis, A. africanus*, and *A. afarensis*) have lighter post-cranial skeletons and smaller cheek teeth and chewing apparatus than the robust forms (Conroy 1997: 133).

The robust line of Australopithecines appears near the end of the Late Pliocene period with emergence of ***Australopithecus aethiopicus***, the tentative name given specimen WT 17000 from Lake Turkana. *Aethiopicus* dates to 2.5 mya and appears to be a primitive form of ***Australopithecus bosei,*** a robust Australopithecine present in eastern Africa between 2.1 to 2 mya and extinct by 1 mya and perhaps also to ***Australopithecus robustus***, a robust species that appears in South Africa about 2.5 mya.

The independent status of these various robust forms remains controversial, however. Yet to be resolved is the question of whether they represent separate genuses, species, subspecies, races, or even merely sexual dimorphism and the variation to be expected within a single lineage evolving over more than two million years (Pilbeam 1984: 94-95; Wolpoff 1980: 156).

Unlike the slender and more lightly built gracile forms, the robust Paranthropus species have large teeth and heavy jaws combined with pronounced sagittal crests across the mid line of the their skulls. Modern gorillas also have sagittal crests to which are affixed powerful jaw and facial muscles. Partly because of this parallel, it has been suggested that, like the gorilla, the diet of robust Australopithecines centered on leaves, roots and other coarse plant materials.

The volume or capacity of the crania of the various forms of Australopithecine range from 370 to 515 cubic centimeters (Conroy *et al* 1998). By comparison, the cranial capacity of modern chimpanzees ranges between 282 and 500 cc., that of modern *Homo sapiens sapiens* between 1,000 and 2,000 cc. or more (Campbell 1982: 148). After comparing the cranial capacity of the various forms of

Australopithecines to their estimated body weight, McHenry (1982), concluded that between ca. 4 to 2 mya, only insignificant changes occurred in absolute and relative brain size.

> what emerges is an image of a creature with a brain size much more like any other member of the suborder *Anthropoidea*...than like *Homo sapiens*. These were creatures with all the constraints of the animal world without language or agriculture. It is a false bias to treat them as humans (McHenry 1982: 161).

"Human" or not, the genus *Australopithecus* was a profoundly successful one whose members persisted in South and East Africa for more than 3.4 million years only to disappear in the middle of the Early Pleistocene epoch.

Genus *Homo* A new, larger-brained form of hominid appears in the late Pliocene fossil record after about 2.5 mya. This new form, present in East Africa until at least 1.8 mya, coexisted with the later gracile and robust forms of *Australopithecus*. It is important to note that, depending upon the characteristics one chooses to emphasize, this first member of the genus *Homo* could as easily be considered an advanced Australopithecine that culminated the developmental history of that early genus (Leakey, Tobias, Napier 1964; Robinson 1965; Tobias 1965; Cronin, Boas, Stringer, Rak 1981: 113).

This taxonomic ambiguity stems in part from the mixture of traits the new hominid exhibits. On one hand, its cranial architecture and dental features indicates that it continues the powerful chewing adaptation developed by the Australopithecines (Wolpoff 1980: 180). On the other hand, its teeth are generally smaller than those of the Australopithecines while its cranial capacity, which ranges from 600 to 752 cc., is dramatically larger (Campbell 1982: 148; but cf., Wolpoff 1981).

To Blumenberg (1983: 590), the markedly larger cranial volume of these new forms indicates that the brain of these hominids had a more complex internal structure and cognitive organization than the brain of the Australopithecines. In his view, early *Homo* is a "big brained" or "advanced" form of hominid whose appearance in the fossil record marks a critical watershed in human cognitive development: from this point forward, the nature of hominid mental capabilities are fundamentally distinguishable from those of either the great apes or the Australopithecines. In any case, at this writing, paleoanthropologists generally class these earliest large-brained hominids as members of the genus *Homo* and divide them into, at minimum, two species: ***Homo rudolfensis***, which first appears in East Africa after 2.5 mya, and ***Homo habilis***, present in both southern and eastern Africa between ca. 2 and 1.5 mya.

The evolution of these earliest members of the genus *Homo* between 2.5 and 1.5 mya is marked by three key events in prehistory: the enlargement of the hominid brain, the first appearance of post-cranial skeletons closely resembling those of modern *Homo sapiens* and, as we shall see a little later in this chapter, the earliest evidence of hominid tool making.

***Homo erectus* Clade or Grade** The next major event in the human evolutionary career is the appearance of *Homo erectus*, a taxon that first appears in the fossil record of East Africa about 1.9 mya and persisted there for at least one million years. While the lineal connections between *Homo rudolfensis* aud *Homo habilis* seem fairly clear, paleoanthropologists are less certain about the relationship between these taxons and *Homo erectus*. A **clade** consists of species demonstrably related to one another through a common ancestor, while a **grade** is merely a arbitrarily defined level of evolutionary development. Is the *Homo erectus* taxon best considered the form or the latter? Paleoanthropologists remain divided. At current levels of knowledge, we can not be certain that the numerous *Homo erectus* fossils recovered worldwide are similar because they have derived their shared traits through descent from a common ancestor or not. What seems certain is that, during the first part of its tenure on earth, *Homo erectus* appears to have co-existed with *Homo habilis* and a robust form of Australopithecine, *Australopithecus boisei* (Walker and Leakey 1978). Whatever the evolutionary relations between it and the earliest species of *Homo* in Africa, the *Homo erectus* population there continued to evolve in both physical and cultural terms in a fashion that allowed it eventually to spread out into Europe and Asia.

The most compelling evidence concerning the size and general morphology of *Homo erectus* was recovered at the **Nariokotome** site on the east side of Lake Turkana in Kenya from a context dating to 1.6 my in age (Leakey and Walker 1988). This site yield the fossilized remains of the most complete skeleton of an early hominid ever found. The skeleton, called ***Homo ergaster***, by its excavators belongs within the *Homo erectus* clade or grade. Precise taxonomic status aside, what is most important about the Nariokotome find is what it tells us about the size and stature of *Homo erectus*. In modern humans it is possible to determine an the age of an individual's skeleton at death by examining the pattern of his tooth eruptions and degree of closure of the long bone **epiphyses**, that is, the ossified ends of the growing bones that fuse when growth is complete. Based on these methods, the Nariokotome skeleton has been estimated to be that of a boy who died when he was between age 11 and 13 years old. This is a striking result since, when he was alive, the boy was between 1.64 and 1.68 meters (5'3' and 5'5") tall, a height that would be considered remarkable for a modern boy of that age. Further, it is estimated that the Nariokotome boy would have grown to a height of 1.85 meters (6'1") and a weight of 71 kilograms (156 pounds) had he lived to full maturity. If the Nariokotome boy's potential height can be taken as representative, the statue of *Homo erectus* likely fell within the range of modern Americans and Europeans.

Despite the similarities between *Homo erectus* and modern humans from the neck down, the morphology from the neck up was is quite another story. *Homo erectus* cranial capacity ranges between 775 and 1,225 cc. which is substantially below that of modern human (Campbell 1982: 148). Further, the general cranial architecture of the species remains distinctly archaic in character. In general *Homo erectus* crania are platycephalic in shape and have thick walls. The verticles of the skulls tend to be marked by sagital ridges and the eye sockets have massive supra-orbital tori above them. The teeth tend to be large and in some specimens the canines project and slightly interlock with a small **diastema** in the upper dentition (Clark 1964:115). A diastema is a gap in a row of teeth that fits one or more teeth in the opposite row when the mouth is closed. From the perspective of our species, these characteristics appear primitive. This is pure hindsight, however. It is important to remember that *Homo erectus* was a successful evolutionary form that persisted on earth for nearly 1.5 million years. Our species should do so well!.

In sum, the evolution of the hominid line during the Lower Pleistocene is marked by major morphological changes including a reduction in the size of the teeth and and the face together with an increase in cranial capacity to 50 to 100 percent above that characteristic of either the pongids or the gracile Australopithecines. More important still, these morphological changes are associated with important behavioral and cognitive transformations (Mann 1976:63; Blumenberg 1983: 590).

The Great Dispersal Begins At the present time, only Africa south of the Sahara has produced fossils that indisputably belong to the Australopithecine and early *Homo* taxa. Fossil evidence that early *Homo* lived outside of Subsaharan Africa in the early Pleistocene is fractional, incomplete and tantalizing. Only two highly fragmentary fossils recovered in Asia are serious candidates for inclusion in one or the other of these two early hominid taxa: one from the **Djetis Beds** in Java and the second at **Longgupo Cave** in China (Larick and Ciochen 1996).

In the absence of actual fossil remains, early stone tools recovered from outside Africa have been taken to suggest the presence of early *Homo* there. For example, the site of **'Ubeidiya** in Jordan Valley of Israel produced crude tools from an occupation level dated to ca. 1.4 mya. Farther north and west, the **Isernia** site in central Italy has produced early cultural materials in a context dated by potassium/argon and paleomagneticism to ca. 780,000 BP. The early date of Isernia makes it one of the oldest sites in Europe. Isernia shows us how deep in the Early Pleistocene the dispersal of the hominids began (Schick and Toth 1993: 257-8). More problematic than either of these sites are very early Pleistocene occupations reported from **Vallonet Cave**, the **Chilhac** site in France and at several other localities in southwestern Europe (Bordes and Thibault 1977). Taken together, the nature of these finds suggests that, if there were early *Homo* populations in the lands beyond Africa during the early Pleistocene, their presence there was at best tentative, faltering and highly circumscribed.

That *Homo erectus* was not so constrained is clearly reflected in the number of fossils of that species that have been recovered beyond Africa. Sites as far afield as **Ternifine** in Algeria, the Djetis and Trinal beds in Java, and at the **Chou-k'ou-tien** (or Zhoukoudian) site and at various localities in the Lantian District in China have produced *Homo erectus* specimens. The wide geographical distribution of these sites indicates that, by at least a million years ago, *Homo erectus* was capable of accommodating himself to a varied and demanding range of environments. With the appearance of this species, the adaptive radiation of the hominid line had begun in earnest. In the course of this radiation, *Homo erectus* apparently began to differentiate into a number of forms. It is gone from the African fossil record by 1.4 mya but apparently persists into the late Pleistocene in Southeast Asia.

The crucial succession from *Homo erectus* to *Homo sapiens* is poorly understood, in part because of a crucial gap in the African hominid fossil record between the last known *Homo erectus* dating to ca. 1.4 mya at Olduvai Gorge, Tanzania and the earliest known *Homo sapiens* fossils. Until quite recently, that earliest example was from **Bodo** in Ethiopia and dated to ca. 600,000 BP. However, the **Danakil** cranium, a fossil skull recently been recovered in the

Danakil Depression of Eritrea, was found in strata whose paleomagnetic signature suggests a date of ca. 1 mya (Gibbons 1998; Abbate *et al* 1998). Thus, in Africa at least, the time gap between the disappearance of *Homo erectus* and the appearance of the earliest *Homo sapiens* has begun to narrow.

The earliest *Homo sapiens* are represented by a diverse group of fossils. In addition to the Danakil and Bodo speciments, possible examples of transitional specimens come, **Florisbad Spring** in South Africa, **Broken Hill** in Zambia, **Arago Cave** (also known as Tautavel cave) and **Fontechevade** in southwestern France, the **Swanscombe** quarry near London, the **Steinheim** and **Mauer** sites in Germany, and **Petralona** in Greece. The hominid fossils from these sites in general are characterized by robust cranial features *and* enlarged cranial capacity. Both their physical morphology and temporal position situate them intermediately between *Homo erectus*, with their robust cranial features and relatively modest cranial capacity, and early forms of *Homo sapiens*, that exhibit cranial capacity in the modern range and cranial architecture of a somewhat more gracile form.

Until quite recently, these intermediate variants were considered to be early forms of our own species and were classed together as **Archaic *Homo sapiens***. Although many paleoanthropologists still favor this categorization, others have come to recognize at least three separate species between *Homo erectus* and *Homo sapiens* in the interval between 1 mya and 200,000 years BP. In Africa, such post-*Homo erectus* fossils have tentatively been classified as ***Homo rhodesiensis***. This label has been applied to the fossils from Florisbad Spring and Broken Hill as well as other sites in Africa.

In Europe, two such forms are now generally recognized: ***Homo antecessor***, a European form that appears shortly after 800,000 BP. and ***Homo heidelbergensis***, a label now being applied to European fossils dating to the 500,000 to 200,000 BP. interval. ***Homo antecessor***, is both the oldest of these fossil forms and the most recently discovered. The loci of their recovery are the limestone grottos of **Atapuerca Hill** in northeast Spain. Atapuerrca Hill promises to be for the middle Pleistocene in Europe what the Great Rift has been for the early Pleistocene in Africa. This region has already produced nearly 80 separate individual *Homo antecessor* fossils thus far. Importantly, these fossils have been recovered in geological contexts exhibiting clear magnetostratigraphy that allows the sites to be dated with reference to the known sequence of worldwide changes in the Earth's magnetic field (Carbonell *et al* 1995).

The *Homo heidelbergensis* label is now applied to the scattered and fragmentary fossils that indicate a hominid presence in Europe during the middle Pleistocene. The most notable site of this age is **Vertessollos** in Hungary where fragmentary hominid skeletal remains together with traces of fire, animal bone and abundant stone tools have been recovered. Definite, if fragmentary, evidence for the presence of such hominids in Europe in the middle Pleistocene have also been reported from southwestern France, central Italy, Czechoslovakia and elsewhere. The aforementioned specimens from Arago Cave, Fontechevade, Swanscombe, Steinheim, Mauer and Petralona have tentatively been placed in this new taxon.

Lower Paleolithic	2,500,000-100,000 B.P.
Middle Paleolithic	100,000-40,000 B.P.
Upper Paleolithic	40,000-12,000 B.P.

Sometime between ca. 128,000 and 90,000 BP., *Homo Heidelbergensis*, is replaced in western Eurasia by a new hominid form, ***Homo sapiens neanderthalensis*** (Trinkaus and Howells 1979: 125). We restrict the term *Homo sapiens neanderthalensis* to the "classic" **Neanderthal** of former usage. This form is characterized by a low-vaulted skull with prominent brow ridges, sloping forehead and nose. Further, the jaw and teeth of Neanderthal was positioned markedly forward with regard to the cranial vault. Neanderthal crania are markedly robust, a trait reminiscent of much like Homo erectus. However, unlike that earlier hominid form, the cranial capacity of Neanderthal falls well within, and often exceeds, the size range of anatomically modern human populations. The post-cranial skeleton of Neanderthals were also robust: both males and females were bulky, solidly built, and heavily muscled. In general, they appear to have had arms, shoulders and hands that were more powerful than those of most anatomically modern peoples (Trinkaus and Howells 1979: 127-128). The Neanderthal anatomical pattern is apparently limited to Europe and southwestern Asia. Of the contemporary fossils from outside of this region, the earlier ones are generally best considered archaic Homo sapiens; the later ones, in parts of Africa and Australia at least, as *Homo sapiens sapiens* (Trinkaus and Howells 1979:125-131; Beaumont, de Villers and Vogel 1978: 416). Both the Neanderthals and the anatomically modern *Homo sapiens sapiens* who replace them during the Late Pleistocene will be discussed in the next chapter.

The Paleolithic Period

The terms "Pliocene and Pleistocene geological epochs" and "**Paleolithic archaeological period**" refer to closely related, yet distinct, phenomena. It is not surprising therefore, that the two phrases are commonly confused with another. The term Paleolithic is derived from the Greek and means literally "old stone." In Europe, Asia and North Africa the period is referred to by this Greek name. In Africa, however, the period is referred to simply as the "Stone Age."

While the Pliocene and Pleistocene are defined using geological strata and chronometric dating techniques, the Paleolithic period is a unit of time used by prehistorians "to organize archaeological evidence so that it can be interpreted in terms of cultural change" (Rowe 1962: 40). The period begins more than 2.5 mya ago with the earliest evidence of hominid cultural activity in Africa and closes at around 10,000 to 12,000 years before the present by which time evidence of a human present. By the end of the Paleolithic period, evidence of human present is found over most of the Earth.

Archaeological Systematics Subdivisions of the Paleolithic period into an early, middle and upper portions are generally made with reference to changes in stone tool form and technology which have occurred through time in the archaeological record. The detailed charting of these changes through time has been called "archaeological systematics" (Laville, Rigaud and Sackett 1980: 31). The development of such systematics has been called

> one of the most impressive achievements of modern prehistoric research. It has been called for refined techniques of excavation to ensure that the artifact assemblages are precisely segregated in terms of individual stratigraphic units of occupation, the design of more refined typologies for classifying stone tools according to their form, and the adoption of quantitative methods whereby artifact assemblages may be compared and then classified into industries not simply in terms of the presence and absence of these artifact types but also according to their relative frequencies (Laville, Rigaud, and Sackett 1980: 31).

Despite this achievement, it is still difficult to correlate and cross-date existing regional stone tool sequences accurately within and among continents in the Old World. Further, many regions remain virtually unknown to archaeology. Understanding of the Paleolithic period is also hampered by our limited knowledge of the function and cultural significance of many of these stone tool types. In fact, it is difficult to draw inferences about most aspects of culture during the Paleolithic since, at least during the early, and perhaps even the middle, portion of the period. After all, the tools and other remains upon which such inferences must be based were the product of hominids whose species (and presumably mental processes) were very different from our own.

Yet with the foregoing difficulties in mind, it may be said that in general terms, studies of archaeological systematics of the Old World Paleolithic period indicate that stone tool technology, and hominid culture as a whole, has moved from simplicity towards increasing complexity through time. Painfully slow at first, technical development in the Paleolithic continued throughout the period and began to accelerate dramatically by its Late or Upper portion.

Naturefacts into Artefacts The selective pressures that lead to the evolution of their characteristic upright stature and bipedal gait freed the hominid arms and hands for tool use. Precisely what those selective pressures were remains in dispute. Did the precursors of the hominids come to walk upright *because* they used tools or was the freeing of the hands for such manipulative activities a secondary consequence of natural selection that favored upright bipedalism for some other reason? We will probably never know.

Whatever the first cause of tool use, however, hominds are not the only primates who do it. Long term studies of free-ranging chimpanzees *(Pan troglodytes)* have revealed that, in their natural habitats, our closest relative in the animal kingdom are also tool users. For example, some chimpanzees have been observed in the wild gathering ants to eat by digging into ant hills with their hands or with digging sticks, then luring some of the furious ants who emerge from the nest onto smaller twigs they have cut for this purpose. Some chimpanzees gather termites in a similar manner. Other chimpanzees have been observed using stones as hammers and anvils to crack open nuts. Finally, chimpanzees sometimes extract the water from the hollows of trees by crumpling leaves in their mouths and using them as sponges (Sugiyama 1997). These examples indicate that *Pan troglodytes* is a systematic users of rudimentary tools. Thus, it appears that tool use qualifies as yet another homology or similarity that chimpanzees and hominids share due to joint descent from a common ancestor. *Pan troglodytes* thus provides us with a good model for use in reconstructing the behavior of this last common ancestor of both apes and hominids.

Contemporary chimpanzee tool use may also provide us with insight into the technology of the earliest hominids. As *Pan troglodytes* does today, our earliest hominid ancestors must first have used tools that were little more than "naturefacts," that is, sticks, bones or stones taken up and, with little modification, applied directly to a task. Although naturefacts probably constituted the earliest hominid tools, they are difficult or impossible to recognize in the archaeological record. We are forced, therefore, to begin the Early Paleolithic period with the first definite evidence for the intentional *modification* of such natural materials into "artifacts."

Early Stone Age Africa At the present writing, the earliest indisputable examples of stone tools yet recovered are from Gona in Ethiopia in a context that dates to between ca. 2.8 and 2.5 mya (Semaw *et al* 1997). However, the recovery of a possible tool from Miocene deposits dated between 9 and 10 mya. in India (Prasad 1983) suggests that an even earlier industrial traditions remain to be discovered and that 2.8 mya date for the onset of the Paleolithic period is probably just a minimum age.

Archaeologists first recovered evidence of the earliest intentional modification of stone–and therefore, for the beginning of the Early Stone– at the site of Olduvai Gorge in Tanzania. Correspondingly, this earliest stone tool tradition was named the **Oldowan** after its type site. The Oldowan tradition is characterized by crudely made, elementary core and flake tools which show a very limited typological variability. Mary Leakey (1971) recognizes a scant six distinct tool types in the earliest Oldowan assemblages from Beds I and II at Olduvai Gorge and only ten in the more evolved **Developed Oldowan** from Upper Bed II. The center piece of the industry is the so called unifacial "pebble tool" or chopper. Oldowan choppers are merely cobble-sized stones on which a cutting edge has been created through the removal or "subtraction" of several flakes from one face by striking the cove against a stone anvil or hitting it directly with a hammerstone. In addition to unifacial choppers, Oldowan industries include a limited variety of other kinds of core artifacts including spheroids, discoids, polyhedrals and bifacially-flaked chopping tools and proto-bifaces. All of these tools are characterized by informal trimming. Little appears to have been expended in creating symmetrical or balanced tools; Oldowan stone knappers simply sought to modify the edges of the stones they used as raw material.

Precisely who made the Oldowan tools and what they used them for remains unclear. It has been a long standing habit among archaeologists to credit the manufacture of these objects to *Homo habilis*. However, Oldowan tools are now known to predate the appearance of *Homo habilis* and these earlier examples presumably were made by *Homo rudolfensis* or some form of Australopithecine. Oldowan tools from the Hadar region may have been made by the roughly contemporary *Australopithecus afarensis*. Other East African locations yielding Oldowan assemblages date to between 2.1 and 1.6 mya., a time range that corresponds fairly closely with the appearance of both *Australopithecus bosei* and *Homo habilis* (Klein 1983: 26). In addition, Developed Oldowan artifacts have been recovered in southern Africa at the sites of Sterkfontein and Swaartkrans. Sterkfontein has produced *Australopithecus robustus* skeletal material, while the remains of early *Homo* have been recovered at both sites (Freeman 1975: 244; Klein 1983: 27). As to the identity or identities of these first stone toolmakers, the jury, it would seem, is still out. Given that tool use appears to be a chimpanzee-hominid homology, it is likely that many, perhaps even all, of the various species of Australopithecines made them.

First Culinary Revolution Whoever made them, it is often assumed that these tools, and the scattered animal bone debris associated with them at Olduvai Gorge and elsewhere, indicate that the hominids of the early Paleolithic period were already hunting horses, pigs, bovids or even game animals as large as *Elephas reckii*. The appearance of stone tools in the archaeological record has often been taken as heralding a shift away from the gathering of vegetal foodstuffs towards a primary reliance on hunting. Washburn and Avis (1958) went ever further by suggesting that this new hunting subsistence lead to the development of a sexual division of labor, the disappearance of the estrus cycle in hominid females, the emergence of the nuclear family and ultimately with the increased size of the brain of early *Homo*. Hunting, in this scenario, is what made us human.

However, studies of predator and scavenger behavior, and of the bone debris such behavior leaves behind, lead Brain (1981) and Binford (1981, 1983) to the conclusion that early hominids were not hunters on the African savannah-grasslands at all. Instead, they probably filled a niche as scavengers around the water holes there. To these scholars, the early hominids were more likely to fall prey to larger animals than to act as predators themselves. However, based on his study of predatory animals on the Serengeti plains of East Africa, Schaller (1972:68) thinks it unlikely that any large species could subsist there exclusively by scavenging. Early hominids might have been able occasionally to drive predators like lions or hyenas off their kills or managed to kill newborn, young or sick individual game animals for themselves. Schaller suggests that early hominids would also probably have had to combine such activities with gathering. But how much gathering? Unlike the teeth of the Australopithecines, the diminished size of their premolars and molars of early *Homo* suggests that such creatures were not primarily vegetarians (Pilbeam 1984: 96). The small size of their teeth would have limited their ability to reduce coarse plant material by grinding it with their mouths. This does not mean that early *Homo* was therefore a carnivore, however. Larick and Ciochon (1996: 544) suggest the early hominids used Oldowan stone tools to do such grinding in place of their teeth. It is their view that tool using hominids would have been better able to process the more intractable animal and vegetal materials of the savannah. In their view, the crude Oldowan implements were chiefly used as a means of "pre-digesting" hard foods by breaking, crushing, spliting and cutting them before ingesting them. "Rather than being implements of predation," say these authors, "early tools underlay a technology that essentially added a new first stage to digestion. Such "pre-digestive" capacity afforded to early *Homo* by their tools thereby gave them access to a whole new range of edible foodstuffs. If their interpretation is correct, the first tools represent the first of a series of "**culinary revolutions**" in the human career whereby human beings profoundly altered and expanded their food niche by cultural means.

Handaxes and Their Makers Their utter simplicity notwithstanding, the tools of the Oldowan industrial tradition were apparently entirely serviceable. After all, little changed, they remained in use for 1.1 mya or more years. The persistence of these tools suggests that they performed adequately such tasks as their early hominid makers were capable of undertaking with them. Then, rather abruptly around 1.4 mya, larger, broader, more symmetrical tools appear in the archaeological record at Olduvai Gorge and other sites in the Great Rift Valley. These large tools were part of a new industrial tradition called the **Acheulean**. Although the oldest manifestations of it yet recovered are from Africa, the tradition bears the name of the French site of St. Acheul where it was first recognized and described.

The center piece of Acheulean assemblages--although by no means their most abundant tools–is the large stone tool type known as the **handaxe**. Handaxes are generally made on grapefruit-size or larger stone cores. In plan view, handaxes range from broadly ovate or tear-shaped to thin and lanceolate in form. Although unifacial handaxes occur, most are bifacially worked, that is, they are flaked on both sides, and have sharp, symmetrical cutting edges around their circumferences. The tips of some tear-shaped handaxes end in sharpened points while those of others are blunt. Other morphological variants include triangularly-shaped handaxes as well as axe-shaped implements known as "**cleavers**." In addition to these bifacially-flaked core tools, Acheulean assemblages include large numbers of both expedient and formal cutting and scrapping tools made on flakes.

Acheulean handaxes appear to have been made exclusively by direct percussion. The earliest examples were made by the **hard hammer ("stone-on-stone") percussive technique.** In contrast, the thinner, sharper profiles of many later handaxes indicate that, by the late Acheulean, knappers sometimes finished shaping these implements using "soft" hammers made of wood, horn or antler. In the hands of a skilled knapper, **soft hammer percussion** results in the removal of thiner, flatter flakes. It also provides the knapper with greater control over the flaking process.

Late in the Early Stone Age in Africa ca. 200,000 BP some Acheulean knappers mastered the **prepared core** or **Levallois technique** (Bordes 1970: 31). If possessed of sufficient skill, a knapper can use this technique to shape a core in such a fashion that the form and thickness of flakes struck from it can be predetermined before their removal. In this way, whole series of sharp-pointed and more-or-less uniform Levallois flakes can be repeatedly and rapidly removed from a single core before its exhaustion. Sparingly used by knappers of the Acheulean tradition, the Levallois technique became the technical hallmark of lithic assemblages of the subsequent Middle Stone Age in Africa and Middle Paleolithic in Europe and Eurasia. No doubt it popularity was due both to the greater efficiency of the technique and to its greater economy of raw materials and human energy.

Between about 150,000 and 100,000 BP, late Acheulean tool assemblages began to be characterized by tools, both core and flake tools, generally smaller in size than heretofore. In part, this size reduction was a by-product of the greater control and economy gained though use of the Levallois technique and its variants (McBurney 1973: 22). However, this change may also reflect increased tool specialization during the Late Acheulian and the correspondingly greater tool efficiency such specialization produced.

Yet, beyond the rudiments of technology, we have only the barest understanding of the uses to which the handaxes were put or to the nature of earliest Acheulean subsistence and social life during the later millennia of the Early Stone Age in Africa. We do know, however, that the industry was produced by *Homo erectus* grade or clade and that, after its early development in Subsaharan Africa, the Acheulean industrial tradition appears to have spread rapidly, first to North Africa and thence to southwestern Asia. By the end of the Lower Paleolithic period, the Acheulean industrial tradition was distributed over a vast region of the Old World from the tip of southern Africa to the unglaciated regions of Europe. Eastward, the tradition feathers out and disappears at the so-called "Movius line" in the middle of the Indian subcontinent. Beyond India, Early Paleolithic industries in Asia lack the characteristic Acheulean handaxe and are considered to represent a different technological tradition.

Handaxe Makers Across Eurasia The Eurasian landmass, the largest of the Earth's six continents, stretches east from the Atlantic edge of Europe to the Pacific edge of Asia. At its northern end, Eurasia crosses the Arctic Circle. At its southern limits it skirt the Mediterranean and Red seas, the Indian ocean, the Bay of Bengal, the South China sea and the western Pacific Ocean. For reasons more historical and cultural than geographic, Eurasia is generally subdivided into a European and an Asian segment along an line that is generally, if quite arbitrarily, drawn northwards from the Caspian Sea to the Arctic Ocean along the Ural Mountains.

The spread of the people who bore the Acheulean industry outside the subtropics into the temperate regions of Eurasia was no doubt facilitated by the mastery of the use of fire. Compelling–if not certain--evidence of the intentional use of fire has been recovered at the site of Chou-k'ou-tien (or Zhoukoudian) near Beijing in northern China in contexts dated to 500,000 BP. Still less certain evidence of fire has been reported from East Africa and dated to shortly after one million ago.

In and of itself, fire is probably not enough. To be fully effective in northerly latitudes of Eurasia, fire must be combined with shelter. Clusters of postmolds and fire hearths exposed at the **Terra Amata** site in Nice, France suggests that, by the late Early Paleolithic period ca 380,000 BP, these hominids may have put up simple shelters and warmed them artificially (deLumley 1969). Further, archaeological remains at sites like **Torralba** and **Ambrona** in north central Spain (dated to between ca. 330,000 to 270,000 BP) suggests that hominids of the late Early Paleolithic period had begun hunting animals as large as the elephant, horse, cattle and steppe rhinoceros by means of cooperative fire drives (Howells 1965:85-100; Freeman 1975:247-248; but cf., Binford 1984, Binford and Ho 1985: 428). If big game hunting was in fact practiced at sites like Torralba and Ambrona, it suggests that, at least towards the end of the Early Paleolithic period, hominids no longer simply relied on "targets of opportunity" in their subsistence pursuits but had begun to focus their subsistence more narrowly. Such hunting demands more than a modest investment in foresight, planning and cooperative effort.

A discovery that tends to confirm this interpretation has been made further north eastern Germany at the site of **Bilzingsleben**. Blizingsleben is a large Early Paleolithic site located along an ancient lake shore that dates to between ca. 350,000 and 230,000 BP. Exposure of the surface of this site through large scale excavation has produced evidence of what appear to be fire hearths associated with oval soil stains about three-by-four meters in size. These large soil stains probably mark the former location of the floors of small huts. Even more dramatic evidence of adaptation to the climate of the north has been recovered recently in central Siberia at the site of **Diring Yuriakh**. Sediments from this site dated by thermoluminescence places the earliest occupation of Diring Yuriakh at or before 260,000 BP. Surprisingly, no evidence of the use of fire has yet been recovered at this site (Waters, Forman and Pierson 1997). Both the early age of this site and its surprisingly crude stone tool repertoire, suggests that we have underestimated the resilience of Early Paleolithic period hominids and their capacities to adapt to the harsh environments of the Northern Hemisphere even to the edge of the Arctic Circle.

Early Paleolithic Asia

The eastern or Asian segment of the Eurasian landmass is generally divided into southern, central, southeastern, eastern and northeastern subregions. Geologists and archaeologists have been systematically working in the vastness of eastern Eurasia since early in this century. Their success, although partial, have been remarkable nonetheless. Unlike Europe, most of Asia was never covered by continental scale glaciation. The absence of a clear glacial stratigraphy there has meant that, until quite recently, the Plio-Pleistocene stratigraphic sequence based largely on biostratigraphy, that is, on the cross-dating of deposits by means of the animal fossils they contain (Jia and Huang 1991: 519). In light of this difficulty, it is not surprising that early rersearchers there attempted to fit the Asian Pleistocene stratigraphic sequences into the simple 4 or 5 glacial cycle model developed to describe the stratigraphic succession in the Alps of Europe. Unfortunately, these attempts at intercontinental cross-correlation between Europe and Asia were premature in both areas. The rejection of the classic European Alpine sequence and the subsequent re-examination of the Chinese succession has revealed that the Pleistocene sequences in the two areas cannot be brought into such easy conformity and that much work remains do be done.

Nonetheless, contemporary deep-sea coring in the North Pacific suggests that, although Asian climates and environmental changes were locally highly variable, their fluctuations were generally synchronous with those experienced elsewhere in the world (Luchterhand 1978: 377). Yet even given such gross, worldwide climatic synchrony local stratigraphic sequences must be constructed before inter-area correlations can be done.

Just as attempts were made to fit the Pleistocene succession into a European or Alpine mold in the first half of the century, so too did scholars tried to impose the classificatory scheme developed to describe tools of the European and African Paleolithic period on Asian stone tools assemblages. Most notable in this regard was the path breaking work of the American archaeologist **Hallam L. Movius, Jr.** (1944). Movius was among the first to recognize that early Asian Paleolithic industries from Pakistan to Java and north to China formed a cultural province distinct from that found at that period in Africa and Europe. His analysis revealed that the stone tool industries of South and East Asia lacked the two hallmarks of the Acheulean industries of the early Paleolithic period in Europe and Africa: (1) bifacially-flaked core tools including handaxes and cleavers, and (2) flakes struck from prepared cores by means of the Levaillois technique.

By contrast, these early Asian industries were characterized by crude implements made on large cobbles through the removal of but a few flakes. According to Movius (1944), such industries as the Soan of Pakistan, the Anyathian of upper Burma, the Patjitanian of Java, and the Choukoutienian of Beijing were seen as collectively forming the "**Chopper-Chopping Tool Complex**. The imaginary "Movius Line" was seen as circumscribing much of South, Southeast and East Asia during the Early Paleolithic period. North and west of the line, handaxes are present. Beyond it to the east, handaxes drop out and stone tool industries centering on the large, minimally-shaped chopper-chopping tools appear.

The over all simplicity of the Chopper-Chopping Tool Complex industries has long been a staple of archaeological interpretation. In the past, it has been common for archaeologist to point to this supposed backwardness as evidence that, in cultural or even cognitive terms, early hominids in Asia lagged behind their Early Paleolithic contemporary in Europe and Africa. However, more recent analysis has revealed that the oft-asserted simplicity, monotonousness, and conservatism of these Asian industries had been greatly exaggerated (Olsen 1987: 135; Chang 1977: 51). Senshui (1985: 184) reports that the most important class of tools in the so-called Chopper-Chopping Complex assemblages is not the large pebble choppers. In fact, as such artifacts comprise only about two to six percent of the total number of tools at most sites. Rather, the dominant implements are small flake tools of considerable sophistication that appear to be scrapers. Further, in a limited number of cases, Chopper-Chopping Tool assemblages have been found to contain a few bifacially-flaked handaxe-like implements. Finally, in South Asia at least, extensive field survey has demonstrated that bifacial handaxes and related tools do occasionally occur beyond the Movius line in the region (Sankalia 1978: 112).

But what if we have altogether over estimated the importance of stone in Early Paleolithic technology? Geoffrey Pope (1989) begins with just that premise in rejecting the idea that the Chopper-Chopping Complex is an inferior technological tradition. In Pope's view, far from representing a retarded cultural assemblage, the Chopper-Chopping Tool Complex constitutes the material remnants of a very effective cultural adaptation to Asian environmental circumstances. Pope notes that the distribution of the Chopper-Chopping Tool Complex coincides in a general way with that of the native bamboo forests of South and East Asia. He hypotheses that, during the early Paleolithic period, Asia hominids made the bulk of their tools out of bamboo rather than stone. In his opinion, bamboo would not simply have been used when stone was scarce. On the contrary, its superior qualities would have made it the raw material of choice. Pope supports this idea by noting that, in historic and contemporary in Southeast Asia, a huge array of useful tools are made from bamboo

> Cooking and storage containers, knives, spears, heavy and light projectile points, elaborate traps, rope, fasteners, clothing, and even entire villages can be manufactured from bamboo. In addition, to the stalks, which a source of raw material for the manufacture of a variety of artifacts, the seeds and shoots of many species can be eaten. In historical times, bamboo has been to Asian civilization what the olive tree was to the Greeks (Pope 1989: 53).

It is Pope's view that, far from being the most important implements in their tool kit, the crude stone choppers were merely the most resistant to decay. Pope's idea is a plausible one and is very likely correct. Unfortunately, it suffers the defect of so many plausible hypotheses about the Early Paleolithic period: there is at present no evidence to support it.

In sum, archaeologists still find Movius' Chopper-Chopping Tool Complex to be useful in interpreting the Early Paleolithic in eastern Eurasia. However, they no longer assume that Chopper-Chopping tools were inferior to those of the contemporary Early Paleolithic assemblages in western Eurasia and Africa. Rather, like Pope, they seek other explanations for these contrasts.

Early Paleolithic South Asia. South Asia is the collective term applied to the territory occupied today by modern India, eastern Pakistan, Bangladesh, Assam, western Burma and Sri Lanka, and western Burma. Although the mountain glaciers in the Himalayas on its northern boundary of South Asia expanded and contracted during the glacial and interglacial stages, South Asia remained unglaciated throughout the Pleistocene epoch. This is not to say that climate and vegetation in South Asia were unaffected by the cycles glaciation and interglaciation taking place in the Northern Hemisphere. This impact is most clearly evident in the distribution of fossil sand dunes in northern India. Analysis of the dunes suggest that, during glacial stages, the desert of that region expanded markedly at the expense of the tropical rainforests and savannah grasslands. Correspondingly, contractions of the desert belt there, reflected in the stabilization of the dunes, appears to have been synchronous with glacial recessions (Goudie 1983: 73). Since the desert regions of Pakistan and northern Indian have progressively expanded in the modern, post-glacial times, such an interpretation seems counter-intuitive. However, contemporary desertification there is probably due largely to deforestation and overgrazing rather than environmentally-caused desiccation.

The Paleolithic sequence is less certainly established in South Asia than it is in East Africa. Nonetheless, recent work in northern Pakistan suggests that its beginning maybe as ancient. The **Siwalik Hills** have produced early Miocene hominoid fossils, including *Ramapithecus* and *Gigantopithecus*. Until quite recently, these forms were regarded as likely ancestors of the hominids but are now seen as leading to the gibbons. More direct evidence of an early hominid presence in South Asia has been presented recently by a group of British archaeologists. They report the recovery of "hominid-struck artifacts and/or debitage" that were embedded in a gritstone outcrop that may be more than two million years in age (Dennell *et al.*, 1988: 495-498).

Although no skeletal remains of early hominids have been reported from South Asia, the presence of humankind there during the Paleolithic has been demonstrated in virtually every region there that has been subjected to systematic archaeological site survey (Agrawal 1982: 12; Sankalia 1978: 101).

Early Paleolithic Central Eurasia The mountainous interior of the Eurasia landmass from the Tien Shan Mountains south to the edge of the Iranian Plateau is generally referred to as Central Eurasia. This subregion includes the rugged Tibetan Plateau and ends were the south slopes of the Himalaya Mountains meet the Ganges River plain in India. The modern states of the region include Iran, Afghanistan, western Pakistan, Nepal, Kashmir, Tibet, and the western provinces of China. Knowledge of the Paleolithic period in this vast region remains remarkably undeveloped.

Early Paleolithic Southeast Asia. Southeast Asia includes eastern Burma, Thailand, Laos, Vietnam, Cambodia, Malaysia, Indonesia, the Philippines and New Guinea. In terms of climate and geography, South China might be included in Southeast Asia. However, for clarity of presentation, we discuss the archaeology of that region with East Asia. Scientific understanding of the Paleolithic period of Southeast Asia is fragmentary at best. The political turmoil and economic difficulties that have beset Southeast Asia since World War II have rendered systematic archaeological work in much of the region extremely difficult. Until this situation stabilizes, our understanding of the archaeology of Southeast Asia is not likely to improve.

The entirety of Southeast Asia falls within the tropics. As a consequence, temperatures in the region are, with little variation, hot throughout the day or from season to season. Even here, however, temperature declines with elevation. Therefore, at least in the mountains of New Guinea, permanent, if very small, glaciers occur (Bellwood 1997: 9).

The great fluctuations in temperature and moisture of the Pleistocene appear less noticeable in Southeast Asia than in the rest of Asia. Nonetheless, during glacial stages, the tropical climate of Southeast Asia on the whole became drier and its great forest contracted (Bellwood 1997: 19). More important still, the periodic rising and lower of worldwide sea level occasioned by the climatic fluctuations of the epoch had significant impact there. Malaysia and the islands of Indonesia including Sumatra, Borneo and Java sit astride a continental platform known as the Sunda shelf. Declining sea level during periods of continental glaciation exposed vast area of sea floor on this shelf. During each successive glacial stage, many of the numerous islands that today form the maritime nations of Indonesia, Malaysia and the Philippines were connected to the mainland of Southeast Asia. Geologist refer to this continuous land mass as **Sundaland**. Its periodic exposure during the Pleistocene allowed plants and animals from the Asian mainland to migrate across it. The most recent submergence of Sundaland came at the end of the Pleistocene. The isolation of these islands during the Holocene epoch resulted in the evolution of local races of elephant, tiger, leopard, and dhole on many of them (Goudie 1983: 67).

Presumably Sundaland was the land bridge from Asia that permitted early hominids to reach as far south as the Indonesian island of Java. We know of their arrival largely because of the pioneering research of the Dutch colonial army surgeon, **Eugene Dubois**. Dubois' excavations in the **Trinil Beds** on the Solo River of Java at the end of the 19th century resulted in the first scientific recovery of remains of *Homo erectus*. Stratigraphically earlier than the Trinil Beds, the **Djetis Beds** in Java have produced *Homo erectus* material from contexts recently dating to 1.9 ± 0.4 million years ago (Jacobs 1972: 148). At the same time, the lower Djetis Beds produced a still more primitive looking hominid cranial fragment called *Meganthropus paleojavanicus* by its excavator. As mentioned earlier, if this cranial fragment is in fact that of an advanced australopithecine, *Homo erectus* was preceded by even earlier hominid form in Southeast Asia.

Archaeological work in Java begun so auspiciously by by Dubois in 1891 continued into the 1930s. At that time, deposits geologically younger than the Trinil Beds were found downstream on a high or Ngandong terrace of the Solo River. This location, called the **Ngandong site**, was dug in a rather sloppy manner during the early 1930s. Nonetheless, a large body of Pleistocene-age animal fossil where recovered there together with fully twelve hominid crania fragments. Testimony varies as to whether or not these crania were associated with a handful of stone, antler and bone tools at the site (Bellwood 1997: 51-52, 65-66). The relative robustness of these crania together with their moderately large cranial capacity has made it difficult to simply pigeon-hole them in one taxon or another. Consequently, the Ngandong skulls have variously been classed as both *Homo erectus* and Archaic *Homo sapiens*, although its membership in the former taxon now seems most likely.

If this designation is correct, the Ngandong hominids are among the morphologically most advanced form of *Homo erectus* yet recovered. However, the real surprise afforded by these skulls is not their form but their apparent age. Electron spin resonance (ESR) and mass spectrometric Uranium/thorium dates from animal teeth recovered from the hominid level in the terrace produced gave a mean age range of between $53{,}000 \pm 4{,}000$ and $27{,}000 \pm 2{,}000$ years BP. (Swisher *et al.* 1996; but cf., Grün and Thorne 1997). If these dates are correct, they indicate that *Homo erectus* persisted longer in Southeast Asia than anywhere else we are aware of and may have co-existed there with anatomically modern humans. In fact, anatomically modern *Homo sapiens sapiens* are clearly in Sundaland by the middle of this time range. On the Indonesia island of Borneo, for example, modern hominid remains have been recovered at **Niah Cave** from a context dating to 40,000 BP. or earlier (Harrison 1957, 1959.

Elsewhere in Southeast Asia, fossil record for the early and middle Pleistocene is even more fragmentary. Scattered skeletal material referable to *Homo erectus* have also been reported from Burma and Vietnam dating to 500,000 and 250,000 years ago respectively. Isolated data points hardly allow the construct of a coherent picture of the region's early evolutionary prehistory.

Fragmentary or not, the fossil record of Southeast Asia is far richer than its cultural record. This is largely because, throughout the entire region, a clear and indisputable association between fossil hominids and Paleolithic tools has yet to be found (Higham (1996:9). In fact, the Southeast Asian Paleolithic sequence is still based in large part on Movius' (1944) pioneering work on the Chopper-Chopping Complex there. Subsequent surface survey and occasional archaeological excavation work in Southeast Asia has resulted in the refinement of Movius's scheme and in the recovery of tools referable to this Early Paleolithic tradition from many parts of the region. For example, the **Anyathian** tradition of the central Irrawaddy river valley of Burma is a moderately well-documented local expression of this larger areal tradition.

The distinction between tool traditions of the Early and Middle Paleolithic periods, so carefully drawn on formal and technological grounds in Europe, are less easily made in Southeast Asia. For example, the **Cabalwanian** industry of northern Luzon in the Philippines assemblages contains a few bifacial tools and others that appear scrapper-like in function and appears to fall in the time frame of the Middle Paleolithic period. However, the bulk of Cabawanian assemblages consist of crude cobble choppers and implements made on unifacial flakes (Fox 1978: 73-79). Altogether, the industry shows little of the technical sophistication of the Mousterian of Eurasia and is best considered to be a late expression the

Chopper-Chopping Complex.

Early Paleolithic China East Asia includes China, Mongolia, Korea, Japan and Taiwan. Clearly this a region of great physical and climatic diversity the unity of which reflects cultural, rather than physical, reality. China is the largest East Asia nation. Geographically, it is divisible into a northern and southern segment the boundary of which is generally placed at the Tsingling mountains, an east-west trending range located near the middle of China at roughly 33°North latitude.

North China is located between the Tsingling mountains and the Amur river of Mongolia. Within this vast region a further distinction is made between the northern forest and steppe land zone of Manchuria and the great valley of the Huang Ho river. The modern climate of North China is continental in character with cold winters, hot summers and low to moderate rainfall. Evidence that the configuration of climate during was very different in China during the Pleistocene is found in the spectacular loess deposits that have blanketed some 119,000 square miles of North China. Loess deposits vary in size, but some have been found to be nearly 100 meters thick. These bluff-colored, homogenous and generally unstratified deposits are made of windblown dust that is largely of Pleistocene age. They appear to have accumulated in the dry, cold conditions that characterized worldwide stages of glacial advance. Conversely, loess beds stabilized in the warmer and more moist conditions and soil horizons formed on their surfaces. These soils were buried in turn when declining temperatures and rainfall and the renewal of glaciation further north and changing wind patterns once again triggered loess formation. the alternating pattern of sediments and buried soil horizons or paleosols apparent in the loess beds constitute a key proxy measure of Pleistocene climate. The North China loess beds have generally been found to contain a pattern of twelve alterations between soil formation and soil stability during the early and middle Pleistocene epoch. This indicates that, although North China was not directly glaciated, its climate was strongly affected by the cycles of glaciation and interglaciation that occurred in other parts of the Northern Hemisphere (Goudie 1983: 39).

Of the two regions of China, the north has not only produced the most detailed Pleistocene climatic record, it is also the locus of the most spectacular Paleolithic archaeological discoveries. Interestingly, many of these discoveries have been made in a most unusual manner. The unusual quality of early Paleolithic archaeology in the north stems from the informal reliance many prehistorians placed on the purveyors of traditional Chinese folk medicines. For hundreds, perhaps thousands, of years, Chinese apothecaries have collected and ground up fossil bones and teeth to use as ingredients in various remedies and poltices. These practices have no doubt resulted in the destruction of vast numbers of paleontological and archaeological specimens, the mere mention of which is guaranteed to raise the blood pressure of most geologists and prehistorians. Yet,.

in one of the richer ironies in the history of archaeology, archaeologists relied on bone miners to provide the key that first unlocked China's prehistory. For example, in the late 1930s, a Hong Kong drugstore sold Ralph von Koenigswald several extremely primitive hominid teeth (now thought to be those of a robust australopithecine) that alerted that great paleontologist to the presence of early man in East Asia (Chang 1977: 41-42; but cf., Wolpoff 1979). Until quite recently with the excavation of Lunggupo Cave in south China, these and other odd drugstore fragments formed the primary evidence of a hominid presence in Asia in the early Pleistocene and Pliocene epochs.

However, even more important was the role of bone miners in the discovery of the legendary deeply-stratified cave site of Chou-k'ou-tien (or Zhoukoudian) near Beijing. Inquiries that eventually lead to the scientific investigation of the site began after a Pa-tung apothecary sold members of the Academica Sinica a molar that now appears to be from a gracile Australopithecine. Whether Australopithecines ever occupied Chou-k'ou-tien remains still uncertain. Nonetheless, scientific excavations have been conducted on-and-off at the cave since 1927 and this work has produced the remains of more than 40 *Homo erectus* individuals together with a great number of stone tools, floral and faunal remains in contexts dated by a variety of absolute techniques to between ca. 460,000 and 230,000 years BP. This material was apparently associated with intentionally-constructed fire hearths (but cf., Binford and Ho 1985). Further, late occupation of the site produced the remains of anatomically modern hominids dating to the Late Pleistocene epoch.

In addition to Chou-k'ou-tien, definite, well-provenanced archaeological evidence of Pleistocene hominids has been recovered from a number of other locations in the Yellow River region of north China. The sites of **Lan-t'ien** and **Kung-wang-ling** have produced fossil remains of *Homo erectus* and these and numerous other localities contain abundant stone tools referable to the later end of the Early Paleolithic period on both stratigraphic and typological grounds. Recent chronometric dating indicates that all of these sites are younger than ca. 700,000 BP. and therefore are of middle Pleistocene age (Xinzhi and Linghong 1985: 35-41).

South China begins below the Tsinglings and extends from the Tibetan Plateau to the Pacific ocean. The climate there changes from subtropical to tropical as one moves southward and is dominated by the annual **monsoon.** It is this seasonal monsoon cycle of dry months and wet, stormy ones that shape the climate of coastal East Africa, India, the Indo-Malaysian islands and mainland Southeast Asia.

Archaeological attention has been focused on south China only recently. As a consequence, the framework that has used to interpret the region's prehistory has been derived largely from the north. Nonetheless, the archaeological sites of the south are significant in their own right and appear to demonstrate a human presence in the south throughout the Pleistocene. For example, the lowest occupation at Lunggupo Cave may date to the Plio-Pleistocene boundary at ca. 2 mya. while the faunal remains recovered at **Guanyindong cave,** including *Ailuropoda* and the extinct proboscidean, *Stegodon*, suggest that the occupation layers containing them date from the middle to the early Pleistocene epoch. The lithic assemblages recovered from these layers in Guanyindong cave include retouched flakes and core tools that appear on morphological ground to be of Early Paleolithic age. Missing from these lithic assemblages are the heavy pebble and core tools characteristic of the Chopper-Chopping Tool tradition (Olsen and Miller Antonio 1992: 141-142, 153-54).

Early Paleolithic Korea, Japan and Taiwan

Like the rest of East Asia, neither Japan, Korea nor Taiwan were subject to continental glaciation during the Pleistocene epoch. In Korea and Japan, glaciers in higher mountain ranges expanded and contracted in response to worldwide shifts in global temperature. Beyond this little more can be said as the early Pleistocene paleoclimatic sequence is poorly known in all three regions (Nelson 1963: 22-23).

During glacial stages, declining sea levels either entirely exposed the floor of the Yellow Sea or turned that body of water into an inland freshwater lake. Although Taiwan remained an island during these episodes, Japan was periodically connected to Korea and the Asian mainland across a bridge of land that is today beneath the Tsushima Straits (Nelson 1993: 23). The Japanese fossil record indicates that this bridge enabled flora and fauna to emigrate there from the mainland at least by the lower Pleistocene and perhaps even earlier. Since *Homo erectus* are known to be in China by the early middle Pleistocene, it is likely that human colonization of Korea and Japan had begun by that time also.

In central Korea, a fragmentary skull cap recovered at **Kumgul Cave** on the Namhan river has tentatively been classed at that of a *Homo erectus*. Several other skulls recovered from cave sites in both north and south Korea have been similarly classified but these remain to be scientifically described. A modest number of other Korean sites localities have produced crude stone nodules and flakes that suggest human workmanship. On typological grounds these tools have been classed as Early Paleolithic in age and taken as evidence that the peninsula was occupied at this early date. Unfortunately, specialists disagree as to whether these specimens were produced by human craftsmanship or through the action of natural geological processes (Yi and Clark 1983).

More sophisticated tools of indisputable hominid workmanship have been recovered at **Chon'gongni** (sometimes spelled as Chonggok-Ni), an open site found in the Pleistocene-age terraces of the Hantan River in central Korea. The lithic assemblage from this site includes tools referable to the Chopper-Chopping tradition together with a modest number of Acheulean-like handaxes. Of course there can be no doubt that hominids made these tools but which hominids and when? No skeletal remains were recovered at Chon'gongni and the geological age of the terrace has yet to be precisely established. As a further complicating factor, excavations were undertaken there only after the cultural deposits had been bulldozed and possibly mixed in the course of clay mining (Nelson 1993: 30-39).

In Japan, stone objects of definite human manufacture and apparently early date have also been recovered. As at Chon'gongni, some of the Japanese sites have produced handaxes and other implements typologically similar to those from Early Paleolithic sites in Europe and Africa. What is surprising and problematical about these Japanese artifacts is that they come from contexts apparently *less* than 50,000 years old. If these late dates are correct, they reflect a striking technological conservatism in this region of East Asia during the Paleolithic period (Aikens and Higuchi 1982: 88). They further demonstrate that archaeological sites in this part of East Asia cannot be securely dated based solely on the form of the artifacts they produce. In fact, until the geological context of these early archaeological sites have been dated using multiple absolute dating techniques, the presence of hominids in Korea and Japan during the Early Pleistocene remains uncertain. In fact, even a hominid occupation during the Middle Pleistoene must be considered "not proven."

Finally, the first evidence of a hominid presence in Taiwan appears to be no earlier than the late Pleistocene. Fossilized *Homo sapiens sapiens* cranial fragments and teeth were recovered at the **Chouqu** and **Gangzilin** sites where down cutting by the Cailiao river exposed a series of vertebrate fossil beds (Olsen and Miller-Antonio 1992: 152-153).

Revolutions Technological, Culinary and Mental

Overall, what are we to make of the culture of the Early Paleolithic period? First and foremost, we must not over estimate its power or capacity. At best, archaeological evidence suggests that hominid culture, like hominid behavior in other spheres of activity, was deeply conservative, slow to change, difficult to convey and almost incapable of innovative recombination. The glacial pace of tool change suggests learning was effected only at great pain and then entirely by imitation. Jelinek (1977: 28) characterizes the cultural life of the period as one "of almost unimaginable monotony." In his view, the archaeological record reflects "a *paleocultural* behavior which differed significantly from the cultural behavior of modern man" perhaps due to the absence of language.

Yet, as different as this paleoculture appears to be from our own, the archaeological record indicate that the Early Paleolithic period witnessed a series of truly revolutionary changes in hominid technology, diet and cognition.

Transmutation of Tools and Facilities according to Philip Wagner (1960:92-97), technology consists of two key elements: *tools*, artifacts designed to transfer energy, and *facilities*, artifacts intended to contain, store or inhibit the loss of energy. In the course of the technological revolution of the Early Paleolithic period, both elements of technology emerged and changed.

British archaeologist Grahame Clark (1977) distinguished six successive "modes" or systems of stone tool manufacture in the history of technology. The technologies of the Early Paleolithic period centered on two of Clark's six modes; its end is marked by the succession of these first two modes by Mode 3 technologies.

Mode 1, the earliest and simplest means of stone tool making, consist simply of removing flakes from cores by direct, hard hammer percussion. The tool kits that resulted from manufacture in this mode contain little more than hammerstones and unifacially flaked cobbles presumed to have been used as choppers and scrappers. These early Mode 1 tool kits were superceded during the Early Paleolithic period by Mode 2 assemblages. In Mode 2, both hard and soft hammer percussion were employed in the fabricating of symmetrically-shaped handaxes, picks and cleavers made by removing flakes bifacially from cores or large boulder flakes.

In contrast to those of the Early Paleolithic period, the technologies of the succeeding Middle Paleolithic period were made in the third technological mode. The succession of Modes 1 and 2 by Mode 3 is a technological transformation or "rollover" of great moment in the history of humankind. In the Old World Paleolithic, this technological rollover was a **time transgressive** event. That is, it began at different times in different regions but nowhere did it occur earlier than ca. 200,000 BP.

In Mode 3, stone tool making centered on the production of smaller tools exhibited more a varied range of types. Tools of this mode were generally made on flakes. However, Mode 3 technology did not simply involve a change of emphasis from the core to the flake tools. It was also characterized by the development of new stone working methods that made it possible to shape or prefabricate the core in a manner that produced uniform flakes of predetermined size and shape. Preeminent among these characteristic new methods of core preparation were the Levallois technique. Although it appears near the end of the Acheulean, Mode 3 technology lays at the heart of the **Mousterian** industrial tradition. We will encounter the Lavallois technique (together with its descendant, the Disc Core technique)in our discussion of the Middle and Late Paleolithic periods in the next chapter. What is important here is that, during the Early Paleolithic period, early hominids increased their physical power and capacity to do work. They did this by learning to make tools that enabled them to transfer the energy of their bodies more efficiently.

By the close of the Early Paleolithic period, hominids had also begun to construct and use facilities, the second of Wagner's elements of technology. The distinctive archaeological features excavated at the sites of Terra Amata in France and Blizingsleben in Germany appear to be the floors of huts or shelters built around fire hearths. If these features were in fact walled and roofed structures, they are perfect examples of facilities as Wagner defines the term. Such huts would have inhibited the loss of heat from the bodies of their residents while at the same time impeded the diffusion of the heat given off by the fire built in their centers. No less than a diving bell or a space capsule, the interior of these early huts would have contained an artificial interior environment insulating their residents from their surroundings.

Culinary Revolutions We do not eat as our close hominoid relatives eat. This distinction between us and them is a key element in our nature. It is therefore no exaggeration to say that the history of humankind can be charted as a series of culinary transformations. The first such transformation was the use of Oldowan tools to render hard foodstuffs eatable by cutting and breaking them down and assisting in their digestion. The second culinary revolution occurred with the harnessing of fire. Following this event, heat could be applied to foodstuffs as a further means of breaking it down and assisting in its digestion. The use of tools and fire in food processing increased the degree to which foodstuffs were "pre-digested" prior to being eaten. Such cultural pre-digestion broaden the range of vegetal and animal material that could be consumed by hominids to a level far beyond that of any other primate. Further, archaeological evidence from sites like Torralba and Ambrona indicate that these new culinary techniques were combined with more efficient methods of group organization and cooperation subsistence pursuits. Taken together, the archaeological record indicates that the Early Paleolithic period witnessed a remarkable cultural expansion and transformation of the hominid food niche. The importance of these transformations is reflected in the marked expansion in geographic distribution of the hominid line that had occurred by the end of the Early Paleolithic period. As noted earlier, while the Australopithecines appear to have been limited to the tropical and subtropical regions of Africa, *Homo erectus* fossils and sites dating to the last million years have been reported from the arid regions of South Africa, the North China plain, Java, southern and southeast Asia and unglaciated Europe and Eurasia from the Mediterranean basin to the edge of the Arctic Circle in Siberia.

Presumably, the combination of (1) an increasingly efficient tool kit, (2) the ability to create artificial "interior environments" by building shelter facilities and warming them with fire, and (3) widening of the hominid food niche combined with a nascent, but expanding, ability to organize and coordinate group efforts in common tasks like hunting led to an expansion in both the population size and in the range of species. Just as modern people adapt to hostile environments like the surface of the moon or the bottom of the ocean by cultural means, so the hominids of the Early Paleolithic period accommodated themselves to an increasingly diverse range of circumstances largely by cultural, rather than biological, means.

Paleolithic *Mentalités* The archaeological evidence for the foregoing technological and culinary transformations suggest something further occurred during the Early Paleolithic period. They imply that, with the emergence of *Homo erectus* some time between 1.9 and 1.5 mya., an important threshold in the development of the intellect had been crossed. Measuring intellect is a difficult task for the archaeologist. Nonetheless, Wynn (1979) attempts to assess the intellectual capacity of *Homo erectus* using "**genetic epistemology**," a theory of intelligence developed by Jean Piaget. Piaget's theory posits four kinds of "operational spatial organization" as characteristic of the thought processes of modern adults: whole-part relations, qualitative displacement, spatio-temporal substitution, and symmetry. These operational modes of thought make it possible for individuals to work towards preconceived goals. In Piaget's view, children, and presumably, our prehistoric ancestors, have "organizational repertoire of preoperational thought." That is, they do not have a clear picture of the end product in mind before they begin a creative task; consequently they work largely by trial and error until they achieve an acceptable result.

Wynn, interested in determining at what point modern "operational thought" came to characterize hominid thinking, examined the "geometry" of the stone tools from Oldowan and Developed Oldowan assemblages recovered at Olduvai Gorge in Bed I and lower Bed II respectively, and from the late Acheulian site of Isimila in Tanzania in terms of Piaget's four kinds of operational spatial organization. He concludes that while the makers of these late Acheulian implements employed "operational concepts of space" identical to those of modern *Homo sapiens*, the "Oldowan artifacts...did not require operational intelligence for their manufacture (Wynn 1979: 385, 388)." That is, the *Homo habilis* or Australopithecine fabricators of the Oldowan tool kits did not achieve standard shapes, regular cross sections, or symmetry in their implements because they were working essentially at the "preoperational level" of thought. Beds I and II at Olduvai date to between 1.6 and 1.5 mya., the Isimila site dates somewhere between 330,000 and 170,000 years ago. In Wynn's (1979: 390) view, the emergence of modern operational modes of thought may have occurred a million or more years ago with the appearance of *Homo erectus*.

Realities Beyond Mere Needs. Significantly, it is also among *Homo erectus* that we begin to find some of the earliest evidence of what Edwards (1978: 135) calls a "concern with realities transcending mere biological needs." The evidence of this concerned cited by Edwards includes (1) the intentional collection and use of ochre or other red colored rocks and minerals, (2) the special treatment of human and perhaps animal skulls, (3) the refinement of utilitarian objects beyond the simple requirements of their purpose and (4) the possible appearance of rudimentary symbolic or artistic engraving on bone. Perhaps the most dramatic indication that *Homo erectus* was concerned with more than simple survival is the evidence that these hominids intentionally collected and used ochre or other red colored rocks and minerals. In recent years a number of scholars have called attention to the presence of "**red ochre**" in early sites and have suggested that such specimens offer important insight into the development of human cognition (for example, Blanc 1961: 123; Edwards 1978: 136; Wreschner 1980; Schmandt-Besserat 1980; Marshack 1981). The term "ochre" refers to certain common iron oxides such as hematite, goethite and limonite which occur naturally in a wide range of colors including red, yellow, brown, black, orange, even violet. Ochre has some wonderful physical properties that no doubt caught the attention of Paleolithic era people. First, it is generally do soft that, when it is rubbed against a harder surface, it leaves a richly colored streak. Ochre makes an excellent stain, paint or dye when it is ground into powder and either used directly or mixed with a medium like water or animal fat. Finally, the heat of an ordinary camp fire is sufficient to cause most ochres to dramatically change color. Significantly, specimens of ochre have been reported from some of the oldest occupation or activity sites known from the Early Paleolithic period in the Old World, including Bed II at Olduvai Gorge in Tanzania, Ambrona in Spain, Terra Amata in France, and Becov in Czechoslovakia. The use of ochre apparently increases during the Middle Paleolithic period in the Mousterian tradition and becomes common in the Upper Paleolithic period.

Ochre has no apparent "practical" or technological use until the development of iron metallurgy sometime in the second millennium before Christ when it becomes a principal ore for iron smelting. Nonetheless, many of the Paleolithic period ochre specimens show evidence of having been worked or utilized in some fashion. For example, the two lumps of ochre recovered at Olduvai Gorge show signs of having been struck directly by hammerstone blows (M. Leakey 1971). Howell (1966: 129) states that the ochre specimen recovered at Ambrona showed evidence of shaping and trimming, although Butzer (1980: 635) asserts this may only be natural cleavage. Still the ochre comes from the same horizon as the famous linear arrangement of elephant tusks and bones and was probably brought to the site by the hominids who are thought to have killed and butchered elephants there.

At Terra Amata, which was occupied around 300,000 BP., deLumley (1969:49) reports a number of ochre specimens recovered from the two occupation layers associated with the pole structures uncovered at the site. Specimens of red, yellow and brown were recovered and the range of color variations suggests the ochre may have been heated. DeLumley also reports apparent use-marks on some of the specimens: the ends of some of the specimens were worn smooth suggesting to him that they had been used in body painting.

Clearer evidence of its use comes from Becov in Czechoslovakia. This cave site, occupied ca. 250,000 BP., yielded a specimen of red ochre that was striated on two faces with marks of abrasion together with a flat rubbing stone with a granular crystalline surface that had been abraded in the center possibly during the preparation of ochre power (Marshack 1981: 138). Whether or not the rubbing stone was actually used in the preparation of ochre powder is uncertain, but a wide area of the occupation floor from which the ochre lump had been recovered was stained with red ochre powder. It is important to note that at least some of the specimens referred to as ochre in the archaeological literature are merely red stones that lack the special qualities of iron oxides. Despite the specific mineralogy of these specimens however, their very presence in Paleolithic contexts is important evidence that they were collected by early hominids capable of thought and action divorced from what we would consider exclusively practical considerations. Even if the red stones may not be classed mineralogically as ochre, they had no apparent practical use. The collection of these red rocks and minerals are among the first signs in the archaeological record that early hominids were, in fact, developing a "concern with realities transcending mere biological needs."

In any event, the recovery of ochre from Paleolithic sites has generally been taken to mean that from very early times, hominids needed, sought, and used pigments. Of course we can only speculate on the purposes to which these colors were put. Marshack suggests they

> could have been used in ritual body decoration, in body decoration to mark rank, status, age, or sex, to decorate tools or symbolic artifacts, to color habitation floors or walls, to delimit symbolic spaces (ritual areas, healing grounds, grave, or altars), or simply as a marking material... (Marshack 1981:190).

His list could easily be expanded, but if these ochre lumps can be taken as evidence of early body ornamentation and the use of cosmetics, they are particularly significant. Lewis Mumford (1967: 108-111) notes the human universality of what he calls "**technical narcissism.**" By this he means body and facial decoration, the use of masks, costumes, wigs, tattooing, scarification and so forth. He suggests that all of these are part of humankind's

> effort to establish a human identity, a human significance, a human purpose. Without that, all other acts and labors would be performed in vain (Mumford 1967: 110).

Even more striking is his assertion that such technical narcissism (of which body and face painting are essential parts) indicates that

> primitive man's first attack upon his "environment" was probably an "attack" upon his own body; and that his first efforts at magical control were visited upon himself. As if life were not hard enough under these rude conditions, he toughened himself further in such grotesque ordeals of beautification. Whether decoration or surgery was involved, none of these practices had any direct contribution to make to physical survival. They count rather as the earliest evidence of an even deeper tendency in man: to impose his own conditions, however ill-conceived, upon nature. Yet what they point to even more significantly is a conscious effort toward self-mastery and self-actualization; and even though often exhibited in perverse, irrational ways, at self-perfection (Mumford 1967: 110).

The presence of worked ochre in Bed II at Olduvai Gorge suggests that the beginning of this "attack" may even predate the appearance of *Homo erectus* and begin instead with *Homo habilis* or the Australopithecines more than 1.5 mya..

Another form of evidence that *Homo erectus* had begun to develop concerns beyond simple survival is the special treatment thought to have been accorded human, and perhaps animal, skulls by these hominids. It is in the context of the celebrated *Homo erectus* cave site of Chou-k'ou-tien in China that we find the earliest evidence of what is referred to as the Paleolithic "Cult of Skulls." The Early Paleolithic period deposits at this site produced the fragmentary remains of some 44 individuals, including 14 crania (Chang 1977: 47). All 14 of these crania show signs of what has been interpreted as systematic mutilation: the bases are broken in around the foramen magnum, the facial regions of most of the cranial were damaged or missing, and the mandibles exhibit a consistent pattern of breakage. The systematic nature of this damage, coupled with the apparent absence of all but a few fragmentary hominid postcranial bones in the deposits, has lead many scholars to conclude

> that these skulls were trophies of head hunters, and furthermore, that said hunters usually bashed in the bases of the skull when fresh, presumably to eat the brains therein contained. Many crania also show that their owners met their deaths as a result of skull fractures induced by heavy blows (Hooten 1949: 304).

Over the last twenty-five years, this interpretation has become enshrined in archaeological orthodoxy. However, the venerable view that head hunting and ritual cannibalism were practiced at Chou-k'ou-tien by *Homo erectus* has been challenged by Binford and Ho (1985). The systematic breakage patterns alone, they note, are not sufficient proof that the damage to the Chou-k'ou-tien crania was inflicted by other hominids. Scavenging animals commonly eat the facial portions of skulls, that rolling and tumbling skulls in streams often results in breakage in the foramen magnum region, and that mandibles naturally break due to structural weakness in the manner of the Chou-k'ou-tien specimens. Since (1) recent taphonomic studies indicate scavenging animals and water rolling consistently produce similar patterns of damage to crania, (2) since the crania were recovered in secondary contexts at the site, and (3) these contexts also contained bone that had been gnawed by animals, the possibility that the crania had been deposited there by animals and by natural processes cannot be rejected.

The third class of evidence of nascent "aesthetic" interest among Early Paleolithic hominids is less easily dismissed (Edwards 1978: 136). Certain classes of utilitarian stone tools show refinement beyond the simple requirements of their purpose and such refinement as indicate that these early hominids appreciated form and color for their own sake. Following Oakley (1957: 127) and Clark (1974: 190), Edwards points particularly to the finely made Acheulian handaxes and perfectly shaped spheroids from Olduvai Gorge and Kalambo Falls in Africa as examples of this type of behavior.

Finally, Edwards accepts Marshack's (1976: 278; 1977:289-292) contention that the fine engraved lines on an ox rib from the Acheulian layer at the site of

Pech de l'Aze in France appear to be similar to the "meander" symbol or iconographic unit of notation that become an important element in the cave art of the Upper Paleolithic period. If this engraving is a meander symbol, it suggests that the complex cognitive development claimed for *Homo sapiens* near the end of the Pleistocene epoch was presaged in the mental and cultural lives of *Homo erectus* sometime before 300,000 BP. Recent work at the Bilzinglleben in eastern Germany provides some confirmation of this hypothesis. There, animal bones exhibiting patterns of incised lines arranged in rhythmic sequences have been recovered in a context dating to between ca. 350,000 and 230,000 BP. The purpose of these incised markings are unknown (Mania and Mania 1988). While no one would mistake these engravings for the Mona Lisa, they clearly reflect an artistic intention.

The End of Early Prehistory

Did they think like us, these hominids of the Early Paleolithic period? If a time traveler from our era were to encounter them, he would probably conclude that he they did not. Face to face with the pace and manner of their lives, their habits, their modes of learning and their manner of communication (or lack of it), our wary visitor from the future might be forgiven for concluding he had fallen amongst an altogether alien genus.

Yet he would be wrong. The articulation of their bodies and, as Wynn and others would have it, the structure of their intellects, demonstrate otherwise. Rather, by the close of the Early Paleolithic period, it is fair to say that the technological, culinary and mental foundations of modern humanity were in place.

Major Points Raised in Chapter Eight

1. The chapter begins with a discussion of the chronology of the Last Interglacial Stage and the **Last Glacial Stage** that occur at this time.

2. Hominid evolution during the Late Pleistocene is discussed with special attention given to the emergence of anatomically modern *Homo sapiens sapiens*. Reviewed here is the debate between proponents of the **Multiregional model** and those favoring the **Rapid Displacement model**.

3. The Middle Stone Ages in Africa is defined and discussed with special attention given to the key sites at **Klaisies River Mouth** in South Africa. The East African rockshelter sites of **Shurmai** and **Kakwa Lelash** are used to evaluate the hypothesis that modern behavior in Africa predates the Later Stone Age.

4. The Middle Paleolithic period in Europe and Southwest Asia is defined and technological changes that characterize it are discussed. The nature of the relationship between this Neanderthals and anatomically modern *Homo sapiens sapiens,* is treated in the fourth section of this chapter. Special attention is paid to the understanding the role played by taphonomy in the interpretation of Middle Paleolithic period sites.

5. The archaeological evidence for expanded hominid mental capacity and religious activity during the Middle Paleolithic is critically discussed. The evolution or arrival of anatomically-modern *Homo sapiens sapiens* after ca. 35,000 B.P. around the end of the Middle Paleolithic period is noted in this connection.

6. The Upper Paleolithic period is defined and the cultural sequence in western and northern Europe reviewed. The technological changes that distinguish it from the preceding Middle Paleolithic include the predominance of blade tools, the appearance of new tool forms, the development of composite tools, and the widespread use of ivory, bone, horn, antler, and wood.

7. Upper Paleolithic settlement pattern is examined for clues to the subsistence and economic activity of the period. The possibly that peoples of Western Europe specialized in reindeer hunting while those of the Eurasian steppes focused their economies on mammoth-hunting is examined.

8. Upper Paleolithic mortuary evidence is examined for clues to the nature of social life in Europe and Eurasia at this time. The chapter closes with a brief consideration of the remarkable tradition of mobiliary and cave art that appeared in Europe during the Upper Paleolithic period.

9. The final section of this chapter examines the Pleistocene and Paleolithic beyond Europe and Eurasia in Africa South of the Sahara, in East, South, Southeast Asia and in Australia and the Pacific.

CHAPTER 8

THE LATE PLEISTOCENE EPOCH IN THE OLD WORLD

Introduction

The Late Pleistocene which dates to between 128,000 BP. and 10,000 BP. In geological terms it was but a brief moment. In the 24-hour model of deep time, it begins slightly more than two and one-half seconds before the end of the final hour and ends a spare one-fifth of a second short of midnight. Precisely because of its relative proximity to the present, the geological and archaeological record of the late Pleistocene is far more complete and well understood than that of the Pliocene and the early and middle Pleistocene treated in the previous chapter. The Late Pleistocene epoch encompasses the final interglacial/glacial cycle of the Quaternary period. The **Last Interglacial Stage** begins ca. 128,000 BP. and ends with the onset of the **Final Glacial Stage** ca. 118,000 BP. The Final Glacial Stages reaches a peak ca. 18,000 BP. and closes around 10,000 years ago.

It was against the background of the Late Pleistocene that the great human cultural and evolutionary events of the later Paleolithic period unfolded in Old World lands of Eurasia, Africa, the Indo-Malaysian Archipelago and Greater Australia. The record of these events, and of the Late Pleistocene generally, has been particularly well studied in eastern and southern Africa and western Eurasia. We therefore begin our discussion of the period in those regions.

The Late Pleistocene in Eurasia

The Eurasian landmass is located well north of the Equator and most of it falls within the modern Arctic, Subarctic and Midlatitude latitudinal geographic zones. Only at its extreme southern and southeastern ends is the continent characterized by subtropical and tropical climates. These will be discussed together with those of Africa, Australia and the insular Pacific in the succeeding section.

Western Eurasia consists of Europe and the Middle East. This sub-region occupies the land between Great Britain and the Atlantic Ocean on the west and the Ural Mountains on the east. On the north it begins at the Baltic and Barents seas of the Arctic Ocean and then stretches southward to the Mediterranean and the Red Sea. The paleoclimatic sequence of western Eurasia has been under close study in one form or the other for the last two hundred years. Although the study of paleoclimate of Asia at the eastern end of the landmass began only in the 20th century, it is now well underway also. As a result, an exceptionally clear picture of the sequence of Eurasian paleoclimate has begun to emerge.

The scientific breakthrough that laid the ground work for that sequence was the successful application of the Uniformitarian assumption to the interpretation of the so-called Drift Deposits found on many northern European land surfaces. In the 19th century these sediments were recognized as identical to those found in association with modern glaciers in the Alps, in Greenland and the Antarctic. This recognition led to the conclusion that such sediments were glacial in origin and therefore evidence that. western Eurasia was colder in prehistory and that glacial ice expanded there on a continental scale. This insight was further enriched by the realization that, while many of the landforms and deposits in modern Europe could only have been produced by the movement of glacial ice, still others were **periglacial** in origin, that is, produced under conditions of extreme cold in areas peripheral or beyond the actual ice masses. The foundation of the modern view of the Pleistocene epoch was established when detailed stratigraphic analysis of these newly identified glacial sediments revealed that not one but many glacial expansions and contractions occurred.

Paleontology was, and continues to be, the second

great source of insight into Eurasian paleoclimate. Such study has demonstrated that, through time, the nature and distributions of Eurasia plant and animal species has changed dramatically. Paleontologists have recognized and defined a series of land mammal ages that follow one another in Eurasia during the Pliocene and Pleistocene epochs. These land mammal ages have in turn been correlated with the cycles of continental glacial advance and retreat that occurred there.

Finally, the third important source of paleo-limatic information has been the **pollen spectra** recovered from sediment cores drilled into the numerous bogs and other sedimentary locations in Eurasia. A bog is an area of water-logged ground whose sediments are composed in large part by plant material in various stages of decomposition. Although bog is a perfectly good Anglo-Saxon word, we hardly hear it anymore in this environmentally-conscious world. Rather, those wet, spongy, plant-choked, in-between worlds, are now referred to as **wetlands**. But whether bogs or wetlands, northern Europe has thousands of them. Sediment cores extracted from deep ones like the **Grand Pile Bog** in France and numerous others have proven rich in both preserved plant fragments, pollen and insight into the environments that produced them.

Pollen has proven to be particularly useful in the reconstruction of ancient environments. Seed plants, which include both the gymnosperms (conifers) and angiosperms (flowering plants), reproduce by means of pollen and pollination. A pollen grain is a microspore containing a male reproductive cell. In order to foster reproduction, such a microspores must be united with a female reproductive structure in order to germinate and produce seeds. Female reproductive organs is the **ovule** in gymnosperms, the **stigma** of the carpel in angiosperms. In some species, pollen is carried to the ovule or stigma by insects, in others, by water or the wind. More rarely, plants effect the contact needed for germination through *self*-pollination.

Different plant taxa produce recognizably different types of pollen. These differences enable **palynologists**, students of pollen, to identify the genus and sometimes even the species of the plant that produced the microspore. In this way, palynology offers a means of reconstructing ancient flora by identifying and counting the different genus and species of pollen grains (and other spores) represented in ancient soils, peat bogs or other kinds of sedimentary deposits. Palynologists extract plant pollen from samples from soils and sediments where it was deposited and examine it microscopically. The percentages of each genus or species present in a pollen sample is plotted on a **pollen diagram**. Such a diagram represents the relative contribution of the various plant types to the pollen rain during the sediment's formation and--indirectly at least--the proportion of each type of plant in the environment at that time.

Pollen samples are especially useful when they have been extracted from sequential layers in stratified sediment columns like peat bogs. The depth below surface of each strata is plotted in order along the y-axis beginning with the oldest at the bottom. The percentage of the various plant taxa contained in the samples drawn from each strata is plotted on the x-axis of the diagram. In this way one obtains a graph that represents the changes that occurred in the vegetation around the collection site through time. A relative sequence of vegetative change of this kind becomes an absolute one when the associated organic material in the sediments are radiocarbon dated. By identifying the fossil pollen taxa incorporated in the stratified layers of countless peat bogs in northwestern Europe, palynologists have been able to reconstruct the sequence of vegetative change that characterized that region during the Late Pleistocene.

Finally, the Eurasian paleoclimatic sequence assembled using palynology, glacial studies and palenotology has been correlated with--nd corrected by--the marine oxygen isotope stages derived from the worldwide coring of deep ocean sediments. This recent correlation of marine and terrestrial phenomena has greatly enlarged our understanding of the paleoclimate of ancient Eurasia during its successive interglacial and glacial phases.

Last Interglacial Stage The final interglacial stage before the present is referred to as the Eem in northern Europe, the Riss/Wûrm in the Alps, the Mikulino in European Russia. More generally it is simply referred to as the "Last Interglacial." The beginning of the Last Interglacial is conventionally placed at the boundary between stages 6 and 5 of the deep ocean oxygen isotope curve which dates to 128,000 BP. (Edwards, Chen Ku and Wasserburg 1987). Much disagreement surrounds the date of its close, but the boundary between oxygen isotope stages 5e and 5d at 118,000 BP. is used here (Kukla, Mathews and Mitchell 1972; but cf., Woilland, G.M. and W.G. Mook).

In western Eurasia the onset of this interglacial is marked by a rise in temperature and humidity reflected in the changing pollen spectra recovered form the Grand Pile Bog and other peat bogs through northwestern Europe. Changing plant pollen frequencies in the successive strata of these bogs indicate that the open tundra vegetation of the previous glacial stage began to be replaced by a subarctic boreal forests dominated by birch (*Betula* sp.), pine (*Pinus* sp.), and spruce (*Picea* sp.). Boreal forests were replaced in turn by temperate woodland by the middle of the interglacial. The temperate woodlands were first dominated by hazel (*Corylus*) but eventually came to support a climax deciduous forest dominated by oaks (*Quercus* sp.) and such warm-loving taxa as, elm (*Ulmus*), limes (*Tilia*), and hornbeam (*Carpinus*) (van der Hammen *et al.*, 1971: 395-396). Mean annual temperatures in Europe at the height of the Eem interglacial are presumed to have been appreciably higher than those of the Holocene epoch. Key Eemian fauna in western Europe during these warmer conditions include hippopotamus, straight-tusked elephant, giant deer, cave hyena and cave lion (Kurten 1968: 13, 29-30, 134-135).

The Final Glacial Stage The Final Glacial Stage began ca. 118,000 BP. It is known as the Weichsel in the European stratigraphic column, the Würm in the Alpine sequence, and the Valdai stage in European Russia. During this last stage the mean annual temperature of the Earth declined, the climate became drier and continental-scale glaciation came to hold vast tracts of the Northern Hemisphere in its icy grip. After some 108,000 years, the Final Glacial Stage closed around 10,000 BP.

The arrival of the colder, drier conditions of the Final Glacial Stage are clearly reflected in pollen spectra from the Grand Pile Bog. At the end of the Eem Interglacial there, the percentage of warm-adapted tree species began to decline relative to cool-adapted tree species. At the same time, the percentage of arboreal species pollen began to decline relative to non-arboreal herb species in the pollen spectra. These changing taxa ratios are taken as indicating declining mean annual temperature and the gradual replacement of the temperate climax forest vegetation, first with by an open boreal forest of fir, spruce, then pine and birch, then by grassland, moorland and peat bogs, and finally by open tundra vegetation. The appearance of open arctic and alpine vegetation in France signaled the onset of the first major stadial of continental glaciation during the Weichsel stage (Woillard, 1978: 6-7). Further north in the Netherlands, even this tundra vegetation disappears during full glacial stadia (van der Hammen *et al.*, 1971: 398).

In response to this worldwide drop in temperature, cold-adapted species like the woolly rhinoceros, woolly mammoth, cave bear, and reindeer return to prominence in western Eurasia. The earliest known appearance at this time of that quintessential arctic form, the polar bear is further evidence of the declining temperatures at the end of the Eem (Kurten 1968: 254). There are two even more interesting species changes in western Eurasia at this time: the hominid form know as Neanderthal (*Homo sapiens neanderthalensis)*replaces more archaic forms (now collectively referred to as *Homo heidelbergensis)* sometime during the early Weichsel stage. In turn, the Neanderthals appear to have been supplanted by *Homo sapiens sapiens* in western Eurasia there after ca. 35,000 BP.

Although the return of continental glaciation in the northern hemisphere resulted in a drop in mean annual temperatures worldwide, the Last Glacial Stage was not one of unbroken cold and darkness. Temperature curves derived from deep ocean cores and Caribbean marine terraces levels to indicate that the glacial climatic cycle oscillated in a "saw tooth" or wave-like fashion. That is, deep sea cores dating to the Final Glacial stage reflect a series of temperature declines separated by periods of temperature increases. The periods of low temperatures averaged about 20,000 years in length and were presumably glacial stadia or growth phases. The intervening periods of warmer temperatures, which averaged only about one thousand years apiece, were interstadial periods characterized by glacial retreat (Broecker and van Donk 1970).

Such a saw-toothed pattern also appears to characterize the terrestrial record of Final Glacial Stage. Pollen and other evidence suggests that, in northern Europe at least, the Weichsel stage was punctuated by at least eight, relatively-brief interstadials of moderate warmth and re-vegetation (van der Hammen *et al.*, 1971: 395; Bowen 1985: 20, 26-27, 30). Each of these short-term interstadial climate amelioration was followed by a stadial characterized by dropping annual temperature, vegetational impoverishment and glacial re-advance. Presumably these advances and remissions occurred roughly at the same time throughout the northern hemisphere, but much stratigraphic correlations between regions and continents remains to be done. However, given the current state of our understanding of the Quaternary, only the last two of these interstadial/stadial cycles can be correlated with great precision to similar late glacial stage events in North America. In Western Europe,

Last Glacial Maximum in Eurasia. By 18,000 BP., worldwide temperatures had dropped to their lowest levels: an average reduction of ground temperature some 5.3° Celsius below those of the modern northern hemisphere and as much as 17° Celsius in some mountainous regions (CLIMAP 1976: 1131-1137). Even more lower temperatures were reached in Greenland at this time. Recent analysis of glacial ice cores from that continent indicated that temperature dipped to about 23° Celsius there (Jensen *et al.* 1998) Correspondingly, continental glaciation reached its maximum expansion around this time. By 18,000 BP. a more-or-less continuous glacial ice mass stretched from Ireland to Siberia in the Old World. At the same time, the glaciers that had filled the mountain valleys of the rugged Alps had expanded and coalesced to form a huge glacial ice complex that virtually engulfed all but the higher peaks of that range. In a similar manner, mountain glaciers throughout Eurasia expanded. Yet we must recall that, even during this **Late Glacial Maximum**, most of the Northern Hemisphere remained unglaciated. This was also true of virtually all of the Southern Hemisphere excepting the higher mountain ranges of the region and the continent of Antarctica.

In the areas of permanent inland ice, summer temperatures remained close to freezing virtually year round. Areas close to the ice front appear to have been "polar deserts" that resembled the high arctic barrens of the modern world in the sparseness of their vegetation. The polar desert stretched across the entire North Sea Basin which was exposed due to lowered sea level and connected Great Britain to the continent during glacial substages.

Rigorous climatic conditions extended eastward from Europe across the Urals into Siberia in northeastern Eurasia. During the Last Interglacial and subsequent interstadial intervals this region was moist and cold; during ice advances of Late Glacial stage, Siberia while remaining largely unglaciated, became intensely cold and arid. In fact, climatic conditions may have become so severe as to render human habitation there impossible throughout all of northeastern Eurasia at least during the millennia of the Last Glacial Maximum ca.18,000 BP. (Davis 1998:177).

While the rigors of Siberian climate is not surprising, the the hydrological impact of these extreme glacial conditions there certainly are so. The Russian Quaternary geologists M.G. Grosswald (1980) sees evidence that the huge Fennoscandinavian continental ice sheet that covered much of western Siberia at various times during the Late Pleistocene interrupted the northward drainage of key rivers like the Irtysh, the Ob', and Enisei that flow from the Siberian interior into the Arctic Ocean. In his view, such blockage would have impounded vast quantities of fresh water behind the ice sheet and created a network of proglacial lakes across the vast interior of the lowlands that would have connected the Enisei and Ob' rivers in the west to Lakes Baikal and the Aral Sea in the eastern Siberia. More dramatic still is the evidence that, at the end of the LGM these glacial ice dams were breeched with catastrophic results. It appears that a huge wall of water 400 to 500 meters deep burst loose and spread rapidly over northern Siberia carrying everything in its path (Baker *et al.* 1993). The record of this event, still evident in the surface geology of parts of northeastern Eurasia, indicates the flood was even greater than the one produced in North America. In that terminal Pleistocene flood, glacial Lake Missoula burst loose in a catastrophic outpouring that created the rugged "scablands" of the Columbia Plateau of the northwestern United States. The impact of these sudden catastrophic events on human life and settlement can only be imagined.

The Late Glacial Steppes of Eurasia Disagreement exists as to the nature of the environment that lay south of the ice during during late glacial times. The sediments dating to this time range from cores drilled into beds of former periglacial lakes generally are generally found to contain very low quantities of plant pollen. The apparent sparsest of pollen influx has led a number of palynologists to conclude that region was vegetatively barren due to the extreme harshest of the climate. Some suggest that at this time the vast periglacial regions of Eurasia were virtually lifeless polar deserts.

The paleontologist R. Dale Guthrie has proposed a very different picture of what the region must have been like at Last Glacial Maximum. In his view, the abundant fossil remains of large Pleistocene mammals from there are a more reliable paleoclimatic indicator than the pollen spectra. These data suggest that the region was in fact a rich grassland setting that supported an immense biomass. Late Glacial Stage ungulates from the region are generally much larger than their Holocene descents. Some, like the woolly mammoth, reached truly gigantic proportions. Further, many species sported elaborate horns and other "social ornamentation" connected with mating competition. All this suggests that, far from scratching for survival in a polar desert, the fauna there had access to high protein diets and long annual growth periods.

Guthrie's explanation for this is simple: the modern Arctic is a poor analogy for the environment of the Late Glacial steppes. First, most ungulates fossil forms recovered from this region lack the spayed hooves that adapt grazing animals to regions of heavy snow cover. He hypotheses that their small hooves indicates that the snow cover on the steppes was lighter there, melted earlier in the summer and returned later in the autumn than it does in the modern Arctic. How can this be? According to Guthrie, the modern Arctic latitudinal geographic zone is located at latitudes some 10° to 15° farther north than this region. The Sun's radiation strikes the Earth at a higher and more favorable angle at this part of the globe. Combined with this more favorable energy balance, the region would not have experienced the curtailment of plant growth that occurs during the six months of winter darkness within the Arctic Circle. Protein rich herbs and sedges would have grown in abundance there in Late Glacial times, Guthrie assets, and this abundance would have turned the vast mid-continental steppes into one of the richest grasslands in the history of the Earth. After the most dramatic member of this assemblage, Guthrie christens this biotope, the **Mammoth Steppe**s.

If Guthrie is correct, the Mammoth Steppe was a rolling tree-less tundra, that is a region of permanent subsurface frost, that supported a rich herbaceous vegetation. This steppe would have stretched from Britain across northern France to the central European loess belt of Germany and Czechoslovakia, on to the East European Plains of Poland and Ukraine. From there it would have reached beyond the Ural Mountains to the steppes of central and far eastern Asia and thence across the Bering Straits landbridge into Alaska. No doubt the portions of this vast plain pinched between the Scandinavian ice sheet and the Alpine glaciers must have experienced particularly severe climatic conditions. A forest tundra or forest steppes vegetational assemblages was apparently located immediately south of the tundra zone. Still farther south, the land was freed of permafrost. Most of the Mediterranean basin and the southern shores of the Caspian sea supported a cold-temperate woodland mixture of coniferous and deciduous tree species (Butzer 1971: 278, 281, 283, 293).

As paleoecological research brings the nature of the ancient Mammoth Steppe into sharper focus, archaeologists are beginning to appreciate the stimulus this biotope gave to Paleolithic cultural evolution. The harsh land of the steppe would only have yielded its

Substages, Stadia, and Interstadials of the Weichsel Glacial Stage in Northern Europe

Substage	Age	Stadial	Interstadial
Upper	10,000 BP	Younger Dryas	
			Allerød
		Older Dryas	
			Bølling
		Oldest Dryas	
Middle	26,000 BP		Denekamp
		5th Stadial	
			Hengelo
		4th Stadial	
			Moershoofd
Lower	50,000 BP	3rd Stadial	
			Oddeerade
		2nd Stadial	
			Brørup/Amersfoort
		1st Stadial	
	118,000 BP		

The period in the Upper Weichsel glacial substage between about 13,000 and 10,000 BP. that includes the Bolling and Allerod interstadials is commonly referred to as the "Late Glacial" substage or the terminal Pleistocene.

endless treasure of game to peoples capable of mastering its cold, its aridity and its vastness. The wooden spear found embedded in the ribs of a straight-tusked elephant at the Lehringen Mousterian site in Saxony testifies to is mute testimony to the success of Paleolithic peoples in meeting and overcoming these challenges–no small task in the comfortless world of the Late Glacial Stage Earth. Thus, despite its rigorous climate, the game herds on the rich tundra grasses and sedges of the Mammoth Steppes appear to have been a magnet for hunting peoples. The cultural basis of their successful adaptation to this part of Eurasia will be discussed later in the chapter.

In any case, after 18,000 BP., world climate began to warm towards the end of the Oldest Dryas stadial. The Bølling interstadial began shortly after 12,400 BP. and lasted until ca. 12,100 BP. when cold conditions returned briefly during the Older Dryas stadial between ca. 12,100 and 11,850 BP. Another period of warming temperatures, called the Allerød interstadial, followed the Older Dryas between ca. 11,850 BP. and 10,900 BP. and then, once again in Europe, and throughout the northern hemisphere, temperatures deteriorated sharply between about 10,900 and 10,100 BP. during the Younger Dryas period. However, this brief return to full-glacial conditions at the very end of the Pleistocene was merely a prologue to the rapid, full retreat of continental glaciation and to the massive mammalian extinctions that marked the close of the epoch. By ca. 6,000 BP. in the Old World, the continental ice sheet had retreated essentially to its modern size and location in Scandinavia, the Alpine glacial complex had shrunk once again into many individual mountain glaciers, and the mountain glaciers disappeared altogether from the Pyrenees. At approximately the same time in the North America, the last remnants of the formerly massive Laurentide continental ice sheet all but disappeared from the vicinity of Hudson's Bay. The Ice Age was over.

The Late Pleistocene in Tropical and Subtropical Africa and Asia

Considerable debate surrounds reconstructions of Late Pleistocene period temperature and precipitation change in the tropics and subtropics during the. This is because in the past, detailed and well-studied terrestrial records of paleoclimatic change during the epoch in Africa, Southeast Asia, and South China have rarely been reported. Fortunately, this situation now appears to be changing.

Africa's Ancient Climates Africa, the second largest continent on Earth, is virtually bisected by the equator. As a consequence, modern temperatures are high over most of the continent and about 90 percent of the continent's land area falls within the tropical or subtropical latitudinal geographic zones. This is a higher percentage of such climates than any other continent (Hance 1975: 46). Nonetheless, the environments of modern Africa are far from uniform. At its extreme northern and southern ends, the giant continent is characterized by midlatitude climates of "Mediterranean" character. Departing from these Mediterranean fringes and venturing into the interior, one encounters true desert, the Sahara in the north and the Kalahari in the south. South of the Sahara and north of the Kalahari, the desert ecology grades into semi-arid steppe or savannah grasslands. These two savannah belts parallel the equatorial rainforest zone that straddles the Equator. Thess four environmental bands extend in rough symmetry across the continent on both sides of the Equator. Only the notable exception of the East African highlands breaks the mirror image zonation of climate and environment across the continent.

Of course, the current configuration of climatic and environmental zones is a relatively recent phenomena. As elsewhere on Earth, the placement of these boundaries and the character of the zones themselves have shifted widely through time. Unfortunately, the precise nature of these shifts is less well understood in Africa and in the Earth's lower or southern latitudes than that of its higher ones. This is due to the difficulty of locating detailed, long-term proxy measures of rainfall change. The best long-term climatic resource presently available from the region are the deep-sea cores recovered from low-latitudes in the Atlantic and Indian oceans. These marine cores have been found to contain windblown dust that originated in northeast Africa and Arabia. During the peak months of the summer monsoon in June, July and August, a huge plume carries dust from these lands out over the oceans. The dust in this plume is from soils that dried out in the early summer after the close of the winter rains and were picked up from the land surfaces by the strong monsoon winds of mid-summer. The size of the plume diminishes in years of heavy rainfall and increases in years when the rains have been poor. Thus, the thickness of each annual dust layer observable in deep-sea core removed from the adjacent sea beds provide a relative measure of the amount of rainfall that fell during the monsoon wet season preceding the layer's deposition (deMenocal 1995; deMenocal and Bloemendal 1995: 263-264).

Using aeolian sediments recovered from tropical Atlantic cores, paleoclimatologists have concluded that

(1) at least two contrastive modes of long-term climate variability were operating in tropical Africa during the last 150,000 years and (2) that these two modes can be correlated in a general way with climatic events in the Northern Hemisphere (Pokras and Mix 1985: 137). The first mode was characteristic of tropical northwest Africa (that is, the southwestern Sahara and the Sahel). This region was driest when the Northern Hemisphere was experiencing glaciation and wetter than it is at present during interglacials and interstadials. Increased aridity there was probably due to the displacement of the westerlies and subtropical high southwards towards the equator by the continent ice sheets in the north. The second climatic mode is evident in equatorial Central Africa southeast of the Sahara during the Late Pleistocene. This region appears driest during high latitude interstadial periods *and* during times of ice growth in the north. Conversely, when the Northern Hemisphere was experiencing deglaciation, Central Africa was at its most humid.

East African climate is considerably more complex than that of West and Central Africa due to the greater topographic variability of that region. Nonetheless, the evidence of changing lake levels in the Rift Valley and elsewhere has proven useful in reconstruction paleoclimate there. Presuming that lake levels provide a relative measure of rainfall and temperature in the Ethiopian and Kenyan highlands, Scholz and Rosendahl (1988: 1647) seismically mapped and dated the ancient shorelines around Lake Malawi and Lake Tanganyika. They concluded that the high levels in these lakes between ca. 25,000/20,000 and 12,000 BP. indicated that East Africa was notably cooler and drier at that time and these conditions had been established in rough parallel with the final stage of late Pleistocene glaciation in the Northern Hemisphere. The sedimentary records of Lakes Victoria and Cheshi also indicate that the timing of local climatic change there occurred more -or-less in concert with events in the Northern Hemisphere. These lakes reached their maximum shrinkage between ca. 15,000 and 12,400 BP. In fact, recent seismic reflection and coring work done at Lake Victoria indicates that this, the largest lake in Africa, dried up completely near the end of this arid interval (Johnson *et al.* 1996) and only began to refill and expand again near the end of the Late Pleistocene between 12,000 and 10,000 years ago. Pollen spectra from East Africa is consistent with this picture as well. These data suggest that the mean annual rainfall in that region decreased by as much as 30% during the Last Glacial Stage and that mean annual temperatures there were 4 ± 2°Celsius lower there than at present (Bonnefile, Roeland and Guiot 1990).

Lest anyone think such changes would have only a trivial impact on the equatorial lands of East Africa, let them climb the side of Mount Kenya. There, at a point literally on the Equator, one can look upon the shrinking remnants of once mighty glaciers. Late Pleistocene temperatures were sufficiently low to allow glacial ice to form, accumulate and push glacical ice out of Mount Kenya's high valleys and down its mountainous sides almost to the plains below.

Subtropical and Tropical Asia As one moves southward across modern Asia, climate becomes warmer and more humid. Much of South China and Southeast Asia is dominated by the annual monsoon cycle of alternating dry and wet months. The seasonal monsoon shapes the climate of coastal East Africa, India, the Indo-Malaysian islands and mainland Southeast and South China. Unfortunately there are only a few continuous palaeoclimatic records available from this vast region. Records that do exist suggest that both temperature, rainfall and the intensity of the monsoon in the region fluctuated widely during the Late Pleistocene and Early Holocene (Gasse *et al* 1991). As noted in Chapter 7, sea level changes that occurred in concert with these climatic shifts intermittently dropped the level of the South China Sea and exposed the sea floor on the Sunda Shelf.

The nature of the environment that came to be established on this exposed sea floor is in dispute. It has been suggested that, during full glacial periods, Sundaland supported an open savanna not unlike the .the Late Pleistocene grasslands of central Eurasia. However, the central Eurasian grasslands supported a distinctive array herbivores that included equids like *Hipparion* or *Equus*, camelids like *Paracamelus* and *Camelus*, and giraffoids like *Sivatherium*. Such taxa are altogether absent from the fossil record of Java and the other Indo-Malaysian islands. Instead, Late Pleistocene fossil deposits on these islands contain orangutan, gibbon, macaque, tapir and other taxa that appear to have been tropical forests dwellers. These fossil data, coupled with pollen cores from the region, suggest the presence of a multi-canopied tropical forest flora. It now appears that Sundaland and other parts of Southeast Asia were dominated by tropical forests throughout the Late Pleistocene (Pope 1985).

The valleys of Southeast Asia generally trend north-south. If these valleys remained tropical throughout the late Pleistocene, they would have provided continuously open dispersal routes across Asia for tropical plants, animals, and of course, hominids, throughout epoch. It is to the question of the origin and dispersal of these Late Pleistocene hominids that we now turn.

Hominid Evolution in the Late Pleistocene

When do the hominids become recognizably human? Plausibly, this may be said to have occurred with the appearance of the various lines of archaic *Homo sapiens*. After all, despite their primitive cranial architecture, these forms have larger cranial capacities than *Homo erectus*. This characteristic above all has lead taxonomists to include them within our species, *sapiens*. Thus, if the Danakil cranium recovered in Eritrea is in fact correctly dated, recognizable humans are on Earth's evolutionary stage as early as 1 mya during the middle Pleistocene period. However, even through they share our genus and species, these archaic *Homo sapiens* are different from us in some important ways. Most importantly, their cranial architecture, though large, is more robust than that of modern humans. Further as we shall see, being "recognizably human" (or even "anatomically modern") and practicing "modern" forms of behavior are not necessarily the same thing.

In any case, "anatomically modern" *Homo sapiens sapiens* appear in the hominid evolutionary stream some time near the beginning of the Late Pleistocene epoch. Modern humans are distinguished by the high-vaulted, globular shape of their skulls the widest part of which is located high on the sides. In contrast to the modern archaic hominids like *Homo erectus* or the Neanderthals, anatomically modern peoples have small brows, high foreheads, and little facial prognathism. The teeth of modern peoples also offer interesting contrasts with those of the archaics. On average, modern front and rear teeth are smaller and are set in more lightly built jaws. Modern peoples also have pronounced chins. Post-cranial contrasts are also present. Anatomically moderns are generally more gracile than their hominid predecessors--most particularly, Neanderthal--and overall have longer legs and short arms.

One Center or Many? The nature of the origins or anatomically modern human has been hotly debated in recent decades. Current reading of the fossil and molecular evidence has resulted in the formulation of two competing models accounting for their Late Pleistocene emergence: the **"Out of Africa" model** (also known as the Rapid Displacement Model or the Single Origin Model) and the **Multiregional model**.

Groups of organisms characterized by a high level of interbreeding are referred to as **breeding populations**. Human breeding populations are usually isolated or defined by geographical or cultural boundaries that separate them and impede interaction between their members. As a consequence of their isolation from one another, breeding populations come to be characterized by distinctive genes frequency. The range and variety of genes present within such population is referred to as its **gene pool** Proponents of the Out of Africa Model see anatomically modern humans as the result of a "significant evolutionary event" that took place within a single breeding population in Africa and altered its gene pool.

Following this evolutionary event, modern humans dispersed from Africa throughout the Old World. In the course of this dispersal, they replaced the indigenous archaic humans populations in each region they colonized. As a consequence, all modern humans are decedents of the same African ancestral population. The Out of Africa Model was developed largely out a series of provocative studies of **mitochondrial DNA (mtDNA)**. Mitochondrial DNA is transmitted strictly through the maternal line. If any variant appears in a group of lineages, it is interpreted to mean that a mutation occurred in the ancestral lineage. It is assumed that each individual is homogeneous for its multiple mtDNA genomes. Therefore, a tree diagram relating the various types of human mitochondrial DNA can also be considered a genealogy linking maternal lineages in modern human population to a common ancestral female.

Recent studies of a large sample of modern mitochondrial DNA have suggested two important conclusions: first, that Africa is the source of the modern human mitochondrial gene pool and second, that the date range for the evolution of the common ancestor of all modern humans can be estimated using the differences that exist within the human genome. To calculate this date range it was first necessary to assume that divergences within the mitochondrial DNA sequence accumulated at a constant rate. Next, estimates of the amount of time required for modern humans to disperse from Africa to New Guinea, Australia, and the New World were formulated. Using these peopling, rates, researchers calculated that mitochondrial DNA changes within modern human populations at a rate of 2 to 4 percent every million years. Analysis of the modern genome indicates that mitochondrial DNA types had diverged from the parent gene pool by an average of nearly 0.57 percent. At the rate of 2 to 4 percent change per million years, fully modern *Homo sapiens sapiens* would have evolved out of African archaic *Homo sapiens* between ca. 290,000 and 140,000 years BP. (but compare Stoneking and Cann 1989:17-28). Proponents of the Out of Africa Model generally agree that, whatever the date of their first appearance in Africa, anatomically modern humans had begun their migration out of the continent by at least 100,000 years ago. When modern humans encountered indigenous archaic peoples during their dispersal, they out-competed and replaced them without interbreeding (Stringer and Andrew 1988). For example, proponents of the Out of Africa interpret the disappearance of Neanderthals *(Homo sapiens neanderthalensis)* from the fossil record of western Eurasian after ca. 35,000 BP. as reflecting the extinction of that taxa and its replacement by anatomically modern peoples of an ultimately African genetic origins.

The alternative to the Out of Africa Model, the Multiregional Model, is based primarily on the human fossil record. Its proponents recognize Africa as the original center of hominid evolution and dispersal. However, they reject the view that anatomically modern human evolved there exclusively a single speciation event. Rather, they see modern humans evolving from *Homo erectus* or archaic *Homo sapiens* populations in a number of different regions of the Old World. They see the hominids of the Middle Pleistocene as belonging to a single species that, although widely dispersed across the Old World, formed an interconnected network of interbreeding populations. In their view, the anatomical features associated with modern humanity evolved locally in different parts of this network at different times and then spread throughout. In their model, there is no single event that defined modern humans and no single regional ancestral population. Rather, anatomically modern populations descend from the *multiple* interbreeding ancestral populations that made up this network (Wolpoff 1989; Wolpoff and Thorne 1991, Lahr 1996). For example, proponents of the Multiregional model do not consider Neanderthal to have "disappeared" from western Eurasia. Instead, they conclude that the Neanderthal population there was connected to populations throughout the Old World and was, as a consequence, transformed through interbreeding into the anatomically modern people found to inhabit the region after ca. 35,000 BP. Multiregionalist consider it likely that Neanderthal genetic material may still be present in modern European breeding populations.

Proponents of both models work on large canvasses: millions of years, hundreds of thousands of generations, vast distances. Expecting fallibility for hypotheses on such a scale is perhaps too optimistic. Although neither model can be categorically rejected at the present time, the weight of evidence (some would say, circumstantial evidence) favors the Out of Africa hypothesis. With that in mind, we begin our discussion of the human cultural sequence in the Late Pleistocene with the African Stone Age.

Cultural Evolution in the Late Pleistocene

The question posed at the beginning of the foregoing section has an important cultural dimension. Probably the most distinguishing characteristic of modern humanity is the manner in which we learn, create and share culture. Therefore, the appearance of recognizably modern humans was primarily a product of *cultural,* rather than a biological, evolution. With that in mind, we turn in this section to an examination of the sequence of cultural evolution that took place in the Old World from the middle of the Pleistocene until its close. In this section we pay particular attention to the events of the Middle and Later Stone Ages in Africa and the Middle and Upper Paleolithic in Eurasia.

Africa During the Middle Stone Age

Sometimes our certainty about a subject turns out to be inversely related to the quantity and quality of our evidence. That is, the less we actually know about a subject, the more certain we are that we understand it. This appears to be the case with regard to the later prehistory of Africa. Not too long ago, most prehistorians accepted the notion that, as the Middle Pleistocene wore to a close, Africa ceased to be a center of human cultural progress. According to this view, at the close of the Early Stone Age, Africa became "a kind of cultural museum" where archaic techniques survived long after they had been discarded elsewhere (Clark 1971: 181). Recent research has forced the rejection of this view by demonstrating

> that technological innovations like flake tools produced on prepared cores, punch-struck blades, and burins appear in Africa at about the same time they did in Europe. Far from remaining a "cultural backwater," (Further) it now appears possible that microlithic technology, pottery, and cattle domestication were indigenous African developments (McIntosh and McIntosh 1983: 215).

Of course, when we consider that similar revolutionary changes have occurred in our view of Africa's place in the human evolutionary career, this new perspective on Africa's cultural role in human prehistory is less surprising.

In Africa south of the Sahara, the Earlier Stone Age (ESA) begins with first appearance of the stone artifacts sometime ca. 2.8 mya The stone artifact assemblages of the early Earlier Stone Age consist of irregularly-flaked pebbles, cobbles, slabs, blocks, flakes and angular fragments. Knappers of the Earlier

Stone Age appear simply to have sought to produce flakes with shape cutting edges. Correspondingly, their greatest effort was directed at modifying the edges of their tools and they gave little attention to carefully trimming and shaping the artifacts further. However, the variety of primary forms in the Early Stone Age began to be changes after ca. 1.6 to 1.5 mya with the appearance of the Acheulean industrial tradition. From this point on, African stone artifact assemblages began to be characterized by an increase in retouched and greater attention to form. The bifacially- flaked tool forms known as the handaxe and the cleaver appear at this time. Sometime after ca. 300.000 BP, the prepared core of Levallois technique is added to the Acheulean tradition in Africa. The lithic assemblages of the following African **Middle Stone Age (MSA)** clearly developed out of those of the preceding Acheulean but there are important distinction between them.

The break between the Earlier and the Middle Stone Ages is not a sharp one. Consequently, assigning a precise date to the end of the Earlier Stone Age and the beginning of the Middle Stone Age is difficult. Nonetheless, between about 200,000 and 100,000 year ago, tool assemblages in Africa began to be characterized by core and flake tools that are generally smaller in size than heretofore. The Middle Stone Age may be said to be fully underway when the large bifacial cutting tools like handaxes and cleavers disappear from the African tool repertoire. Prehistorians working in southern and eastern Africa conventionally assign of dates of between ca. 200,000 and 180,000 BP. for the onset of the MSA and ca. 40,000 BP. for it's close. The culture of the MSA did not remain static during this vast period. Correspondingly, change evident in tool form combined with stratigraphic evidence allows the MSA to be divided into at least four sequential phases labeled MSA I, II, III, IV. Nonetheless, dating and interpretive problems persist and, as we will see below, the Middle and Later Stone Age sequence of Africa can not at present be simply or easily calibrated with that of Europe or Asia.

Lacking the large bifacial core tools like handaxes and cleavers, MSA assemblages are dominated instead by much smaller tools like side scrappers, points, denticulates, notches and backed blades fashioned on well-made flakes and blade commonly produced by the Levallois technique. In part, this reduction in the size of tools in the MSA was a by-product of the greater control and economy gained through use of the Levallois technique and its variants (McBurney 1973: 22). As we will see, the use of the Levallois techniques to produce flakes and blades is only one of a number of attributes present both in the MSA of Arica and in the later Mousterian industrial tradition of the Middle Paleolithic period in Europe and parts of Asia.

The reduction in tool size during the MSA may also reflect increased specialization and the correspondingly greater tool efficiency such specialization produces. For example, the use of smaller flakes may be associated with the development of **hafting**, that is, the mounting of stone flakes on handles or shafts. Flakes were probably wedged into grooves cut into the shaft and bound to the handle with twine made of animal sinew or plant fiber. Mastic made of plant resins may have been employed to hold the flake fast under the binding. A flake mounted on a shaft is a more formidable and effective tool or weapon than one merely held in the hand as the shaft allows the tool user to muster greater leverage in his task. Further, a flake mounted on a spear shaft enables its user to strike prey from a distance by throwing or thrusting with greater force and accuracy. The replacement of the large bifacial implements like handaxes and cleavers probably signals a shift from hand-held tools to shaft-mounted or "composite" tools.

The trend towards reduced flake size that began in the MSA was carried farther during the subsequent Later Stone Age when tool inventories in Africa were dominated by particularly diminutive stone artifacts know as **microliths**.

Who Made the MSA Tools? Only a few African sites have produced hominid skeletal material in direct association with Middle Stone Age stone tool assemblages. For example, a gracile archaic *Homo sapiens* (perhaps best classed as *Homo rhodesiensis*) was recovered in association with a small assemblage of apparently early Middle Stone Age artifacts at the Florisbad Spring site in South Africa. This site has recently been re-dated to between 200,000 and 100,000 BP.(Kuman 1986: 123). MSA material has also been found in association (1) with four anatomically-modern *Homo sapiens sapiens* at the **Border Cave** site in South Africa, (2) with Omo I, a skull of modern aspect, recovered in the lower Omo River valley in southwestern Ethiopia (Klein 1983: 34-35), and (3) with modern human teeth at **Die Kelders Cave** in South Africa (Avery *et al* 1997). Taken together, these data indicate that tools of the Middle Stone Age industrial tradition were made and used by *both* archaic and modern humans.

On reflection the association between the MSA and a wide range of Late Pleistocene human types is not surprising. Chronologically, the MSA spans the crucial period of time when *Homo sapiens* were evolving from their archaic to their modern anatomical form. This dual association thus seems fully consistent with the transitional position of the MSA in African prehistory.

In any case, a more detailed understanding of the African MSA will clearly be necessary if we are to apprehend the crucial interplay in prehistory between the evolution of anatomically modern people and the development of recognizably "modern" forms of behavior and culture.

The Klasies River Mouth Sites The best archaeological data on the Middle and Later Stone Ages in Africa comes from the large group of sites at the mouth of the Klasies River in the Republic of South Africa. The **Klasies River Mouth sites** is a complex of interrelated caves and rockshelters located at the very tip of the southern African continent near the point where the Klasies River flows into the Indian Ocean. In Shelter IB some eighteen meters of sedimentary deposits were excavated to bedrock. Large, if less dramatic, accumulations of sediments have also been excavated in the adjoining caves and rockshelters of the complex (Singer and Wymer 1982: 26-27). As thick as these and other deposits at the site are, only a few sediment lenses in it are primarily of nonhuman origin. The accumulation of the thick, sedimentary units at the site were the result of human activities including stone tool making, food processing, waste discard, fecal disposal, and the introduction of sand and soil into the site on the moist feet of the site's occupants (Butzer 1982: 41).

Since the deposits there are largely the result of human activity in a fairly limited number of local environments, the sedimentary sequence provides little information that can be directly or easily correlated with the worldwide climate sequence. This has proven to be an interpretive problem. Fortunately, the Klasies River Mouth sites is located on the shore of the Indian Ocean. The cave deposits reflect the changing level of the sea relative to this portion of the coast and can therefore be tentatively correlated with the worldwide record of glacial-eustatic fluctuations. The sand component in the site deposits is particularly sensitive to the level of the sea with regard to the site. The presence of typical foreshore aeolian sand and artificially introduced beach cobbles found in the sites' deposits are taken as indicating periods of moderately high sea level when the beach was close to the sites. Destruction of deposits at the sites when storm-driven waves washed inside the caves and rock shelters indicates especially high sea level. On the other hand, the recovery of "regressional eolianites" in the deposits, that is, dune sand cemented below groundwater by calcite, in the deposits record a falling sea level.

The level of the sea is also grossly reflected in the fluctuating number and nature of shells and other marine faunal remains found through time within the deposits. Oxygen isotope ratios found within the marine shells in the various strata was correlated with the oxygen isotope ratio curve based marine deep sea

Chronology of Paleolithic Period in Europe and Eurasia and the Stone Age in Africa

Years Before Present (BP)	Paleolithic Period	Stone Age	Years Before Present (BP)
10,000	Neolithic Mesolithic	Neolithic	3,000 (?)
46-35,000	Upper Paleolithic	Later Stone Age	40,000
90,000	Middle Paleolithic	Middle Stone Age	200,000
2,900,000	Lower Paleolithic	Earlier Stone Age	2,900,000

sediment cores collected worldwide. Radiocarbon dates from the Late Stone Age components at the sites helped in dating the stratigraphic sequence. Based on this calibration work, the date ranges of the Middle Stone Age occupation phases at Klasies River Mouth have been estimated by Butzer (1982: 39, 41-42) as follows:

Dates	Period
80,000 to 70,000 BP.	MSA IV
95,000 to 80,000 BP.	MSA III
ca. 95,000 BP.	Howieson's Poort
120,000 to 95,000 BP.	MSA II
ca. 120,000 ± 5,000 BP.	MSA I

If these date assignments are correct, the sediment sequence at the Klasies River Mouth sites contain the remains of some 50,000 years of intermittent human occupation and provides a dated stratigraphic context for one of the longest and most complete succession of Middle Stone Age occupations yet discovered (Butzer 1982: 42). The Klasies River stratigraphy also contains a long record of human settlement and subsistence activity. Especially important to this record is the large number of animal bones and mollusk shells recovered at the sites. The percentage of various species present in this faunal record changed in a pronounced fashion during the 50,000 years that the caves remained in use. Two changes in species frequency there are especially suggestive. First, the percentage of the bones of large bovids in the deposits declined over time while those of small bovids increased. Second, the percentage of adult seals represented in the bone remains decreased relative to those of young seals.

The excavators of the Klasies River Mouth sites concluded that the shelters there generally served as a "home base" for foraging peoples who systematically hunted bovids and seals and gathered mollusks in its vicinity and then returned to the site to process, cook and share these foodstuffs. Therefore, they interpret the changes in the ratios of large to small bovids and seals in the faunal record as representing climatically-induced environmental changes in the vicinity of the sites. For example, they see the decline in large bovids as representing a habitat shift in which open grasslands were replaced in the vicinity of the sites by the bushy habitats favored by the smaller bovids (Klein 1974). In their view, MSA peoples were hunters who, through time, used the same hunting techniques to confront a changing array of prey species in the catchment area that surrounded their habitation sites.

A Dissenting View After studying the patterns of frequencies of different bones, bone breakage, stone tool cuts, and animal gnaw marks on the faunal remains from the sites, Lewis Binford (1984) offers a radical alternative interpretation of the Middle Stone Age at the Klasies River Mouth sites that rejects the excavators' interpretations on all counts. In his view, (1) the caves and rockshelters there never served as home bases and (2) the inverse relationships in the bone frequencies of large and small bovids and seal is evidence of human *behavioral* changes during the MSA rather than environmental shifts.

Binford (1985) has long been a proponent of the view that early hominids were scavengers first and hunters only reluctantly and later. He bases this conclusion on two sorts of data: (1) the relative abundance of different skeletal parts in the faunal assemblages from early hominid archaeological sites, and (2) studies of the patterns of cut marks and carnivore teeth marks on the surfaces of these bones. Binford's examination of hominid faunal assemblages lead to the recognition that head and foot bones tended to be the most common animal skeletal parts recovered at most Early and Middle Stone Age sites. Since neither head bones nor foot bones carry much flesh, they are far less desirable parts than long bones or ribs. Binford presumes that whoever has earliest access to a carcass will chose to remove the "meat joints" first and concludes therefore that faunal assemblages dominated by head and foot bones were the work of *scavengers,* not hunters. Actualistic experiments done by Blumenshine (1986) with animal kills in contemporary East Africa has largely confirmed this conclusion. Binford's studies of the patterns of surface-modification on bones from early homind sites further strengthened his conclusion. In his view, the teeth marks of large carnivores generally appear to have been put on these bones *before* the cut-mark incisions were made on them by hominids using stone tools. Such an order suggests that the hominids were late on the scene and obliged to scavenge the remains of kills left behind by the big cats. Analysis of the bones recovered at the Klasies River Mouth sites revealed such a head-and-foot-dominated skeletal-element pattern (Klein 1976).

The presence of this pattern coupled with his own analysis of the patterns of cut marks and carnivore marks on the bones lead Binford to the conclude that three different exploitative strategies were in practice at Klasies River Mouth sites during the MSA: (1) the scavenging of the carcasses of larger bovid species killed by nonhuman predators, (2) the hunting of small bovids and the young of large bovids, and (3) the collecting of shellfish and scavenging of beached seals from the adjacent beach (Binford 1984: 201). These three tactics were not practiced in the same proportion to one another overtime, however. Binford (1986: 60) asserts that slowly the hominids using the sites (1) increased their hunting of small bovid species and the young of larger ones and simultaneously decreased the scavenging of marginal parts from carcasses of large bovids killed by carnivores and (2) began to hunt young seals while simultaneously decreasing their scavenging of beached adults. In his view, it is only near the end of the MSA that true hunting begins to supplant scavenging among *Homo sapiens* (Binford 1984: 226-227).

Binford's revisionist view of prehistory is by no means universally accepted by the archaeological profession. In fact, it is fair to say that most prehistorians, are prepared to extend the beginnings of hunting well back into the Middle Pleistocene or earlier. Nonetheless, Binford's interpretations have challenged archaeologists to look more closely at their data in order to resolve this interpretive dispute.

For example, Marean and Kim (1998: S79) reexamined the excavation and laboratory techniques used at Klasies River Mouth sites and other key sites that exhibited the "head-and-foot-dominated" skeletal pattern. They found that at these sites, the faunal assemblages were "either biased assemblages, shaft fragments having been discarded by the excavators, or unbiased ones in which shaft fragments were not included in the zooarchaeologist's analysis."

Milo (1998) took this process one step further by microscopically re-examining the bone modification in the identified bovid assemblage from Cave 1 at Klasies River Mouth sites. His analysis appeared to contract Binford's on every point and suggested that (1) occupants of the sites had been able to obtain the choicest parts of bovids in all size classes, and (2) the carnivore damage to these bones appeared to be ephemeral. Milo concluded that his data falsified the hypothesis that the carcasses of large bovids had been stripped of their prime meat before they were apropriated by hominids or that a large number of the smaller bovids in the faunal assemblage were carnivore kills. Milo's study lead him to conclude that

> hominids were the sole, regular accumulators of bovids in all size classes. That many of those bovids were obtained by active hunting is suggested by the tip of a stone point embedded in a cervical vertebra of the extinct giant buffalo, *Pelorovis antiquus* (Milo's 1998: 99).

Nonetheless, the level of predatory competence of early modern humans remains an open question in archaeology. We therefore return to this question again later in this chapter with our discussion of the Neanderthal in Europe.

The Anomalous Howieson's Poort Perhaps the most intriguing element in the Klasies River Mouth sequence is the so-called **Howieson's Poort** industry. The Howieson's Poort industry at the sites dates to around 95,000 BP. between the MSA II and III phases in layers composed of thin layers of hearths and black carbon-rich soil. These layers were interpreted by the excavators as evidence of a long term, semi-permanent occupation of the shelter during this phase (Singer and Wymer 1982: 87). Howieson's Poort is noteworthy because its lithic industry is based on the production of small blades flakes. The great age of these small blade tools makes them the earliest known precursors of the Later Stone Age and Mesolithic period tool known as the **microlith**. Microliths were produced by systematically breaking a standard-sized blade flake iuto a number oif bladelets of a regular, often geometric, form. These bladelets were then delicately trimmed to produce a series of specialized tool forms. Taken together, the contrasts in artifactual debris and occupation debris suggests that the Howieson's Poort occupation represents change in settlement, hunting or other behavioral patterns by the human occupants of the site at this period.

In sum, the Howieson's Poort occupation appears to be an "site-unit intrusion," that is, an intrusive cultural assemblage with no developmental antecedents at the site where it is found. Equally striking is the fact that the industry has no technological descendants at the Klasies River Mouth sites either. It was followed there by a rather conventional Middle Stone Age III phase occupation that lacks microliths. Altogether, the appearance and date of the Howieson's Poort occupation at the Klasies River Mouth sites appear revolutionary. In terms of our current understanding of Stone Age prehistory, its technology is at least 60,000 years *too early*. Given his view of the retarded nature of MSA subsistence patterns, it is not surprising that Lewis Binford is unwilling to accept the Howieson Poort occupation at face value. As Binford notes, the assemblage contains

> tool forms that are generally unknown worldwide from assemblages dating prior to 35,000. Even more surprising is the occurrence of some tool forms that are generally unknown from most other parts of the world prior to 20,000 years ago (Binford 1986: 61).

For these and other reasons, Binford questions the

validity of the dates assigned by Butzer to the Klasies River Mouth sequence. Who is right? Does the Howieson Poort industry at the Klasies River sites disprove the conventional view that microliths developed late in the Stone Age? Or have the Klasies River Mouth sites been mis-dated? Only the discovery of Howieson Poort occupations in MSA sites elsewhere or the development of new and more direct ways of dating Klasies River Mouth sites or both will settle the issue. For now the Howieson Poort industry remains an intriguing prehistoric anomaly.

Around 80,000 BP., the Middle Stone Age III occupation gives way to the Middle Stone Age IV. MSA IV people appear to use the sites intermittently for some 10,000 years before abandoning it altogether after ca. 70,000 BP. (Singer and Wymer 1982: 27). Butzer suggests that use of these caves and rockshelters ceased because the sea shore had retreated from their vicinity with the worldwide drop in sea level that occurred after 70,000 BP. in concert with the return of continental glaciation in the northern hemisphere. Rising sea level after the close of the Pleistocene appears to have made the sites an attractive location once more. After ca. 5,000 BP. the sites was briefly used and sporadically occupied by Later Stone Age, historic and modern peoples.

Africa's Later Stone Age

In Africa south of the Sahara, the end of the Paleolithic period is called the **Later Stone Age (LSA)**. This final period is commonly thought to have begun between ca. 40,000 and 30,000 BP., although some evidence suggests that a substantially earlier date may be correct (Ambrose 1998: 379-380). The MSA/LSA transition is marked by a technological change. The characteristic MSA technique of manufacturing stone tools on levallois flakes removed from prepared cores disappears in the LSA. In place of this technique, flakes removed from plain platform cores and subsequently broken to form microliths became the norm. Associated with this change in lithic technology is the appearance of tools and ornaments of bone, ivory, horn and shell that were shaped by grinding and were often perforated for suspension on twine and worn as jewelry. Still later, LSA peoples began to produce rock paintings, carving and other forms of plastic arts. Most importantly, "modern" patterns of hunting, gathering, resource exploitation, trade and social organization are thought to have come on line during the LSA (Binford 1989; Klein 1995).

The LSA period comes to an end when these stone using peoples were contacted or displaced by iron-using peoples. This contact happened at very different times in different parts of Africa but seems to have been underway at least by ca. 4000 B.C.

LSA Technology and Subsistence Later Stone Age assemblages continue the trend towards size reduction that began in the MSA. As noted above, the most distinctive LSA stone tool was the very small stone bladelet known as the microlith. Microliths were commonly set in wood or bone handles or shafts to form composite tools or weapons, such as barbed arrows or fish-spears (Oakley 1957: 108-109) The manner in which such composite microlithic tools were manufactured has been reconstructed by examining the limited number of such tools with microlithis still in place recovered from dry caves and other contexts with special preservation. The bone or wood shaft was generally grooved or split to received the microlith. However, before the stone was inserted in the groove, one of its edges had to be "backed," that is, trimmed or dulled to prevent it from splitting the haft. Generally, vegetal resin or mastic was used to fasten the bladelet in place in the groove (Phillipson 1985: 58-59). At least some of the Later Stone Age backed microliths were used as arrow points as were the polished bone points that have very occasionally been recovered along with bow and arrow fragments in dry caves (Deacon 1984: 221-222). However, microliths can also be "employed in composite tools for plant gathering, harvesting, slicing, grating, plant-fiber processing for lines, snares, nets and traps, shell openers, bow-drill points and awls" (Clarke 1977: 452).

Later Stone Age technical repertoires were also characterized by a profusion of bone tools including needles and awls. These tools, together with the abundant and varied stone scrapers and the occasional fragments of sewn leather, suggest that skin working was an important craft in the Later Stone Age presumably for the manufacture leather clothing and carrying bags. Finally, as in the roughly synchronous Upper Paleolithic period of Europe, Later Stone Age peoples appeared to delight in adorning themselves. To this end, they used beads, pendants, and amulets of freshwater and marine shells and ostrich eggshells. They also buried their dead, sometimes with grave goods, in a formal manner (Deacon 1984: 221-222).

Perhaps most important, the Later Stone Age in southern Africa witnessed changes in human subsistence. Among the most notable change is the evidence that fish and marine birds, apparently ignored by Middle Stone Age peoples, were exploited systematically at that time. In addition, faunal remains from sites of the period indicate that the big game hunting ability of these later people "represented a quantum advance over the preceding Middle Stone Age." As it did (and continues to do) in many parts of the Earth, increased human technological capacity in southern Africa appears to have contributed to the extinction of game species as giant buffalo, giant hartebeest, southern springbok, giant Cape horse, and other late Pleistocene species (Klein 1983: 43-44).

Still Another Stone Age Anomaly Late Stone Age peoples were most certainly anatomically fully-modern *Homo sapiens sapiens*. This assertion is based both on the association of anatomically modern remains with LSA assemblages and by the fact that the Later Stone Age is characterized by the appearance of such elements of "modern" material culture such as bone and ivory carving, rock painting, formal burial, and so on. Of course, these elements are considered modern because they have continued in use almost to the present day among African hunting and gathering peoples (Deacon 1984:221; Volman 1984:174). As we will see, however, their is some dispute as to whether these elements--or at least all of them--appeared first in the LSA.

Beyond question the peoples of the Later Stone Age practiced a "modern" form of behavior that we ourselves would find generally familiar. But were they the first to initiate it? Until quite recently most African archaeologists argued that modern forms of behavior first appeared in Africa during the Later Stone Age or at the very end of the MSA (for example, cf. Klein 1995). However, this date has been challenged by recent archaeological work done at an MSA site on the **Semliki River** in Zaire. These sites have yielded sophisticated bone harpoons and other tools previously found only in LSA contexts. The stratigraphic layer containing these tools has been dated by thermolumiscence (which provides an age for quartz grains in the soil) and electron spin resonance methods and found to be at least as early as 90,000 years old. Based on these dates, the excavators assert that sophisticated tools (and by implication, "modern" behavior) appeared in prehistory during the Middle Stone Age (Brooks *et al* 1995; Yellen *et al* 1995).

The apparent early date of the these tools coupled with their advanced nature is, like the Howieson's Poort occupation at the Klasies River Mouth sites, an anomaly. If these dates and their interpretations are correct, modern behavior began in Africa much earlier than anywhere elsewhere in the world. Such a conclusion would force a substantial revision in our current view of the prehistory of the Old Stone Age. However, one site is never sufficient to occasion such a interpretive revolution. Before the question can be resolved, the finding from the Semliki River site must be replicated MSA sites elsewhere in Africa.

While by no means able to resolve this question, work done by one of your authors (Dickson) and his colleagues at the **Shurmai** and **Kakwa Lelash** rockshelters in East Africa perhaps offer some insights into it. These two sites, located near one another at the edge of the remote and rugged Mukogodo Forest in central Kenya, were tested excavated during the early 1990s (Kuehn and Dickson 1998). Shurmai Rockshelter (GnJm 1) produced 4,782 stone artifacts and was occupied during both the MSA and the LSA. Kakwa Lelash (GnJm 2) yielded 7,862 stone artifacts and appears to have been occupied no earlier than the LSA. All of these lithic materials received as detailed analysis (Gang 1998; Dickson, Gang and Kuehn 1998). At both sites, the earliest occupation date to the end of the MSA and continue into the LSA and later.

The area in which the two shelters are located is rich in quartz and volcanic rocks including basalt, granite, gneiss, and schist. Unfortunately, all of these locally available types of rocks make very poor raw materials for stone tool making. Excavations revealed that the inhabitants of the two sites procured higher quality of raw materials like chert and obsidian from more remote sources. Given the present state of knowledge, we cannot be sure where the nearest source of high quality chert is located. However, for a variety of geological reasons, such sources are very likely located no closer than 10 and no farther than 50 kilometers from the two sites (Gang 1997: 84). The chert recovered from these two sites was considered to be "non-local" in origin. The nearest known obsidian source is located on Mount Kenya about 70 kilometers from the two sites (Merrick and Brown 1984: 129-135). Therefore, obsidian recovered at Shurmai and Kakwa Lelash can be considered to be "exotic."

This classification was used to compare the raw material procurement strategies of the Middle and Later Stone ages at the two sites. The number of stone tools, flakes and cores made of local, non-local and exotic raw materials at the sties were tallied. Chi square analysis of these tallies indicate that distribution of the three raw material types in LSA occupations at Shurmai and Kakwa Lelash are statistically identical and that the raw material distribution in the MSA occupation at Shrumai is distinctly different from both LSA occupations. This comparison suggests that raw materials were procured quite differently by the MSA and LSA occupants of the two sites. The MSA occupants of the Shurmai rockshelter appear to have relied almost entirely on local basalt and quartz despite the low quality of such raw material for flint knapping. Although the LSA people continued to use local basalt and quartz, they appear to have relied just as heavily on non-local chert and, to a much lesser extent, on exotic obsidian.

In addition to differences in the types of raw materials preferred, there are also apparent differences in the way MSA and LSA knappers transported these raw materials to their sites. The amount of cortex on lithic material from the MSA occupation at Shurmai suggests that their locally available raw materials were taken directly to the site as nodules and were not first roughed out at their source. In contrast, the lower quantity of cortex on lithic materials in the two LSA occupations suggests the opposite: LSA knappers apparently roughed out their raw material into blanks before transporting it to the site. Of course, such behavior reflects the effect of differential "transportation costs" and simple least-effort calculations. MSA raw materials had only a short distance to be carried, while transportation of the non-local and exotic materials preferred during the LSA were at a substantial remove

The small amount of non-local and exotic lithic materials recovered from the MSA component at the Shurmai Rockshelter suggests that the MSA knappers probably did not procure non-local and exotic raw material in a systematic manner. Rather it seems most likely that MSA occupants of Shurmai Rockshelter obtained high quality raw materials more-or-less inadvertently while hunting, collecting, or engaging in seasonal movements. On the other hand, since non-local and exotic raw materials constitute over 50 percent of the lithics recovered in the LSA components at both Shurmai and Kakwa Lelash rockshelters. It would thus appear that LSA knappers did Western in fact procure high quality lithic raw materials in an intentional and systematic manner. In Binford's (1979:259-261) terms, the raw material procurement strategy of MSA peoples appears to have been "embedded," that of LSA people "curated."

Thus it would seem that, as compared to their MSA predecessors, the LSA inhabitants of our two sites gave greater consideration to the quality of their raw materials, preferred to use high quality raw materials for tool manufacture, made a systematic effort at acquiring such materials, and attempted to minimize the "cost" or effort of transporting these raw materials to their occupation sites. Such acquisition and manufacturing strategies suggest at least two additional behavioral differences between the knappers of the two periods. First, compared to the peoples who occupied Shurmai Rockshelter during the MSA, the LSA peoples probably had better information about areas beyond their own immediate territory or had better trade connections with surrounding peoples or both. Second, it seems probable that the people of the LSA had higher performance requirements for their tools than did their MSA predecessors. .

Both LSA and MSA knappers at the two sites used the same primary technique for removing flakes for tool making: direct, hard hammer percussion. Neither MSA nor LSA knappers made use of the Levallois technique. Despite these technical similarities, LSA knappers produced flakes that were shorter, thinner and more varied in shape than those of the MSA. Correspondingly, the average size of LSA cores is also smaller. At the same time, the cores recovered from the LSA occupations are more apt to be "exhausted" than those from the MSA occupation. An exhausted core is one that has be used so throughly that virtually no more flakes can be removed from it. This exhaustion may be due to the relative scarcity of cores of non-local material. Their value must have induced LSA knappers to use them "down to the nub" before reluctantly discarding them.

There are also a number of stylistic contrasts between the lithics of the LSA AND MSA components at the two sites. Core tools such as handaxe or cleavers were not recovered at either of our sites. In both the MSA and the LSA components, all tools were made on flakes except hammer stones (Gang 1997: 244). Beyond this, there are few other stylistic similarities between the components. The MSA tool assemblage from the Shurmai Rockshelter includes blades of irregular or rectangular shape and flake tools including points, denticulates and naturally backed flake "knives." Tools in the LSA components at Shurmai and Kakwa Lelash rockshelters consist largely of microlithic bladelets and small flakes with pronounced use wear. The production of naturally backed flake knives and convergent flakes was also low in the LSA components at both sites. This change probably indicates that the simple, naturally backed flake knife was replaced during the LSA by more sophisticated composite tools made by inserting microlithic bladelets into wooden handles.

The work at Shurmai and Kakwa Lelash rockshelters does not falsifying either of these conflicting hypotheses. However, the greater sophistication, systematization and efficiency evident in the patterns of resource use, tool manufacture and style in the LSA components at the two sites is consistent with the older view that the origins of modern human behavior are to be found in the Later, rather than the Middle, Stone Age in Africa.

The end of the LSA period takes place when these stone using peoples were contacted or displaced by iron-using peoples. This contact happened at very different times in different parts of Africa. The differential nature of this contact and diffusion in later

African prehistory created what McIntosh and McIntosh (1983: 215) refer to as a distinctively "mosaic" pattern of technological stability and change on the continent. In the western and central areas of the continent, flaked stone repertoires were enriched after ca. 4,000/3,000 B.C with ground stone tools and pottery which presumably diffused in the company of farming peoples and domesticated plants and animals. By 1,000 B.C., iron-working penetrated into these areas and iron tools began to replace flaked stone implements. However, in southern Africa, displacement by iron-using peoples did not begin until around the time of Christ and the San (Bushman) and the Khoi (Hottentot) peoples were still making and using stone tools of the Later Stone Age tradition when they were encountered by Europeans in the 17th century (Deacon 1984: 222).

Western Eurasia in the Middle Paleolithic Period

The Late Pleistocene epoch in western Eurasia encompasses the later portion of the Middle Paleolithic period (ca. 250,000 to 35,000 BP.) and the entirety of the Late or Upper Paleolithic period (ca. 35,000 to 10,000 BP.). In a very general way, the these two subdivisions of the Paleolithic period can be correlated with the Earlier, Middle and Later Stone ages of Africa. The Middle Paleolithic period begins when the Acheulian industrial assemblages there are replaced in western Eurasia by a new industrial tradition, the **Mousterian**. Dating this replacement has presented prehistorians with real problems as it occurred at a time beyond the effective range of radiocarbon dating and after that of the potassium/argon method. As a consequence, most dates for the beginning of the Middle Paleolithic were, until quite recently, extrapolations from archaeological stratigraphy. From these stratigraphic extrapolations, it was concluded that the Mousterian arose in western Eurasia during the early millennia of the Weichsel glacial stage between ca. 115,00 and 90,000 BP. Thus, by these lights, the Middle Paleolithic period was see as lasting no more than about 80,000 years in western Eurasia and closing when the Mousterian tradition gave place to the earliest stone tool industries of the Late or Upper Paleolithic period.

In recent decades, the widespread application of techniques like thermoluminescence, electron spin resonance and U-series dating to Mousterian sediments and artifacts has complicated the Middle Paleolithic age picture considerably and forced a revision of the Lower and Middle Paleolithic period chronology in western Eurasia. It is now apparent that the key elements of the Mousterian tradition are present in both Europe and the Middle East by at least 250,000 BP. or perhaps even 100,000 years earlier (Falguères, Bahain and Saleki 1997; Blackwell, Schwarcz and Debenath 1983; Delpech and Laville 1988; Mercier *et al.* 1995: 506-507). It is also apparent that the replacement of the Acheulian by the Mousterian was time transgressive to a decidedly unexpected degree. That it, it occurs in different places in western Eurasia at radically different times. The end of the Middle Paleolithic is also time transgressive; it comes to an end in western Eurasia between 46,000 and 35,000 BP.

Mousterian Tool Form and Function As might be imagined, the Mousterian label has been assigned to a widely varied range of stone tool types and complexes in the Old World. However, in this book we use the term in its narrowest sense to refer exclusively to the Middle Paleolithic tool assemblages that follow the Acheulean and precede the Upper Paleolithic industrial traditions in western Eurasia. Beyond western Eurasia, Middle and Late Pleistocene tool assemblages from Africa that exhibit many of these attributes are classed as Middle Stone Age industries. Those of a similar nature recovered in eastern Eurasia are best referred to as "Mousteriod" rather than Mousterian in the strict sense. This latter distinction is more than a mere semantic nicety as at least some of the Asian Mousteroid assemblages appear to be much younger than those of classic Mousterian of western Eurasia.

Recall that, over most of the Old World west of the Movius line, the centerpieces of the LowerPaleolithic period Acheulian industrial tradition were bifacially-flaked tools like handaxes, picks and cleavers made on cores. In contrast, the succeeding Mousterian industry there is characterized by:

1. the use of the prepared core or Levallois stone working technique tin the production of flakes for retouching into tools, and
2. the presence of four primary types of tools: side-scrappers, denticulates, notched pieces, and perforators.

Some Mousterian assemblages in western Eurasia have also been found to contain handaxes, that is, core tools descended from the Acheulean handaxe but generally smaller and perhaps more specialized in function that those ancestral forms (Bordaz 1970: 38). Much less commonly, Mousterian industries have produced triangular flakes often identified as "projectile points." As traces of an adhesive or mastic have been found on a few such triangular flakes, it appears that at least some of them were in fact hafted on shafts (Boëda, Connan, Dessort *et al.* 1996). Nonetheless,

Mousterian points show nothing like the range of stylistic variation found on projectiles points made by later, behaviorally-modern peoples (Holdaway and Cosgrove 1997: 69). Finally, late in the Middle Paleolithic period, tools like burins and endscrapers, are present in small quantities in some Mousterian assemblages. While these tools first appear in the Mousterian, they rise to prominence only later in the industries of the subsequent Upper Paleolithic period.

Nearly forty years ago the French prehistorian François Bordes, assessed the then-chaotic state of Mousterian taxonomy and developed as magisterial classifactory scheme to remedy the situation. Bordes meticulously defined some sixty-three separate classes and subclasses of Mousterian tools and considered these tools in each type to have been consciously manufactured in accord with strict stylistic constraints and formal criteria. Bordes descriptive efforts introduced a long overdue rigor into the study of Middle Paleolithic technology. However, he went further than mere description.

Using differences in the ratios of key tool types, Bordes recognizes six major European variants or facies of the Mousterian tradition (Bordes 1968: 98-105). Unfortunately, these facies do not follow one another in a fixed, successive order that would indicate that they developed one out of another. Rather, at key sites like the **Combe Grenal rockshelter** in southwestern France, the facies occur, disappear and then recur almost randomly through the successive layers in the stratified deposits. Bordes (1972: 146-149, 1973; Bordes and de Sonneville-Bordes 1970) interpret the stratigraphic alteration of these facies in southwestern Europe as evidence for the persistence of "parallel phyla" during the Middle Paleolithic period. In his view, the six Mousterian variants represent six separate cultural groups like "tribes" that co-existed in southwestern Europe for 65,000 years or more without materially affecting the independence of each other's technical traditions. Alternatively, it has been argued that this is evidence of functional rather than cultural variation and that the six Mousterian facies represent distinctive tool kits used by essentially the same people to accomplish different sets of tasks at different times or under different environmental conditions (Binford and Binford 1966, 1969; Binford 1973, 1983: 95-108; Freeman 1966, 1972b). However, neither interpretation has proven to be adequate explanation of Mousterian tool form variation.

In recent years much of the formal variation between tool types which Bordes took to be the result of conscious choices between styles has been found to be due to differences in the amount of edge re-sharpening to which the individual tools had been subjected (Dibble 1987). Further, the extent to which individual tools were reworked and re-sharpened by their users appears to be a function of the distance to raw material sources. That is, the amount of re-sharpening evident on Mousterian tools increases as the distance from the raw material source from which they were drawn lengthen (Rolland and Dibble 1990). Therefore, Bordes six Mousterian facies may reflect variation in access to raw materials rather than political or ethnic differences.

However, Kuhn (1995) sees still another relationship in this data. He contends that the degree of tool re-sharpening is greater in Mousterian assemblages associated with fauna remains showing signs of having been scavenged rather than hunted. In his view, both re-sharpening and scavenging are expedient or opportunistic behavior. When found to have occurred together at a site, they are evidence that the particular location had been used only briefly. In contrast, he sees long term occupations as having occurred at those Mousterian sites that have produced fewer re-sharpened tools and faunal remains of hunted, rather than scavenged, game. In his view, long term camps would more likely have been well supplied with stocks of raw materials for stone tool making rendering tool re-working less critical.

In any case, the Mousterian industrial tradition must have offered its users certain adaptive advantages over the late Upper Acheulian industries. We may infer this from the evidence that the rigorous pro-glacial environment of European Russia was successfully exploited by Mousterian peoples before the Last Interglacial. This penetration was no doubt also made possible through the use of artificial shelters. Features uncovered at Moldova in the Ukraine indicate that mammoth bones were used to construct frameworks which probably were then covered with hides or earth. More limited Mousterian structural remains have been recovered in southwestern Europe. Remnants of dry masonry walls have been found in association with Mousterian components within the French rockshelter of **Pech de l'Aze** and at the Spanish rockshelter of **Cueva Morin**. Cobble stone pavings, uncovered in the Mousterian levels at **La Ferrassie** and Combe Grenal rockshelters in the French Perigord region, perhaps once floored long-vanished wood or hide structures (Freeman 1980: 84-85).

Splitting Heirs: Are We Kin to the Neanderthal? The great dispersal of the hominid line in western Eurasia was begun by *Homo erectus* during the Early and Middle Pleistocene; it was continued there during the Late Pleistocene by *Homo sapiens neanderthalensis* or Neanderthal (Trinkaus and Howells 1979: 125). This hominid form, which is apparently limited to

Europe and southwestern Asia, first appears in the fossil record there during the Last Interglacial Stage. In its classic form, Neanderthal is a striking amalgam. Its crania combine large interior brain capacity with "primitive" features like large brow ridges, sloping foreheads and low cranial vaults. Further, Neanderthals had remarkably robust post-cranial skeletons of the sort one would only find today on athletes intensely training for power sports.

We can only speculate as to what suite of environmental conditions favored this combination of characteristics but since these hominids were able to penetrate the rigorous subarctic environments of western Eurasia during the Late Glacial Stage, it has been Howell (1957) contends that the classic Neanderthal physical pattern was a genetic response to conditions of extreme cold. He sees the pattern becoming established in European hominid breeding populations because of their geographical isolation during the Last Glacial Stage. In contrast, Brace (1962, 1979) interprets the robust Neanderthal facial morphology as largely the product of cultural behavior; specifically, the extensive and heavy use of the teeth and jaws by Neanderthals in chewing tough foods and as tools for the processing of hides and the performance of other tasks. In Brace's view, the hominid face began to assume its gracile modern form when improvements in technology during the Late Paleolithic period lead to (1) a reduction in the use of the teeth and jaws as tools and (2) a still another "culinary revolution" characterized by more efficient food preparation techniques. Lieberman and Crelin (1971) and Lieberman 1976) on the other hand, suggests that the emergence of the modern cranial morphology reflects the evolution of the capacity for speech in our species (and its absence in Neanderthal and earlier hominids).

The origin of fully modern hominids and the relationship between these forms and the Neanderthals have been persistent problems in archaeology. The question of the nature of the replacement of pre-existing Neanderthal populations in Western Eurasia by *Homo sapiens sapiens* still remains to be answered. Three competing hypotheses have been proposed. As framed by ApSimon these include:

1. straightforward evolutionary transformation *in situ* of Neanderthal into anatomically-modern people;
2. anatomically-modern people immigration followed by interbreeding in in which the distinctive characters of the Neanderthals were lost;
3. immigration of anatomically-modern peoples into Western Eurasia leading to the extinction of the Neanderthals (ApSimon 1980: 271).

Since the resolution of this question hinges on the degree of anatomically similarity or difference that exists between these two taxons, let us briefly compare the crania, pelvis, the degree of sexual dimorphism and the DNA of Neanderthals and *Homo sapiens sapiens* (after Howells 1975; Trinkaus and Howells 1979; Stringer 1974, 1982; Stringer and Andrews 1988; Rak and Arensberg 1987).

Neanderthal Crania The most pronounced differences between *Homo sapiens sapiens* and Neanderthals are in the shape of their skulls or crania. Although the average cranial capacity of the Neanderthals equals or slightly exceeds that of modern man, the shapes of the crania of these two forms are quite different. Modern crania tend to be rounded and characterized by high vertical foreheads. Maximum breadth of these crania is high on the parietals, that is, at the two paired membrane bones that form the back and sides of the skull and join together at the sagittal suture. This shape is classed as **dolichocephalic**, or long headed. In contrast, the maximum breadth of Neanderthal crania tends to be in the middle of the parietals. These wide middle parietal bone dimensions give Neanderthal crania a **platycephalic** or wide, low-vaulted, barrel-shape.

Another contrast between the Neanderthal crania and those of *Homo sapiens sapiens* is their relative facial angle. When viewed in profile, hominid faces range from nearly vertical or flat to markedly protruding, that is from **orthognathous** to **prognathous**. The extent of this variation can be measured and expressed as the **gnathic index**. In anatomically modern populations, faces range from orthognathous to moderately prognathous, that is, from low to moderately high gnatic index values. Neanderthal crania, on the other hand, are invariably displaced forward in a pronounced fashion and have very high gnathic indices rarely found in modern individuals. At the same time, the mandibles of modern people are thinner and lighter, the chin more pronounced and pointed and the teeth smaller than those of the Neanderthal.

Neanderthal crania also exhibit strongly-marked ridges for muscular attachments to a degree that is generally absent on those of *Homo sapiens sapiens*. For example, in the occipital region at the back and base of Neanderthal skull have pronounced "buns" and relatively small nuchal area that form the nape of the neck. However, the most greatest contrast between the crania of Neanderthals and modern peoples is the absence among the latter of the massive supra-orbital ridges or tori over the eye sockets. Surely these orbital ridges must have been the most arresting elements in the countenance of living Neanderthals (Clark 1964: 50-51; Kennedy 1980: 385 Stringer and Andrews 1988: 1261).

Neanderthal Pelvic Structure The post-cranial skeletons of the two taxa are also contrastive. The morphologies of Neanderthal and *Homo sapiens sapiens* female pelvises are different. Although the birth canals are approximately the same diameter, they of somewhat different shape Pilbeam (1984a: 96) takes these different shapes as indicating that birthing is harder for anatomically-modern females. He even suggests that the Neanderthal gestation may have been as long as eleven months since such a period is more "in line with predictions based on general mammalian relations among maternal body size, fetal size and length of gestation." However, Rak and Arensberg (1987), contend that the size, shape, and structure of the Neanderthal pelvis reflect important differences in locomotion and posture between the two taxons rather than contrasts in gestation.

Reductions in Sexual Dimorphism Anatomically-modern skeletons tend to be lighter, more slender and less powerful than those of Neanderthals. In addition to this general difference in skeletal robustness, the two groups show very different patterns of sexual dimorphism. Anatomists use the phrase sexual dimorphism to refer to size or behavioral differences between the males and females of a species that are not directly related to their contrastive reproductive functions. Size difference between males and females is far less pronounced in late Pleistocene *Homo sapiens sapiens* populations as compared with those of the Neanderthals. And, as Frayer (1978) notes, sexual dimorphism in our species continues to decline following the close of the Pleistocene epoch in the subsequent Mesolithic period. He states that while late Pleistocene males show consistently larger dimensions than their Mesolithic descendants, females from the two periods are relatively stable in post-cranial dimensions.

Frayer relates this reduction in male robustness to changing hunting patterns. He hypothesizes that there occurred at this time (1) climate change and extinction which lead to a general reduction in the size and aggressiveness of the game animals being hunted in Europe, and (2) an improvement in the effectiveness of the tools and weapons used to hunt them. He suggests that such changes would result in the development of new hunting strategies which placed less reliance upon simple physical strength. Assuming that later Paleolithic peoples divided labor by sex, such changes would reduce the selective advantage of large body size in males. Since female activities would be unaffected by these developments, little change would be expected in their body sizes. Frayer's model is perhaps even more plausible as an explanation of the reduced sexual dimorphism in *Homo sapiens sapiens* populations as compared to Neanderthals, since the technical contrasts between the cultures associated with the two forms are even more dramatic than those between the cultures of the terminal Pleistocene and the early Holocene epochs. In summary, the skeletal remains of *Homo sapiens sapiens* suggest that anatomically modern people are less similar to "classic" Neanderthals than to certain late archaic *Homo sapiens*.

Neanderthal DNA In an interesting case of scientific symmetry, the skeletal remains that lead to the scientific recognition of this existence of Neanderthal are now contributing to the clarification of its phylogeny. The first Neanderthal was discovered in a cave in the Neander valley of Germany in 1856. Now, 142 years after this discovery, the arm bone of this skeleton has yielded a small quantity of ancient mitochondrial DNA. Mitochondrial DNA comes only from one's mother. It is not subject to genetic recombination Therefore, the order and character of the long chains of nucleotides strung together to form specific mitochondrial DNA chains remains unchanged from generation to generation except when mutations occur s within them. As a consequence, geneticists count the number of differences between nucleotides of two mitochondrial sequences in order to access the degree of evolutionary distance between them. When the mitochondrial DNA sequence extracted from the Neander valley specimen was compared to sequences from 986 distinct living humans from around the world, it was discovered that, on average, there was

> three times more difference between the Neanderthal and modern human sequence than between pairs of modern humans. Specifically, pairs of modern human sequence differed at an average of only eight positions, while human-Neanderthal pairs differed at an average of 25.6 position. And the range only barely overlapped: The most divergent modern humans differed in only 24 nucleotides, while the closest modern-Neanderthal pair had 20 differences. Also, the type of nucleotide substitutions and their locations were different. These data put the Neanderthal sequence outside the statistical range of modern human variation (Kahn and

Gibbons 1997: 177).

These genetic comparisons make it appear highly unlikely that Neanderthal (the individual from the Neander Valley, at least) contributed to the genome of anatomically modern humans. Neanderthal does not even appear genetically closest to Europeans but is roughly equidistant from all races of modern humanity. Thus, the DNA data appears to falsify the hypothesis that the immigration of anatomically-modern people into Europe lead to interbreeding that lead to the disappearance of the distinctive Neanderthal genetic traits. Finally, the number of mutations that have accumulated over time in the Neanderthal mitochondria can be used to calculate the approximate number of years that have passed since their lineage and that of anatomically modern *Homo sapiens sapiens* differentiated from their last common ancestor. Using these data, the authors estimate that this split occurred between 550,000 to 690,000 years ago (Kahn and Gibbons 1997: 187). If this estimate is correct, it would indicate that Neanderthal and modern humans evolved separately from *Homo erectus* or from different forms of Archaic *Homo sapiens*. A likely candidate for such a last common ancestor has recently been recovered in the Atapuerca Hills of Spain and dated to ca. 800,000 BP. This form has been tentatively called *Homo antecessor*. Thus, both the skeletal and genetic dissimilarity between Neanderthals and modern humans renders it unlikely that this taxon was the direct ancestor of modern people (ApSimon 1980: 271; Stringer 1982; Stringer, Hublin, and Vandermeersch 1984: 115-124). Further, such contrasts makes it unlikely that Neanderthals and *Homo sapiens sapiens* simply merged into a single population. Thus, current physical evidence would seem to favor the third hypothesis, that *Homo sapiens sapiens* evolved outside Eurasia and then immigrated there, replacing the Neanderthals and forcing them into extinction (but cf., Smith 1984: 192-198). Nonetheless, the question does not end with the physical contrasts between Neanderthals and modern humans.

Taphonomy and the Emergence of Mind The most important issue pertaining to the differences between Neanderthals and modern people does not concern their contrasting skeletal morphologies or DNA. Rather, it hinges on the relative intellectual capabilities and mental processes of the two types of hominids. Did the Neanderthals think and act like anatomically modern people or did the appearance of *Homo sapiens sapiens* represent an important intellectual watershed in human prehistory? Mousterian sites have in fact produced data that is often taken as evidence of an emerging human cognitive or intellectual complexity among the Neanderthals. For example, red ochre or hematite from the Mousterian occupation at Pech de l'Aze shows evidence of

> having been scratched with flint tools to produce a black powder which could be smeared with the fingers, while others were either rounded smooth by wear or had been shaped into rectangular or pointed crayons (Schmandt-Besserat 1980: 128).

Evidence of ochre use of this nature have been found at a number of European Mousterian sites. However, only isolated instances of art or intentional engraving has been found in Mousterian context. Freeman (1980:85) notes that an incised cross was found scratched on a fossil at Tata in Hungary, that bones with linear "tally marks" have been found in the Charente, and that the site of Cueva Morin in Spain has produced a deer bone with regularly spaced curvilinear grooves (but, cf. Chase and Dibble 1987 for a contrary opinion). Klein concludes that this rare and scattered evidence

> indicates prefigurations of the behavior of fully modern man, but their production seems to have been confined to a few isolated and perhaps unusual individuals (Klein 1980: 85).

Yet these scattered markings are not the most significant "prefigurations" of modern behavior to appear at this time. That title must be reserved for the intentional burials that Neanderthals began to accord some of their number during the Middle Paleolithic period. Intentionally buried Neanderthal remains have been reported from a number of western Eurasian Middle Paleolithic sites including: Teshik-Tash in Uzbekistan, Shanidar in Iraq, Mt. Carmel in Israel and La Chapelle-aux-Saints, Regourdou, Le Moustier, La Ferrassie, and Correze in southwestern Europe. The occasional recovery of objects like red ochre, stone tools and animal bones apparently associated with these remains has been taken as evidence that the deceased were sometimes buried with mortuary offerings, personal possessions or food.

In addition to isolated burials, complexes of mortuary remains and features have been reported from La Ferrassie and the controversial site of Regourdou in France. At La Ferrassie, burials of two adults and six children are associated with a series of regularly spaced mounds and trenches. The bulk of the excavations at La Ferrassie were done by Peyrony and Capitan before

World War I. However, Delporte's (1976) limited test excavation and reanalysis of the site has confirmed most of Peyrony and Capitan's findings (Stringer, Hublin, and Vandermeersch 1984: 98). At the Regourdou site, a single individual was buried along with grave goods under a low mound or cairn of rocks. In the vicinity of the mound, and perhaps associated with it, a number of trenches were found as well as stone cysts containing brown bear (*Ursus arctos*) remains.

The discovery of these Neanderthal graves provides rare insight into the expansion of the mind taking place during the Paleolithic Following the physical demands for food, sleep, and sex, no other aspect of life shapes the consciousness of our species more than the recognition of the inevitability of our death. Although, we rebel against its immutability in countless ways, the utter simplicity of this horrendous fact ultimately colors all human choices. The death of other's recalls our own fate to us and forces us to prepare for it in some fashion. For these, and a thousand other reasons, we cannot allow the death of those close to us to pass unmarked. Therefore, these Middle Paleolithic period burials are *prima facie* evidence that, at least on this very fundamental topic, the Neanderthals thought as we think and acted as we act. As Freeman notes:

> Whatever their interpretation, these burials suggest a concern for the proper treatment and well-being of members of society beyond death's frontiers and the beginnings of complex ideological and social practices like those of fully modern man (Freeman 1980).

But are these finds legitimate or were the supposed intentional burials really accidents of the taphonomic process? That is, did the excavators at these sites confuse the results of the operation of natural, environmental formation processes with intentional human activity? Robert Gargett (1989) reviewed the excavation records from key Mousterian burial sites and in fact concluded that the archaeological evidence of purposeful disposal of the dead in the Middle Paleolithic period is insufficient to reject the hypothesis of accidental burial by natural process. Gargett makes a strong case for rejecting a number of skeletons reported to have been intentionally buried and is most convincing when discussing sites dug early in the 20th century under very less control. He casts his net too far in rejecting all such reports, however for, as Trinkaus notes, he is unable to explain how a number of Neanderthal skeletons:

> managed to be found in highly accessible upper Pleistocene rockshelters and caves in near-anatomical position and overall skeletal-part frequencies identical to those of recent cemetery samples. These partial skeletons retain many fragile elements largely intact, despite the ubiquitous presence of carcass-destroying carnivores...and rodents in the vicinities of the sites, the lack of evidence in most cases for sufficiently rapid natural sedimentation rates to shield them from scavengers, and the absence of comparably preserved nonhominid skeletons in similarly accessible upper Pleistocene locales (Trinkaus 1989: 183-184).

Thus, there is currently sufficient evidence to justify the conclusion that at least *some* Neanderthals began burying their dead in the Middle Paleolithic period. Beyond this, however, the ice becomes thinner. For example, in some instances, Neanderthal graves (as well as those of later *Homo sapiens sapiens*) have been found to contain skeletal remains tightly constricted into the fetal position. It has been suggested that the corpses of the individuals must have been tied and bound before being placed in the grave. We remain forever uncertain as to whether this binding was merely to enable the survivors to fit a large corpse into a small grave pit or was meant instead to prevent the dead person from leaving his grave and returning to the land of the living. It has often been asserted that this constricted burial position was meant by the Neanderthals to represent the position of the fetus in the womb and thus to suggest that they viewed burial to presage some form of rebirth. It seems unlikely however, that by the Middle Paleolithic period people would possess any such understanding of the position of the human fetus *in utero* (James 1957: 29).

Evidence for even more complex ideological development and ritual practice by Neanderthals was formerly thought to be present at the cave site of Drachenloch in the Swiss Alps. Early in this century, Emil Bachler reported the recovery of a series of cave bear (*Ursus spelaeus*) skulls stacked in wall niches and in what he supposed was a man-made stone cyst at the site. He interpreted these data as evidence of the worship and ritual hunting of the cave bear by Mousterian peoples. The drama of such a cult, together with its supposed survival in the bear worship of certain Siberian peoples and the Ainu of Japan, has resulted in the widespread acceptance of Bachler's interpretation. Bachler's primitive techniques of excavation and recording render it difficult to check his findings. However, recent comparative analysis of similar bone material from cave bear sites that were never occupied by hominids has lead Kurten (1976: 83-107) to the conclusion that Drachenloch was never anything but a cave bear lair and Bachler's piles of cave bear "trophy heads" probably accumulated naturally in the cavern *without* human agency. Binford (1983), studying bone attrition and patterns of bone accumulation by both modern people and scavenging animals, rejects Bachler's Neanderthal cave bear cult hypothesis for similar reasons.

Other evidence of ritual activity during the late Middle Paleolithic period has also come under suspicion for taphonomic reasons. The celebrated Burial No. 1 at the cave site of Monte Circeo near Rome is often cited as evidence for the existence of a "Cult of Skulls" among the Neanderthals of Europe. On the floor of an inner gallery of this site a Neanderthal cranium was discovered *in situ* surrounded by a circle of stones. The skull is described by its excavator (Blanc 1961: 125-126), as exhibiting conspicuous damage to the right temporal region appeared to have been caused by violent blow or blows suggesting the individual died a violent death. More interestingly, the skull appears to exhibit careful and symmetric incising around the periphery of its foramen magnum. The foramen magnum is the aperture at the base of the skull which connects it to the spinal cord. These "incisions" completely destroyed the foramen magnum and produced a sub-circular opening about 10-12 centimeters in a diameter. Blanc found that trophy skulls collected among head hunting peoples of Melanesia and Borneo exhibited similar mutilations around the foramen magnum, and concluded that the Monte Circeo cranium was such a trophy. He speculated that the Neanderthal had been ritually murdered, the base of his skull broken open, and his brain extracted and eaten in a ritual feast. Blanc even suggested that, since the skull rested in situ on its forehead with its mutilated base turned upwards, it may have served as a drinking vessel or container during the grisly ceremony. We will surely never know whether or not Blanc was correct. However, at the Monte Circeo site, as at Chou-kou-tien and Drachenloch, we cannot rule out the operation of natural, postmortem, taphonomic processes as the source of the pattern skull breakage and therefore must remain skeptical of Blanc's scenario.

A Neanderthal Flute? In asking whether the Neanderthals behaved as we do, are we not are really asking if they could *think* as we do? While this is certainly true, it puts the archaeologist interested in this question in a bad position. After all, thought is elusive. Armed only with screen and trowel, archaeologists, seldom recover the commodity in the course of an ordinary excavation. But of course, some excavations are extraordinary. The dry cave site of **Divje Babe I** appears to fall in this latter, extraordinary category. Divje Babe I, excavated recently by the Slovenian archaeologist Ivan Turk, is located in the Slovenia, a newly-independent republic that once formed the northern portion of Yugoslavia. From Stratum 8, which has been radiocarbon dated to 43,100 ± 700 BP., Turk recovered a heavily-gnawed fragment of a bear thigh bone which exhibited two complete holes along its cylindrical shaft and two partial holes on its gnawed distal and proximal ends.. The regular spacing of these holes suggested human workmanship and the excavators concluded that the enigmatic object had once been part of a bone flute.After being shown a photograph of the fragment, the musicologist Robert Fink concurred with this identification and took it one step further. Fink noted that the distance between the second and third holes on the bone is twice that between the third and the (partial) fourth. In musical terms, this distance means that the first interval represents a whole tone one, the second interval a half tone. Such a three note tone progression was most surely diatonic Although the fragment only contained four holes, Fink hypothesized that from their placement that, when complete, the instrument could produce four notes corresponding with the third, fourth, fifth, and sixth notes of a minor diatonic scale. A diatonic scale is composed of five whole tones and and two half tones. The familiar musical scale, *doe-re-mi-fa-so-la-te-doe,* is diatonic Most of the world's music is composed in this seven note diatonic idiom (Anon 1997:205).

If Fink is correctly identified, the Divje Babe I site has yielded the world's oldest known musical instrument and that instrument was made and used by Neanderthals. If the Neanderthals played flutes, is it too much to suggest that they also made music by vocalizing? This question will never be answered but whether they sang along with the flutist or not, the morphology of this instrument suggests that the Neanderthals used the same seven-note scale as do most anatomically modern people today. The Divje Babe I specimen is a flute is thus provocative evidence that the Neanderthals perceived musical tones and harmony as we do. His find thus begs an important question: if Neanderthal brains processed musical sounds and patterns as ours do, is it plausible to suppose that, in other crucial areas such as speech, our brain functions were entirely different?

Thus, if it is a flute, the Divje Babe I bone fragment has much to tell us about the nature of Neanderthal thought. But is it in fact a flute? Things are rarely so simple in archaeology. Once again, solutions to those elusive questions of mind and cognition must wait for the resolution of the more mundane issues of taphonomy. But the taphonomic evidence is ambiguous. The bone was clearly heavily chewed by a carnivore, probably a wolf, before its deposition but such chewing does not exclude the possibility of either previous or subsequent use of the bone as a flute by Neanderthals. Nonetheless, after carefully examining the specimen, Philip Chase and April Nowell (1998) reject the hypothesis that it is an artifact. In their view, the presence of carnivore gnawing combined with the absence of clear evidence of human workmanship such as tool marks on the specimen is decisive. With regard to the Bivje Babe I bone fragment at least, the question of whether Neanderthals made music can be pursued no further. If we are to accept that they did, we must first discover other, less ambiguous, bone flutes in a Mousterian context.

Western Eurasia in the Upper Paleolithic Period

The Mousterian industrial tradition disappears from western Eurasia in a time transgressive fashion between ca. 46,000 and 35,000 BP.:

1. In the Middle East, the transition occurs between 46,000 and 42,000 BP.at the site of **Kebara Cave** on Mount Carmel in Israel (Bar-Josef *et al.* 1996)
2. Westward across North Africa at **Haua Fteah** and other cave sites the Middle Paleolithic occupation disappears shortly before 40,000 B.P (McBurney 1967).
3. In Spain at the sites of **El Castillo** and **Abric Romaní** and in France at **L'Arbreda Cave** the Middle/Upper Paleolithic appears to have taken place between ca. 43,000 ± 1,000 and 37,000 ± 2,000 B.P. with the intrusion of the Aurignacian industrial tradition (Bischoff *et al.* 1994: 541-542).
4. In other parts of southwestern Europe, the Mousterian lingers until ca. 35,000 BP. with the appearance of the **Chatelperonian** or **Lower Perigordian** industrial tradition..
5. On the Russian end of the European Plain, the earliest Upper Paleolithic occupation appears to be no older than ca. 40,000 BP. (Anikovich 1992).

The final dissolution of the Mousterian industrial tradition marks the end of the Middle Paleolithic period and is correlated with the disappearance of the Neanderthals. The beginning of the subsequent Late or Upper Paleolithic period in western Eurasia is distinguished by dramatic change in stone tool form. Relatively brief compared to the two earlier subdivisions of the Paleolithic, the Upper Paleolithic period lasts a mere 25 to 35,000 years to close at the end of the Last Glacial Stage at around 10,000 BP. The period is roughly contemporary with the Later Stone Age of Africa and the Upper Paleolithic of eastern Eurasia. The earliest definite evidence of humankind in the New World occurs during the Paleo-Indian period (ca. 12,000 BP. to 10,000 BP.) and is contemporary only with the very end of the Upper Paleolithic in the Old World.

Despite its brevity, the Upper Paleolithic period was one of accelerating technology change and of profound transformation of human behavior and intellectual life in western Eurasia. During the Upper Paleolithic period, anatomically-modern people came into biological ascendancy and the pace of cultural change dramatically accelerated. The technology of the period, although still centered on stone, cames to include a host of new raw materials including bone, ivory, and horn. New techniques and tools were added to the technical repertoire of the period and, for the first time in prehistory, such tools were accompanied by numerous carved and decorated objects that qualify as the oldest surviving artistic tradition in the archaeological record. Portable art objects dating to the Upper Paleolithic period have been reported from most parts of the Old World. An even more dramatic aesthetic outpouring was the appearance of brilliantly painted cave murals at that time. Upper Paleolithic period cave paintings, which constitute the first great artistic tradition in human prehistory, are concentrated in the **Franco-Cantabrian region** of southwestern Europe. Finally, near the end of the Pleistocene epoch, massive animal extinctions took place. These extinctions had a profound impact on human subsistence as the Upper Paleolithic period came to a close.

Middle Paleolithic/Upper Paleolithic Transition Often in science, initial research quickly produces a simple solution only to have it dissolve into ambiguity as subsequent work reveals the true dimensions of the problem. Based on research done mostly in southwestern France early in this century, it was concluded that a linear developmental sequence could be traced from the Mousterian through the Upper Paleolithic industrial tradition called the Aurignacian. The Aurignacian was viewed as an indigenous European development out of which the later Solutrean and Magdalenian industrial traditions developed. Coeval cultures in North Africa were thought to have resulted from migrations of "Aurignacian" peoples out of France during the Upper Paleolithic period.

However, work in Europe over the last fifty years has revealed that the term "Aurignacian" as it was formerly used included a number of distinctive cultural traditions under a single heading (Peyrony 1933; Bricker 1976:136-137). It is now widely accepted that at least three major regional cultural traditions existed, roughly contemporary with and yet independent of one another, in western Eurasia at various times during the early Upper Paleolithic period. These three traditions include the Perigordian in northern and western Europe, the Gravettian in the central and eastern, and the Aurignacian present in some form across most of Europe except the extreme eastern regions of Russia. It now appears that the Lower Perigordian or Chatelperronian, not the Aurignacian, is the final expression of the Mousterian tradition in western Europe. The Aurignacian did not develop *in situ* in western Europe but arrived there fully-formed from somewhere outside shortly after the appearance of the Lower Perigordian (Bordes 1968: 155, 220).

Upper Paleolithic Southwestern Europe The Upper Paleolithic period begins in western Eurasia with the appearance of a new stone working tradition called the Aurignacian. Although traditionally dated to between about 34,000 and 33,000 BP., recent dates from the sites of **El Castillo** and **Abric Romaní** and in France at **L'Arbreda Cave** indicate that it began as early as between ca. 43,000 ± 1,000 and 37,000 ± 2,000 B.P. (Bischoff *et al.* 1994: 541-542). These new, early dates present interpretive problems as, in the Perigord, components of the typical Aurignacian appear stratigraphically directly atop both Mousterian and the much later Chatelperronian layers as well as on virgin bedrock in rockshelters newly formed in the middle of the Last Glacial Stage. These chronological problems have yet to be resolved.

Typical Aurignacian components have been recognized over a wide area of southwestern Europe and it is often suggested that these components have genetic or historical connections with the Aurignacian traditions of central and eastern Europe and the Near East. These similarities have lead Denise de Sonneville-Bordes (1973: 48-49) to conclude that "at this period France belonged to a sort of European community of which it formed only the Atlantic province."

The Chatelperronian The Lower Perigordian or Chatelperronian stone tool making tradition in southwestern Europe sometime between about 35,000 to 34,000 BP. The tradition, which is apparently unknown outside of Franco-Cantabrica, lasts between 2,000 and 3,000 years and then disappears entirely from the archaeological record after ca. 32,000 BP. Since the Mousterian industrial tradition is the product of Neanderthals, the transition to the Upper Paleolithic traditions is generally assumed to have been correlated with the appearance of *Homo sapiens sapiens* in Europe. Given the transitional position that the Chatelperronian tradition occupies in the archaeological sequence, the question whether or not anatomically-modern humans or Neanderthals were responsible for its manufacture is an important one. Unfortunately, controversy surrounds the association and identity of the scattered human remains recovered in the past from such Chatelperronian sites as Roc de Combe-Capelle and Grimaldi. However, the discovery in 1979 of the partial skeleton of a Neanderthal female buried in the Chatelperronian level at Saint-Cesaire in the Charente-Maritime district (Leveque and Vandermeersch 1981), suggests that these questions may be on the verge of resolution. After all, we now know of exceptions to both of the old equations of anatomically modern people with Upper Paleolithic tools, and Neanderthals with Mousterian ones in western Eurasia (Harrold 1981: 38). The Neanderthals apparently did not disappear at the end of the Middle Paleolithic but persisted into the early Upper Paleolithic period in southwestern France. Further, their persistence was not in some thinly populated peripheral area; "Saint-Cesaire is, in fact, situated in a region densely occupied during the middle and Upper Paleolithic" (Stringer, Hublin, and Vandermeersch 1984: 106). Like the Middle Stone Age industries in Africa, the Chatelperronian in Europe seems to represent a transition in human biology as well as technology.

It thus appears that the transition from the Middle to the Upper Paleolithic period was not a throughgoing cultural revolution that moved like the wind and everywhere swept away the old and replaced it with the new. Rather than producing a stark uniformity of homind form and cultural tradition, the transition seems at first to have created a biological and cultural **mosaic** of diverse and overlapping hominids and traditions in western Eurasia (Straus 1995). Recall, that in the discussion of the African Stone Age earlier we noted that both archaic and anatomically modern hominids have been found associated with MSA assemblages, a pattern that suggests a mosaic patterned emerged in the MSA/LSA transition there as well.

Chatelperronian Technology. Whatever the species of its fabricators, the Chatelperronian stone working technology definitely developed out of the preceding Mousterian industrial tradition and retained a number of its important elements. The Levallois technique continued to be used and Mousterian-style scrapers, points, and saw-toothed denticulates remained important tools (Delporte 1970; Bordes 1968: 147-148; Mellars 1973). However, a number of distinctive new tool types appeared in the industry including "backed" blades, in which the back of the blade is blunted by steep retouching, burins, endscrapers, perforators, and broad, tapered blades with curved, blunted backs called Chatelperronian backed knifes or points (Harrold 1981: 17-19).

Aside from the small quantity of ornamental carved bone pendants and teeth drilled or grooved for suspension recovered at the Arcy-sur-Cure site east of Paris (Leroi-Gourhan and Leroi-Gourhan 1965: 40), Chatelperronian occupations have not produced much more worked bone than the Mousterian that lie beneath them. However ochre appears to have been used far more lavishly than it was in the Mousterian (Harrold 1981: 37). Limited and fragmentary as it is, this is essentially the extent of the evidence of "non-utilitarian" behavior during the Chatelperronian. To date, none of the complex representational or symbolic art of the later Upper Paleolithic period has been attributed to this complex.

The Aurignacian As noted, the Aurignacian occupation of southwestern France apparently overlaps with the final phases of the Chatelperronian tradition there. It is supposed by Bordes (1968: 157-158) and others that the two industrial traditions existed side by side and yet independent of one another for perhaps a millennium. Bordes supposes that such isolation and insularity could exist between contemporary and proximate cultures because the artificers of the two assemblages belonged to different "tribes" or ethnic groups. Bordes Upper Paleolithic version of his Mousterian "parallel phyla" theory presents the same theoretical and practical problems for the Upper Paleolithic sequence that it did for the middle. Binford (1973) and others have sought to resolve by suggesting that the formal contrasts between such assemblages are due to "functional" or "seasonal" differences between the sites from which they have been recovered but the dispute remains unresolved.

Aurignacian Technology. Aurignacian tool assemblages contrast markedly with those of the Chatelperronian and later Perigordian industries in their high percentages of blades and fairly low percentages of burins and by the extensive, often elaborate, use of bone, ivory, horn and antler for tools, weapons, and jewelry. More significantly, the earliest cave art in prehistory was the work of the Aurignacians. They were apparently the first people on earth to express their aesthetic impulses through the drawing and carving of representational, schematic and symbolic images and they did so with powerful and vivid strokes of creative genius as if sensing the monumental edifice of human art that was to follow them. The media they chose were varied: limestone and other rock, bone, ivory and perhaps less permanent materials that have since vanished. Aurignacian art centered on three subjects: vulvas, human figures (generally female) and animals (Collins and Onians 1978: 11). Bas relief carvings or engravings of what appear to be human vulvas on stone slabs and cavern walls seem to be among the earliest in southwestern Europe (Wymer 1982: 255). Enigmatic "cup mark" depressions and deeply incised lines are often associated with these vulvae-shaped reliefs. A limited number of red and black painted vulvae, figures and animals are also attributed to the Aurignacians there as well (Sonneville-Bordes 1973: 47). Carved figurines appear earliest in the Aurignacian components of central Europe. The Vogelherd site in Germany produced the first animal sculptures carved in the round out of mammoth ivory. The Aurignacian components at the Hohlenstein-Stadel and Bruno II sites each produced fragmentary male statuettes (Hahn 1972: 262-263). The Aurignacian industries of Europe developed a complex system of signs and produced large quantities of jewelry and portable art objects. The jewelry and art objects are often recovered in Aurignacian burials associated with red ochre and other evidence of an increasingly elaborate burial ritual and religious ideology emerging within the tradition (Hahn 1972: 252). After supplanting the Chatelperronian in southwestern Europe about 32,000 BP., the Aurignacian tradition is replaced in its turn in western Europe by the middle and late Perigordian some time after 29,000 BP.

The Perigordian. The Perigordian IV phase apparently spanned the years from 29,000 BP. to 28,000 BP. in the Perigord during a climatic phase characterized by relatively mild and humid conditions (Laville, Rigaud and Sackett 1980: 228, 275). According to Bordes (1968: 152), it is characterized by (1) the disappearance of virtually all the Mousterian tool forms retained in the Chatelperronian with the exception of some types of side scrapers and (2) a great increase in the numbers of burins of various types, end scrapers on blades, and Gravette points ranging in size from large down to small "microgravettes" bladelets. While the Perigordian flint workers excelled at the various techniques for detaching blades from cores, they neglected to retouch finished tools or did so only poorly. Their bone industry was also undistinguished

(Sonneville-Bordes 1973: 47).

The most original element of the later Perigordian tradition are the the so-called **Venus figurines** which appear within it. These figurines depict nude or semi-nude females of a variety of ages and shapes. The exaggerated character of the breasts and buttocks of some of these statuettes has lead many scholars to conclude that they are "fertility" images but they are just as likely to represent more complex female deities or to be portraits of real, flesh and blood women. It is also entirely possible that, although the Perigordians didn't know much about art, they indeed knew what they liked.

Reported dates for the Venus figurines range from 29,000 to 14,000 BP. However, a critical review of the evidence by Gamble (1981: 97) suggests that the bulk of them date to a much narrower range between 25,000 and 23,000 BP. The figurines are carved in a variety of media and are reported from western, central and eastern Europe. The fact that these figurines are so similar across this vast area is striking and suggests that the Upper Perigordian tradition of southwestern Europe was part of the larger "Gravettian" cultural province centered in Western Eurasia.

The tradition of cave painting begun in the Aurignacian continued in the Perigordian period in western Europe although, according to Collins and Onians (1978: 4), the emphasis shifted away from female form and sexual apparatus towards the realistic representation of animals. Engravings and carvings of ivory, antler and horn still continued to be made in great quantity. By the final phase of the Perigordian, bone tools were particularly well; characteristic bone tools include grooved and un-grooved wands, points with hooked butts, small eyeless needles, and the ***baton-de-commandement***, or bone handle with a hole or socket at one end. The function of this tool is unknown but it is thought to have served as a kind of wrench used in straightening wooden dart shafts.

Despite it ubiquity during early times, the final phase of the Perigordian tradition (Perigordian VII or the Proto-Magdalenian) is known at only a few sites in western Europe. The tradition disappeared altogether from the area sometime between 22,000 and 21,000 BP. (cf., Movius 1974: 95; Laville, Rigaud and Sackett 1980: Figure 8.1, 220, 277).

The Solutrean. A new cultural tradition emerges in southwestern France sometime after about 21,000 BP. This industry is termed the Solutrean after Solutre, an isolated open air site located near Lyon on a tributary of the lower Rhone river that produced large quantities of horse bone in association with the Solutrean component. The Solutrean tradition is of relatively short duration; it disappears from the region around 17,500 BP. after only 3,500 to 4,000 years. Its geographic distribution is also limited when compared with traditions like the Aurignacian, Perigordian and the Magdalenian. Unlike those earlier and later traditions, the classic Solutrean has been found almost entirely within an area bounded on the the south by the Spanish Pyrenees, and on the north and east by the Sarthe the lower Rhone river valleys in France.

The Solutrean stone tool repertoire is characterized by a high proportion of specialized blade tools including endscrapers, and the celebrated unifacial and bifacial **foliates** or leaf-shaped projectile points and blades. These implements were made on both unifacial and bifacial blades and finished by means of a highly regular form of flaking called **Solutrean retouch**. This retouch resulted in flat, narrow, parallel flake scars on the surface of the blades and was probably done by pressure flaking, that is, by applying pressure to the blade surface with a bone tool, rather than by direct percussion with a bone baton. Through time foliates were made in a variety of forms: laurel leaf-shaped, willow leaf-shaped and in an asymmetrical shape with a "shoulder" on one side only. Foliates are large objects; the largest yet reported is nearly 14 inches long (Bordaz 1970: 76). Certainly this specimen, and others nearly as long and thin, was simply too big and too fragile to be put to any practical use. Instead, it may have served as a token of status or wealth much as did the widely-traded shell jewelry of the Trobriand Island Kula Ring described by Malinowski. Alternatively, these such blades may simply been made as an expression of bravura craftsmanship by master stone workers. Whatever their original purpose, the spectacular foliate blades make the Solutrean the most distinctive lithic tradition in European prehistory. Excepting the foliates, however, Solutrean stone tools are rather undistinguished and this has led Smith (1965: 390) to assert that most Solutrean assemblages consist of two basic classes: a rather conservative pool of "ordinary" artifacts like endscrapers and perforators and a highly flexible "experimental" or "mutating" group of specialized artifacts.

The technical schizophrenia created by the presence of the foliates and the special retouch has been taken as evidence by both sides of two conflicting interpretation of the Solutrean. One side argues that the tradition was introduced into southwestern Europe by foreign invaders from say, the contemporary blade cultures of North Africa or Central Europe. The alternative is that the tradition was produced by indigenous peoples who merely inducted a few distinctly foreign ideas and practices into their preexisting local cultural repertoires (Smith 1964:47). Careful analysis of the stratigraphy and tool systematics at the key French Solutrean site of Laugerie-Haute indicates that both the foliates and the distinctive retouch have an in situ developmental history within the Perigord itself (Smith 1964:53). It would thus appear unnecessary to seek developmental prototypes or donor traditions for the Solutrean to explain its invasion or adoption and diffusion in southwestern Europe.

Solutrean craftsmen worked with bone, horn, and ivory to create bracelets, pendants and other decorative items. They also made bone **"spatulas"** that often bear enigmatic patterns of grooves or notches which French prehistorians have traditionally referred to as *marques-de-chasse* or hunting tallies. Recently, at least some of these marks have been interpreted as part of a system of calendric reckoning and record keeping.

The most important bone or ivory implement made by the Solutreans was the eyed-needle. This tool was apparently invented by the them and its significance can hardly be over-estimated. Without it, how difficult must have been the production of weatherproof, form-fitted clothing. Further, the small size of many of these needles suggests that they were used for tailoring and fine-stitching clothing, for decorative needle craft or fancy stitchery, or for sewing decorative elements like shells or ivory buttons on garments (Smith 1964). The elaborate shell covered cloak, of which traces were found on a burial of an apparently high status individual at the Sungir site in the Soviet Union, is but one example of the manner in which the needle could be used to produce garments capable of keeping out the cold and at the same time marking the special status of the individual who wore it.

Solutrean in Spain The Solutrean tradition in the Spanish Pyrenees differs somewhat from that in France. In addition to appearing later there, the Spanish Solutrean is characterized by a number of unique stone tool forms not found in the French heartland. Some of these forms include bifacial foliate points with concave bases or asymmetrical blades and small barbed and tanged points. The distribution of these blades and points appears to be variable. Concave-based points are restricted in distribution to the narrow coastal zone of northern Spain between the Cantabrian Sea (Bay of Biscay) and the Cordillera, from Central Asturias to the French border (Straus 1977: 34). The small barbed and tanged points on the other hand, are generally found further east in the Pyrenees and south along the Mediterranean coast as far as Parapallo cave in Valencia (Smith 1973: 68). Recent radiocarbon dating of collagen from bone samples associated with the upper Solutrean occupation at Parapallo place the appearance of these points at around 18,080 BP. (Davidson 1974). These have been interpreted as arrow points and compared with supposedly similar specimens from the Aterian in North Africa.

The brilliant Solutrean culture disappeared round or after about 18,000 BP. in the cold, dry climatic conditions of the worldwide glacial maxima for reasons that are unknown. The oldest levels of the next tradition, the Magdalenian, often overlie the Solutrean without a break so much that it is difficult to trace the dividing line between these two levels except by the typology. But within the limits of what we know about the Paleolithic culture, it is impossible to think that the one culture might have evolved into the other. The implements are completely different. If the Solutrean represents one of the zeniths of stone working, the lowest Magdalenian is certainly one of its nadirs (Bordes 1968: 159-161).

The Magdalenian. The last great industrial tradition of the Upper Paleolithic period in northern and western Europe appears in the Perigord between 17,500 to 17,000 BP. The tradition disappears more than 7,000 years later near the close of the Pleistocene epoch. The Magdalenian tradition, named after its type site of La Madeleine rockshelter in France, appears to be limited to south central portion of that country in its earliest phases. Later, the industry spread east across the Saone and Rhone rivers and southward deep into Spain. By late glacial times, the Magdalenian was present as far north as Belgium and the lower Rhine (Sonneville-Bordes 1973: 52-53). As the Pleistocene epoch closed, a Magdalenian-like hunting tradition called the Hamburgian appeared in the wake of the retreating continental ice mass in Germany. The Hamburgian developed into the Ahrensburgian tradition there as the ice, and the Ice Age world, disappeared.

Magdalenian Technology When compared with the virtuosi stone tools of the preceding Solutrean, Magdalenian stone work appears crude and carelessly made. Magdalenian craftsmen preferred to apply their ingenuity and skill to crafting tools of antler, bone, ivory and horn. It was in these media that they achieved their greatest successes. Two tool types commonly made of antler or bone, the **harpoon** and the **spear point**, were particularly popular among the Magdalenians. Both implements were used to tip throwing or thrusting spearshafts. However, harpoon points are distinguished from spear points by their barbed heads and the presence of a small hole at their base to which a raw hide cord was presumably affixed. A cord attached to this hole would have allowed the hunter to retrieve the harpoon head, preferably with the animal still impaled on it, after he had thrust or thrown the shaft into its body. Harpoons are especially useful in hunting sea mammals like whales and walruses but can be used against land animals as well.

Magdalenian harpoon and spear points have received a great deal of scholarly attention from prehistorians. In addition to being objects of great beauty and interest, the style of these implements changed over time and their formal changes have proven useful in dating Magdalenian occupations relative to one another by seriation. Not surprisingly, therefore, studies of these implements have generally focused on their stylistic and aesthetic attributes. Yet, surely the popularity of these implements among the Magdalenian and the fact that they replaced projectile points made of stone must have been due to their effectiveness rather than simply their beauty. Convinced of this, James Pokines set out to measure the functional and technical characteristics of Magdalenian antler projectiles by experimental means. Pokines made twenty replicas of Cantabrian Lower Magdalenian antler spear points, hafted them to wooden shafts and then proceeded to systematically throw these implements at a dead animal target until they broke. His replication experiment conclusively demonstrated that antler points were highly durable and that he concluded that:

> their great strength and flexibility translate into a longer use life and an ability to penetrate (by grazing past or punching through thin bone) into body cavities housing vital organs. Their durability far surpasses that of experimentally reproduced and propelled stone points. While antler points represent a larger time investment, their greater reliability correlates with the dominant hunting strategy of the Cantabrian Lower Magdalenian: a shift towards the planned harvest of large herds of red deer and ibex at one time (Pokines 1998: 884).

Thus, Pokines work suggests that the Magdalenian neglected stone working because they had found antler projectile points to be functionally superior. As noted, however, Magdalenian craftsmen were not content with simply producing durable and efficient *plain* tools. Rather, they commonly chose to decorate these antler objects (as well as others of ivory, horn, bone and wood) with elaborate engravings and carvings. At the same time, they used these media to made jewelry, clothing decorations, palettes, and a variety of other objects whose purpose is lost to us. But the devotion of the Magdalenians to beauty did not end with merely with the decoration of portable objects or the manufacture of jewelry. They did more, far more and it to their greatest chievements in art that we now turn.

Magdalenian Aesthetics and Religion. Craftsmanship aside, the clearest expression of the Magdalenian genius is found in their splendid polychrome cave paintings, wall engravings, sculptured friezes, and, their rare modeled clay statues. Upper Paleolithic period art, especially cave wall painting, reached its apogee during the Magdalenian occupation of Altamira, Lascaux, Tito Bustillo, Isturitz, Angles-sur-l'Anglin and numerous other sites in southwestern Europe (Dickson 1990). The richness of the aesthetic impulse and the complexity of the intellectual life during the Magdalenian is reflected in other ways as well. Magdalenian burials are far more common than those of earlier periods and the ochre and grave goods often included with these burials hint at an elaborate cosmology, grave side ritual and mortuary cult. Further, traces of the intricate system of notation and calendric reckoning thought to have been present in the Solutrean appears to have continued to be used in the Magdalenian (Marshack 1972).

Perhaps the most striking hint of the complexity of the spiritual life of the Magdalenians has come from the site of **Cuevo del Juyo** in the Spanish province of Santander. Freeman and Gonzalez Echegaray (1981) discovered an elaborate mound constructed of stone, clay and sand within this cave. Associated with the mound were large quantities of animal bones and a large block of stone which appeared, to the excavators, at least, to have been crudely sculpted into a feline face. The placement of the mound in the cave's interior, the central position of the stone "face" and other unusual features, suggests that the site served as a shrine or sanctuary. The date of 13,920 $\pm$ 240 years BP. from a contemporary layer in the cave's main gallery places the mound's construction near the end of the Magdalenian sequence.

The final phase of the Magdalenian closes with the onset of the Holocene epoch sometime before ca. 10,000 BP. Close analysis of cave stratigraphy and radiocarbon dates from Cantabria leads Butzer (1981: 177-178) to conclude that, in northern Spain, at least, the final Magdalenian phases overlapped the subsequent Epi-Paleolithic or Mesolithic Azilian tradition by as much as 2,000 years before disappearing.

The Azilian. The magisterial terminal phase of

the Magdalenian tradition disappears from northern and western Europe with the waning of the Late Glacial period. In both northern Spain and southwestern France it is overlapped and gradually succeeded by the Epi-Paleolithic or Mesolithic tradition known as the Azilian. The Azilian appears sometime after about 12,500 BP. during the Bolling Interstadial oscillation and fades after 9500 BP. with the end of the Younger Dryas (Butzer 1981: 177). Azilian occupations are generally found directly on top of Magdalenian occupations. The technical base of the Azilian clearly descended from the earlier complex and thus was derived from a cultural adaptation to the rigorous environments of the Late Pleistocene. Nonetheless, new elements in the Azilian technical repertoire suggest that the tradition was not just a continuation of the earlier ways; rather, it represents a transition and transformation of Upper Paleolithic life in response to the more temperate and humid condition of the post-glacial epoch.

The most dramatic cultural change to take place in the Azilian is the disappearance of the magnificent Upper Paleolithic parietal art tradition. Azilian art, at least what has survived into our time, consists of small, smooth pebbles painted (and occasionally engraved) with red and black lines, dots, and borderings. Less common designs include simple geometric elements such as crosses and chevrons. Prehistorians have often made invidious comparisons between the magnificent polychromatic cave painting and engraving of the Magdalenians and the paltry pebble art of the Azilians (cf., for example, Clark and Piggott 1965: 144; Sonneville-Bordes 1973: 53). Suggestions that the pebbles merely served as gaming pieces merely adds to the image of Azilian decline; instead of getting back to the serious business of cave painting, they just sat around and gambled!

However, recent work by Couraud suggest that these painted pebbles with their enigmatic symbols are in fact the material remains of a system of counting or notation. Couraud undertook a systematic reexamination of some 1,967 Azilian pebbles. Within this corpus, which included virtually all of the artifacts available for study in Europe, he identified 16 different signs. Potentially these 16 signs could be combined into 246 binary combinations. However, only 41 such combinations were actually used on the pebbles. This appears too low to be the result of chance. Rather it suggests that formal rules governed the placement and combination of the symbols. That is, there appears that a "grammar" of some kind governed the combination, placement, and order of the symbols on the Azilian pebbles. It is Couraud view that the meaning of a symbol was largely dependent upon the position it occupied on the pebble. Yet the meaning of the messages remains elusive. The possibility that the dots and lines were numbers has been recognized virtually since the discovery of the painted pebbles in the late nineteenth century. However, Couraud discovered an apparent pattern in these "numbers." Fully 85 percent of the pebbles represented digits between one and four. Some 87 pebbles contained symbols that could be read as amounts ranging from 21 to 29 or their multiples, numbers that may correspond to lunar phases or lunations and this has been taken as further evidence that systems for recording the changing phases of the moon were developed during the late Upper Paleolithic (Bahn and Couraud 1984: 157-158). In any event, with the northward retreat or extinction of the great herds, the attendant reforestation of Europe, and the rising worldwide sea level, the Upper Paleolithic period comes to a close in the Azilian. But whatever else the Azilian pebbles signified, they mark the end of the Stone Age world in northwestern Europe.

Upper Paleolithic Tools and Facilities

Although still a stone technology, the tool assemblages of the Upper Paleolithic period in western Eurasia contrasts with the industries of the Middle Paleolithic period there on a number of critical dimensions. As contrasted with the earlier period, stone assemblages of the Upper Paleolithic contain:

1. tools made predominantly on blade flakes,
2. new and often highly standardized tool forms,
3. composite or multi-component tools,
4. artifacts made of shaped bone, ivory, wood, and antler.

Blade Tools A stone blade is merely a flake whose length is twice its width. Flakes struck off stone cores as long, linear blades can be quickly transformed into such useful tools such as end scrapers, backed blades, and burins with little additional flaking. Blades are found in limited numbers in the assemblages of earlier periods and tools continue to be made on ordinary, non-elongated flakes in the Upper Paleolithic period. However, blade tools clearly came to dominate technical repertoire at the late end of the Paleolithic Experimental studies os stone tool manufacture indicates that the long, this sharp, blades of the period must have been struck from their cores with a "soft" hammer or wood, bone, or antler billet rather than another stone. However, the exact technique used in Upper Paleolithic glade production cannot be known with certainty since experimental work indicates that such blades may be produced either by "direct percussion," in which a hand-held core is struck directly with a billet, or by "indirect percussion," in which a punch or chisel made of wood, antler, or bone positioned on the edge of the core is struck with a hammer (Newcomer 1975). Techniques of retouch remain larger unchanged during the Upper Paleolithic (Mellars 1973; White 1982: 169).

New Tool Forms. Upper Paleolithic assemblages contained new, often highly standardized, tool forms (Mellars 1973; but cf., Binford 1973; White 1982: 169). Most commonly these new tool forms fall into three general classes: burins, backed blades and various types of endscrapers on blades (Bordaz 1970: 57). Of these three, the beaked burin or graver was probably the most significant since this tool enabled craftsmen to incise deep grooves in such resistant material as bone, antler, ivory or horn and pry up and remove long strips of it by the so-called "groove and splinter method" (Bordaz 1970: 71). In this fashion, Upper Paleolithic peoples could utilize a virtually inexhaustible set of resources in a new way. I n addition, artifact types become standardized for the first time during the Upper Paleolithic period. For example, Solutrean levels we have produced laurel and willow leaf points that are almost exact duplicates of one another. The same is true for some Magdalenian antler or bone harpoons. Unprecedented control was being exerted over morphological and metrical attributes of these implements (Freeman 1975: 253).

Composite or Multi-component Tools The spear thrower or **atlatl** is present at least by the Magdalenian period at the end of the Upper Paleolithic period in western Europe. The spear thrower is a rudimentary tool for increasing the propulsion of spears or darts. It consists of two basic parts: a stick or board generally less than two feet in length equipped with a notch or hook to engage a projectile shaft, and a "dart" or spear shaft generally 50 to 70 percent longer than the board which cradles and then propels it. The spear shaft is first engaged to the back of the throwing board and then both are held in the user's hand. The spear is then cast from the board with a sharp, over hand motion and released with a snap of the wrist. The purpose of the throwing board

> is to prolong contact between the spearman and his missile and thus allow him to concentrate a greater percentage of the total force of his throw behind the dart. In using the throwing board, a spearman is in effect lengthening his reach and therefore applying force to the dart for a moment longer than he would be able to do with his arm alone (Dickson 1985: 10, after Howard 1974: 102).

Experiments with the spear thrower suggest it increases the range of a spearman by as much as 40 percent over tosses with the unaided arm.

Two other key composite tools present in the Upper Paleolithic include: (1) small, tanged projectile points found in the Levantine Solutrean assemblages that probably indicate the presence of the bow-and-arrow at that early date, and (2) the Magdalenian antler and bone harpoons and spears points used with detachable foreshafts in discussed earlier.

Artifacts of Bone, Ivory, Wood, and Antler Materials like bone, antler, and ivory must have been abundant and readily available to hunting and scavenging peoples throughout the Paleolithic period and Freeman (1978: 32) and White (1982: 171-172) suggest that archaeologists may often have failed to recognize evidence of crude but intentional modification of bone during the excavation of Middle Paleolithic period sites. But, if the peoples of the preceding Lowerand Middle Paleolithic made use such materials in tool manufacture, their utilization appears was apparently neither extensive nor systematic. The situation changes dramatically in the Upper Paleolithic Period and bone, ivory, wood, and antler artifacts of all kinds become ubiquitous. This abrupt increase is presumably the result of the development of the burin and the grove and splinter method of obtaining strips of workable bone, ivory and antler.

Finally, bone, ivory, antler, animal teeth, shell, stone and other materials were perforated for suspension and were probably used as pendants, necklaces, and other items of adornment. These objects are the first and oldest definite jewelry in the archaeological record. Another Upper Paleolithic period development that was perhaps related to the appearance of personal ornaments is the first appearance of amber, flint, marine shells, ocher and other materials from exotic or non-local sources (Mellars 1973). Such exotics could be obtained either through trade or travel across long distances but, in any case, neither kind of movement appears to have been common during the Middle Paleolithic period (White 1982: 172).

Facilities and Shelters The most abundant archaeological evidence of Upper Paleolithic shelters and settlements in the Old World is concentrated in the Franco-Cantabrian region of southwestern Europe. Interestingly, archaeological excavation and survey, underway in Franco-Cantabria since before the turn of the century, have turned up significantly more Upper Paleolithic period sites than those dating to any of the previous periods (Mellars 1973: 268-270; Straus 1985: 504). The increase in the gross number of sites has often been taken as evidence human population grew markedly during the Upper Paleolithic period. However, White (1982: 172) demurs on the grounds that simple site tallies do not give sufficient weight to multi-component sites, that some of the temporal assignments of some of the sites have proven to be in error, and that the effects of sampling error cannot be discounted.

Upper Paleolithic sites are also larger in area when compared to sites of the previous Middle Paleolithic period. This increase in site size has also been taken as indicating Upper Paleolithic populations were growing based on the assumption that an increase in the gross numbers of hunting and gathering people utilizing a location will result in the discard debris over a larger area, and therefore, in the creation of larger sites in the archaeological record (Mellars 1973: 271-272). In fact, support for this hypothesis was provided by Yellen (1977) who mapped and counted the occupants of camps occupied by !Kung Bushmen, a contemporary hunting and gathering people living in southern Africa. However, the same study reveals that larger debris scatters also result when the *same* number of individuals occupy the same location for a *longer* periods of time. It is thus possible that the increased size of Upper Paleolithic sites resulted from longer occupancy.

At least some of the larger sites dating to the period probably represent seasonal aggregation sites (White 1980, 1982: 171-172; Conkey 1981). A common pattern among historically-known hunting and gathering bands was to live as dispersed, independent groups during most of the year but to congregate seasonally or occasionally in large, joint camps. Commonly these camps were in proximity to some abundant wild resources. This pattern may have begun to emerge in the Upper Paleolithic period. If so, it suggests that the subsistence and social organization (the "socio-ecology") was becoming more complex, cosmopolitan, and effecient than that of the Middle Paleolithic period.

Abri Pataud A rockshelter, what the French call an *abri sous-roche*, is an undercut reentrant along the base of a bedrock cliff (Farrand 1975: 2). Recall that we provided an example of the archaeological interpretation of a the Coxcatlan rockshelter was provided in Chapter 5. The rockshelters of Franco-Cantabria are largely of Plio-Pleistocene age and appear to have been formed both by solution erosion and frost-shattering. The chemical and mechanical erosion of limestone cliff faces created overhanging rockshelter ledges. Once formed, the rockshelters generally began to accumulate sediments. Commonly, these sediments were insulated to some degree from weathering agents active in the exterior environment and thereby preserved. Humans and animals have sought refuge under the overhanging ledges of the Franco-Cantabrian rockshelters for millennia and have left debris behind in the accumulating sediments. As a consequence, these rockshelters are excellent sources of archaeological and ecological data, a fact not lost on either archaeologists or pothunters who between them have excavated nearly two hundred such *abris sous-roches* in the Department of the Dordogne of southwest France alone (Sackett and Gaussen 1976)!

The site of Abri Pataud in the Dordogne region of southwestern France is a good example of a rockshelter occupation site. Abri Pataud is located in the village of les Eyzies about 100 kilometers (about 62 miles) from the coast of the Atlantic ocean. It is nearby the celebrated Paleolithic sites of **Lascaux cave** and **La Ferrassie** and **Laugerie** rockshelters. The diverse and complex deposits at Abri Pataud reach some 9.25 meters in depth and have been divided by excavators in to fourteen major human occupation horizons or levels called *couches* in French. Virtually the entire Upper Paleolithic cultural sequence of the Franco-Cantabrian region from the Aurignacian to the Proto-Magdalenian is represented in the fourteen *couches* at this site. Radiocarbon dates from these deposits indicate that the occupation of the rockshelter fell between ca. 34,250 ± 675 and 20,400 ± 450 years BP. (Mansouri, El Fouikar and Saint-Martin 1996).

Systematic excavation undertaken at Abri Pataud in the 1960s revealed, among other things, a series of superimposed sediments consisting of fragmentary limestone debris (or *eboulis*) that resulted from the frost-shattering of the surrounding bedrock, windblown sediments carried into the shelter from the nearby river flood plain, and stone tools and tool making debris, worked and unworked animal bone remains, and human skeletal remains introduced into the shelter by its' human occupants (Farrand 1975: 16). The human debris was found on living floors and in association with fire hearth dating from the Early Aurignacian through the Perigordian period to either the Proto-Magdalenian or the Lower Solutrean between about 34,000 BP. and 22,000 BP. (Movius 1975). Chemical and physical analysis of the sediments that formed these *couches* enabled the excavators to reconstruct the nature of the environment during each of the successive human occupations of Abri Pataud (Farrand 1975: 23-24).

The configuration of fire hearths in this sequence of floors is particularly interesting. The excavator, Hallam Movius, noted that the layers dating to the Early Aurignacian (*Couches* 14 to 7) tended to be small and closely spaced, while those in the later Perigordian IV layer (*Couche* 3) were somewhat larger and were laid out at roughly two meter intervals paralleling the rear wall of the rock shelter. Movius interpreted these contrasting hearth patterns as reflecting markedly different community organizations between the Early Aurignacian and the Late Perigordian IV. He suggested that the residence unit at Abri Pataud during the Early Aurignacian was perhaps limited to one extended nuclear family. In contrast he concluded that the linear alignment and even spacing of the hearths in *Couche* 3 indicated that a "longhouse" form of community pattern probably characterized by a lineage kinship system was present during the Perigordian VI occupation of the rockshelter. He viewed other, larger "bonfire-sized" hearths in the Perigordian as consistent with the view that a larger social aggregate was present at Abri Pataud near the later end of its occupancy (Movius 1966: 320-321).

However, in a recent discussion of the use of ethnographic analogy in archaeological interpretation, Lewis Binford (1983: 162-163) suggested that these configurations of hearths might be seen in a different light. Binford (1983: 144-149) claims to have isolated a number of uniform patterns of spatial use in the ethnographic record. According to him, these patterns have remained more-or-less unchanged through time and across culture because the size, shape and functioning of the human body has remained roughly the same through much of prehistory. For example, he presents the dimensions of a "standard bed sizes" which he asserts is essentially the same in all cultural systems and all time periods. When he superimposed the area of his standard, one-person bed size in the spaces between the five small Early Aurignacian fire hearths located behind a cluster of roasting pits along the back wall of the shelter in *Couche* 11 at Abri Pataud. The spacing of these hearths corresponds closely to this standard. In addition, Binford notes that,

> among modern groups...the alteration of hearths and single beds typifies the sleeping patterns of all male hunting camps. The combination of this form of bed distribution with the presence of roasting pits located in front of the seeping area suggests to me that, at this date, the site was not used for residential purposes (as Movius proposed), but was a temporary camp (Binford 1983: 163).

The pattern of hearths in *Couche* 3 which Movius took as evidence of a longhouse community pattern also was more likely the result of the spacing of hearths in a hunter's camp. The fact that these hearths were larger and were closer to the back of the rockshelter than those of the Early Aurignacian *Couche* 11 probably merely indicated that the Perigordian hunters camped at Abri Pataud during the winter and therefore built bigger fires and aligned themselves parallel to the back wall of the shelter rather than perpendicular to it (see Figure 15, from Binford 1983: 162). If Binford is correct, his analysis suggests that Abri Pataud, and perhaps many of the other rockshelters of Franco-Cantabria, may only have been used by small task groups as seasonal camps or as limited activity areas.

Modfied Rockshelters Upper Paleolithic peoples in Franco-Cantabrian did not always remain content with the natural features of the rockshelters and sometimes altered them. For example, the Early Aurignacian occupants of Cueva Morin near the Cantabrian coast of northern Spain undertook fairly extensive modifications of their site. They first cleared a great deal of accumulated occupational debris and cave sediments from the floor of the cave in order to create a level building surface and then dug a 2 by 4 meter rectangular pit to the depth of 1.7 meters or more. In the eastern corner of the pit they scooped out a shallow basin for a hearth and cut a narrow slit or draft hole in the side of the pit to allow the smoke to escape. A hut may have been built around or within the pit but no postmolds or roofing material was recovered from the excavation of the pit so we can only speculate that walls and a roof once covered it. Outside the pit a line of six postmolds were recognized. These posts may have formed a wooden wall or fence that extended out from the hut at its northeast corner. Two burial tumuli were located several meters southeast of the pit outside of the area enclosed by the wooden fence. The excavators concluded on stratigraphic grounds that these tumuli were contemporary with the fence and pit (Freeman and Gonzales Echegaray 1970). In addition to Cueva Morin, a number of other Franco-Cantabrian rockshelters were modified and improved through the construction of stone flooring, stone-lined hearths, low stone walls, wind-breaks, and perhaps even erection of small tented structures within them (Wymer 1982: 233-234).

The rockshelters and caves of France and Spain certainly contain dramatic and convincing evidence of human settlement and use but their excavation has enshrined the people of the Paleolithic period forever as "cave men" in the popular imagination. Countless books, movies, and cartoons, depict these people as little more than rude skulkers in the dark places under mountain ledges. As we have seen, Paleolithic peoples certainly used rockshelters and caves extensively in the karst regions where such landforms occur. Yet, as Binford's interpretation of Abri Pataud suggests, even in those regions, "cave dwelling" may have been but one part of more complex settlement patterns that, for a variety of reasons, have proven difficult to reconstruct from the archaeological record. Although rare, convincing archaeological evidence of artificially-constructed, open-air dwellings has been recovered in recent years in southwest Europe. For example, the site of Solvieux in the Isle River Valley of southwestern France appears to contain 14 separate open-air occupations dating to the Upper Paleolithic period. The remains at the site are not spectacular: they consist of roughly rectangular scatters of cobbles. Many of these cobbles show signs of fire cracking, suggesting they were part of hearths. These rock scatters have been interpreted as "pavements" which may have served as the floors for temporary structures of wood, skins or brush (Sackett and Gaussen 1976; Taranik 1977). The open-air location of the occupations at Solvieux immediately exposed any the organic debris around the site to decay. Consequently, few clues to the diet of the site's inhabitants or the season of their occupation have been recovered. It seems plausible that these sites and others like them were occupied during the spring or summer by peoples seeking to exploit the local plant and animal resources of the Isle valley, possibly by peoples who utilized the nearby rockshelters during the winter.

Life on the Steppes The unglaciated regions north and east of Franco-Cantabria across the great loess plains of the Mammoth Steppes of central Europe to the Ukraine and Siberia are not blest with numerous caves and rockshelters. To survive in these regions, Paleolithic peoples had to build entirely artificial dwelling facilities. Unfortunately, the nature of the facilities used over much of the territory beyond Franco-Cantabria is unknown. Perhaps this is because Upper Paleolithic hunting peoples there made insubstantial structures or used portable skin tents that would be invisible in the archaeological record.

Evidence of open-air structures may also have been destroyed by the actions of E-transforms. The central European loess plain was characterized by rigorous glacial and periglacial environments for prolonged periods during the late Pleistocene epoch. Much of the permanently-frozen ground there was subject to such arctic and subarctic soil processes as solifluction (the slow flowage of water saturated soils and sediments downslope over a frozen substratum) and frost heaving (the lifting of surface deposits by the internal action of frost). Both of these processes distort or destroy archaeological remains. However, the most probable reason structures at open sites of the period are not more abundant in the archaeological record is that many are too deeply buried to be located in ordinary archaeological surface reconnaissances. For example, the remains recovered at Mezin, Mezhirich and other sites in the Dnepr river valley of the Ukraine were found beneath loess deposits more than 10 meters thick (Gladkih, Kornietz and Soffer 1984: 164). The oldest levels at Moldova on the Dniester river valley occurred between 6 and 8 meters below the current land surface. In most cases, the original discovery of these sites occurred following their partial exposure by natural erosion and river downcutting or by human activity like road building or clay quarrying (Klein 1974).

The most striking feature of the structures recovered at these and other sites in the Ukraine and Siberia is the use of large bones, mostly of reindeer and mammoth, in the construction of large, oval-shaped foundations, lower walls, and frameworks. Particularly elaborate structures were built at Mezhirich. There

> the bones were not just packed into the structure at random. Instead, their geometry was exploited as an element of design. Skulls were placed at regular intervals in an arc or a full circle to form the foundation of the interior base wall (and the) foundation wall was extended upward and outward with bones arranged in an intriguing architectural anatomical pattern (Gladkih, Kornietz and Soffer 1984: 167).

A settlement of five such oval mammoth bone structures varying in size from four to seven meters in floor area have been recovered at Mezhirich. The builders used different kind of mammoth bones and arranged them in unique, symmetrical geometric patterns in the upper parts of the walls of each of the five buildings.

At other central Eurasian sites, the bone frameworks were built around pits and surrounded with earthen embankments and stone slabs. Storage pits and permanent stone heaths have sometimes been found within some of these structures (Shimkin 1978: 226).

No direct evidence concerning the nature of the roofs of these structures has been recovered. However, most of them lack postmolds in their floors and this suggests that skins, brush or other light material was stretched or piled over the bone frameworks.

The massive quantities of bone used in the construction of these frameworks has sometimes been interpreted as the gristly remains of a single huge hunt. This is not the correct. Bones recovered from the same structures commonly return different radiocarbon ages. This indicates that the builders collected, and probably stockpiled and reused, bones of mammoth that had died or been killed on *many* separate occasions. But a major question remains unsettled: what is the degree of sedentism represented by the Ukrainian and Siberian mammoth bone structure? Their construction surely involved a great investment of time and energy. Does this investment indicate that the structures were occupied more-or-less permanently on a year round basis? Many scholars assume that it does but, at present we cannot be certain.

Economy of the Upper Paleolithic Period

Reindeer Hunters By the end of the Upper Paleolithic period, subsistence systems of peoples like the Magdalenians in western and northern Europe appear to have centered on large herd animals such as reindeer, bison, horse, and wild cattle, supplemented with ibex, chamois, red deer, pig, salmon and other fish, birds of various species as well as bear and other fur-bearing carnivores. Salmon and other fish may also have been important.

The correlation between the classic late Magdalenian tradition and the rigorous climatic regime of the late upper Weichsel stage has led to the conclusion that tradition developed as close adaptation to full glacial conditions. Since historic arctic and subarctic hunting peoples were adapted to broadly similar circumstances, they have often been used, somewhat uncritically, as analogues for the Magdalenian tradition (Stringer, Hublin, and Vandermeersch 1984: 111). However, certain difficulties attend the drawing of any strict interpretive parallels between the Magdalenian culture and that of the historic Eskimo of the North American Arctic. First of all, the pro-glacial environment of southwestern Europe during the Pleistocene would by no means be identical with the arctic and subarctic regions of the modern world. The position of Europe on the globe means that it would not have been subject to the gloomy six months of darkness characteristic of the polar regions. Further, the incidence at which the sun's radiation strikes the earth in Europe is very different than in the polar regions. This difference, coupled with the absence of the lengthy arctic night, means that, even under full glacial conditions, mid-latitude Europe supported a much larger biomass than does the modern arctic (Butzer 1971: 463). In addition to a larger overall quantity of game, Upper Paleolithic period peoples were hunting a diversity of large species that has simply not been present in northern climates since the close of the Pleistocene (Spiess 1979: 184).

The presence of rockshelters in southwestern Europe and their absence in the modern arctic and subarctic further complicates the analogy (Smith 1965: 393). It seems likely that neither the carrying capacity of the late Pleistocene environment in Europe nor the subsistence strategies developed by Magdalenian hunters to exploit that environment have not been duplicated on Earth since the beginning of the Holocene epoch. As a consequence, the precise nature of Magdalenian subsistence strategies remain a subject of some dispute among prehistorians.

Much research has centered on determining whether the great rockshelters sites of western Europe were occupied on a seasonal or on a more-or-less permanent basis. To answer this question, the patterns of tooth eruptions and tooth wear evident in the reindeer mandibles from Magdalenian sites have been given close scrutiny. Reindeer calves are born in the spring and their teeth erupt in a systematic fashion throughout the first year of their growth. The presence of manidbles with eruption patterns from all seasons would indicate that the shelters were permanently occupied; the systematic absence of some eruption stages in the mandibles would suggest the sites were abandoned during that season. Recent analysis of the faunal remains at Abri Pataud and other sites reflect the latter pattern and indicate the *abris* that line the steep cliffs of the Dordogne, Lot, and Vezere river valleys of the Perigord were apparently utilized during the late fall, winter and perhaps early spring seasons (Binford 1973: 238-244; Spiess 1979: 234-235).

Where did the Magdalenians go in the season that spent beyond the rockshelters in the interior uplands of the Masif Central, Calabria, and the Pyrenees. Bahn (1977) concludes that the Magdalenians followed the fall movement of the reindeer herds as they migrated westward down the narrow, east-west trending river valleys that led to the coast of Europe. Bahn presumes that the reindeer returned to the uplands in the spring or summer with the hunters in their train. However, studies of modern arctic and subarctic caribou and reindeer herds indicate that hunters cannot and do not follow the herds rather, they contrive to intersect and ambush them at various points along their seasonal migration routes. White and other have concluded that late Pleistocene hunters in western Europe operated in a similar fashion and points to the strong association between Magdalenian sites and fords and natural topographic traps in the Perigord as support of his view. He argues that such a strategy would force the hunters to develop a heavy reliance on plant and animal species besides the reindeer during a large part of the year (White 1980: 100-101; also cf., Burch 1972). The open-air Magdalenian sites of the west central Dordogne with their cobble stone pavements and dense, often highly specialized, lithic debris (Sackett and Gaussen 1976; Taranik 1977: 252-255), may represent summer settlements in such a seasonally-timed subsistence strategy.

In addition, a number of scholars have suggested that "anadromous" fish, specifically the Atlantic salmon, could plausibly have complemented a transhumant subsistence system centered on the reindeer. Anadromous fish spend the bulk of their lives in saltwater but annually swim up freshwater rivers and streams in the springtime in order to spawn. Prior to river pollution and the construction of hydroelectric power dams, many of the western European rivers that empty into the Atlantic were characterized by great spring spawning runs along hundreds of miles of their lower courses (Netboy 1968).

Sieveking (1976: 586-587) notes that Magdalenian sites are clustered along rivers which had some of the richest seasonal salmon runs in Europe as late as the 19th century. Jochim (1983: 215-216) also makes a strong inferential case for the importance of the Atlantic salmon in late Upper Paleolithic subsistence. He notes that modern populations of the fish are limited to waters between two and 18° Celsius. The CLIMAP reconstruction of the currents and water temperatures of the eastern Atlantic indicate that, during the last glaciation in Western Europe, such conditions would have been met only between southern England and the Straits of Gibraltar. Presumably salmon runs at the northern end of this range would have been quite short due to seasonal freezing and the shortage of food; in the south, runs would also have been short due to the large number of predators and competing species. However Jochim concludes that in southwestern Europe (the presumed central portion of the range) the runs probably tended to last for a much longer period of time. As a consequence, neither intensive harvesting nor storage would have been necessary in order for these fish to be a significant part of the diet. The northern coast of Spain and the southern Atlantic watershed of France contained the potential for significant exploitation of salmon without technological elaboration (Jochim 1983: 215).

Unfortunately, actual salmon remains are only rarely reported from French Magdalenian sites, although suggests that this absence may be due to poor preservation and poor recovery techniques, combined with the difficulty of locating open-air Magdalenian sites. Sieveking calls attention to an more indirect evidence of Magdalenian familiarity with and interest in the salmon: the occasional appearance of salmon and other aquatic animals in the art of the period. One of the most dramatic examples of such artistic evidence is Marshack's (1970; 1972: 169-174) identification of a spawning salmon as part of the decoration of a Magdalenian *baton-de-commandement* from the site of Montgaudier in department of Charente in southwest France. He also recognizes engravings of a flower, sprouting plant and other images of spring flora and fauna including snakes and a pair of seals on the baton. If his identifications are correct, they suggest that the engraver was concerned with timing of the spring salmon run and sought to represent the signs associated with its arrival. As Marshack put it,

> the plants and animals carved on the baton were time-factored and storied images of creatures whose comings and goings and seasonal habits were known,...they represented the birth of the 'new year,' if not calendrically and arithmetically at least observationally and probably in story (Marshack 1972: 173

Mellars (1985) disputes the claim that salmon harvesting was an important subsistence activity during the Upper Paleolithic period. He notes that fish bones are an important part of the faunal assemblages from terminal Magdelenian and Azilian sites and are found in fairly large quantities. In contrast, only a small number of salmon bones, along with the bones of other fish species, have been reported from late Upper Paleolithic period sites, furether, when they do occur, salmon represents a smalller percentage of the total fish bones than do species such as river trout and pike. On the basis of these data, Mellars concludes that the dearth of salmon remains from earlier occupations is not an artifact of poor preservation or recovery techniques, but evidence that salmon was not a primary part of the diet during the Magdelenian, and that fishing became an important element only in postglacial times.

Mellars (1985:279) also asserts that Jochim has been misled by ethnographic accounts of historic reindeer hunting peoples in Siberia and the North

American Arctic. In those regions reindeer migrate over vast distances and can use a number of alternative routes between their summer and winter grazing areas, and reindeer-hunting is indeed a precarious undertaking that scatters hunting groups and places a high premium on shared information. However, Mellars contends that these conditions did not pertain in southwestern Europe during the late Pleistocene. The geography of that region instead would have necessitated only a short migration between the coastal wintering areas and the summer pastures of the Masif Central, Calabria, and the Pyrenees. Further, these migratory poles are connected by a set of narrow, east-west trending river valleys. Since it is precisely within the lower foothill zones cut by these river valleys that the greatest concentration of Upper Paleolithic sites are found, Mellars concludes that peoples of this period had located in these places to be strategically placed for the interception of the annual migration of the reindeer herds. Such placement meant that the reindeer herds "were probably never very far removed from the main centers of Upper Paleolithic settlement over the full range of their annual cycle (Mellars 1985:279)". Finally, Mellars notes that even in seasons in which reindeer were relatively scarce, the high animal biomass of full glacial southwestern Europe probably meant that a host of alternative game species would be available.

In sum, gaps in the data prevent the reconstruction of the subsistence and seasonal movements of the Magdalenians with any precision. However, we can be virtually certain that the great herds of Late Pleistocene herbivores in southwestern Europe migrated seasonally from highlands to lowlands in response to changing temperatures. It does not seem unreasonable to suppose therefore, that the subsistence schedule of the Magdalenians was in general synchronization with the movements of these herds. Further, hunter-gatherers wintering on the lower reaches of the west-flowing rivers of Franco-Cantabria would be ideally situated to harvest the salmon toiling upstream towards them from the Atlantic in the springtime.

Mammoth Hunters? Since the discovery of the mammoth bone structures in the river valleys of the Dnestr, the Don and Dnepr and other rivers of the Ukraine beginning in the 1920s, prehistorians have assumed that the vast loess plains of eastern Europe and western Russia supported a form of subsistence centered on mammoth hunting markedly in contrast with the reindeer hunting economies of western Europe.

Despite the harsh environment that prevailed of there, humankind was drawn to these Mammoth Steppes by the immense richness and variety of its animal life. The excavation of Mousterian sites in the Ukraine indicate that the penetration of the tundra steppes began as early as the Middle Paleolithic period. However, Upper Paleolithic peoples seemed to have had achieved far greater adaptive success in the region than the Mousterian peoples. According to Klein (1974: 104):

> This conclusion is supported by the distribution of Mousterian and Upper Paleolithic sites. Systematic reconnaissance in those northern parts of the U.S.S.R. where glaciation has not been obliterated all evidence of ice-age man has failed to locate even one Mousterian site north of latitude of 54 ° Celsius . Upper Paleolithic sites however, have been found even above the Arctic Circle. The much greater number of Upper Paleolithic sites, particularly ones with "ruins" suggesting a settled or semisettled existence, implies higher overall population numbers and greater population density for Upper Paleolithic peoples.

This Upper Paleolithic adaptive success on the Mammoth Steppes was probably due in part to the large heat retaining effectiveness of the mammoth bone structures built in the Ukraine at this time. Additionally, the probably made effective use of animal fur in the manufacture of form-fitting clothing. Not only have the remains of fur bearing species as marmot, beaver, and hare have been recovered from Ukrainian sites (Soffer 1985: 286-289), skeletons of wolves and arctic foxes lacking paws have been recovered in some of these contexts. These paw-less skeletons are particularly significant in this connection as, among modern hunters, paws are commonly removed and kept with the hide when an animals is skinned for its fur (Gladkih, Kornietz and Soffer 1984: 164).

However, heat efficient facilities and warm fur clothing were probably only part of this adaptive success; very likely, new and more effective organizational strategies in big game hunting activities were probably even more important. The immense quantities of mammoth bones recovered from these sites has lead prehistorians to the conclusion that the site's occupants were mammoth hunters. However, in her recent extensive review and analysis of the archaeological data from the central Russian plain, Olga Soffer doubts that mammoths constituted the **primary** prey of the Upper Paleolithic peoples who lived there, and that ancient hunters killed whole herds of mammoth in mass drives. As noted, radiocarbon dating indicates that the bones used in the construction of the mammoth bone structures are of very different ages. These differences indicate that Upper Paleolithic peoples did not assemble the huts from the bones of a single season's hunt but rather collected, used, and then reused the bones over many generations. No doubt the short summers and low annual temperatures on the plain that inhibited the decay of these large bones would have made this possible. Since mammoth bones were "curated," rather than rapidly used and discarded, the vast numbers of bones in the Ukrainian sites can not be taken as *prima facie* evidence that the mammoth hunting provided these peoples with correspondingly vast stores of meat.

Second, Soffer carefully tabulated the numbers of individual animals represented at each site together with the numbers of individuals represented in each age and sex category. She compared these data with similar demographic profiles from New World mammoth kill sites, from a large fossil site or "cemetery" where mammoths presumably had died of natural causes rather than human predation, and from a living African elephant herds. Soffer predicted that, if hunters had killed entire herds, the demographic profiles at the kill sites would match those found in the living African herd and contrast with those from the cemetery. In fact, she found **no** statistically-significant differences between the demographic structure of the majority of both Old and New World mammoth assemblages and both the fossil cemeteries and living African herds. These results suggest that two very different archaeological site formation processes, natural mortality on one hand and human hunting and butchering practices on the other, create the same demographic patterns in the archaeological record, and that the resulting demographic pattern is not statistically different from the population structure of living elephant herds. Since different formation processes apparently can produce the same results, Soffer concluded that her statistical analysis reduced the *probability* that the ancient hunters took whole herds of mammoth to a mere *possibility* (Soffer 1985: 303-307).

Finally, Soffer assumed that mammoth would have been more abundant in the southern Ukraine, where the late Weichsel climate was less rigorous, than in the northern Ukraine, and that southern sites would therefore contain greater quantities of mammoth bones than northern ones. In fact, she found no statistically-significant differences between northern and southern sites. On the basis of these results, Soffer (1985: 281) concluded

> that mammoth did not play the major role in subsistence that has been widely attributed to them in both Soviet and Western literature. While some mammoth were undoubtedly hunted, how many remains a question for future research, as does determination of the frequency of mammoth kills, whether they were a seasonal or annual event, and if harvesting occurred at long intervals at frequencies similar to those postulated by Steward (1938) for Shoshone antelope drives (every 10 years) or by Frison (1978) for prehistoric bison kills (every 25 years or so).

Less uncertainty surrounds the remains of other big-game species recovered at these Ukrainian sites. For the most part, those bone remains are recovered from middens and appear to be the discards of a hunting camps rather than carefully curated building materials reused from generation to generation. Analysis of the bone remains reveal that the sites' occupants concentrated on the smaller gregarious herbivores of the late Pleistocene such steppe bison, musk-ox, aurochs, horse, and reindeer and occasionally took woolly rhinoceros, wild boar, birds and fish where possible (Soffer 1985: 297-301). Were these ancient Ukrainian steppe dwellers mammoth hunters? No doubt they were but, as Scotty MacNeish said of the late Pleistocene hunters in the Valley of Mexico, if they bagged one, they probably never stopped talking about it! Meanwhile, the bulk of their sustenance was probably supplied by the other game animals of the late Pleistocene north lands much as it was in western Europe.

New Forms of Social Life One of the most dramatic contrasts between the middle and the Upper Paleolithic periods in the European archaeological record is found in the burial or "death assemblages" of the two time periods. These assemblages reflect dramatic changes in the numbers of burials and in the nature and composition of burial populations. Lewis Binford's (1971: 23) widely quoted dicta states that, in general, the form and structure which characterizes the mortuary practices of any society are conditioned by the form and complexity of the organizational characteristics of the society itself.

In line with this view, Sally Binford (1968), Harrold (1980), Quechon (1976), Vandermeersch (1976) and others have sought clues to the nature of Upper Paleolithic society by comparing its burial populations with those of the preceding Middle Paleolithic period. The differences that these scholars have found between the two burial cohorts have interesting social implications. First, a comparison of the raw numbers of burials reported for each period suggests that the Upper Paleolithic period witnessed a marked increase in the number of individuals who received funeral rites. Of course, this apparent increase may simply have been the result of a higher population density during the Upper Paleolithic period or their "shorter period of exposure to destructive forces before recovery (Harrold 1980: 204)." Secondly, as compared to burials from the Middle Paleolithic, those of the Upper Paleolithic period tend to exhibit far greater quantity and diversity of grave goods such as tools, art objects, shells and ochre (Harrold 1980: 206). This distinction cannot simply be attributed to differences in population or in the relative ages of the burials. Instead, it suggests either greater material well being in the later period or a more complex, well developed funereal ritual and ideology, or both. Thirdly, females buried with grave goods become common for the first time during the Upper Paleolithic. This perhaps suggests that women were allowed a greater degree of participation in the status system than had been the case in the Middle Paleolithic period. Harrold (1980: 207) concludes that: "apparently at least some females in the societies represented by Upper Paleolithic burials enjoyed statuses as complex as those available to males." Despite the willingness of Upper Paleolithic peoples to occasionally bury women with grave goods, in the burial population examined by Harrold (1980: 202), male burials outnumber females two to one. This preponderance of males could merely reflect sampling error or the well documented tendency of researchers to class as male all those specimens with intermediate characteristics. But it may indicate that males were simply more likely than females to receive this kind of burial treatment in the cultures of the Upper Paleolithic period. In contrast, Harrold's (1980: 204-205) analysis suggests that individuals of different ages do not seem to have been treated significantly different from one another at death.

Unlike those of the Middle Paleolithic, Upper Paleolithic period mortuary remains show a great deal of intra- and inter-regional variability with regard to the position of the body, associated grave goods or features and so forth. Following Binford (1971), Harrold (1980: 204) states that mortuary customs function less to mark cultural boundaries than to symbolize status distinctions within societies. He interprets the variability in Upper Paleolithic mortuary remains as reflecting a greater number of social distinctions and social complexity in the societies of the period.

Finally, the occurrence of multiple burials, that is, the simultaneous interment of two or more individuals, is significantly more likely in Upper Paleolithic period contexts. Harrold (1980: 207) interprets the more common occurrence of these multiple burials as indirect evidence of greater population density and increased sociocultural complexity since, "the most likely causes of such multiple deaths, such as warfare, communicable disease, or ritual sacrifice, would seem likelier under (such) conditions..." However, Vallois's (1961) analysis of 76 skeletons from Upper Paleolithic period burials provides a sobering insight into the mortality rates of the period. Less than half of the individuals in his sample were older than 21, only 12 percent were older than 40 years of age. The females in his sample all died before they were 30. Butzer (1971: 482) concludes on the basis of this data that the mortality rates of the Upper Paleolithic period represented no improvement over that those of the preceding Middle Paleolithic period.

Caves and Calendars The Upper Paleolithic period witnesses the first appearance of cave or "parietal" art, portable or "mobiliary" art and the possible development of complex notational and calendric systems. The Upper Paleolithic period seems to have witnessed a powerful outpouring of aesthetic and intellectual energy. This outpouring is clearly mirrored in the representational cave and portable art of the period, and in the contemporary symbolic or nonrepresentational systems interpreted by some scholars as expressing sophisticated information, computations or calendric reckoning. The implications of these data for our understanding of Upper Paleolithic period religion is of course immense; here at last is a window into the mind of Ice Age humanity. Further, the window opens onto a mental universe that seems foreign to us in content but not in structure; a universe created and dwelt in, after all, by people like ourselves. The appearance of art and symbol in the Upper Paleolithic period thus represents a crucial step in what Teilhard de Chardin calls the process of "hominisation." And hominisation, or the gradual emergence of thought and self-consciousness, is itself but one part of the "grand cosmic phenomenon of the vitalization of matter (Teilhard de Chardin 1964: 208)."

Some Comparisons The Upper Paleolithic period in southwestern Europe lasted less than 25,000 years and was characterized by nine or ten distinct techno-industrial traditions. These traditions for the most part succeeded one another in time and persisted on the average for 2,500 years apiece. In contrast, the

European Mousterian of the Middle Paleolithic period was characterized by six distinct facies some, perhaps most, of which are contemporary with one another rather than successive. These facies spanned phenomenally long time ranges of 25 or even 70 millennia. In effect, the half-life of the Mousterian industrial variants was of an order of magnitude greater than that of the Upper Paleolithic industries (Butzer (1981: 178). The duration of the Mousterian is remarkable. More remarkable still is the apparent capacity of the various Mousterian facies to maintain their separate identities despite each others' contemporary presence. The human cultures known to anthropology today and in historical times simply do not operate in this manner. It would appear that the archaeological record of the transition from the middle to the Upper Paleolithic period records a radical change in the dynamics of human culture.

Eastern Eurasia in the Middle and Upper Paleolithic Periods

The Middle Paleolithic in eastern Eurasia is a generally poorly understood period. Chronometric dates are scarce compared with both Africa and western Eurasia. Nonetheless, progress has been made over many areas of eastern Eurasia and our understanding of the prehistory of the later periods of the Paleolithic period is improving.

China The Middle Paleolithic in China is, as elsewhere in eastern Eurasia, a poorly understood period. Chronometric dates are scarce and, compared with both the Lowerand Upper Paleolithic sequences of China, Middle Paleolithic sites are few in number. Nonetheless, the modest corpus of sites dating to the Middle Paleolithic period contains localities from both north and south China and this widespread distribution is an encouraging sign that many more sites are yet to be found. The presence of choppers and chopping tools that are retouched on one face only in most of the stone tool assemblages of the middle period suggest a developmental continuity with the industrial tradition of the LowerPaleolithic period in China. Further, clear technological improvement is reflected in a number of new artifact types that appear that this time. These include points, scrapers, bifacially-flaked handaxe-like implements, and in some cases, stone spheroids interpreted as bola stones (Zhongland 1985).

Sites dating to the Upper Paleolithic are somewhat better represented in the archaeological record, particularly in north China and Mongolia. By the Upper Paleolithic period, the distinctiveness of culture of East Asia as compared to western Eurasia appears to have been reduced.

A good of example of the increasing similarity of the archaeology of the two regions can be found in the tool assemblage of the **Ordosian culture**, the earliest Upper Paleolithic tradition on the north China plain. The later Pleistocene deposits in the Ordos region containing the type site of this tradition have produced fossil antlers and mammalian teeth dated by the uranium series method to sometime between 50,000 and 30,000 BP. (Xinzhi and Linghong 1985: 43). Ordosian lithic industries exhibit specialized blade tool types not unlike those of the western Europe. As the Ordosian tradition begins to wane, microlithic blades begin to appear in its lithic assemblages. However, these Chinese early Upper Paleolithic industries are also surprisingly conservative in some ways. Tools on flakes produced by the prepared or Levallois core technique continued to be made and pebble tools of the older Asian mode were apparently still used; occasionally Ordosian repertories are even found to contain a few choppers and chopping tools (Chang 1977: 61-65).

By the end of the Upper Paleolithic period, stone tool assemblages of north China exhibit a great deal of regional diversity (Chang 1977: 65). This heightened technical diversity may reflect increasingly specialized subsistence adaptations to contrasting local environments or a reduction in pan-regional interaction, or both. In southwestern China.

Japan, Korea and Taiwan The archaeological record of the Japanese Paleolithic becomes clearer after approximately 30,000 BP. Between roughly between 30,000 and 20,000 BP., early peoples in Japan produced a stone tool industry characterized by large pebble handaxes-like tools and end and side scrappers on flakes. The presence of these tool classes led Aikens and Higuchi (1982: 88) to characterize industries of this period as "Mousterian-like." However, they also note that the classic element of the European Mousterian, the prepared core or Levallois flaking technique, was apparently not practiced in Japan at this time.

The Japanese analogue for western Eurasian Upper Paleolithic period culture dates to between ca. 20,000 and 13,000 BP. As in other parts of Eurasia, stone tool assemblages come to be dominated by tools made on blades struck from carefully prepared cores. A specialized tool made with steep retouch over one side of thick blade flakes is common in industries of this period. Although all of the tools of this form are thought to have functioned as knives, the particular style of their manufacture exhibits a good deal of regional variation across Japan. Late in the Paleolithic, bifacially-flaked projectile points made by flaking the entire surface of blade flakes develop out of the manufacture of these knives. As in much of the other parts of the world, the end of Paleolithic period in Japan witnesses a reduction in the size of stone tools. This trend, which begins after ca. 15,000 BP., culminates in the subsequent period with the appearance of true microliths. The terminal Paleolithic in Japan dates to between 14,000/13,000 and 12,700/12,500 years BP. During these late millennia, stone tool technology in Japan came to be dominated by the manufacture of microlithic tools (Aikens and Higuchi 1982: 89-91). It is within this terminal Paleolithic period technological setting that the earliest ceramic-making cultures of the **Jomon tradition** begin their rise to prominence. Although its origins are in the Terminal Pleistocene, the Jomon tradition will be discussed in the next chapter.

Southeast Asia The distinction between tool traditions of the Lower and Middle Paleolithic periods, so carefully drawn on formal and technological grounds in Europe, are less easily made in Southeast Asia. For example, the **Cabalwanian industry** of northern Luzon in the Philipines appears to fall in the time frame of the Middle Paleolithic and its assemblages contains a few bifacial tools and others that appear scrapper-like in function. However, the bulk of the assemblages consist of crude cobble choppers and implements made on unifacial flakes (Fox 1978: 73-79). The industry lacks little of the technical sophistication of the Mousterian of Western Eurasia and is best considered to be a late expression the Chopper/Chopping tradition.

The Upper Paleolithic period in Southeast Asia continues to reflects what Shutler and Shutler (1975: 19) call the "remarkable persistence of lower Paleolithic tool types into late Pleistocene times." In addition to the continued use of large unifacial tools, these authors note that Upper Paleolithic stone technology throughout the area is characterized by various types of small flake tools. Two deeply stratified cave sites: **Niah cave** in Borneo and **Tabon cave** on the Philippine island of Palawan (Fox 1970) have provided a basic idea of life during this final cultural period in Southeast Asia. Scientific understanding of most of the Paleolithic period in Southeast Asia is fragmentary at best. Fortunately, due to a series of well-conducted excavations at **Spirit Cave** in Thailand and a number of other sites of the **Hoabinhian tradition** in Philippines and Malaysia, the prehistory of the terminal Pleistocene and early Holocene is better understood. This tradition is examined in the next chapter.

Australia, New Guinea, and Tasmania Australia and the continental islands of New Guinea and Tasmania are collectively referred to as Greater Australia. The impact of late Pleistocene climatic change on the lands of Greater Australia appears to have been fairly modest. Here, as elsewhere, glaciers in the mountain ranges of the area formed or expanded during periods of glacial maxima and a general correlation appears to have existed between northern hemispheric glaciation and cooler, drier conditions in the southern hemisphere. Of course, the same cautious skepticism must be applied to these correlations as to those alleged for the Pleistocene in Africa.

Although the period after ca. 60,000 is thought to have been generally colder and more arid than at present, the establishment and expansion of lakes between 30,000 and 25,000 BP. in southern Australia indicates that moister conditions and reduced rates of evapo-transpiration rates prevailed over a 5,000 year period in the Late Pleistocene. The effects of worldwide warming near the end of the Pleistocene were not uniformly felt throughout Greater Australia. Although valley glaciers began to recede and tree lines began to move to higher elevations by ca. 20,000 BP. in the Australia mountain ranges, similar changes did not occur in the mountains of New Guinea and Tasmania until after ca. 15,000 BP. The onset of the Holocene after ca. 10,000 BP. initiated a period of increased precipitation and greater warmth throughout the entire area (Lourandos 1987: 148).

Australia, New Guinea, and Tasmania sit astride a feature of the Earth's crust known as the **Sahul Platform.** This platform is analogous to the Sunda Shelf that supports the Indo-Malaysian Archipelago to the north of it. Sea level recessions during Plio-Pleistocene glacial stages exposed portions of the ocean floor on both of these crustal features. However, the elevation of the Sahul Platform is lower than the Sunda Shelf. Consequently, the sea level drops that occurred during Pleistocene were never sufficient to entirely expose the Sahul Platform nor fully connected its islands to the adjacent Sunda Shelf by dry land bridges. Although at various times during the Pleistocene, the sea between the Sahul Platform and the Sunda Shelf shrunk to a relatively narrow channel, that water barrier was apparently deep and formidable to create one of the Earth's most pronounced and important zoo-geographic boundaries, 'Wallace's Line'" (Goudie 1983: 67). Most of the Earth lies beyond Wallace's Line and there marsupials and monotremes are rare or absent. This is because these mammal taxa were out-competed by placental mammals and driven to extinction long ago. Isolated from such competition by the water barrier that forms Wallace's line, marsupials and monotremes thrive in Greater Australia. This sea barrier also apparently prevented *Homo erectus* from reaching the lands on the Suhul Platform. The sea was not a serious impediment to anatomically modern humans, whose remains have been recovered from Late Pleistocene contexts in Australia, nor perhaps to archaic *Homo sapiens* whose presence there is less certain.

Modern hominids were definitely in Australia sometime between 35,000 and 40,000 BP. If the recent dates from several sites in **Deaf Adder Gorge** in northern Australia are correct, they were there between 50,000 and 60,000 BP. Much earlier dates of ca. 116,000 to 176,000 years BP. have been obtained from the occupation of the **Jinmium Rockshelter** in northern Australia were obtained by thermoluminescence method. However, these dates have proven problematical for a number of technical and methodological reasons and are not now generally accepted (O'Connell and Allen 199-). Continued cultural and biological interchanged across the Sahul Platform must have taken place throughout the terminal Pleistocene in Greater Australia. However, the land bridge between Australia and Tasmania was severed by 12,000 BP. and that between Australia and New Guinea by 8,000 BP. Thereafter, contacts between these regions would have to have been made strictly by water craft (Lourandos 1987: 148). While these severances were no doubt significant, the impediments they presented to contact and migration in the terminal Pleistocene should not be exaggerated. After all, evidence for successful early sea voyaging of a far more challenging character than merely crossing the Sunda Straits has accumulated in recent years. For example, radiocarbon dates from cave sites recently excavated to the east of New Guinea on New Ireland indicate that modern peoples arrived on that distant Pacific island by 33,000 BP. (Allen *et al.* 1988).

Whatever the method of transport and the date of the earliest human arrival in Australia, the impact of this new and cunning predator upon the conservative and formerly isolated animal species below the Wallace Line must surely have been pronounced. The late Pleistocene bestiaries of Australian, Tasmania and New Guinea contrast markedly with the relatively impoverished assemblages that characterize those lands in the present. Further, the final stages of the extinction process that led to this impoverishment took place during humankind's tenure in the area. Here as elsewhere in the world, students of the Quaternary period disagree as to whether this correlation is spurious, proves profligate human hunting practices like fire driving game ("fire stick farming") caused the massive faunal changes of the last 40,000 years, or reflects the interaction of hunting with terminal Pleistocene climate change in effecting these extinctions.

Of course, animal extinctions have followed in the wake of humankind throughout the world. Therefore, the correlation between the impoverishment of the Australian fauna and the arrival of humankind down under is not a surprising one. What is unexpected, however, is the simplicity and homogeneity of the stone tool kits used by these early immigrants. Current fossil evidence indicates that, while the migrants who crossed the Sunda Shelf were anatomically modern, their tools kits were decidedly archaic. This is particularly true of the peoples who eventually came to occupy Tasmania, the island just off the south coast of Australia. In western Tasmania, for example, stone technology is

> dominated by unmodified flakes rather than shaped tools. Tools that might be argued to conform to a particular style are limited to so-called thumbnail scrapers--small flakes that have been shaped at one end. Other modified tools forms would not look out of place in Eurasian Middle Paleolithic assemblages. Stone projectile tips are missing, the only possible candidate being a series of bone points that may, however, have functioned as awls (Holdaway and Cosgrove 1997: 70)

Further, only low-quality local raw materials were used in these Tasmanian stone industries. Little effort was expended at securing better raw lithic materials from afar. Finally, there are no objects of art or adornment associated with these rudimentary stone assemblages as is the case in Eurasia and Africa. It would appear that in Greater Australia, as elsewhere in the Old World, the transition to behavioral modernity created a biological and cultural mosaic rather than a uniform transformation.

Conclusions

After untold millennia of inertia and virtual immobility in the early and middle portions of the Paleolithic, the pace of cultural and technical change heightened dramatically as the period drew to a close. Compared to these earlier periods, the Upper Paleolithic is a time of accelerating technical development, demographic expansion and social change. Humankind has entered into the cultural trajectory that will carry it in to the present. However, this transition was not effected with out cost. The cultural and technological acceleration of the Upper Paleolithic period appears to have been purchased with the squandering of the vast faunal resources of the late Pleistocene epoch. Perhaps there is a cautionary tale in the archaeological record of these stone age hunting-and-gathering peoples that modern technologically-sophisticated humankind could ponder with profit.

Major Points Raised in Chapter Nine:

1. The chapter begins with a discussion of the chronology and paleoenvironment of the last (Sangamon) interglacial stage and the last (Wisconsin) glacial stage in the New World.

2. The Plio-Pleistocene land mammal ages in North and South America are examined and the role played by the land bridges at the Bering Straits and Panama in shaping these mammal ages discussed.

3. The questions of the earliest human migration to the New World ("early Early Man") is examined. The Laurentide ice sheet, the Cordillerian glacial complex, and the possible ice free corridor between them are considered in discussing possible routes from Beringia into the New World that might have been taken by these early migrants

4. The controversy surrounding the date of the earliest colonization of the New World is noted and key archaeological sites used as evidence in support of their views by radical, liberal, and conservative prehistorians are evaluated.

5. The Paleo-Indian, the earliest indisputable cultural tradition in the New World, is described. Three sequential complexes make up this tradition: the Clovis or Llano complex, the Folsom complex and the Plano complex. These three complexes are discussed from the perspective of their technology and subsistence. Archaeological evidence and ethnographic analogy are used to reconstruct Paleo-Indian strategy for hunting late Pleistocene big game or "mega-fauna." Speculations are made about the nature of Paleo-Indian society, economy, and religion.

6. The chapter ends with a discussion of the mammalian extinctions–particularly megafauna extinctions–that took place throughout the world at the end of the Pleistocene epoch. The question of humankind's complicity in "mega-cide" is discussed.

CHAPTER 9

THE LATE PLEISTOCENE EPOCH IN THE NEW WORLD

Introduction. In this chapter we turn to the archaeological and geological evidence of the human career in the New World. Unlike the Old World, indisputable archaeological evidence of human presence is no earlier than the late Pleistocene. In fact, the archaeological community is deeply divided over the date of the earliest appearance of humankind in the Americas. But the dispute over the date of arrival is but one of the major foci of this chapter and other archaeological bones of contention are considered including the nature of the big game hunting Paleo-Indian life and the role of humankind in the extinction of the late Pleistocene megafauna.

Time and Environment in the Late Pleistocene New World

Chronology. Although the earliest records of glaciation in North America are of Miocene age, the first major stage of continental glaciation there, referred to as the Nebraskan, begins around the Plio-Pleistocene boundary some 1.8 to 1.6 mya (Barry 19–: 394). During the Pleistocene the New World experienced cycles in which middle-latitude continental glacial advances–punctuated by interstadial climatic and glacial remissions–alternated with interglacials characterized by abrupt warming and the full-scale retreat of continental glaciation.

The late Pleistocene is defined as the period of the last interglacial and last glacial stages. In North America, the final interglacial stage is referred to as the Sangamon and dated to between about 125,000 and 118,000 B.P. The final glacial stage there, referred to as the Wisconsin, begins at the end of the Sangamon and lasts until ca. 10,000 B.P. The New World glacial-interglacial cycles have been generally correlated with those of the Old, the late Pleistocene epoch in both hemispheres through absolute dating, stratigraphic correlation and geomorphology. Thus, the Sangamon is temporally equivalent to the Eem-Riss/Würm-Mikulimo integlacial stage, the Wisconsin to the Weichsel-Würm-Valdai glacial stage in the Old World.

The Wisconsin stage is exceptionally well-known in Illinois and in the eastern Great Lakes region around Lakes Huron, Erie and Ontario. Based on the glacial stratigraphic succession in these regions, the Wisconsin stage can be divided into early, middle and late periods and into a series of stadial and interstadial substages (Bowen 1985: 51-53). The early Wisconsin occurred between ca. 118,000 B.P. to 53,000 B.P. It is characterized by the gradual build-up of the continental ice-sheet until about 75,000 B.P. Between ca. 75,000 and 70,000 B.P., glaciation spread southward rapidly, then entered a period of retreat known as the St. Pierre interstadial, sometime between ca. 65,000 and 60,000 B.P. The continental ice mass began to enlarge again roughly between 60,000 and 53,000 B.P. and reached its maximum southward expansion during that time. The middle Wisconsin period occurred between ca. 53,000 and 23,000 B.P. In contrast to the extensive ice advances of the early Wisconsin, this period was characterized chiefly by the dwindling of the ice mass during a series of long, cool interstadials known as the Port Talbot I, Port Talbot II and the Plum Point. The dates of these interstadials have not been established with great precision, although the Port Talbot II seems to have occurred between 48,000 and 42,000 B.P. and the Plum Point interstadial sometime between ca. 30,000 and the end of the middle Wisconsin.

The late Wisconsin period occurred between 23,000 and 10,000 B.P. In what is called the Woodfordian substage or stadial, the continental ice mass rapidly enlarged in a series of southward surges that culminated throughout the northern hemisphere in the maximum glacial advance about 18,000 B.P. (Barry 19–: 394-395). Deglaciation began after ca.

15,000 B.P. and the Woodfordian substage ended around ca. 12,000 B.P. in the warm conditions and extensive deglaciation of the Two Creeks interstadial (ca. 12,000 B.P. to 11,000 B.P.). The end of the Two Creekan is marked by the sudden but limited re-advance of glacial ice and a brief return of glacial climate during the Valders (or Great Lakes) event between 11,000 and 10,000 or 9,500 B.P. This last event, clearly worldwide in scope, is closely correlated with the Younger Dryas event in Europe and the El Abra event in South America (Ruddiman and Duplessy 1983).

Late Pleistocene climate and vegetation in the New World The climate of North America during the Sangamon interglacial stage appear to have been generally moister than the present and characterized by higher winter and lower summer temperatures. This conclusion is based in large measure on the biogeographic differences between the last interglacial and the Holocene. For example, North American pollen spectra indicate that a number of warm-loving tree species were growing well north of their present boundaries during the Sangamon (Boardman 1985: 69-70) and fauna comparable to that of the modern Gulf Coast have been recovered from Sangamon contexts as far north as southern Kansas (Hibbard 1970). The pronounced reddish color of certain Sangamon paleosols and the fact that most have been more intensively weathered and strongly developed than soils formed in the Holocene is also cited as evidence that, at least by the period of thermal maximum, the climate of North America was warmer and wetter during the last interglacial than it is at present (Butzer 1971: 390). Boardman (1984) however, cautions us to remember that, since the Holocene is relatively short in comparison to earlier interglaciations, **time** rather than **temperature** may account for difference in character between its soils and those of earlier glacial remissions.

The return of cooler conditions at the end of the Sangamon around 118,000 B.P. and the cyclical advance and retreat of continental glacial ice at the middle latitudes after 75,000 B.P. profoundly altered those climatic and environmental conditions. By the late glacial maximum around 18,000 B.P., a huge glacial ice mass–larger than that on the modern Antarctica–covered parts of Alaska and most of Canada and reached far down into the northern midwestern United States. As in the Old World, the presence of this middle latitude ice mass effectively depressed or displaced the major climatic belts southward. With those belts, the major New World vegetation zones were also pushed south and distorted. The results of this depression were not identical throughout the northern hemisphere.

By 18,000 B.P., the unglaciated portions of Alaska supported a tundra vegetation but only a narrow band of tundra climate and vegetation appears to have been located immediately south of the ice sheet. The eastern United States supported a coniferous woodland or boreal forest dominated by and spruce (*Picea*) jack pine (*Pinus banksiana*) from the mid-Atlantic region to the lower Great Lakes in central Midwest. The southern limit of this woodland may have reached as far south as southern Tennessee and perhaps upper Georgia and was thus more than 1,200 kilometers south of the modern lower boundary of the boreal forests in Manitoba and Ontario. In the very limited area of the southeastern United States below the boreal forest, warmer conditions supported mixed coniferous and hardwood forests and provided a North American refugia for warm-loving plants and animals during full-glacial times (Delcourt and Delcourt 1985: 15, 13).

West of the great boreal forest, a permafrost zone characterized by an open park land vegetation stretched along the margin of the ice front from the western Great Lakes region across the northern Great Plains to Montana. This permafrost and open park land vegetation gave way to a non-permafrost, montane grassland in the western Great Plains along the Rocky Mountain piedmont south of the Dakotas. At the same time, the vast deposits of aeolian glacial sediments or *loess* that had accumulated in the center of the continent in what is today Illinois, Iowa, Missouri, and northern Kansas, appear to have supported either grasslands or park lands interfingered with woodlands in the river valleys (Butzer 1971: 374-375). Like the tundra park lands of the European and Russian steppes, these open plains of the central United States contained one of the most opulent and raucous, varied and roughscuff assemblage of animal life ever seen on this continent.

Across the Rocky Mountains much of the Pacific Northwest supported a subalpine park land. In the Great Basin and the Southwest, vast playa lakes reached their highest levels in the structural basins between the mountain ranges. At glacial maxima, forests descended to their lowest elevations on the mountain piedmont in the Southwest and a mixed-woodland assemblage of piñon (*Pinus* sp.), or juniper (*Juniperus* sp.) or both with a perennial grass understory dominated what today is desert grassland in New Mexico and Arizona (Van Devender and Spaulding 1979: 701; Van Devender, Betancourt, and Wimberly 1984: 344).

Correlation of the Pleistocene stratigraphy of North America with that of South America is as difficult to effect as that between Europe and Africa. However, great headway has been made in recent

Wisconsin Glacial Stage in the Great Lakes Region of North America

Period	Date Range	Substages: Stadial	Substages: Interstadial
	10,000 B.P.		
Upper		Valders	
			Two Creeks
		Woodford	
	23,000 B.P.		
			Plum Point
Middle		4th Stadial	
			Port Talbot I
		3rd Stadial	
			Port Talbot II
	53,000 B.P.		
Lower		2nd Stadial	
			St. Pierre
		1st Stadial	
	118,000 B.P.		

years. Based on work in the Colombian Andes, van der Hammen (1974) asserts that the stadial and interstadial events of the Wisconsin glacial stage in North America are reflected in rainfall and temperature oscillations in highland South America. He sees a general and progressive decline in rainfall after 20,000 B.P. reflected there in a lowered treeline and snowline. During this glacial stage, grasslands expand dramatically in the Andes and Fairbridge (1976; also cf., Danmuth and Fairbridge 1970) argues that the much of lowland and northeastern South America was also characterized by an arid to semiarid climate that was much in contrast with the warm, moist conditions in the region in interglacial times and at present. Near the end of the Wisconsin stage in South America, warmer, moister conditions reappear briefly and then fade in rough temporal concert with the Two Creek interstadial in the north. The return of cooler, drier conditions in the subsequent El Abra stadial makes the end of the Wisconsin stage in South America just as the contemporary Valders event does in North America. Of course, the impact of these climate and vegetational changes on the animal life of the New World was great indeed and it is to this animal life that we now turn.

The New World land mammal ages Like those of Europe and Eurasia, the land mammal ages of North and South America are well defined. Figure 12 lists the Plio-Pleistocene and Holocene land mammal ages for the New World. Note that North and South American land mammal ages have different names and different time boundaries. This is because the continent of North America was connected by land bridges with Europe or Asia at various times during the Cenozoic and the faunal interchange that occurred across these bridges resulted in the uniting of North America, Asia, and Europe in a single, vast biotic province. In contrast, from the Late Cretaceous until the late Pliocene, South America was an isolated island continent characterized, like Australia, by a unique biotic assemblage. The continent's isolation ended around three million years ago when the Bolivar Trough, the ocean barrier between South and Central America, disappeared due to structural change and dropping sea level. The interconnection of South, Central and North America over the Panamanian land bridge resulted in a dramatic mixing of North and South American fauna known as the "Great American Interchange" (Marshall 1988: 380).

Blancan, Irvingtonian, and Rancholabrean land mammal ages in North America North American fauna dating to the middle and late Pliocene fall into the Blancan land mammal age (see chart). The Blancan age dates to between ca. 5.3 and 1.9 mya and begins slightly later then the corresponding early Villafranchian land mammal age in Europe; both the Blancan and the middle Villafranchian come to an end around the Plio-Pleistocene boundary.

A number of important taxa indigenous North American appear during the the early Blancan including the great predatory stabbing or saber-toothed cats of the genus *Smilodon*, the genus *Equus*, that includes all modern horses, asses, and zebras, and the primitive equine subgenus, *Dolichohippus*, the genus

Tapirus, which included a host of tapir species that came to spread throughout the world, *Mylohyus* or long-nosed peccary, the ancestral camel, *Camelops*, the four-horned pronghorn, *Capromeryx*, and the white-tailed deer (*Odocoileus*) (Anderson 1984: 59, 65, 66, 69, 71, 75). The most unusual taxa to appear in North America during this age is the giant beaver (*Castoroides* sp.). Although this huge rodent was as large as a black bear, it was probably muskrat-like in its living habits and mercifully neither constructed dams nor felled trees (Anderson 1984: 61).

Many of the land mammals characteristic of the Blancan age in North America were of Asia derivation. The most likely route for the arrival of such species is the Bering Straits between Siberia and Alaska. Not only is the Asian mainland closest to the North America at the Bering Straits, the region now submerged beneath the Bering Sea has been periodically exposed as a dry land bridge between the two continents. The repeated emergence of a land bridge in the region beginning at least by the Miocene facilitated the exchange of biota between Siberia and Alaska over the last seven million years. Conversely, the disappearance of the bridge has periodically isolated fauna on either side of the Straits. The land beneath the Straits today has been sea floor only since the end of the late Pleistocene epoch. Before that, it formed a relatively narrow strip of dry land called "Beringia" by geologists and assumed by them to have generally supported a tundra-grassland and boreal forest biome across which Asia and New World animals species could wander dry shod. As Flint (1971: 773) points out, the Beringian causeway must have served as a filter "that permitted the transit of mammals of primarily grazing habit which could withstand boreal temperatures." And of course, the predators, including hominids, that preyed upon the grass-eaters could cross in their company.

In most instances, the Old World forms that crossed the Bering Straits land bridge during the Blancan seemed to have a great competitive advantage over indigenous competitors advantage and the consequent establishment in North America of such Old World immigrants as the brown bear (*Ursus arctos*) and various members of the genus *Panthera*, including lions, leopards, tigers and jaguars gave the early Blancan age a decidedly Villafranchian character. Of course earlier, the spread of *Hipparion*, the primitive, three-toed ancestor of the modern horse, across the Old World with the expanding grasslands of the Miocene and the later dispersal of various equid and camelid species, indicates that such transfers were not altogether in one direction. Further, once across the land bridge, Old World forms often evolved in dramatically new directions. For example, Stegomastodont, the chief proboscidean in North America during the early Blancan, evolved in North America from more primitive Old World stock.

The late Blancan begins with the arrival in Central America and southern North America of seven genera of South American land mammals and one large ground bird shortly after the emergence of the Panamanian land bridge. Included among these genera were a form of porcupine, several armadillo species, a giant armadillo-like animal known as *Glyptotherium*, and two forms of ground sloth (*Glossotherium* and *Nothrotheriops*). By far the most imposing immigrant was the huge, flightless bird known as *Titanis*. *Titanis* stood over three meters high and–judging by its great size, powerful legs, and formidable beak–preyed upon land mammals. Recovered from late Blancan age deposits in Florida, bones of *Titanis* indicate that, had this carnivorous bird not become extinct in the later Pleistocene, shopping malls and theme parks would never have become so common in this part of the Sun Belt.

The close of the late Blancan near the end of the Pliocene around 1.9 or 1.8 mya ushers in the Irvingtonian land mammal age. The Irvingtonian is contemporary with the late Villafranchian and the Cromerian/Mosbachian land mammal ages in Europe. The age is characterized by the arrival of additional Old World immigrant species most important of which were: the wapiti or elk (*Cervus elaphus*), the caribou or reindeer (*Rangifer caribou* [Gmelin]), and the mammoth (*Mammuthus primigenius* [Blumenbach]). The Columbian mammoth (*M. columbi*) is an indigenous North American descendant of *primigenius*. Competition from the Columbian mammoth apparently led to the extinction of the indigenous Stegomastodont in North America during the Irvingtonian, but the inability of the mammoth to penetrate south America allowed the Steogmastodont to persist there until nearly the end of the Pleistocene (Anderson 1984: 85). The giant Shasta groundsloth (*Nothrotheriops*) evolves indigenously in North America at this time. The Irvingtonian age closes after about 400,000 B.P.

The faunal assemblage of late glacial America is referred to as the Rancholabrean land mammal age, after the celebrated fossil beds of the La Brea tar pits in Los Angeles County, California. The age begins ca. 400,000? B.P. and closes about 10,000 years ago with the great wave of extinctions that mark the final millennia of the Pleistocene epoch. A number new taxa indigenous to North America appeared during the Rancholabrean land mammal age among the most important of which was the moose elk or "stag-moose" (*Cervalces scotti* [Lydekker]). At the same time, various species of old, native North American taxas including pronghorn, peccary, tapir, and the ancestors of the modern horse continue to flourish. The last species of the camelops line, *Camelops hesternus* (Leidy), appears during the Rancholabrean. *C. hesernus* probably resembled the living dromedary but had longer limbs and a narrower muzzle. It has been recovered in a few kill sites and appears to have become extinct around 10,000 B.P. (Anderson 1974: 71). Finally, a distinctly North American form of forest-dwelling proboscidean, the Mastodont (*Mammut americanum* [Kerr]), evolved during the this age.

Species continued to immigrate from South America during the Rancholabrean. Among the most impressive of these is the six meter long giant groundsloth or *Eremotherium* and the dire wolf (*Canis dirus*) a predator that was half again the size of a modern wolf (Anderson 1984: 53, 55). The hallmark or directory fossil of the Rancholabrean, however, is an Old World immigrant, the steppe bison or wisent (*Bison priscus*). This species reached Alaska at the onset of the age and was the source population from which evolved the giant North American *Bison latifrons* and later *Bison bison antiquus* and *Bison bison occidentalis* and, at the end of the Pleistocene, the modern form, *Bison bison* (Anderson 1984: 76-77). Other Old World species to reach North America during the Rancholabrean age were the shrub oxen (*Euceratherium*), the musk ox (*Ovibos*), the moose (*Alces alces* [Linnaeus]), and mountain sheep (*Ovis*) (Anderson pp 78, 79, 72-73, 80). Finally, the domesticated dog (*Canis familiaris*) probably crossed the Bering Straits land bridge and entered late Pleistocene America in the company of early human immigrants (Anderson 1984: 55; also cf., Olsen 1985). In sum, North America near the close of the Pleistocene was endowed with animals in an abundance and variety unknown on the continent anytime since. Mammoth, horse, camelops, several species of bison and pronghorn and a host of other herd animals–many of them giants when compared to their modern descendants–dotted the plains in every direction. Moose, shrub oxen, caribou, mastodont, deer, bison, peccary, tapir, giant beaver, and numerous other species filled the forests. In the southern regions of the continent, giant groundsloths, anteaters, and other warm-loving animals were abundant. A proportionate abundance of truly fearsome predators like *Smilidon*, dire wolf, lion, and panther stalked the land. Late Pleistocene North America was surely a zoologist's and a hunter's paradise–but a paradise **lost**. Early 19th century travelers on the North American Great Plains spoke glowingly of game in breathtaking and uncountable abundance, yet they glimpsed only a depauperated remnant of the late glacial world. To experience anything remotely like the richness of the New World in late glacial times, one must nowadays leave the continent and journey to the Serengetti Plains of Africa–and even that remnant corner of the Pleistocene world is in decline.

Montehermosan, Chapadmalalan, Uquian, Ensenadan, and Lujanian land mammal ages in South America In South America the land mammal age of the late Miocene and Pliocene epoch is called the Montehermosan. With the exception of a few "waif" species of Central American origin that probably washed ashore in South America on vegetation rafts, the fauna of the Montehermosan had evolved in locally and in isolation since the late Cretaceous. However, the unique mammals of the Montehermosan age represent the final epoch of South American insularity. The age ends shortly after three million years ago, and the succeding Chapadmalalan age (ca. 2.8 to 2.5 mya) is marked by the first appearance of North American immigrant species that must have walked dry shod across the Panamanian land bridge. These North American immigrants include such early Blancan forms as: skunk (*Conepatus*) and peccary (*Platygonus*). However, the largest North American species penetration occurs in the to South American appears during the succeeding Uquian age (ca. 2.5 to 1.5 mya?). the primitive horse, *Hippidion*, makes its appearance together with *Smilidon*, the saber tooth cats, bears, tapirs, camels–including the ancestors of the modern llamas and alpaca, deer, and a variety of other North American taxa (Marshall 1988: 382-383; Anderson 1984). The transfer continued during the Ensenadan and Lujanian land mammal ages of the middle and late Pleistocene. As a result, by the Holocene epoch, "nearly half of the families and genera now on the South American continent belong to groups that emigrated from North America during the last three million years" (Marshall 1988: 387).

Extinctions have occurred throughout geological time and the Plio-Pleistocene land mammal ages provide no exception to this generalization. However, while the scale of New World extinctions over the last three million years remained fairly modest, at the end of the late Pleistocene, a massive wave of extinctions took place among mammalian taxa in both South and North America. These extinctions–and the questions they raise about possible human role in their occurrence–will be discussed at the end of the chapter. We turn now to the of humanity's arrival portentous entry into the natural richness of the New World.

The Problem of "Early Man" in the New World

As we have seen, archaeological and geological

work undertaken over the last 150 years in the Old World has successfully demonstrated that humankind evolved there during the Plio-Pleistocene. Conversely, at least a century of similar scientific diligence in New World has revealed neither australopithecine nor *Homo erectus*, Archaic *Homo sapiens* nor Neanderthal. Arguing from negative evidence is always risky in archaeology, but given such concerted efforts at finding it, the absence any credible evidence of earlier forms of hominids strongly indicates that only anatomically modern forms of humankind (e.g., *Homo sapiens sapiens*) were ever present in the New World. Thus, scholars have generally concluded that hominids did not evolve in the New World but migrated there from an Old World evolutionary hearth in Asia.

Based on his examination of the teeth of some 9,000 individual burials, Christy Turner concludes that this hearth was in northeastern Asia. Turner's study is based on the fact that there are over two dozen traits of the human tooth crowns and roots and that these traits are largely independent of one another genetically. Populations differ in there expression of these traits and Turner found that American Indians tended to share a distinct tooth root and crown pattern with modern East Asians. Turner calls this pattern "sinodonty" and distinguishes it from a related southeast Asian pattern he calls "sundadonty." The earliest archaeological evidence for the evolution of sinodonty comes from burials dated to ca. 18,000 B.P. from the site of Chou kou tien but Turner suspects the pattern emerged perhaps 30,000 or more years ago.

In the New World, Turner finds that American Indians exhibit three distinct varieties of sinodonty that correspond to three great geographic and cultural groupings: (1) Eskimos and Aleuts, (2) Northwest Coast Indians and Indians of the Alaskan interior, (3) Indians of North and South America. He hypothesizes that these three patterns reflect three separate immigrations of peoples into the New World from northeast Asia after about 16,000 B.P. (Turner 1987; Greenberg, Turner and Zegura ????; but cf., Gruhn 1987). As we saw in the preceding section, the most likely route for this migration from northeast Asia is the Bering Straits land bridge between Siberia and Alaska. The periodic emergence of Beringia provided the route for the entrance of hominids into the New World, but once there, they were confronted by a second problem. Beringia was exposed only during periods of low worldwide sea consequent on the expansion of continental glaciation in the northern hemisphere and geological evidence indicates the bridge was open during most of the 115,000 years of the Wisconsin glacial stage (Hopkins 1986). Once across the landbridge, hominid populations would have found that only the central plateau region of Alaska and the adjacent lowlands of the Arctic coast unglaciated–and provocative evidence from Bluefish Cave in Alaska suggests that people may have crossed the Straits and entered this unglaciated region as early as 25,000 years ago (Lewin 1987: 1230). Beyond these regions to the south and east they would have been confronted by the ice mass of the North America continental glaciation.

During the late Wisconsin stage of Pleistocene glaciation, two ice masses formed in northern North America: the Laurentide ice sheet in the east and the Cordilleran glacier complex in the west. The Cordilleran glacier complex extended at its maximum from Alaska to the Columbian Plateau in the state of Washington. It was centered on the Coastal Range, the Cascades, the northern Rockies, and the Aleutian Islands and had emerged from the coalescence of numerous individual mountain glaciers as well as the ice sheet that covered the Aleutians. Through much of the late Pleistocene, the Cordilleran glacial complex calved ice burgs into the northern Pacific from the Aleutians as far south as Puget Sound (Butzer 1972: 492).

The Laurentide ice sheet covered the eastern side of the continent from the edge of the Rockies across Canada and the northern United States to Greenland. Its southern most extension is marked today by the drift border, a line of glacial morraines, erratic boulders and other features that stretch across the Midwest on a generally oblique angle from Ohio to Nebraska. Geological evidence indicates that, at least during the period of maximum glacial expansion ca. 18,000 B.P., the Cordilleran and the Laurentide joined at a number of points on the continent to create a continuous ice mass across North America from the Atlantic to the Pacific (Flint 1972: 463). This ice mass must have blocked–or at least complicated–passage south into lower North America and South America from the ice free refugium in central Alaska (but cf., Reeves, Stryd and Smith 1971 for the view that the ice never formed such a barrier).

However, before 18,000 B.P. and definitely after ca. 14,500 B.P., the two ice masses parted slightly to create a gap or "ice-free corridor" between them along the eastern margins of the Rocky Mountains (Butzer 1972: 493). If early people entered the continental United States on this eastern route, they would have been led directly to the central and southern Great Plains. From the Great Plains, they would have found further natural routes downstream along the river valleys that course eastward across the Great Plains to the Missouri and then to the Mississippi rivers. Once across the Mississippi, early peoples could have followed its major eastern tributaries upstream into the interior of the continent (Sauer 1944: 555).

Early peoples might also have taken the intermontane passage through the interior of British Columbia, along the valleys between the Coastal Range and the Rockies (Wedel 1961: 48). This route, passable only during periods of deglaciation, would have brought them to the Columbian Plateau and the Great Basin which in turn would have been a conduit into the southwestern United States, Mexico, and ultimately South America.

However, some scholars have held that, even when the ice-free corridor was open, conditions within it would have been too harsh and game too limited to have allowed poeple to traverse it south. Such objections have led to a search for alternatives to the inland route. Fladmark (1979) cites recent archaeological and geological data indicating that, during the late Pleistocene, a chain of ice-free plant and animal refugia existed along the North Pacific coast on the continental shelf exposed in full glacial stages by lowered sea level. He suggests that peoples practicing a maritime fishing and gathering subsistence might have leap-frogged south along this chain of refugia and become the first to penetrate the New World. Fladmark's thesis is an original one. Conventional wishdom–and current evidence–favors the view that coastal-maritime subsistence first developed at the end of the Pleistocene in response to Holocene climate changes or to population pressure in the Mesolithic period. Unfortunately, Fladmark's thesis is difficult to test since the rise in sea level in post-Pleistocene times has reflooded the continental shelf.

Finally, one result of the transit across the Straits to the New World had repercussion not felt until 1492. This was fact that the passage acted as a disease filter. The most damaging infectious diseases that have plagued humankind–typhus, malaria, cholera, typhoid, diphtheria, to name but a few–appear to have evolved relatively late in the human career. In general, these diseases require dense population as reservoirs for infection and thrive in the filthy, crowded settlements. These conditions emerged in prehistory only with the expansion of population and increased sedentism consequent on the development of agriculture and the rise of archaic states. Hunting and gathering subsistence systems provide poor setting for the persistence and spread of many diseases. Food collecting societies are generally characterized by low population densities and the necessity of moving settlements at various times during the year in response to the shifting availability of wild plants and animals. It is therefore more difficult for disease to become permanently endemic among small bands in the first place and the relative isolation bands one from another prevents acts to slow the spread of epidemics.

Further, the passage of the early emigrants from Asia across the Bering Straits into the New World probably filtered out whatever diseases were endemic in the parent populations from which the emigrants sprang. Cold conditions inhibit the establishment and spread of certain infectious diseases and, if the passage was rigorous, it would have tended to favor healthy individuals and eliminated sick ones. Once in North America, the immigrants would have found relatively disease free circumstances and the severance of the land passage between Siberia and Alaska by the rising sea level in post-Pleistocene would have cut the link that might otherwise have allowed the epidemic diseases of later times to enter the New World.

Of course, isolation from infectious disease also meant that Amerindian populations in the New World developed little disease immunity and this sad fact left them vulnerable to ghastly scourges of sickness and death which beset them with the arrival of the Europeans and Africans. No evidence of the independence and isolation of the New World from the Old before 1492 is stronger than the utter vulnerability of American Indian populations to the endemic and epidemic diseases of Europe and Africa in the early post-contact period (For example, Clark *et al* 1987; but cf., El-Najjar 1979).

The date of the initial colonization of the New World: conservative, liberal, and radical views The most controversial and intractable issue surrounding the study of the colonization New World is **date** of the arrival of humankind in the northern hemisphere. At present, scholarly opinion on the matter currently falls into radical, liberal and conservative camps distinguished by the maximum arrival date the membership of each camp is willing to accept and by the kinds of archaeological data considered acceptable evidence of human presence.

The earliest dates for the arrival of humankind in the New World are championed by "radicals" like the late Lewis Leakey, Ruth Simpson, George Carter and Richard "Scotty" MacNeish, who place no *a priori* or theoretical limit on the date of arrival of humankind in the New World. In the view of many radicals, hominids may be present there prior to beginning of the Wisconsin glacial stage and perhaps thousands of years earlier. One putative early Early Man site, Toca de Esperança in Brazil, has even been assigned a date of 3 mya by its excavators!

Radicals assert that, as a consequence of the great age of the first arrival, most of the evidence of early human presence has been either deeply buried or destroyed by erosion. Nonetheless, members of the radical school assert that a vast alluvial fan in the arid Calico Mountains south of San Diego, California contains crude artifacts indicating a hominid presence there 50,000 to 80,000 years ago (Leakey, *et al.* 1968). The matter was complicated recently by a geological study of the Calico alluvial fan that suggests a minimum age of 200,000 B.P. for the site (Budinger 1983).

The objects recovered at Calico Hills have been characterized by their excavators as being part of a core or chopper-scraper tool complex that was widespread in the New World. According to Carter (1980), this complex centers on the manufacture of extremely crude tools made from cobble-size blocks of quartzite, basalt, or porphyry. It must be noted that these objects appear no cruder than some of the specimens that have been recovered from Plio-Pleistocene strata at Omo, Olduvai Gorge and other localities in East Africa, generally accepted to be of hominid manufacture, and assigned to the Oldowan industrial tradition. However, unlike their African counterparts, these New World objects lack any clear archaeological associations with human remains or with features of definite human construction or modification. Most are surface finds, others have been recovered from sites, that for one reason or another, can be considered problematic. Consequently, these objects have yet to receive wide acceptance as artifacts outside of radical circles. Haynes (1973: 305-310), for example, considers the supposed core tools recovered at the Calico Hills site to be "geofacts" produced without human agency by the natural processes of stone breakage and shattering that accompany slope erosion in desert regions. Proponents of the Calico sites assert that this argument is merely a restatement of the old statistical canard that infinite monkeys with infinite typewriters would ultimately produce the works of Shakespeare purely by chance. In their view, nature simply could not produce and reproduce the flaking patterns evident in the Calico collections. Typological questions aside, only the recovery of such objects in an indisputable archaeological context will settle the early, Early Man controversy.

In fact, the radical and liberal cases for the presence of an early Early Man in the New World was strengthened briefly during the 1970s when the new absolute dating technique known as "aspartic acid racemization" returned dates ranging from 20,000 to 40,000 B.P. on a series of skeletons recovered from Early Man sites in California and elsewhere in western North America. However, recent efforts at re-dating these skeletons using radiocarbon dating by means of accelerator mass spectrometry has forced a revision of these late Pleistocene dates for these skeletons. Based on these latest dates, it is now clear that the amino acid racemization technique had been in error and that these skeletons are of post-Pleistocene age (Taylor *et al.* 1985; Bada 1985).

By being less heroic in the magnitude of the dates they postulate, scholars like James Adovasio, T. D. Dillehay, Alan Bryan, Niede Guidon, G. Delibrias and others constitute the "liberal" camp on the question of humankind's New World arrival. While this group argues that humans entered North America well before 11,500 B.P., unlike the radicals, they do not consider current evidence to indicate that the occupation of the New World began before the last glaciation. Rather, members of the liberal camp currently place that beginning some time around 33,000 B.P. or slightly earlier. The archaeological materials recovered from sites in this time block are intriguing and may eventually be accepted as demonstrating an early human presence in the New World. At present, however, they do not form a consistent or entirely convincing pattern.

The site of Old Crow Flats in Alaska produced a caribou bone flesher or scraping tool which was initially dated to more than 27,000±3,000 B.P. from a deposit of redeposited fossil bones on the Porcupine River (Irving and Harrington 1973). This object was considered strong evidence until recent re-dating revealed it to be of late Holocene age (Nelson *et al.* 1986). Currently, the best evidence for the presence of people in the New World during this early time block comes from Meadowcroft rockshelter, a deeply-stratified and meticulously excavated cave in Pennsylvania. Stratum IIa, the lowest occupation layer at the site, produced fire hearths and stone artifacts and chipping debris and has been radiocarbon-dated to between 19,600 and 8,000 B.P. (Carlisle and Adovasio 1982). However, flora and faunal remains recovered from the stratum contained only species subsequently present in the Holocene. For example, oak and walnut macrofossils were recovered in contexts that returned radiocarbon dates as early as 16,000 B.P. This is shortly after the maximum glacial advance. The Laurentide ice mass was less than 100 kilometers north of the site. Surely the region must have been too cold for such a warm-loving forest assemblage. Are the dates wrong? Perhaps. Outcrops of coal and lens of black organic carbon are present within the shelter near its base prompting critics to suggest that the fire hearth charcoal used to date Stratum IIa may have been contaminated by older carbon (Haynes 1980; Mead 1980). Doubts about the validity of the early date for Stratum IIa were vigorously denied in the course of an acrimonious exchange between critics and excavators (Adovasio *et al.* 1980; Adovasio and Carlisle 1988:

713-714), the questions raised probably will have to be tested by further excavation work at the site.

Similarly, various technical objections have been raised regarding either the validity of the early radiocarbon dates or the association between dates and artifacts at the Valsequillo site in Mexico (Irwin-Williams 1967), Pikimachay Cave in the Ayacucho region of Peru (MacNeish 1979), and the other sites used in support of a "liberal" New World arrival date (Waters 1985: 132-137).

However, in the mid-1980s two more candidates for early, Early Man status were added to the liberal cohort: the sites of Boquerirao do Sitio da Pedra Furada in Brazil and Monte Verde in Chile. Like Meadowcroft Rockshelter, both of these sites contain evidence of very early, apparently human material in a stratified context. The Boquerirao do Sitio da Pedra Furada is located on a plateau in the state of Piaui in northeastern Brazil. It is one of 200 known painted rockshelters in the limestone cliffs of this region discovered in the early 1970s by a joint Brazilian and French research team. A deep test trench was excavated to a depth of about three meters in the deposits near the mouth of this rockshelter. The preliminary report describes a stratigraphic column of five separate strata that produced abundant charcoal and organic material. The layers also contained circular hearths and large quantities of crude lithic materials including hammerstones, flaked pebbles, flakes and stone fragments. The stratigraphic sequence appears continuous and, on the basis of the radiocarbon-dates from the strata, formed between ca. 6,160$\pm$130 and 32,160$\pm$100 B.P. In addition, a piece of charcoal from a level in which a painted fragment of the wall had been buried returned a date of 17,000$\pm$400 B.P. (Guidon and Delibrias 1986). The preliminary nature of the accounts thus far published about Boquerirao do Sitio da Pedra Furada make this site difficult to evaluate. However, there are at least two reasons for reserving judgement it. First, only five stratigraphic units were designated in the course of excavating three meters of deposits. Adequate modern excavations in dry caves of comparable depth generally recognized ten or twenty times that number. Second, the excavation of this huge test trench seems to have been accomplished in two brief field seasons, a hasty pace given the depth–and importance–of these deposits.

The Monte Verde site in south-central Chile has been described in more detail to date (Dillehay, *et al.* 1982; Dillehay and Collins 1988; Dillehay 1989, 1997). The site, buried in the banks and adjacent terraces of the Chinchhuapi Creek in south central Chile, contains two apparent occupations: one radiocarbon-dated to 13,000 B.P. and an earlier one dating to 33,000 B.P. The upper or latest occupation layer contains evidence of a possible mastodont kill and putative human settlement. The smaller lower and earlier layer consists of burned, hearth-like basins containing specks of charcoal and ash and charred and uncharred seeds and root fragments. Twenty-six objects of non-local stone were found in association with these basins. Although 18 of these specimens appear to have been broken by natural means, eight are considered to be crude tools by the excavators (Dillehay and Collins 1988: 150). Both the stratigraphic interpretation and the radiocarbon dating of this early later seem to be unimpeachable. Only the crudeness and limited number of the stone objects justifies a certain agnosticism about the site's human origins. The question of whether or not they are geofacts rather than artifact has yet to be answered.

If Monte Verde and Boquerirao do Sitio da Pedra Furada have been correctly interpreted and dated, they are the oldest archaeological sites currently known in the Americas. If they withstand the scrutiny they will surely receive from the scientific community in the next few years, they will be the two most important archaeological sites in the hemisphere. As we say in prehistory, time will tell (see also, Gillespie, *et al.* 1984: 15-20).

Finally, members of the conservative camp, led by C. Vance Haynes and Paul S. Martin, asserts that humankind arrived only recently in the New World. Of the three schools of thought, the conservatives have the most stringent requirements for the kinds of archaeological evidence of Early Man that they will accept. According to Haynes (1969), demonstration of the presence of early humans at an archaeological site minimally demands: (1) the presence of human skeletal remains or definite artifacts in a (2) conclusively-dated geological context which is (3) undisturbed–preferably primary–and clearly displays the relationship between the artifacts and stratigraphy. Haynes applies his three criteria to the late Pleistocene archaeological record and concludes that to date the evidence that has been mustered for an early hominid presence in the New World by the radical and liberal schools is defective in terms of one or more of his three criteria (also cf., Waters 1985 for an update of Haynes study). In his view, indisputable evidence for human presence in the New World is no older than about 11,500 B.P. to 12,000 B.P. at sites of the Paleo-Indian tradition such as: Lehrer, Blackwater Draw, Naco, Murray Springs, and Folsom in the American Southwest, Shawnee-Minisink in Pennsylvannia, Debert in Nova Scotia, Olsen-Chubbuck in Colorado and a number of others. Thus, Haynes and other conservatives deliver the Scot's verdict of "Not Proven" in the question of the presence of "early

Early Man" in North America. Of course, scholars in the radical and liberal camps reject Haynes' "legalistic" approach to the archaeological record and defend the acceptance of a broader range of evidence for human presence (Bryan 1986). For example, Adovasio asserts that the unfluted but lanceolate biface in Stratum IIa at Meadowcroft Rockshelter is unequivocally older and therefore possibly ancestral to the Paleo-Indian projectile point forms of post-12,000 B.P. despite Haynes' technical challenges to the early date for the stratum.

New World Upper Paleolithic: the Paleo-Indian Tradition

The earliest indisputable archaeological sites in the Northern Hemisphere are part of what has been termed the Clovis culture or the Llano complex. The Llano complex is the first of three sequential assemblages that form the Paleo-Indian tradition in the New World. The Llano complex is tightly dated to the terminal Pleistocene between ca. 11,500 B.P. and 11,000 B.P. and is relatively well documented. The Folsom complex succeeds the Llano during the terminal millennium of the Pleistocene epoch and is dated in absolute time between about 11,000 B.P. and 10,000 B.P. The Folsom complex is in turn supplanted by the various regional expressions that have been grouped together as the Plano complex and by the Archaic and Desert Archaic cultural complexes at various times in various regions after the close of the Pleistocene epoch around 10,000 B.P. Only at a few rare sites, such as Blackwater Draw in New Mexico (Hester 1972) and the Carter/Kerr-McGee site in Wyoming (Frison 1984) is the Clovis-Folsom-Plano succession–or even portions of it–present in stratigraphic order.

A South American version of the Paleo-Indian tradition–called the Fishtail Projectile Point Horizon or the Old South American Hunting Tradition–appears sometimes after 12,000 B.P. At least by about 9,000 B.P., Paleo-Indian peoples were occupying Fell's Cave in Tierra del Fuego on the southernmost tip of South America. Although the tradition begins to be replaced after 10,000 B.P, it lasts as late as 5,000 B.P. in some parts of the South American continent (Lynch 1978: 457, 487).

The Paleo-Indian stone assemblages throughout the New World are characterized by distinctive lanceolate-shaped projectile points and numerous bifacially-flaked implements including knives, drills, denticulates or saws, and spokeshaves. Unifacial tools in Paleo-Indian tool kits include burins, side scrapers, and steep-edged end scrapers. Rogers (1986) has recently confirmed that the spurred end scraper or perforator made on a flake (variously called "graver," "beak" or "spur") is a particularly diagnostic Paleo-Indian artifact type.

Paleo-Indian peoples seem to have invested a good deal of effort in the manufacture of their stone tool kits. For example, many of these Paleo-Indian stone tools are made of high quality cryptocrystalline rock which often had to have been derived from a quarry source at great distant (Kelly and Todd 1988: 237-238). Further, Paleo-Indian stone knappers commonly seemed to have carefully heated this raw material above 500 degrees Celsius before working with it in order to enhance its flaking properties (Pavlish and Sheppard 1983). Finally, modern stone knappers who had replicated Paleo-Indian points attest that "fluting" them in the characteristic manner is a difficult operation and often breaks the point at the very final stage of its manufacture. Small wonder that Paleo-Indian lithic artifacts commonly show resharpening and other evidence of a long use life (Kelly and Todd 1988: 237-238). In archaeological parlance, these objects were carefully "curated," rather than discarded, after use.

The occasional recovery of needles, rods, wanes, points and other cylindrical, subcylindrical and beveled objects made of ivory and bone from Paleo-Indian sites indicate (1) that perishable raw materials were also extensively utilized in tool manufacture, and (2) the "splinter-and-groove" technique for removing long strips of workable raw material from bone, tusk or horn was known to the Paleo-Indians (Frison and Zeimans 1980). Further, as Cotter (1981) points out, many of these bone and ivory objects are morphologically similar, if not identical, to tools from the Magdalenian and Solutrean traditions of the European upper Paleolithic period. Perhaps the most striking parallel to the Old World upper Paleolithic is the bone handle with a socket recovered at the Murray Springs Llano site in Arizona. Objects of this type, thought to have functioned as dart shaft wrenches or straighteners, are common in European upper Paleolithic assemblages where they are referred to as *baton-de-commandements* (Haynes and Hemmings 1968).

However, the most celebrated artifact of the Llano complex, the "fluted" Clovis projectile point, has no obvious Old World parallels or origins. Clovis points occur in many sizes but are generally quite large. According to Irwin (1970: 26), they average 6.5 centimeters in length and range between 2.5 to 13 centimeters. They are lanceolate in shape and were produced by direct, soft hammer percussion. As the final step in their manufacture, such points were thinned on one side (and sometimes two sides) along their medial axis through the removal of a small flake or "flute." The removal of this flute created a channel scar which runs from the base of the point for about a quarter to a half of the point's length. This removal was the final operation performed in making the Clovis point and the difficulty of doing it correctly resulted in the stone knapper ruining the point perhaps as often as one in every three tries. The base of Clovis points are slightly concave and this, together with the fluted groove, probably facilitated mounting the point tightly on a shaft. The edges around the base of these points usually show signs of having been ground, apparently to blunt them and thus prevent the point from cutting the thongs used to haft it to the shaft. Some Clovis specimens have been retouched along the edges of their blades by pressure flaking either to improve their shape or resharpen them after they had been dulled by use.

The Anzick site in Montana has provided some clues to the manner in which Clovis point were hafted (Lahrens and Bonnichen 1976). Road construction near this site accidentally exposed two sub-adults buried in a collapsed rockshelter. The skeletons, radiocarbon dated to 10,600±300 B.P., are among the oldest certainly-dated human remains in North America (Taylor *et al* 1985: 138). Recovered along with these burials were more than 100 Llano complex stone and bone artifacts including cylindrical or rod-like bone objects with beveled ends in associated with a number of Clovis points. Bone rods recovered at other Llano sites have been interpreted either as projectile points or fleshing tools. Based on their work at the Anzick site, however, Lahrens and Bonnichen (1976) suggest they are the bone foreshafts for holding points on darts or spears. They conclude that these artifacts were used with small wooden splints or shims but such objects have yet to be recovered at a Llano site. In their view, Llano peoples made a composite form of projectile weapon by gluing a Clovis point to a bone foreshaft and wooden splint with resin and then wrapping the three parts with sinew (see figure). They did this because a composite foreshaft and point would have been "more resilient to stress than wood in the actual stabbing operation and would allow deeper penetration of the point into the animal" (Lahrens and Bonnichen 1976: 149). The composite point/foreshaft assembly could have been inserted into a separate main shaft for throwing. If this was done, many such points/foreshaft assemblies could be carried for use in a single lance or *atlatl* dart. Unfortunately, it is not known whether or not Paleo-Indian peoples used the *atlatl* or throwing stick to propel their projectile points. The *atlatl* in certainly present in the Old World by 14,000 B.P but the earliest evidence *atlatl* recovered to date in the New World dates only to 6,500 B.P. (Dickson 1985: 8).

Whatever the means whereby Clovis points were mounted and used, they appear to have been part of an effective hunting technology. Such points have been found in both primary and secondary archaeological association at a variety of "kill sites" with various forms of Rancholabrean fauna megafauna including horse, camelops, bison and, almost invariably, the Columbian mammoth. In addition to kill sites some camp and quarry sites are known to contain Clovis components. Clovis points have been recovered from central Canada to central Mexico and "fishtail" and other variants of the Clovis have been covered as far south as Patagonia in South America. The ubiquity of these points no doubt indicates the presence of Llano peoples over most of unglaciated North America and much of South America during the late Pleistocene. Nonetheless, the largest number and most unambiguous associations between late Pleistocene megafauna and Clovis artifacts has come from the Great Plains and the southwestern United States.

A classic example of a Llano kill site was excavated from a deeply-eroded channel wall of the San Pedro river on the Lehner ranch just north of the Mexican border near Hereford in southeastern Arizona. The original surface of the site had been buried under more than 2.5 meters of alluvial deposits. It was therefore entirely invisible on the modern ground surface and would remained undetected by archaeologists had not natural circumstances intervened. Like most other streams in the Southwest, the San Pedro began eroding its floodplain just before the turn of the century and this erosion removed the overburden and exposed the site in the newly exposed wall of its deepened channel where it was noticed in the 1950s. The original excavation of the site revealed a bed of disarticulated animal bones, several fire hearths, Llano complex stone tools and chipping debris. Most importantly, some 13 Clovis projectile points, some of which were found touching the bones of nine disarticulated mammoths, were recovered from the bone bed. The hearths contained the charred bones of horse, bison and tapir. Carbon-14 dates from the site range in age from 11,600±190 to 10,410±190 B.P. Pollen recovered from the site suggests it was a grassy swale at the time of the kill. Very likely the mammoth and perhaps some of the other animals were driven in to the swale, mired down and dispatched.

The nature of Llano complex kill sites like Lehrer–not to mention the tool kits recovered from them–remind C. Vance Haynes of the late Pleistocene age mammoth sites of the central Russian plain in the Old World and, in his view, the resemblance is more than accidental. For Haynes, the Llano complex represents a late migration of descendants of Siberian mammoth hunters into the New World at the end of the last interstadial of the Wisconsin glacial stage. He hypothesizes that the sudden appearance of the Llano complex in lower North America after 11,500 B.P. followed their successful transit in the ice-free corridor opened between the Cordilleran and Laurentide ice masses around that time. In lower North America, these descendants of the mammoth hunters of the Old World would have found a continent teeming with Rancholabrean age game naive to the threat posed by two leg predators!

How did these new predators operate? We can never know with certainty what techniques were used by the Llano hunters but clues for interpreting the archaeological record are to be found by examining the behavior and nature of modern domesticated and wild elephants as well as ethnographic accounts of peoples like the Akoa and M'buti pygmy peoples of equatorial Africa who hunted modern elephants until quite recently.

Domesticated elephants provide some particularly unexpected insights. Some years ago, P. E. P. Deraniyagala published an interesting compilation of fact and folklore about domesticated Indian elephants based on his own field work and on his interviews with South Asian *mahouts*, or elephant handlers. According to Deraniyagala, whose work was published by the Ceylon National Museum, the elephants most remarkable feature

> are the numerous sensitive spots located all over its body which enable a mahout to force the animal to kneel, raise a limb or trunk, move or to trumpet. These are induced as a series of reflex actions either by pressing, pulling or striking appropriate areas with the mahout's bare person or prodding with a goad. Prodding some of these spots to a depth of a couple of inches can even cause death and consequently this science is a jealously guarded secret (Deraniyagala 1955).

Deraniyagala asserts that the complex nervous system of the elephant provides the *mahout* with fully 90 nervous centers or pressure points near the external surface of their bodies. Knowledge of the location of these points allows the *mahout* to control the beast using only a short stick or goad with a hook on one end and a point on the other. He further asserts that there are another 12 points that can enrage the animal to the point of killing and, 18 points which will cause death to the animal if he is merely prodded on them deeply enough. Is this merely folklore? Deraniyagala himself is uncertain, but he reports it nonetheless. He also cites a Dutch colonial hunter's assertion that an elephant can be killed with a bullet or arrow lodged behind the ears or in the middle of the of forehead and hurt by strikes to vital centers located on the middle of the occiput and on the soles of the feet. In addition to these points of vulnerability, Deraniyagala reports that modern elephants can not travel more than four miles without requiring food and water. Deprived of food for two days, an elephant will fall down but will continue to live for up to five days more. If deprived of water, an elephant will expire in three or four days (Deraniyagala cited in Orr 1978: 79).

Studies of wild modern elephants help complete the picture. Modern African elephants–and by extension their extinct proboscidean cousins–move over large territories in response to season changes in rainfall and vegetation. Once in a locality, however, elephants tend to follow established trails and repeatedly use the same springs (Oliver 1984). The Akoa and M'buti ambush elephants along these trails either (1) by showering poisoned arrows or spears on the beasts in order to injury them at a distance or (2) by sending individual hunters to hamstring an elephant from behind or jab a poisoned spear in its stomach or between its stomach and rib cage (Coon 1976: 135-144). The elephant is seldom killed in the initial attack and the hunters must track it until it falls from exhaustion. Wounded elephants generally flee towards water. A wounded beast able to reach a marsh or river often becomes mired in it and, so entraped, is finished off by the hunters.

Archaeological evidence of the hunting strategy of the Llano mammoth hunters appears consistent with this ethnographic information. Based on their study of Paleo-Indian site placement in the central Rio Grande valley, James Judge and Jerry Dawson recognized a specific constellation of physiographic features which Paleo-Indians sought when choosing the locations for their camps. According to them, sites of this tradition are commonly (1) located on ridges, (2) near *playas* (shallow desert basins that formerly held water during the cooler, wetter conditions of the late Pleistocene) or water courses, and (3) adjacent to areas suitable for grazing. We cannot be certain to what degree this constellation of features can be generalized to North America at large, although a similar playa-ridge-drainage site location pattern has also been observed on the Llano Estacado in Texas (Wendorf and Hester 1962: 166). It is also interesting to note that a high percentage of the kill sites excavated thus far had once been bogs or marshes and that such sites commonly contain artifacts made of lithic material drawn from quarry sources often 100 to 300 kilometers distant.

These geographic and cultural configurations

suggest that Paleo-Indian settlement technology was an adaptation to a animal behavior not unlike that of the modern elephant. For example, the common presence of artifacts made of lithic materials from distance sources in Llano kill sites may indicate these people moved over wide territories in response to seasonal mammoth migrations. The number of Llano kill sites located in springs, marshes, or along boggy sections of stream courses suggests the hunters sought to use locations where the great size of their prey would mired and immobilized them and allowed their easy dispatch. Paleo-Indians may have driven the game into these bogs by means of grass fires. However, it seems just as probable that, like the historic pgymy elephant hunters, Paleo-Indians ambushed individual mammoth or small herds along the routes to their accustomed water hole or feeding ground and then trailed the injured beasts into marshy ground for the kill.

If the Deraniyagala's assertions about modern elephants are true of of their extinct ancestors, then a wounded mammoth, forced to mill about in a box canyon or struggle in a bog would rapidly become exhausted, parched with thirst, and weakened from lack of food. Careful hunters using Clovis points on the tips of thrusting spears could repeatedly jab at the beasts vital spots exposed during the struggle. An abundant supply of the socketed bone foreshafts postulated by Lahrens and Bonnichen would be particularly useful at this stage in the attack. Key resources needed for such hunting include appropriate water conditions, grassland areas capable of supporting ample herds of megafauna, and vantage points high enough to allow these herds to be monitored and stalked from strategic distances.

This makes it all sound so easy and might tempt us to forget that such hunting has always been dangerous. Coon (1976: 137) describes a petroglyph from Saharan Libya which shows a man in the grasp of an elephant's trunk–a scene which no doubt records an ancient fatal encounter. We can only speculate on how commonly such scenes were repeated in prehistory but we know that modern pgymy hunters have often been lost in this way. The intelligence, speed, acute sense of smell–not to say the large tusks and body–of the modern African elephant makes it a formidable prey. The extinct Columbian mammoth, about 20 percent larger than the modern African elephant, was surely every bit its equal. Yet, in recognizing that proboscidean hunting has never been a safe and easy way to make a living, we must not exaggerate the difficulties involved in it. After all, ancient peoples seemed to have engaged in it with great alacrity and effect–and kept it up right to the bitter end of proboscidean extinction in the northern hemisphere.

But given the apparent zeal of the Llano hunters and the ubiquitness of their tools through the Americas, it is interesting to note that Paleo-Indian kill sites are relatively abundant only in the western United States. Elsewhere they are absent or extremely rare. Despite concerted archaeological effort, for example, the first indisputable evidence that Paleo-Indian peoples living in eastern North America hunted mastodont and *B. b. antiquus* was not recovered until the beginning of this decade (Graham, *et al.* 1981; Webb, *et al.* 1984). The *B. b. antiquus* site consisted merely of a bison skull dating to ca. 11,000 B.P. with a chert projectile fragment lodged in it recovered eroding from the bank of the Wacissa river in Florida. The discoverers of the Wacissa bison suggests that the paucity of such sites in the east and their abundance in the west reflects contrasting environmental circumstances and correspondingly different herd dynamics in the two areas rather than different Paleo-Indian lifeways. They suggest three possible explanations for the rarity of megafaunal kill sites in the east:

> First, the herd animals, which were presumably the main quarry, were probably broken up into smaller groups in the predominantly forest settings of the eastern regin. Second, opportunities for ambush and entrapment generally may have presented themselves on a smaller scale than did the grand topographic features of the west. Thus, hunting by Paleo-Indians in eastern North America may have taken place on a smaller scale than it sometimes did in the west. And third, the more humid climate may have obscured such small scale sites under water, vegetation, or soil development, and may have destroyed the evidence by leaching in acidic soil (Webb, *et al.* 1984: 391).

An even more probable reason for the contrasts in the number of kill sites between the east and west stems from differences in the late Pleistocene geological conditions that first buried, then uncovered, Paleo-Indian sites in the two areas. According to Haynes (1982: 385), studies of the Quaternary sedimentary history of streams all over lower North America indicate that most of them had begun to degrade their floodplains after about 12,000 B.P. That is, during the final phase of deglaciation, these water courses had begun to cut into and lower the grade of their channels and incise narrower ones. As the streams cut down into their former floodplains and abandoned them, they created terraces or benches. After 11,500 B.P., terraces near riparian locations were apparently favored as camp site locations by Llano peoples. About 11,000 B.P., this stream degradation appears to have ended. A cycle of aggradation or deposition followed in the west and the stream channels began to fill with sediments. Many Llano sites located on the terraces were buried by clayey silts washed from adjacent slopes during this aggradation phase.

> After 7,000 or 8,000 B.P. many smaller tributaries, particularly in the Southwest, entered a cycle of gully cutting and filling every 2,000 years or so; the cycles tended to be in phase (within the margins of error for radiocarbon dating) over large areas. Because of this alluvial history many Clovis sites still lie buried beneath modern floodplains (Haynes 1982: 385).

Since the 1880s, streams of the southwestern United States, and to a lesser extent, the southern Great Plains, have experienced dramatic downcutting and erosion that has deepened their channels. This regional erosion–perhaps caused and certainly exacerbated by overgrazing and other anthropogenic practices–has exposed huge profile sections of formerly buried floodplain sediments all over the Southwest. Also exposed in these profiles have been animal bones and Paleo-Indian artifacts at sites like Clovis, Blackwater Draw, Folsom in New Mexico and Naco, Murray Springs, Domebo, and of course, Lehrer in Arizona. The vegetational cover and river regimes of the eastern United States are different than those of the Southwest and have not experienced stream channel down cutting on a similarly regional scale and this has greatly impeded the discovery of Paleo-Indian kill sites in the East. Nonetheless, the discovery of the bison kill eroding from the Wacissa river bank and Thunderbird site eroding from the modern floodplains of the Shenandoah River illustrate that Paleo-Indian remains are present in the deep aluvium of the East. Perhaps some truly large Paleo-Indian game drive sites await discovery there, too.

Sites containing indisputable and direct associations between late Pleistocene megafauna and Paleo-Indian remains, although extremely rare, have been excavated in South America. For example, mastodonts and projectile points were found together in the later of the two occupations at Monte Verde, Chile (Dillehay 1989), sloth, *Scelidotherium*, horse, camelid, deer and other fauna were associated with stone tools at Pikimchay Cave in Peru (MacNeish 1979), and horse, mylodon and guanaco at were associated with fishtail points at Fell's Cave in Tierra del Fuego at the tip of the South American continent (Bird 1969). As Lynch (1978: 475) notes, virtually all South American sites of the early South American Paleo-Indian tradition have been reported from (1) regions that would have supported tundra, savannah, steppe or other open habitats during the Pleistocene, (2) from the vicinity of freshwater. He concludes that the distribution of these sites in is clearly related to the habits of the game animals and the hunting methods employed and that, even in the absence of kill sites like those in the north, it is reasonable to conclude that Paleo-Indians there practiced communal hunting and game driving against the South American megafauna. However, none of these South American Paleo-Indian sites found to date are on the scale of the more dramatic western North American kill sites.

The Folsom complex. The Llano complex begins to disappear in the western United States after about 11,000 B.P. and is succeeded in absolute time by the Folsom complex. Radiocarbon dates of the complex cluster between 11,000 and 9,000 B.P. but earlier and later dates for Folsom and Folsom-like material have been reported. Nonetheless, at the Blackwater Draw Locality 2 and Carter/Kerr-McGee sites, the stratigraphic precedence of Clovis before Folsom, Folsom before Plano is clearly demonstrated.

The diagnostic implement of the Folsom complex is the Folsom projectile point. Although lanceolate in shape, points of this type are generally smaller, squatter and more delicately made than Clovis points. Folsom points average only 5 centimeters in length and range between 2.5 and 7.5 centimeters (Irwin 1970: 26). Folsom points were mainly produced by direct percussion, but generally exhibit fine pressure retouch along their edges. They are widest at their mid-sections rather than their bases and have markedly incurvate stems with ear-like projections. Flutes or channel flakes were removed longitudinally along the medial axis of the point from its base almost to its tip. On the classic Folsom, flutes were struck from both sides rending the point bi-concave in cross-section. The flute, which is really a small blade flake, was removed either by indirect percussion or by pressure flaking perhaps using a chest-punch. In addition to these distinctive fluted points, Folsom assemblages commonly include unfluted lanceolate points, gravers, thumbnail scrapers, and long thin blade flakes struck from prism-shaped cores. While the Llano complex and its South American variants appears to have been distributed throughout the Americas, the Folsom complex is, in the strict sense, limited to the North American Great Plains from Montana to southern Texas and to the southwestern United States. Across the Mississippi in the southeastern United States, the

Cumberland, Quad, and Dalton points appear to be formal variants of the Folsom point and were probably part of assemblages contemporary with the Folsom complex.

Like the Llano complex, Folsom assemblages have been found in direct association with extinct forms of late Pleistocene megafauna. However, the Columbian mammoth had largely faded into extinction by 11,000 B.P. Instead of mammoths or mastodonts, *B. b. antiquus* and *B. b. occidentalis* are the most common species found in Folsom kill sites such as the type site at Folsom, New Mexico. Other Rancholabrean forms present at such sites on rare occasions include camelops, horse, and smaller forms like pecarry. Perhaps the replacement of the Clovis point by the Folsom point was related to the changing technical requirements of bison hunting as opposed to elephant hunting. In addition to kill-sites, at least one Folsom camp–the Lindenmier site in Colorado–has been extensively excavated.

Since Folsom hunters targeted different species than their predecessors, how did their hunting strategies differ from those used in earlier times by the Llano peoples? Prehistorians have attempted to answer to this question using both archaeological evidence and analogies drawn from historic accounts of American Indian bison hunting. Modern bison (*B. bison*) are myopic and rely on their highly developed sense of smell to alert them to danger. Like many other herd animals, bison are prone to blind stampedes when frightened. Therefore, historic Plains Indian hunters were able to approach bison herds from the down wind side and either pick off individual animals or, by suddenly making their presence known, induce the entire herd to stampede. Intentionally set grass fires were also used to stampede these animals. When the stampede approach was used, it was generally necessary to build barriers or place other hunters along the flanks of the herd to frighten and direct its path either towards a cliff or deep arroyo–where the fall would kill most of them–or into a box canyon or log corral–where they could be dispatched at will. Bone remains and other archaeological evidence indicate that a number of Folsom sites were terminal points for such bison drives. Key examples include Folsom in New Mexico (Figgins 1927), Bonfire Shelter (Dibble and Lorrain 1968) and Lubbock Lake (Johnson 198-) in Texas, and the Folsom component at the Agate Basin site in Wyoming (Frison and Ziemans 1980: 231). Although *B. b.antiquus* and *B. b. occidentalis* were larger than modern *B. bison*, the fact that Paleo-Indians were able to drive them over cliffs or into steep arroyos suggests these ancient beasts behaved much as their modern descendants do and were hunted in the same manner.

The Plano complex. The final cultural assemblage within the Paleo-Indian tradition is referred to as the Plano complex. This complex appears at different times and places between ca. 10,000 B.P. and 6,000 B.P. The Plano complex apparently evolves out of Folsom complex and reaches it greatest development in the heartland of the older tradition on the Great Plains and in the American Southwest. However, variants of the Plano tradition are also found in the mountains of the northwestern United States and north central Canada, both regions first exposed in early Holocene times by the retreating continental ice masses. To the south, the Lerma complex, a regional variant of the Plano culture appears in northern and central Mexico at about this time. The hallmark of this tradition is the distinctive bipolar Lerma point. Meanwhile, on the margins of the Plano regions in the eastern United States and the expanding Holocene deserts of the west and southwest, a new tradition had begun to emerge more-or-less contemporarily with the later Plano complex. These traditions are known respectively as the Archaic tradition and the Desert Culture or Desert Archaic tradition.

The hallmarks of Early Plano complex tool repertoires are large, generally lanceolate, projectile point and knife types clearly derived from the point making traditions of the earlier Llano and Folsom complexes but distinct from them. While early Plano points exhibit the lanceolate shape and basal thinning of the Clovis and Folsom points, they lack the distinctive basal fluting and exhibit more formal variability than those earlier types from which they presumably developed. Further, in contrast to the uniformity of the Clovis and Folsom points throughout the ranges of the two culture complexes, projectile points of the Plano complex exhibit pronounced formal and regional diversity. A. L. Bryan (cited in Willey 1966: 44), divides Plano point types into two major categories: the Plainview group and the parallel-flaked group. Point types in the former category, such as Plainview, Meserve, Midland, Agate Basin, and Milnesand, are essentially unfluted versions of the earlier Clovis and Folsom points. Types in the parallel-flaked group are less rigidly lanceolate in shape and were manufactured by means of careful pressure flaking in which blade flakes were removed in a continuous, serial fashion across the face of the projectile point blade. This technique (which is also called transverse flaking) results in a series of uniform flake scars running from base to tip parallel to one another but perpendicular to the medial ridge. Types in the second group tend to be thinner, stronger and more regularly-shaped than the more randomly flaked points in the Plainview group. Stemless point types in this group include the Agate Basin and Angostura points. Slightly later, parallel-flaked types in the group, such as the Scottsbluff, Eden, Cody, and Hell's Gap types, exhibit slight shoulders and straight-stemmed hafting areas at their bases. In general, Plano projectile point development proceeded from the early, unfluted, lanceolate forms to later shouldered and stemmed varieties.

At early Plano complex sites, unfluted, lancolate points are associated with late Pleistocene megafauna such as horse and *B.occidentalis*. Later style Plano points are associated with a faunal repertoire that is essentially a modern in character and includes *B.bison*. Abundant archaeological evidence from sites like Plainview in Texas, Olsen-Chubbeck site in Colorado (Wheat 1972), and Casper in Wyoming (Frison 1974) indicate that the Plano peoples continued the use of the jump and the surround techniques employed by the earlier Folsom peoples. Fauna reported from other, later Plano sites include deer, elk, antelope, rodents and other forms best taken by other techniques. The range of game from sites within the culture area of the Plano complex suggests that some of the diversity of Plano projectile point types reflects diverse local specialization in game.

In addition, recent fieldwork in northern Wyoming indicates that, at least shortly after 9,000 B.P., Plano peoples were practicing another form of hunting: animal trapping using nets. The evidence for this is a large net made of juniper bark cordage radiocarbon dated to 8,860±170 B.P. recovered from a dry cave in the Sheep Mountains of northern Wyoming. The net had been folded and carefully stored in a cave in the midst of prime mountain sheep (*Ovis canadensis*) country. It is too fragile to be unfolded, by it appears to be shaped like a modern tennis court net and is estimated to stand two meters high and stretch to a length of 50 or even 65 meters. Both its size and historic accounts of net hunting by American Indians suggests that the net was used to trap large mammals like deer, prognhorn, or mountain sheep. The net was probably stretched on poles across a game trail and animals then driven into the net by beaters. As they fought to free themselves from its coils, the animals were probably clubbed to death by hunters waiting nearby. Net hunting in this manner would leave little tangible evidence behind in the archaeological record. Therefore, although the Sheep Mountain find is the earliest definite evidence of its practice among Paleo-Indians, net hunting may substantially pre-date the Plano complex (Frison *et al.* 1986). Finally, late Plano sites such as the one at Horner, Wyoming, have produced grinding stones or crude manos and metates suggesting a greater dietary role for wild plants also developed at the end of the Paleo-Indian tradition.

Economy and society in late glacial America Although the archaeological record allows us to speak with some confidence about Paleo-Indian big game hunting technology and subsistence, that record tells us very little about their socio-economic organization. Consequently, it has been fashionable to attempt to flesh out the picture of Paleo-Indian society and economy using analogies drawn from accounts of modern or historically-known hunter-gatherers like the !Kung Bushmen of Kalahari desert of southern Africa. Ethnographers have found that foraging peoples like the !Kung depend more heavily on the gathering of plant resources rather than hunting and are organized into bands or family clusters that more-or-less systematically follow a seasonally-timed circuit of movement within a specific territory. The !Kung settle for a period at a water hole location in order to take advantage of the plant and animal resources in its vicinity. They move on when these local resources begin to diminish but return to the locale in the following year. It is also common for modern foraging groups like those of the !Kung to break up and disperse in smaller units in one season only to reunite in the next season in response to increased availability of wild foodstuffs.

Seasonal movement, aggregation and dispersal, and the utilization of a broad-spectrum of resources are practices that, according to Kent Flannery (1968), prevent foraging peoples from over-exploiting the wild resources in their territories and degrading the environment upon which they are dependent. At the same time, modern foragers practice various forms of birth control and infanticide in order to hold their population size below the human carrying capacity of their environment. Finally, in historic and modern times, individual foraging bands are (or were) integrated into extensive extra-band social networks. That is, they are connected to other bands in wide-ranging regional trading, marriage, political, and religious exchange fields. These networks or exchange fields have important adaptive implications. The individual bands within them: (1) share intelligence about game movements or plant availability with one another, (2) arrange marriages or cooperate in hunting, ritual, or social activities through the networks, (3) exchange raw materials, food stuffs, and other commodities, and (4) move temporarily into the territories of other bands in the regional networks during times of poor hunting or local resource shortage in their own territories.

By analogy with the !Kung and other contemporary foragers, many scholars conclude that Paleo-Indians also relied more on plants than animals in their diet, practiced seasonally-scheduled subsistence activity within a fixed territory, developed a variety of social and economic mechanism designed to prevent over-exploitation of the wild resources upon which their subsistence depended, controlled their population growth by cultural means, and were integrated into extra band, regional networks.

For example, Edwin Wilmsen uses a model of this sort in reconstructing the socio-economic system of the ancient inhabitants of the Lindenmeier site, a Folsom complex camp site in Colorado (Wilmsen 1974). Wilmsen did not excavate the site himself; that was done by the celebrated archaeologist Frank Roberts during seven field seasons at the site from in 1934 to 1941. However, Wilmsen used Roberts' detailed field notes and huge collections in performing a painstaking analysis of the stone tool and tool making debris from the site as part of his doctoral dissertation.

Although his analysis focused on the Lindenmeier stone tools and debris, Wilmsen's real interest lay in reconstructing the nature of social organization of the site's inhabitants. He used the former material to reach the latter goal by assuming that increasing degrees of social separation between social groups will be accompanied by increasing differences between the stone tools made by each group. These differences will occur because, as contact and interaction between members of two groups diminish, the probability becomes greater that random differences in techniques of manufacture of the same kinds of tone objects will occur and become fixed in the separate groups. Therefore, separate Folsom bands would, over time, make the same types of tools in slightly different ways; the greater the time or degree of their separation, the more pronounced these differences would become.

When Roberts excavated the Lindenmeier site, he found there were two fairly separate and distinct clusters or concentrations of the stone material on its occupation surface. Wilmsen's analysis of these materials revealed that the populations of stone artifacts and chipping debris collected from the two clusters differed from one another in terms of small–but measurable and statistically significant–differences of shape, style, chipping technique, and raw material type. He concluded that these differences indicated that two contemporary but distinct Folsom social groups had been present during the occupation of the Lindenmeier site. Wilmsen hypothesized that these two groups were probably semi-autonomous foraging bands whose territorial ranges overlapped at Lindenmeier. He goes so far as to suggest that:

> periodically, as conditions permitted, these groups scheduled their movements so that they would meet in this valley. They did so to cooperate in bison hunts–which neither probably could carry out alone–and to continue a series of social transactions upon which both depended for continued existence. Mates could be exchanged, adolescents initiated into adulthood, the sick and recently dead given proper ceremonial attention, and the natural environment given whatever ritual care that may have been thought necessary. The groups need not have remained together for long; a few weeks would have been enough to allow time for all cooperative activities and transactions to be completed. They would then move in their separate directions to exploit other parts of their own territories from other locations. Lindenmeier seems to have been the single largest camp unit in their settlement system, and presumably there were a number of smaller campsites at which fewer people were in temporary residence (Wilmsen 1974: 116).

Wilmsen's lithic analysis of the Lindenmeier lithic collection was both original and exhaustive and I am satisfied that he has isolated genuine differences between the two concentrations of lithic debris found there. However, it is important to note that, even if the two clusters were in fact deposited by two separate social groups, there is not enough resolution in the Lindenmeier archaeological record to be certain that the groups were on the site at the same time. It is equally possible that the clusters were discarded by different groups occupying the site years apart from one another–or even the same group occupying and then re-occupying a slightly different portion of the sites at a later time.

Wilmsen's interpretation of the Lindenmeier as meeting place for two bands is a reasonable interpretation of the pattern of debris there only if Paleo-Indians really were organized along the lines of modern hunter-gatherers like the !Kung. Kelly and Todd (1988: 231), however, suggest that they probably were not so organized. To them, the archaeological picture of Paleo-Indian life that emerges from the sites and the analysis of Paleo-Indian technology suggests an economy and society characterized by a primary dependence not on plants but on hunting (though not exclusively of megafauna), and by high residential, logistical, and territorial mobility rather than a fixed territoriality. They emphasize that, unlike modern foragers, Paleo-Indians groups were apparently pioneers on an empty continent during a time of pronounced climatic change. They conclude, therefore, that the low human population density combined with the environmental instability of the late Pleistocene meant that Paleo-Indians: (1) would have focused on big game hunting because such activities are less heavily dependent on detailed and intimate knowledge of the locale than is plant gathering, and (2) would not have occupied rigidly fixed territories and could not have relied upon neighboring groups with a detailed knowledge of local resource geography in times of stress, and (3) would have had socio-economic organizations that differed from that of modern foragers to the extent that no adequate ethnographic-parallel to it can be found in the modern world. Kelly and Todd interpret the relative homogeneity of the Llano culture all over North America as strong evidence that a generalized big game hunting adaptive strategy was employed by there peoples all over their North Ameican ranges because such a strategy worked well in a host of different local conditions and allowed them to overide or overcome regional environmental differences. At the same time, such a strategy would have allowed them to live more-or-less autonomously with little support from the kinds of regional networks relied by the !Kung and other long-established and environmentally-specialized foraging peoples.

I would add further that the social and economic mechanism which, according to Flannery, restrain modern foraging peoples from over-hunting or over-harvesting wild resources is an adaptation to modern demographic conditions. Modern foraging peoples occupy fixed territories because they are hemmed in and surrounded by the fixed territories of other peoples. The specialized adaptations found by ethnographers among modern foragers–their close adaptations to their local environments, their parsimonious use of wild resources, the control they exercise over their reproduction–are all probably post-Pleistocene developments that emerged as foraging groups came to be packed into their territories by contiguous and competing groups.

If a "generalized" big game hunting strategy did work in most late Pleistocene environmental settings on the essentially empty continent of North America, Paleo-Indians would have faced no immediate or short-run need either to practice restraint in their hunting or in their population growth. Such freedom–rare or absent in later times–may have had severe and unintended consequence for the late Pleistocene "mega-fauna" of the New World upon whom the Paleo-Indians preyed. In fact, the charge of "mega-cide" that has been leveled by some scholars against the Paleo-Indian is examined at the end of this chapter.

In any case, the diversity represented both in the Plano projectile point types and in the faunal repertoires recovered from Plano sites indicate that important changes in technology, subsistence economy, and social organization were overtaking the Paleo-Indian peoples at the end of the Pleistocene. The generalized big game hunting Paleo-Indian adaptation to the environment of late Pleistocene North America perhaps was giving way in the face of the new requirements presented by the environments of the early Holocene epoch. The regionally-specific adaptive solutions that were found for these new, post-glacial environments of North America are referred to in aggregate as the Archaic cultural tradition. Archaic peoples appear to have developed the kinds of seasonally-timed, broad-spectrum hunting and gathering economies represented in modern times by peoples like the !Kung and we turn to an examination of this tradition in the next chapter.

An ancient thirst for visions? Although people do not live by bread alone, archaeologists are seldom able to recover much of the spiritual life of non-literate peoples with their sifting screens and shovels. Yet sometimes archaeological excavation does provide small windows into the religious life of the ancients. For example, the earliest occupations at three deeply-stratified dry rockshelters–Fate Bell Shelter (ca. 9,000 B.P.) and Bonfire Shelter (10,390 to 10,070 B.P.) in the lower Pecos river valley in Texas and Frightful Cave (9,215±85 B.P.) in Coahuila, Mexico–have all produced seeds from a plant called the mescal bush (*Sophora secundiflora*). The mescal bush is a rather unprepossessing shrub found today in canyons and upland forest zones throughout the Sonoran and Chihuahuan deserts of the greater Southwest. In historic times, seeds of this bush–known as red or mescal "beans"–were highly prized by American Indians because of their powerful hallucinogenic properties. Such beans were consequently widely used and traded all over the Southwest and southern Great Plains.

The base religion of historic American Indians from Alaska to Patagonia may be characterized as "ecstatic-visionary shamanism" (LaBarre 1974: 188). The quest for contact with the supernatural through altered states of consciousness is a key element in such shamanism. Divination is done by means of these altered states and, in curing rituals, the shaman enters the other dimension to fight the spirits of disease and misfortune. Often natural psychotropic drugs derived from roots, seeds or mushrooms are used to aid them in achieving these states.

The Paleo-Indians who camped at Fate Bell, Bonfire Shelter, and Frightful Cave surely knew of the powerful hallucinogenic power of the mescal beans and they collected them, one presumes, for use in the shamanic quest for visions. Some of their mescal beans remained behind after thousands of years to be sifted up in the screens of the archaeologists who dug those sites. The visions these beans once conjured, however, the comfort they gave, the truths they taught, the strength they engendered, have flown with the dreams of those ancient people.

The Great Hunt Ends: Terminal Pleistocene Extinctions

As we have seen, the mammalian animal life on Earth during the Pleistocene epoch was rich almost beyond measure. At the epoch's close, however, a massive series of animal extinctions took place. Extinctions have occurred throughout geological history, of course, but at least five aspects of the terminal Pleistocene episode render it unusual. First, a great many of these extinctions were concentrated worldwide in the brief period (geologically-speaking) between ca. 12,000 B.P. and 8,000 B.P. In some areas of the world these extinctions took place during an even narrower time frame. For example, of the 35 Rancholabrean genera that became extinct in North America during the last glacial stage, 22 disappeared between 12,000 and 10,000 B.P. (Grayson 1987).

Second, "megafauna," that is, species with large adult body sizes and body weights over 44 kilograms (100 pounds), were more severely affected than were smaller taxa (Martin 1984: 335). For example, Columbian mammoth, mastodonts, horse, camelops, the giant or Shasta ground sloths, saber-tooth cat and 70 other genera disappeared from North and South America. At the same time in Europe and Eurasia, woolly mammoths, woolly rhinoceros, and giant deer (Irish Elk) became totally extinct, musk oxen, steppe lions, and hyenas disappeared from these areas but survived elsewhere, and the distribution of horses and steppe bison or wisent became greatly restricted. Even Africa, with its seemingly limitless faunal repertoire, lost between 15 and 20 percent of the large mammalian taxa to extinction at the end of the Pleistocene (Martin 1984: 354).

Third, during most earlier mammalian extinction episodes, chiefly taxa at the species level were affected; in the terminal Pleistocene, higher taxonomic categories including genus, families, and orders disappeared as well. Fourth, whereas following most previous extinction episodes, ecological niches of extinct forms were claimed by new taxa, many niches remained empty after the close of the Pleistocene. The total number and diversity of mammalian taxa on most continents were greatly reduced. Compared to that of the late Pleistocene, the terrestrial vertebrate fauna of the early Holocene epoch was markedly impoverished. Fifth, the most unique–and perhaps most significant–difference between the terminal Pleistocene episode and earlier extinctions is that they coincided with the accelerating dispersal of anatomically fully-modern *Homo sapiens sapiens* over much of the Earth.

Anthropogenic Overkill in the Terminal Pleistocene

Attempts to explain the extinctions of the terminal Pleistocene epoch first began in the 19th century, but the development of radiocarbon dating and improved geoarchaeological techniques sparked a vigorous renewal of the debate over causality in the 1960s (Martin and Wright 1967; Grayson 1984: 807-823). Participants in this debate have generally explained extinctions as having been due either to geological and climate change, to human agency, or to an interaction of all these causes. However, in spite of concentrated effort–and sometimes heated debate–a final resolution of the question has been as elusive as ever. Students on both sides of the question have been able to advance only circumstantial evidence, not demonstrate causality. As Paul S. Martin (1984: 355) drily observes, "Late Pleistocene extinctions have proved refractory."

For many scholars, the timing of the climate changes that took place at the close of the Pleistocene, combined with their violence and rapidity, makes them a likely cause of the faunal extinctions. During the last 128,000 years of the late Pleistocene, the Northern Hemisphere had experienced a long stage of continental glaciation, punctuated by short term ice contractions or interstadials. The last stage of Pleistocene glacial expansion terminated abruptly about 12,000 B.P. and was followed by a millennium of rapid deglaciation and warmer, moister climate during the Two Creekan-Allerod interstadial. This last interstadial ended abruptly about 11,000 B.P. and glacial conditions returned to the northern hemisphere for another thousand years during the Valders-Younger Dryas event. There can be little doubt these climate changes affected profoundly affected late Rancholabrean zoogeography. Expansion of the forests in Europe, Eurasia and North America, for example, must have severely curtailed the range and population sizes of plains dwellers like the horse, mammoth, bison, and woolly rhinoceros. However, ice remissions and rapid climatic changes, occurred a number of times during the late Pleistocene without triggering similar waves of extinctions. The extinctions were neither entirely coeval with this final deglaciation nor with each other. Perhaps most curiously, some animals became regionally extinct in

North America at the end of the Pleistocene only to be successfully re-introduced there by man in historic times. The horse is the chief example, although a case can also be made for the camel. One would expect that, if the new environments of North America had proven fatal to the horse at the end of the Pleistocene, the wild descendants of Spanish mustangs would not be able to roam the western United States today with such success. For these and other reasons, a climatic cause for the extinctions is not an altogether satisfactory explanation and a strong case can be made for human hunting on a grand and profligate scale–what Martin calls "anthropogenic overkill"–as a major cause of the extirpation of the megafauna at the end of the Pleistocene epoch.

New World "mega-cide" The most important evidence for anthropogenic overkill as cause is the stratigraphic association between the stone tools of early man and the bones of extinct megafauna found in the terminal Pleistocene age archaeological sites. This evidence is strongest in North America where direct and primary association between Clovis points and megafauna is well-documented at a host of localities. In North America the extinction of key elements of the Pleistocene Rancholabrean megafauna, including ground sloths, camelops, tapirs, mastodons, various species of bison, mammoths, and horses between 12,000 and 10,000 B.P. is clearly correlated with the abrupt appearance of the Llano complex between 11,500 and 11,000 B.P. (Meltzer and Mead 1985: 166). Grayson (1984: 814) recently compiled a list of North American sites reported to contain megafaunal remains that date to after 12,000 B.P. There are 151 sites in this corpus, of which I estimate perhaps ten percent were found to contain Pleistocene megafauna in direct association with human cultural material, chiefly stone tools. Given the problems of sampling error, the shortness of the time involved, and the odds against finding the remains of any short term event in the archaeological record, the evidence of human predation is remarkably abundant.

However, the date of the arrival of human kind in the New World is a crucial element in the hypothesis of an anthropogenic cause of the late Rancholabrean faunal extinctions and this is a subject of great contention, If the specialized Llano big game hunters were in fact the first people to arrive in the New World, they would have found a hunter's paradise filled with game but empty of human competitors. That paradise would have been unlike anything in the Old World as the animals that lived in it, having developed throughout the Plio-Pleistocene without ever having been hunted by hominids, would not have evolved the behavior needed for dealing with cunning, two-legged human predators. As a consequence of their naivete, the megafauna would have been struck hard when such creatures finally arrived on this continent.

As the populations of these early peoples grew, they would have expanded southward and across the continent in a moving wave of advance. By employing profligate and damaging hunting strategies that indiscriminately killed large numbers of game, the Llano and later Paleo-Indian peoples would have inadvertently been waging a kind of *blitzkrieg* against the megafauna, and extinguishing them along the ever-widening front created by their expanding population. Mosimann and Martin (1975), who first proposed use of the term *blitzkrieg* in this context, simulated this scenario by computer. Their simulation led them to conclude that the descendants of 100 people entering the New World in Alaska at 11,500 B.P. could have destroyed 3.3 million mammoth in only 500 years! Simulations by other scholars have generated equally provocative results (for example, cf., Whittington and Dyke 1974; 451-465; Martin, Thompson, and Long 1985: 5-14).

Evidence of profligate, *blitzkrieg* hunting by Paleo-Indians is not hard to come by but the clearest example in the archaeological record is from the late Plano complex Olsen-Chubbuck site. The rather modest Colorado arroyo in which the site was located contained the bones of nearly 200 *B.b.occidentalis*. As noted in Chapter 5, the resolution at this site is quite high and it is possible to clearly demonstrate that (1) an entire bison herd was driven into the arroyo and that (2) only the individual animals near the top of the pile were butchered. The remaining animals, out of reach at the bottom of the arroyo, were left to rot intact. Selective hunting was impossible in such game drives and, consequently, the Olsen-Chubbuck bison population contains adult, immature, and juvenile animals of both sexes (Wheat 1972). It is not too much to suppose that such wasteful hunting, practiced as it was in a time of changing climate, proved fatal to numerous genera. By the time Paleo-Indian peoples reached the tip of South America around 9,000 B.P., the megafauna were gone.

Old World extinctions The record of Pleistocene extinctions in Africa and Europe contrasts with that of North America in a number of ways. Megafaunal extinctions occurred frequently in Africa during the early Pleistocene, but only in modest numbers at the conclusion of the epoch. Although Europe lost more taxa in the terminal Pleistocene than did Africa, there too extinctions were neither as abundant, nor as sudden, as those of the New World. Martin accounts for these differences as due to the fact that megafauna and the hominid line evolved together in the Old World. Fossil evidence indicates that the hominid line first appeared some 4 million years ago in Africa; anatomically fully modern *Homo sapiens sapiens* probably evolved on that continent some 100,000 or more years ago (Stringer and Andrews 1988: 1263-1268). Martin suggests that the behavior of African megafauna evolved during the Plio-Pleistocene partly in response to the selective pressure of human predation; hence the extinction rate of African fauna declined between the beginning and end of the Pleistocene epoch. To support this idea, Martin (1984: 368 following Kurten 1968) notes that eleven of the twelve genera of large mammals that did survive the wave of North American terminal Pleistocene extinctions–including mountain sheep, moose, wapiti, caribou, musk ox, grizzly bears, and a few others–evolved originally in the Old World. Martin concludes that, like contemporary African forms, these taxa had evolved appropriate behavioral responses to human hunting and were able to maintain themselves because of their long previous conditions to man.

At present there is no definite archaeological evidence of anatomically fully-modern *Homo sapiens sapiens* in Europe before about 35,000 years ago. The greater rate of terminal Pleistocene extinctions in Europe as compared to Africa may reflect this shorter period of association between late Pleistocene European game animals and human predators (Martin 1984: 380-384, 386-387). Anatomically-modern people were the first to penetrate European Russia and Siberia, climatically the most rigorous parts of the Old World during the late Pleistocene. Klein suggests that it was probably these areas of Eurasia that served as refugia for cold-adapted fauna like woolly mammoth and woolly rhinoceros during earlier interstadial warming intervals of the late Pleistocene. The appearance of hunting populations in these refugia for the first time beginning after ca. 35,000 may have hastened the extinction of European megafauna (Klein 1984: 568). Further, the vast quantities of mammoth bones recovered from the celebrated kill sites in the Ukraine certainly suggests that Upper Paleolithic foragers slaughtered prodigious numbers of these beasts. In sum. the timing of the arrival of modern peoples in Europe in the late Pleistocene, combined with the comparatively sophisticated stone tool repertories in their employ, may have placed the megafauna of Europe under greater hunting pressure than the megafauna of Africa at a comparable time period.

Conclusion and prospectus Whatever the cause of the wave of extinctions in the terminal Pleistocene, the world of the post-Pleistocene was a very different place. In Chapter 10 three important processes of change in human culture and biology initiated during and after these events are discussed:

(1) Broad-spectrum hunting-and-gathering subsistence systems developed to replace earlier big game hunting economies;
(2) Sexual dimorphism among *Homo sapiens sapiens* began to be reduced; and
(3) Incipient agriculture emerged.

These three processes were in large part adaptive responses to the depauperation of the Earth's environments by the terminal Pleistocene megafaunal extinctions. Thus, whether or not profligate hunting practices caused or hastened these terrible extinctions, humankind was deeply affected by them.

Major Points Raised in Chapter Ten:

1. The Holocene epoch–the most recent unit of geological time–is defined and its chronology discussed in the first section of this chapter. The section begins with the melting of the continental ice masses in the northern hemisphere and in the resulting Flandrian sea level transgression that resulted. The rise of worldwide sea level, the flooding of the continental shelves, and the transformation of river and glacial valleys into *ria* coasts and fjords is discussed. The severance of important land bridges between the New World and Siberia, England from Ireland and France, and Japan from China is noted.

2. Worldwide changes in rainfall, temperature and vegetation during this epoch is examined using paleoenvironmental data drawn from oxygen isotope ratios in tree rings, deep sea cores, and glacial ice cores as well as other sources of data include changing lake levels, and changing plant assemblages and distributions recorded in peat bogs and pack rat dens. The simulation of the terminal Pleistocene and Holocene climatic sequence simulated by the COHMAP group is discussed and special attention is paid to the climatic events known as the **Hypsithermal** period, Neoglaciation, the **Little Climatic Optimum** and **Little Ice Age**. The section ends with a discussion of the impact of Terminal Pleistocene/Early Holocene extinctions and their role in altering the nature and distribution of animal assemblages on Earth.

3. Whatever caused the wave of extinctions in the terminal Pleistocene, three important processes of change in human culture and biology were initiated during and after that time: (a) broad-spectrum hunting-and-gathering subsistence systems of the distinctive "Mesolithic" pattern developed in replacement of earlier big game hunting economies, (b) human sexual dimorphism among began to be reduced, and (c) incipient agriculture emerged. Cohen's thesis of a **food crises** in the early Holocene is discussed in light of these changes.

4. The Mesolithic continuation of late Pleistocene hunting and gathering subsistence adaptations into the Holocene epoch is discussed. The question of human subsistence in the new environmental conditions of the Holocene is first reviewed from a general, worldwide perspective. In discussing Mesolithic technology, the widespread use of microlithic stone tools and of composite weapons like the *atlatl* or the bow-and-arrow in most parts of the world is emphasized. The new elements added to the diets of early Holocene epoch peoples are also discussed. The fact that these new elements tended to be: (a) drawn from low levels on the food chain, (b) predictable in yield and seasonal availability, (c) easily collected by women and children, (d) characterized by procurement and/or processing techniques that rendered them costly in terms of caloric yield per unit of effort, and (e) storable resources that could be accumulated as surplus above immediate subsistence needs.

5. Northwestern Europe has produced the most detailed Mesolithic sequence in the Old World. This sequence is discussed in detail beginning with the **Blytt Sernander model** of European Holocene vegetational change. The **Maglemose-Kongemose-Ertebølle** sequence of southern Scandinavia is discussed with particular attention paid to the sites of **Star Carr** and **Morton Tayport B** in Great Britain.

6. By way of contrast, the Holocene environmental changes and the Mesolithic cultural adaptations and key sites in two regions of the Old World beyond Europe are discussed. These include: the **Aqualithic** tradition of Sahara and Sahel of Africa and the **Jomon** tradition of Japan.

CHAPTER 10:

THE HOLOCENE EPOCH AND THE MESOLITHIC INTERLUDE

Introduction

The most latest unit of Earth history began some 10,000 years ago and continues into the present. It is most commonly referred to as the Holocene epoch but is sometimes called the Recent, Postglacial, Neothermal (after the warming of the world climate during the epoch), even the Flandrian (after the marine transgression or world wide rise in sea level that occurred with the melting of the continental ice masses). The Holocene is the second and final epoch of the Quaternary period; thus and the term technically refers both to time and to strata younger than that of the Pleistocene epoch.

Prior to the publication of Sir Charles Lyell's monumental book, *Principles of Geology* (1833), the term Pliocene was used in reference to the most recent geological strata in the geological column. Lyell subdivided the Pliocene epoch into early, middle and upper blocks and retained the name Pliocene in reference only to the lower or earlier portion. For the middle of this sequence, he coined the term "Pleistocene" and defined its strata as characterized by the presence of animal forms 70% of which survived into the present. He called the upper part of the column the "Recent" and stated that strata referable to this epoch were characterized only by species still alive on Earth today. The term Holocene-which means "wholly modern"-was first proposed for the epoch by Gervais in 1869 and has achieved wide currency ever since. It should be noted that some scholars consider the distinction between the Pleistocene and the Holocene to be an artificial one and prefer to regard the last 10,000 years of geological time as simply the most recent episode of the Pleistocene, that is, as the latest interglacial substage to follow the last stage of glacial advance.

Chronology of the Holocene Epoch The beginning of the Holocene is conventionally placed at the date of the breaching of the Baltic ice lake, an event that, according to de Geer's varve chronology, occurred in 7,912 B.C. This date is generally rounded to 8,000 B.C. (10,000 B.P.). The Holocene epoch continues into the present and is divided into three subdivisions: the early Holocene (ca. 8,000 B.C. to 6,000 B.C.), the middle Holocene (ca. 6,000 B.C. to 3,000 B.C.), and the late Holocene (ca. 3,000 B.C. to the present).

Holocene Environmental Change Paleo-environmental data drawn from oxygen isotope ratios in tree rings, deep sea cores, and glacial ice cores as well as other sources of data include changing lake levels, and changing plant distributions and species assemblages composition recorded in pollen in peat bogs and plant macrofossils collected in pack rat dens have been used to reconstruct worldwide changes in rainfall, temperature and vegetation that occurred during the Holocene epoch. In addition, for the late Holocene, weather records and other forms of historical documentation are available. Collectively, these data indicate that a number of important changes in the environmental of the Earth took place during the Holocene. Although the last 10,000 years have been characterized by an essentially "interglacial" climate, this regime has been punctuated by "numerous short-term fluctuations as well as longer term trends to a cooler or warmer, moister or drier climate" (Butzer 1971:526). It is important to note that as pronounced as some of these fluctuations have been, the climatic shifts of the Holocene have never had the profound ecological impact of the climatic events of the Pleistocene.

Six major environmental events or processes distinguish the Holocene from the late Pleistocene epoch. These include: generally higher levels of global temperature and humidity, the remission of worldwide glaciation, exposure of the formerly glaciated lands in the northern hemisphere, increased volume of water run-off from the continents during the early Holocene, a rise in world sea level, and the return of the forests over much of the Earth. Finally, the Holocene witnessed the transformation of human subsistence, a marked expansion in human population size and increasing geographic range of our species on Earth. Humankind became a significant biological and geological force on Earth during the Holocene–and with that new significance brought an accelerating anthropogenic environmental impact in its wake.

General Increase in Global Temperature and Humidity The Ice Age climate of the Earth began to change in the closing millennium of the Pleistocene epoch between 10,000 B.P. and 9,000 B.C. and these changes continued–albeit in a punctuated fashion–into the Holocene epoch (Lamb 1982: 26). In general, the world climate became warmer after about 8,000 B.C. and this global warming apparently intensified after about 7,000 B.C. By about 6,000 BC, the oxygen isotope ratios in the Camp Century ice core from Greenland indicates that mean annual temperatures were between 1° and 3° C. warmer than at present (Dansgaard, Johnsen and Clausen 1970). This phase in Holocene temperature history, referred to as the **Hypsithermal** or **Altithermal** interval. The Hypsithermal was defined by Deevey and Flint (1957) as a "time-stratigraphic unit." That is, it was presumed to refer to a definite interval of geological time–in this case, to the four phases of the middle Holocene represented in the **Blytt-Sernander European Pollen Zone System** (which we discuss in detail later in the chapter) embracing the Boreal through to the Sub-Boreal between ca. 7000 B.C. and 350 B.C.–and that its beginning and end were "time parallel," that is, were the same everywhere on Earth. However, it now seems apparent to many scholars that the warm interval of the middle Holocene was "time transgressive," that is, the dates of its beginning and end varied from place to place. It is probably best to regard the Hypsithermal as a worldwide climatic episode that occurred primarily between ca. 6,000 B.C. and 3,000 B.C. in the middle Holocene but which varies from location to location in precise date and impact (Wright 1976: 591-593).

After the Hypsithermal peak, worldwide temperature began to decline and Holocene climates entered a phase of "**Neoglaciation**." It is important to note that temperatures oscillated during the Neoglaciation episode and did not remain consistently. Further, its dates and impact also were time transgressive and varied in different parts of the globe. This variation has made dating and worldwide correlation problematical.

Worldwide temperature averages rose again slightly between ca. A.D. 700 and 1200 in a period referred to as the **Little Climatic Optimum**. During this time, Vikings were able to settled in Greenland and wine grapes could be cultivated in Britain (Lamb 1982). Climates fell to a nadir once again between ca. A.D. 1550 and 1850 during the climatic episode known as the Little Ice Age. The **Little Ice Age** is relatively well dated in part because the effects of its occurrence are documented in historic records that been used to supplement the standard methods of geochronology. With the close of the Little Ice Age, worldwide climate entered another relatively warm and humid period between ca. 1850 and the 1940s commonly referred to as the **Climatic Optimum**.

If all things had remained equal, worldwide temperatures would probably have continued to drop in the late 20th century. However, it now appears that increases in the amount of carbon dioxide in the atmosphere due to industrial pollution have created a greenhouse effect on Earth that will reverse this trend and create a warmer, drier worldwide climate by the 21st century (Anon 19–).

Waning and Recession of Glacial Ice in the Early Holocene Followed by Neoglaciation in the Late Holocene In the northern hemisphere, the Fennoscandinavian and the Laurentide continental ice masses and the Alpine, Cordillerean and other glacial ice complexes began to disintegrate or retreat northward during the terminal Pleistocene epoch. Their remission in the northern hemisphere was generally correlated with the retreat of mountain glaciers in the southern hemisphere to higher elevations. Glacial shrinkage intensified during the Hypsithermal climatic episode. The Fenno-scandinavian ice sheet rapidly disintegrated after 5,000 B.C. and the last remnants of the Laurentide faded away in the wake of the worldwide temperatures peak around 4,000 B.C. and had disappeared completely by 3,000 B.C. (Denton and Porter 1967: 101).

After the close of the Hypsithermal, glacial ice once again began to advance world wide, albeit on a much small scale than during the Pleistocene (Denton and Porter 1970 or 1967?). The timing and extent of Neoglaciation varied greatly from area to area and this has made local glacial events difficult to correlate with one another (Grove 1979). Nonetheless, the latest Holocene Neoglacial advances occurred during the Little Ice Age and these ice advances have been firmly dated and cross dated worldwide (for example, cf., Thompson, Mosley-Thompson, Dansgaard, and Grootes 1986).

Exposure of Millions of Square Kilometers of Formerly Glaciated Northern Lands The disappearance of continental glacial ice from much of the northern hemisphere occurred rapidly in geological terms. This disappearance exposed vast tracts of formerly ice bound territory and ameliorated the harsh polar desert climatic conditions that had prevailed in the lands adjacent to the continental ice masses. Thus, de-glaciation vastly expanded the areas available for plants and animals and made possible the human colonization and settlement in far northern latitudes. Further, as Goudie (1977:99) notes, the release of this far northern territory from the grip of the ice expanded the breeding and feeding grounds of the world's migrant waterfowl, an expansion that must have quickly translated into marked increases in the numbers of these creatures worldwide.

Marked Increase in River Discharge Draining the Continents of the Northern Hemisphere The modern valleys of many rivers and streams in the northern hemisphere are much larger than the water courses that flow through them today. In geological parlance, these rivers and streams "**underfit**" their valleys. George Dury (1977) calculated that many of the large valleys in North America were carved by rivers carrying water volumes as much as one hundred times greater than they do at present. This huge increase in discharge occurred in the terminal Pleistocene and early Holocene as meltwater from rapidly disintegrating continental ice masses poured into the water courses draining the northern continents. These great torrents subsided by the middle of the Holocene and, as discharge levels in the northern hemisphere approached modern levels, rivers and streams shrunk back into the floors of their now much enlarged valleys.

Rise in World Wide Sea Level Following the Release of Huge Quantities of Water Formerly Held on Land as Ice **Eustasy** or changes in the worldwide level of sea has a number of causes including: variations in the shape of the globe during its rotation (geoidal eustasy), the introduction of "juvenile" water from the Earth's deep interior onto its surface, the process of mountain building (orogeny) and the progressive **infilling** or **siltation** of the ocean basins. However, the two most important causes of sea level change during the Quaternary period were **glacial eustasy** and **isostatic rebound**. Glacial eustasy refers to changes in sea level caused by glacial advances and retreats. Sea level is lowered during glacial stages when vast quantities of water remain on the continents in glacial ice caps rather returning to the ocean as run off. Conversely, sea levels rise when these ice caps melt during interglacial periods. At the end of the Pleistocene, the rising sea level resulted in a net decrease in the land area exposed on the shelves at the edges of the world's continents and severed a number of key land bridge connections between those land masses.

Although the level of the sea rose uniformly throughout the world, the effects of the coastal innundation were most dramatic and extensive in the coastal regions south of the ice margins. This because coastal regions in the glaciated parts of the northern hemisphere rebounded upwards when the weight of the continental ice masses that depressed them was removed. Such uplift or isostatic rebound partly compensated for the rising sea level and reduced its impact on the coasts of formerly glaciated portions of the northern hemisphere.

In general, the post-Pleistocene rise in sea level changes probably improved conditions for human exploitation along the world's coasts. The mouths and lower valleys of the rivers of the world that flow directly into the ocean were "**drowned**" by rising sea level and these former inland valleys were converted into wetlands, estuaries and ocean bays. Estuaries, bays and coastal wetlands are the places where the freshwater running off the land first mixes with the salt water of the oceans. Water level, temperature, salinity, velocity of flow, degree of mixing, and nutrient content in estuaries fluctuates widely in response to tidal currents, seasonal variations in run off from the land, erosion, deposition and other factors. Generally, animal and plant life is zoned in accord with changes in water salinity. The type, abundance, and relative availability of plant and animal species encountered often changes dramatically from the upper to the lower end of even small estuaries. As a consequence of their remarkably varied and complex ecologies, coastal wetlands and shallow estuaries are areas of particularly high biological productivity. Although they

> constitute but a small proportion of the earth's surface, they are said to produce more living matter per acre than any other part of the world. Per acre they are twice as productive as good forest or grassland, four times as productive as a good cornfield, and fifteen times as productive as the oceans (Hunt 1974: 194).

While there were estuaries throughout the Pleistocene, at least three attributes made the wetlands formed during the early Holocene biologically-richer–especially in northern latitudes–than the late Pleistocene ones they replaced.

First, the combination of their previous geomorphologic history and their recent flooding made early Holocene estuaries more sinuous, topographically varied, and more environmentally diverse than the mature ones of the late Pleistocene. Additional coastal multiformity resulted as the rising sea level turned isolated ridges and other land forms in the flooded river valley mouths into islands. Coastal land forms characterized by exaggerated topographic indentions, wide tidal ranges, and great micro environmental variety are called *rias*. The number of *ria* estuaries and island habitats must have increased geometrically during the early Holocene epoch.

Second, warmer worldwide ocean temperatures promoted the northward migration of ocean species that had been concentrated farther south during the late Pleistocene. The distribution of **anadromous** and **catadromous** fish were particularly affected by these temperature changes. Anadromous fish spend their adult lives in the sea but bred in fresh water rivers. Catadromous fish species have life cycles that are the reverse. They live as adults in fresh water rivers but journey to the sea to spawn. As the Early Holocene oceans grew warmer at more northerly latitudes, the estuaries and rivers along the coasts of northern Asia, Japan, northern Europe, Britain, and North America became the seasonal homes of annually-migratory anadromous and catadromous fish species that had confined to much more southerly locations during the late Pleistocene epoch. The impact of this single shift on human subsistence in the Early Holocene can hardly be overstated.

Third, rising worldwide sea level decreased stream gradient and reduced rate of flow which in turn resulted in greater river meandering and correspondingly greater aggregation and flooding of lower valleys than had characterized the rivers of the late Pleistocene. The slow-moving water and low stream gradients of these new river regimes improved conditions for host of waterfowl, anadromous fish and shellfish species (Turnbaugh 1975: 59).

The microenvironmental heterogeneity and high biological productivity of the newly-formed bays, estuaries, *ria* coasts and islands, and slowly meandering lower river valleys made the Holocene coasts ideal for ***strand loping***, that is, for human food-collecting systems centered on marine and aquatic fish, intertidal zone mollusks, marine mammals and birds, and other resources (Goudie 1977:99; Small 1977: 422). Further, the warmer ocean temperatures made *strand loping* possible and attractive at latitudes much farther north than ever before. However, marine resources are immensely rich in protein and this is a problem as

> a diet in which more than 50 percent of the energy needs are satisfied through protein intake could have lethal consequences, even over the short term. It would have been particularly on the coast to live on the flesh of shellfish, fish, and marine mammals for longer than a few days at a time without the inclusion of a carbohydrate- or fat-rich source of energy...Prehistoric coastal subsistence strategies, and thence diet, must have been developed to avoid over-consumption of protein (Noli and Avery 1988: 399).

Fruits, nuts, and leaves are key sources of such carbohydrates but during the late Pleistocene, the coasts in northern latitudes supported steppe tundra and boreal forests and the vegetation in these assemblages can contribute very little to human subsistence. The return of the pine and hardwood forests to the northern coasts during the early Holocene would have meant that the gathering of terrestrial plants could easily and efficiently be combined with marine *strand loping*. In addition, forest animals are an excellent source of fats.

Evidence that peoples took advantage of the new environmental new richness of the Holocene littoral zones is found in the numerous shell middens and other types of sites dating to the period found along the Atlantic coast in Europe and eastern North America and the Pacific coasts of North America, eastern Asia and Japan (Aikens, Ames, and Sanger 1986). Of course, untold numbers of Mesolithic sites must have disappeared beneath the rising sea level of the early Holocene period. Fortunately, portions of the coasts in the northern hemisphere have were gradually uplifted during the early Holocene as the edge of the continental shelf rebounded isostatically with the remove of its ice burden. This uplifting raised a number of early Holocene period Mesolithic shell middens sites well above the modern sea.

The Return of the Forests During the late Pleistocene, lower worldwide temperature and humidity severely the distribution of the world's forests to a number of relatively limited refugia in both the northern and southern hemispheres. Conversely, in the climatic amelioration of the early Holocene, the world's forest once again began to expand. In Europe and Eurasia, the boreal forests began to re-colonize the vast tracks of land in the northern hemisphere which had been polar desert or dominated by treeless tundra steppe or steppe parkland since before the late Pleistocene glacial maximum around 18,000 B.P. The pine and spruce that initiated this re-colonization presumably came from (1) the limited tracts of boreal forests able to survive late Pleistocene climatic conditions in the sheltered river valleys that cross-cut the steppes, and (2) the forest refugia located south of the Pyrennees on the Iberian peninsula and along the Mediterranean basin on the southeastern fringe of Eurasia (van der Hammern *et* *al.*, 1971).

Unlike the middle latitudes of Europe and Asia, most of Japan was not deforested during the last glacial maximum. Although the northernmost island of Hokkaido supported a tundra and tundra parkland vegetation at that time, most of the remainder of the Japanese archipelago was covered by a heavy coniferous forests. Cold tolerant species in the north gave way to cool temperate forms in the southern (Tsukada, 1983). At the close of the Pleistocene, pollen spectra indicate that the species composition of these forest shifted and the Japanese forest came to be dominated by more temperate hardwoods and pines.

Unlike Japan, the distribution of vegetation of North America was profoundly affected by the full-glacial climate. The vast arid steppe tundra or Mammoth Steppe assemblage of that dominated the interior of Eurasia extended unbroken across the Bering Straits into the unglaciated portions of central and southern Alaska. The unglaciated regions of Alaska represents the farthest, most easterly extension of the Mammoth Steppes. The Laurentide and Cordillerean, the great continental ice complexes of North America, bordered that biotope on the east and south. At 18,000 years ago, this vast glacial ice complex filled virtually the whole of what is today modern Canada. From the edge of the ice southward was a broad band of open and exposed land and ice-front lakes that was constantly swept by cold, fierce winds. This rigorous periglacial environment appears to have been largely, but not entirely, dominated by arctic tundra vegetation. The pollen record suggests that the tundra belt was no identical with that of the modern arctic in its species composition and may have been more like a grassy steppe. whatever the nature of this Full Glacial tundra, it does not seem to have been continuous in its distribution across the continent. Instead, vegetation formed a patchy mosaic in which treeless tundra alternated evergreen forests depending on local conditions. In some parts of the continent, forests apparently grew virtually to the edge of the ice front forming a patchy, vegetation mosaic in which, depending on local conditions, forest alternated with treeless tundra. South of the open ground–or tundra/forest mosaic, an extensive evergreen forests stretched across much of the continent (Pielou 1991: 84, 94).

A vast freshwater body, whose modern remnant can be seen in the Great Salt lake and the Bonneville Salt Flats filled much of the basins of the intermontane region between the Pacific Range and the Rockies. Above the lake levels, unflooded land in the Great Basin supported a relatively lush open parkland vegetational assemblages. The country from the Rocky Mountains east to the Mississippi and south to Mexico was experienced cooler temperatures and greater rainfall in the late Pleistocene. Consequently, tree grew at lower elevations there and the region sustained a more lush vegetation than it does at present. The interior of the Great Plains, although apparently largely devoid of treeless, was dotted with lakes, bogs and ponds and carpeted with grass. The river valleys were heavily wooded. Between the mountains on the west and the forests across the Mississippi in what is today Nebraska, vast sand dunes were formed in the Late Pleistocene . Stabilized remnants of these dunes form the Sandhills of Nebraska and they reflecting the arid nature of the climate during the Late Glacial Maximum as well as the exceptional strength and constancy of the periglacial winds that blew down the plains off the ice front at that time (Pielou 1991: 108).

In Late Glacial times, the Great Plains supported a bestiary made up of some of the largest and most extraordinary mammals ever seen on Earth: mammoths, mastodons, giant moose, endless herds of horse, camels, caribou and bison, beavers the size of modern black bear, dire wolves twice the size of their modern descendants and, of course, saber-tooth tigers and the fearsome short face bear that, were it alive today, would dwarf–and could perhaps outrun–a modern grizzly bear (Anderson 1984).

Then, as now, the climate of North America became steadily more moist from the Rocky Mountains across the Plains to the Atlantic ocean. The unglaciated northeastern portion of the continent there was probably a mix of boreal spruce forest and tundra. The interior of the southeastern United States appears to have been a refugium wherein beech (*Fagus grandifolia*), sugar maple (*Acer saccharum*), oak (*Quercus*), hickory (*Carya*) and a host of other warm-adapted hardwood tree species waited out the dry cold of the LGM (Watts 1983; Delacourt and Delacourt 1985: 13).

When the great Laurentide and Cordillerean ice masses began to retreat, the arid steppe tundra zone was rapidly re-colonized by boreal spruce forest species migrating northward out of the eastern and central portion of North America and northwest along the Pacific coast and Cordilleran slope from refugia in unglaciated southern British Columbia and Washington state (Heusser 1985: 141). At the same time, pine and hardwood tree species from the southeastern United States began to spread north and west to re-colonize mid-latitude North America (Delcourt and Delcourt 1985). Pine and hardwood forests continued to expand northward as Holocene warmth and aridity climbed to a peak during the Hypisthermal. However, in the middle of the continent, the forests contracted on its western margin and was replaced by prairie vegetation. These biogeographical trends reversed in the cooler conditions that followed the onset of neoglaciation (Wright 1976: 581). Although late Pleistocene climates were most severe in the northern hemisphere, the southern hemisphere did not remain unaffected. Current biogeographic and palynological evidence combined with computerized climate simulations indicate that at glacial maximum, the southern hemisphere was substantially colder and drier than at present. As a consequence, the distribution of subtropical and tropical forests there were severely curtailed during the late Pleistocene. This trend began to reverse following the increases in temperature and humidity that occurred during the early Holocene. Sometime after ca. 10,000 B.C., the rainforests of Africa began to expand out of the small refugia in eastern Zaire and in the Cameroon-Gabon area where it had come to be concentrated during the late Pleistocene. The Queensland rainforest at the northern end of Australia–virtually eliminated during glacial maximum–began to regenerate and expand in early Holocene times (Kershaw 1978).

The boundaries of the Amazon also appear to have been markedly reduced during the late Pleistocene. Current evidence indicates that, at least by 35,000 BP, the mighty rain forest had shrunk downward into the lower elevations of its watershed in response to cooler temperatures. With the return of warmer temperatures in the early Holocene, the Amazon rain forest once again began expanding into higher elevations in northern and central South America. Once re-established, the Amazon did not remain stable and unchanged but appears to have been markedly affected in one region or another by rainfall shifts, storms, large scale erosion, and massive fires throughout the Holocene and these "intermediate level disturbances" have contributed to the evolution of the great regional species diversity that characterizes the area at the present time (Colinvaux 1989: 105-106). In Central America.

By late Holocene, the modern floral realms of the Earth were essentially in place. On a global scale, the flora of the Earth is divisible into six large floral assemblages called realms. These realms can be further subdivided into regions (or zones), provinces and districts. The six realms are: (1) the **Holarctic** including North America, Greenland, Europe and the northern part of Asia, (2) the **Neotropical** realm including Central America to the southern tip of South America, (3) the **Palaeotropical** realm including all of Africa south of its Holoarctic fringe along the coast of the Mediterranean, India, southeast Asia and the Islands of the East Indies north of Wallace's line. This boundary is formed by the southward limit of the northern hemisphere coniferous family (Pinaceae) and the northward limit of the palms (*Pamae* and *Arecaceae*), (4) the **Cape** realm limited to the southern tip of Africa, (5) the **Australian** realm and (6) the **Antarctic** realm limited respectively to those continent.

Re-Forestation and Human Adaptations The re-forestation of vast portions of the northern and southern hemispheres during the early and middle Holocene, must at first have had a profoundly negative impact on human subsistence. For example, the replacement of the steppe-tundra biome in Europe and Eurasia with its herds of mammoth, reindeer, and horses was replaced by closed boreal forest vegetation resulted in a substantial reduction in animal biomass available for human exploitation. Compared with the volume of game available on the late Pleistocene landscape, the early Holocene boreal forest is nearly barren; "in the poverty of its biomass, the boreal forest can be compared to a desert or semi-desert" (Jansen 1982:17). Yet, as he goes on to note, seasonal fluctuations in game availability is no doubt less drastic in the boreal forest than on the steppe-tundra and the wild resources in this forest were–if one includes plants–far more diverse. Eventually, humankind was able to overcome the reduction in animal biomass by adapting to the great plant and animal diversity and season stability of the forests and turn these characteristics to an advantage.

The coeval replacement of expansive tracts of temperate forests and savannah grassland by tropical rainforests on the continents of the southern hemisphere at the end of the Pleistocene must have had a profound impact on human subsistence. Since tropical rain forest biomes have the greatest biomass, highest productivity, and greatest species diversity of any terrestrial ecosystem (Sponsel 1986), we might conclude that the expansion of such habitats would have been entirely advantageous for humankind. However, following the lead of Betty Meggers (1973), some many archaeologists have concluded that, despite their great biomass and vegetative productivity, the world's primary or climax rainforests are–or at least were–uncongenial to peoples exclusively dependent upon a hunting and gathering adaptation.

A recent re-statement of the case against the rainforest argues that, despite its apparent richness,

> most of the energy in a tropical rain forest exists in the form of inedible woody tissue. The intense competition for light and ground

> space selects for plants that invest most of their energy in structural support and maintenance. Only 2% of the biomass in central Amazonian forests is in the leaves, and plants in climax forest environments often concentrate toxins in their leaves, which discourages feeding by insects and animals. Very little energy goes into reproductive parts such as flowers, fruits, and seeds, the very parts that are usually the most edible for human foragers and many of their prey species (Bailey *et al.* 1989)

As a result, these scholars contend, edible plant and animal species are widely dispersed both spatially and seasonally and many of the plants present difficult collecting and processing problems and therefore "expensive" in terms of the their ratios of energy expended for food energy returned. Therefore, a strong case can be made for the view that only hunter-gatherers who have developed symbiotic trading partnerships with agricultural peoples **outside** the forest can survive within it (Bailey *et al.*, 1989). If this view of the tropical rain forests is correct, then we must conclude that the amount of territory available for human exploitation in the southern hemisphere was sharply and repeatedly constricted throughout the Quaternary period.

However, recent work in a deeply-stratified rockshelter site in the eastern or Brazilian Amazon called **Caverna de Pedra Pintada** produced clear evidence of a well-established broad-spectrum hunting and gathering adaptation in a tropical rainforest setting dating to more than 10,000 B.P. (Roosevelt *et al.* 1996). At this early date, foragers could not be in symbiotic relationship with farming people. The Caverna de Pedra Pintada site appears to falsified the hypothesis that the tropical rainforest was unfit for pure foraging.

But whether or not human foragers could have adapted to the rainforest or not, there is evidence of more than 20 stages of glacial advance during the Pleistocene and presumably each of these stages caused a great natural deforestation of the Earth like that known to have occurred during the final stage of the late Pleistocene epoch. Correspondingly, the forests returned during each subsequent episode of interglaciation.

However, this ancient natural cycle of forest destruction and regeneration was interrupted beginning in the middle of the Holocene epoch. Whereas previous periods of interglaciation have been characterized by forest stability and ecological dominance, the late Holocene has witnessed worldwide de-forestation. People began intentionally burning the world's forests at least by the early Holocene (Bush 1988). In large measure, this burning appears to be an attempts to create, expand, or reconstitute the edge habitats favored by key game animal animals like deer sought by Mesolithic hunters.

Following the rise of agriculture, farming peoples continued burning the world's forests and accelerated the process of deforestation by combining burning with overgrazing. The pace of this destruction was accelerated further with the rise of civilization and further still by the fuel and raw material demands of modern industry and the unquenchable land hunger of the world's burgeoning Third World populations. Deforestation promises to continue unchecked and unabated into the next century and only run to ground with the disappearance of the world's forests. Yet in the midst of the mindless, catastrophic and seemingly unstoppable destruction of the world's forest going on in our own time–with all of its ominous implications for people yet unborn–the Pleistocene history of the forests may contain a message of hope for the future. After all, we known that the Earth has been deforested due to climate changes many times during the two million years or more. Still, after each of these episodes of glaciation and deforestations, the forests have proven capable of regeneration and return when conditions changed. If we can save remnants of the great forests, perhaps those remnants may prove enough to allow future, more enlightened generations to reforest the Earth that we leave barren.

Conclusion of the Unprecedented Wave of "Mega-Fauna" Late Pleistocene Extinctions The extinctions of Pleistocene megafauna that began after about 14,000 BP (12,000 B.C.) reached a conclusion in the early Holocene by around 10,000 BP (8,000 B.C.). These extinctions–and the current dispute over what caused them–was discussed in the preceding chapter. Suffice-it-to-say that, although most extinctions known to have occurred throughout geological times can be explained as due to climate change, those of the terminal Pleistocene and early Holocene cannot so certainly credit non-human agency. In Europe and Eurasia, the disappearance of the steppe-tundra biotope and the rapid re-colonization of the northern latitudes by forest vegetation may have been crucial events in these extinctions **but** it is curious that the return of the forests during previous interglacials did not similarly trigger extinctions on such a grand scale. In this regard, the correlation between the extinctions of the New World megafauna and the apparently late appearance of humankind in the Americas seems particularly significant. At present, it seems plausible to conclude that anatomically modern humankind–practicing predation on the large and profligate scale reflected at kill sites like Solutré, Doni Vestonice, and Olson-Chubbuck in northern lands–tipped the balance against megafaunal game species already facing the stresses of changing climate. If these sites do indeed reflect human connivance in "**megacide**," they provide important evidence that fundamental human alteration of natural world had already begun by the end of the Pleistocene. Like the prehistoric cycles of forest expansion and decline, the record of late Pleistocene extinctions contains a grave cautionary tale for our own time. Unfortunately, unlike the forests, once destroyed, species do not "regenerate."

Humanity in the New Circumstances of the Holocene The environmental changes that followed the end of the Pleistocene had profound positive and negative effects on humankind but the archaeological record clearly demonstrates that the people of the early Holocene were able to meet and master these new challenges and take advantage of the new opportunities of the epoch. In the millennia that followed, that record bears witness to a remarkable expansion of the geographic range of humankind and a dramatic increase in the size of its population. Unfortunately, prehistory also records a corresponding increase in human environmental impact and destruction.

The Mesolithic Period

The Mesolithic–literally the "middle stone age"–represents a critical period of cultural transition in the prehistory of the northern regions of the Old World. When identifiably Mesolithic cultures first appear in Europe and Eurasia, they represent a continuation of late Pleistocene hunting-and-gathering subsistence adaptations into the new environmental conditions of the terminal Pleistocene and early Holocene epoch. As a consequence, these early Mesolithic traditions are often simply termed **Epi-Paleolithic**. By the close of the period–and it ended at different times in different areas of the Earth–Mesolithic culture seems to have been swept aside or crowded out of existence by the revolutionary new Neolithic life way based on food **production** rather than food collection. Because both its beginning and its end are rather inauspicious, the importance of the Mesolithic period has often been downplayed or misinterpreted by prehistorians. Nonetheless, as David Clarke notes,

> Mesolithic artifacts and subsistence patterns are not incompetent attempts at later Neolithic artifacts and patterns, or degenerate efforts at Upper Paleolithic products; Mesolithic artefacts and subsistence patterns were well designed and well adapted for what could best be done to achieve an adequate food supply from certain environments, using that technology (Clarke 1976: 259).

Throughout the early Holocene, Mesolithic technology and economy–and by inference, society and political organization–changed dramatically. No doubt some of these changes were merely refinements or specialization intended to fit existing Paleolithic patterns more conformably to local conditions. Other changes, however, represented genuine transformations and innovations that offered effective, and often ingenious, solutions to the new problems and challenges presented to humankind in the early Holocene epoch.

While those Old World regions "out of the reach and the consequences of the last glaciation" also experienced significant environmental change during the post-Pleistocene, it was generally of a very different order than that of the north (Zvelebil 1986: 7). Greater similarity exists between the Old World Mesolithic culture and that of early Post-Pleistocene cultural developments in North America. However,

> the mountain ranges in North America are aligned from north to south, contributing to a markedly different distribution of flora and fauna in temperate North America from comparable areas of the Old World. Moreover, North America was colonized by man only recently, late during the last glaciation; consequently, the early postglacial societies in North America must have been facing very different opportunities and constraints from those encountered by their counterparts in the Old World (Zvelebil 1986: 7).

In the New World, the cultural and chronological analogue of the Mesolithic is the Archaic tradition which is discussed in the next section.

Mesolithic Changes in Human Biology and Culture At least seven important changes in human biology and culture occurred in Europe and Eurasia during the Mesolithic period in concert with the profound environmental changes of the terminal Pleistocene and early Holocene.

1. In general the efficiency of Mesolithic stone tool technology comes to surpass that of the late Upper Paleolithic period. This improvement is due to the perfection and widespread use of a number of key upper Paleolithic period implements and stone

working techniques in combination with Mesolithic period technical innovations
2. Broad-spectrum hunting-and-gathering subsistence systems became more common than specialized big game hunting economies of earlier times.
3. Plants became increasingly more important in human subsistence and gathering occupied a correspondingly larger place in Mesolithic settlement and subsistence strategies than it did among those of late Pleistocene foraging peoples.
4. Sexual dimorphism among *Homo sapiens sapiens* began to be reduced.
5. The sexual division of labor, food sharing, and central place foraging characteristic of the society and political economy of historic and modern hunting and gathering peoples like the !Kung Bushman probably developed during the Mesolithic period.
6. The dog (*Canis familiaris* L.) was domesticated in many parts of the world.
7. Incipient agriculture emerged in limited areas of the world.

These seven developments appear in large part to have been adaptive responses to the environmental changes that occurred during the terminal Pleistocene and early Holocene times in the northern latitudes of the Old World. However, while these changes represent human adaptation to environmental conditions coming to approximate those fo the present, we should not therefore conclude that the life ways of Mesolithic peoples were identical with those of ethnographically known hunter-gatherers (Zvelebil 1986: 6).

Key Changes in Mesolithic Stone Technology Nowhere is the legacy of the upper Paleolithic more evidence than in Mesolithic period stone tool technology. This continuity is especially evident in the Mesolithic cultures of Northwestern Europe. In European assemblages (1) blade flakes continued to be backed by steep retouch blows in the manner of the upper Paleolithic, (2) burins remained common, no doubt because they continued to be used in the extraction of bone strips by the groove-and-splinter technique, (3) microlithic blades–that first appeared in Africa in the Middle Stone Age and only much later emerged in southwestern Europe during the Magdalenian–became the characteristic stone implement of the Mesolithic, and (4) the bow-and-arrow, probably present in the Spanish Solutrean by 17,000 BP, became the chief weapon of the period.

However, in northwestern Europe and elsewhere in the Old World, a number of important new tool forms appear after the close of the Paleolithic. These new tool forms suggest that Mesolithic technology–that is, its techniques and the social organization of its production–were undergoing important developmental changes. Principal among the technical innovations to appear at that time include:

1. new composite tools,
2. axes, adzes, and chisels made of flaked stone in early times and on ground stone near the close of the period,
3. mortars, querns, and other stone tools for grinding,
4. fishing and fowling tools and facilities,
5. canoes and skin boats,
6. in northern latitudes, at least, sledges and possibly skis, and
7. in a very limited number of localities, the earliest ceramic vessels.

Until quite recently, prehistorians tended to disparage the technology of the Mesolithic period and suggest that it represented a decline from that of the late upper Paleolithic period. Chief evidence for this supposed decline was the increased importance of stone implements made on microlithic blades. Microlithic blades are on average less than about two inches long (Bordaz 1970: 92), about half the length of the typical upper Paleolithic period blades. Microblades

> were manufactured from very small blade nuclei sometimes no more than an inch high and were worked with a punch and hammer, or perhaps rather by pressure. A later method consisted of fragmenting larger blades by means of a notching technique sometimes called the micro-burin technique. Once flaked or fragmented, the tiny flints were frequently retouched until they assumed standardized geometric forms such as crescents, triangles, and trapezes (Bordaz 1970: 93).

Microliths represent an marked improvement in stone tool manufacturing efficiency. The modest size of the microblades meant that stone knappers could extract more cutting edges from each kilogram of raw material that they processed and could use smaller nodules of flint and chert as cores. However, microblades are generally too small to be effectively used by holding them directly in the hand. Therefore, the increase in cutting edge per unit of raw material that their production entailed would have been of little use had not some means of mounting them on a handle or shaft been developed. The technique generally used to do this is referred to as "insertion." A notch or groove was cut with a burin in a bone shaft. The shaft was then soaked and the microblade inserted in the groove. According to Semenov (1964: 203), "by the action of water bone swells only very slowly, but, owing to the lime in it, it dries very quickly and grips." Bladelets were also fastened to such shaft grooves using tree resins or bitumen from natural oil seeps.

Composite tools of this kind are not easy to make but they have the advantage of being strong and easy to repair. Although a large Solutrean foliate blade provided its user with a long cutting edge, such a tool required a great deal of skill, time and raw material to make, and was easily broken. A cutting tool made of microliths set in a bone handle not only would have been harder to break, when one of its individual bladelets were broken or lost, a new one could readily be inserted into the vacant slot in the handle.

The bow-and-arrow largely replaced the throwing stick or *atlatl* as the primary hunting and military weapon during the Mesolithic period over much of the Old World; the diffusion of the tool to the New World takes place only in the late Holocene (Dickson 1985: 21-24). Consequently, one of the most important Mesolithic composite tool was the arrow. Microliths of varius geometric shapes were used as arrow points by mounting them with resins in grooves cut into the tips of wooden shafts. In the early Mesolithic, additional microliths were sometimes mounted on one or both sides of the shaft just behind the arrow points. Once the arrow point had penetrated the animal, these lateral microliths would have induced the primary wound to hemorrhage and at the same time prevent the arrow shaft from working loose as the animal struggled to escape. (Clark, 1980: 48). This technique seems to have been abandoned in the later Mesolithic, perhaps because improvements in bow design provided penetrating power sufficient to kill the animal with the primary wound alone.

Emergence of Broad-Spectrum Foraging In general, large game animals predominate in the faunal assemblages from Old World archaeological sites dating to the period of maximum continental glacial expansion around 18,000 BP Sites from southwestern France to the Ukraine are dominated by reindeer and/or horse; those from northern Spain across to Italy and Greece by red deer (Dennell 1983: 100; Straus, *et al.* 1980; Jarman 1972), those from the Levant by fallow and roe deer, gazelle or ibex, and bovids (Bar-Josef 1980: 118). In the millennia between glacial maximum and the end of the Pleistocene, mammoth appears to have been central in the subsistence of the Ukraine and other regions of central Europe (Gladkih, Kornietz, and Soffer 1984), while red deer, aurochs and horse provided most of the food in Northern Spain and southwestern France (Dennell 1983: 124), and fallow deer and gazelle or ibex became primary in the Levant (Bar-Josef 1980: 122). Scholars still debate the date of the arrival of humankind in the New World. However, the earliest indisputable peoples, the Paleo-Indians, were there in late glacial times and appear to have focused their hunting on big game. Between 12,000 BP and the end of the Pleistocene in western North America, first mammoth, then various species of bison, were the chief targets of the Paleo-Indian peoples.

The extinction of the megafauna must have forced big-game hunting peoples of the terminal Pleistocene epoch to realign these subsistence adaptations. At first, probably only involved relatively minor adjustments were needed. The various taxa did not become extinct simultaneously, so it was no doubt possible to shift to secondary prey species as primary prey animals disappeared. For example, bison species became favored targets of the Paleo-Indian peoples following the extinction of the mammoth in North America. However, mere substitution became difficult or impossible as the extinctions continued. Survival by foraging in the new conditions of the early Holocene necessitated substantial adaptive changes.

Cohen (1977) has asserted that the worldwide terminal Pleistocene extinctions, the worldwide rising sea level, and the resulting reduction in available game placed great pressure on the populations of early Holocene hunting and gathering peoples. It is his view that this pressure 6initiated a "**food crisis**" of global proportions that led ultimately to the emergence of agriculture. Whether Cohen's view that a Malthusian food and population crisis overtook humankind at the close of the Pleistocene is correct or not, the archaeological record provides ample evidence that subsistence bases in Eurasia, Europe, Africa, and North and Middle America began to broaden considerably to exploit a progressively greater range of species and habitats. Such hitherto neglected animal species as fish, crabs, turtles, mollusks, land snails, migratory waterfowl, rabbits, and other small game became a routine part of the diet. At the same time, plants were exploited more systematically for their fruits, tubers, nuts, and seeds. In general, these additions to the diets of early Holocene epoch peoples tended to be

1. drawn from low levels on the food chain,
2. predictable in yield and seasonal availability (Flannery 1969: 78-79),
3. easily collected by women and children (Flannery 1969: 79),

4. characterized by procurement and/or processing techniques that rendered them "costly" in terms of caloric yield per unit of effort, and
5. storable as surplus above immediate subsistence needs.

The procurement of wild species of these sorts must have necessitated substantial technical and strategic changes on the part of Mesolithic period foragers. Further, the physical demands of broad-spectrum subsistence on Mesolithic people must have been very different than those of the preceding period. Clues to these differences are found in human skeletal biology to which we now turn.

Reduction in Human Sexual Dimorphism Although anatomically modern in most senses, late Pleistocene *Homo sapiens sapiens* populations exhibited more pronounced sexual dimorphism than subsequent human populations. Frayer relates this pronounced dimorphism to subsistence patterns of the Late Pleistocene. Noting that this subsistence centered on large mammal hunting, and assuming that hunting was done primarily by males, Frayer concludes that the hunting of large mammals with only a simple technology would selectively favor large individuals possessed of greater physical strength and stamina. By contrast, gathering and camp maintenance activities–presumably associated with the female role–confer no such advantage on larger, stronger individuals. While late Pleistocene **males** show consistently larger dimensions than their Mesolithic descendants, females from the two periods are similar in postcranial dimensions. The reduced sexual dimorphism in post-Pleistocene hunter-gatherer populations must indicate that selective pressures operating on males had changed while those operating on females had not.

Frayer, and more recently Foley (Lewin 1988), hypothesize that these differences in sexual dimorphism between the two populations were caused by changes in hunting patterns. He notes that the reductions in male robustness follow:

1. the extinction of the Pleistocene megafauna; and
2. a general improvement in the effectiveness of Mesolithic hunting technology.

He suggests that hunting smaller game with an improved technology resulted in the development of new hunting strategies that placed less reliance upon simple physical strength. Thus, the selective advantage that had accrued to males of large body size in the late Pleistocene diminished during the Mesolithic period. Since female activities were unaffected by these developments, little change occurred in the body sizes of women. Recently, Frayer's hypothesis that technical improvements and the end of big game hunting changed selective pressures that had favored large men was tested against the post-cranial skeletal remains of recent Inuit (Eskimo) who specialized in whaling and large marine mammal hunting versus those who practiced salmon fishing (Collier 1993). While Collier found some evidence of greater robustness among the whaler males, he discovered that this was also true of the **females** in the whaler population as well. Thus, the whaler population was more robust, but less sexually dimorphic, than the salmon fisher population leading Collier to conclude that females cannot be seen as constant and unchanging in any adequate model of economic behavior–whether Mesolithic or modern.

Nonetheless, Foley accepts Frayer's argument and concludes further that reduced sexual dimorphism may be taken as evidence, (1) plant gathering was relatively more important in the subsistence strategies of Holocene epoch foragers than in those of the late Pleistocene peoples, and (2) the sexual division of labor, food sharing, and central place foraging characteristic of historic and modern peoples like the !Kung Bushman is of comparatively recent origin (Lewin 1988).

Foraging Band Organization Comes of Age Although direct evidence is probably never be forthcoming, it seems plausible that the sexual division of labor, food sharing, and central place foraging characteristic of the society and political economy of historic and modern hunting and gathering peoples like the !Kung Bushman probably had developed by the end of the Mesolithic period The archaeological record of the Old World indicates that peoples practicing a generalized or broad-spectrum hunting-and-gathering subsistence were well-established in many parts of the Old World between approximately 12,000 and 8,000 BP or 10,000 and 6,000 B.C. There are no direct means of reconstructing the socio-cultural systems of these Mesolithic foragers. Therefore, prehistorians have used studies of historic and contemporary food collectors to arrive, by analogy, at certain generalizations about the infrastructure, cultural ecology, demography, and socio-political organization of societies of broad-spectrum hunting-and-gathering societies of the Mesolithic period.

By analogy, the infrastructure of Mesolithic broad-spectrum foraging societies–that is, their economic base–presumed to have been characterized by (1) simple technology, (2) subsistence systems capable of producing relatively low levels of food energy, (3) a relatively high ratio of output of food energy per unit of labor (Service 1979: 12), (4) diets to which plants contributed a greater percentage of the calories than animals, and (5) a dependence upon wild food resources which tend to be spatially dispersed and to fluctuate seasonally in availability. As the extinctions of the terminal Pleistocene dramatically illustrate, the over-exploitation of a wild resource may yield short term benefits, but in the long term it can damage or destroy the resource base. The absence of megafauna in the comparatively impoverished environments of the early Holocene meant that the results of human over-exploitation of remaining wild resources would translate quickly into human suffering. Therefore, successful broad-spectrum foraging peoples appear to have developed social and economic mechanisms designed to prevent over-exploitation of any given wild resource. Chief among these mechanisms were: (1) the seasonal-timing of the movement of the band from resource to resource, (2) the ready flexibility of organization allowing the band to fission into its constituent family units in the face of resource scarcity and to fuse or reunite in times of abundance, and (3) the use of extensive and generalized (rather than intensive and specialized) strategies of resource exploitation (Flannery 1968; Jochim 1976, Keene 1981).

Given the foregoing infrastructure, the demographic structure of Mesolithic foraging societies was probably characterized by (1) low densities of population per square kilometer, (2) low rates of population growth through time, (3) maximum population sizes determined by the amount of wild food stuffs collectible during the season of **minimum** availability, and (4) an absence of epidemic diseases due to the small sizes and low concentrations of human populations. Small population sizes and relatively modest rates of growth among contemporary–and by extension, Mesolithic–hunter-gatherers may be in part a reflection of the precariousness of the foraging mode of production. However, they are mainly due to conscious population policy instituted by peoples aware of (a) the minimal economic contribution of children's labor to their subsistence, (b) the difficulties of carrying young children for highly mobile populations, (c) the relatively high opportunity costs of female labor in societies where in many cases females contribute the major amount to household production (Harris and Ross 1987; Blanton 1988), and as noted, (d) the need to maintain human population below a level that can be sustained during the least productive season in the annual cycle. Behavioral means of reducing female fertility available to hunters and gatherers include prolonged breast feeding (which reduces female fertility) and the avoidance of sexual intercourse for long periods following child birth. More direct means of population limitation include abortion, infanticide, gerontocide (the killing or abandonment of the aged or infirm) and permanent band fissioning.

The socio-political organization of Mesolithic foraging societies probably was characterized by (1) band (or family cluster) organization, (2) reliance upon kinship as the most important principle of social organization, (3) the band as corporate group holding land resources in common but generally granting unrestricted access to these resources to all members of the band, (4) economic distribution and exchange based on reciprocity, (5) little emphasis on accumulation, (6) an absence of full time specializations beyond the sexual division of labor, (7) an absence of ascribed statuses and roles beyond those of age and sex, (8) feuding but no true warfare, and (9) participation in extra-band, regional networks of rather loosely-defined groups and communities linked by a variety of social relationships including trading fields exchanging goods, marriage fields exchanging personnel, and political and religious fields exchanging information or alliance support. These networks often extended over vast regions and even entire continents and appear to have been critical to the survival of foraging bands.

Finally, it is important to note that a number of contemporary archaeologist, in revolt against the "tyranny of the ethnographic record," reject the view that accumulation, ascribed status, and specialization were absent. These scholars mustered evidence–admittedly limited–that social stratification, sedentism, storage systems, and moderate population density characterized at least some foraging socio-cultural systems dating to the Upper Paleolithic and Mesolithic periods. If correct, their work will force a substantial modification of our picture of the prehistoric foraging adaptation, since such phenomena has heretofore been considered the exclusive attributes of later, food-producing social formations. Key expressions of this point of view are found in Soffer (1985), Price and Brown (1985), Bender (1978, 1981), and Mueller-Wille and Dickson (1991)

In sum, Mesolithic period broad-spectrum hunting-and-gathering was probably a subsistence mode characterized by balance and limitation. Foragers do not control the production of foodstuffs within their environment; being unable to override the cycles of nature, they must ride with them. They do this by limiting the amount of food energy which they extract from their environment and by limiting the size of the human population which they must support within it. Foraging bands prevent the over-exploitation of any given wild resources in their subsistence base and keep the size of their group below the carrying capacity of the environment by scheduling the movement, fissioning and fusing of their social groups in close synchronization with the seasons; by practicing "satisfying," rather than "maximizing," food collecting strategies; by population control; and by pooling foodstuffs through reciprocal food sharing arrangements. In this way, they ameliorate or replace the natural checks that control the population sizes of non-human species by cultural ones. They adapt to nature by mimicking it.

Emergence of Incipient Agriculture "Incipient agriculturalists" was the label applied by Robert Braidwood to the Mesolithic (called the Epi-Paleolithic by most Near Eastern archaeologists), broad-spectrum foraging peoples in southwestern Asia who specialized in gathering and processing wild emmer wheat, wild barley, and other cereal grains at the end of the Pleistocene epoch (Briadwood 1958). Although not food producers themselves, incipient agriculturalists, like the Natufians along the rim of the eastern Mediterranean or the Karim Sharirians located in the foothills of the Zagros Mountains, provided the critical transition between food collecting and food producing.

The socio-ecology of incipient agriculture, like that of broad-spectrum hunting-and-gathering, appears at least in part to be a response to the resource depletion brought about by extinction and environmental change at the end of the Pleistocene. Foley suggests that the two subsistence systems developed more-or-less simultaneously (Lewin 1988). However, while broad-spectrum hunting-and-gathering developed over much of the globe, incipient agriculture emerged only in limited number of areas. The relative rarity of incipient agriculture appears due to the fact that wild species differ in their capacity to be domesticated. Incipient agriculture developed only in those regions where domesticable species where part of nature wildlife array. Thus, broad-spectrum hunting-and-gathering was the mainstream and incipient agriculture was decidedly the minority, solution to the environmental problems of the early Holocene epoch.

Sometime around 10,000 BP (that is, 8,000 B.C.), the wild cereal plant species exploited by the Natufians and other incipient agriculturalists became fully domesticated. The techniques of wild plant procurement, processing, and storage developed by incipient agriculturalists were adapted and used by later, fully-agricultural subsistence systems. In this sense, the practice of incipient agriculture "pre-adapted" foragers to food production and, in doing so, laid the groundwork for the development of village agriculture in the subsequent Neolithic period (Flannery 1969: 77). Incipient agriculture was thus a critical step in the development of full agriculture, and its socio-ecology, like that of broad-spectrum hunting-and-gathering, developed as a response to resource depletion at the end of the Pleistocene. A great deal more will be said of the development of incipient agriculture in the next chapter.

The Early Holocene Epoch and Mesolithic Period in Northwestern Europe

In light of this general discussion, let us now turn to the record of early Holocene epoch environments and Mesolithic cultural adaptations in the region of Old World where it is most fully understood: western Eurasia, specifically, northwestern Europe. Northwestern Europe includes the British Isles, northern Germany, and the modern Scandinavian countries of Danmark, Norway, Sweden, and Finland. By 18,000 B.C. during the final phases of the Würm-Weichsel glacial stage in the late Pleistocene most of northwestern Europe was covered by ice and characterized by harsh climatic conditions at once arid and extremely cold. However, as the Ice Age drew to a close after about 13,000 BP, these conditions began to change. The trajectory of this terminal Pleistocene and Holocene environmental change in the Old World is best understood in northwestern Europe where it has been reconstructed primarily through the analysis of changes in the plant macrofossils and pollen spectra recovered from stratified layers in peat bogs, varve banks, archaeological sites and other sources.

In the late 19th century, two Scandinavian paleobotanists, **A.G. Blytt** and **R. Sernander**, attempted to reconstruct the changes that took place in the climate of northwestern European during the late Pleistocene and Holocene by studying the sequence of plant macrofossils preserved in the stratified layers of the peat bogs of the region. Blytt and Sernander concluded that changing moisture and temperature conditions in northwestern Europe were reflected (a) in the changing composition of the vegetation these layers contained and (b) in the relative thickness and rate of growth of the sequential layers in the peat bogs. Using these data, they recognized and named eight sequential stratigraphic and vegetation "zones" in the peat bogs. The eight-fold scheme and the climatic inferences based on it is called the Blytt-Sernander European Pollen Zone System.

At about the same time that the Blytt-Sernander system was being formulated, European botanists were developing the science of palynology or the study of plant pollen. Palynologists extract plant pollen from samples from soils and sediments where it was deposited and examine it microscopically. Different plant taxa produce recognizably different types of pollen. Such distinctions enable palynologists to identify the genus and sometimes the species of the plant that produced the pollen. The percentages of each genus or species present in a pollen sample is plotted on a **pollen diagram**. Such a diagram directly represents the relative contribution of the various plant types to the pollen rain during the sediment's formation and–indirectly at least–represents the actual number of each type of plant in the environment at that time.

Pollen samples are readily extracted from sequential layers in stratified sediment columns like peat bogs. The number or depth below surface of each strata is plotted in order along the Y-axis beginning with the oldest at the bottom. The percentage of the various plant taxa contained in the samples drawn from each strata is plotted on x-axis of the diagram. A picture of the changes that occurred through time in the vegetation around the collection site can thus be obtained by comparing relative pollen percentages in each layer from bottom to top in stratigraphic order.

By identifying the fossil pollen taxa incorporated in the stratified layers of the peat bogs of northwestern Europe, palynologists were able to further refined the vegetation sequence in the Blytt-Sernander system. Finally, the dates in years in which the eight zones were deposited were determined using de Geer's European varve chronology and these absolute dates have been further refined over the last three decades through radiocarbon dating of the organic material in the bog sediments. These scientific developments converged to give northwestern Europe the first complete, absolutely-dated model of the last 13,000 years of vegetation and climate change.

However, it is now recognized that the changing pollen frequencies characterizing the eight zones of Blytt-Sernander system are not solely due to changes in temperature and rainfall. At least four other factors contributed to the changing pollen species frequencies including: (1) differences in the speed at which the plant species were able to migrate northward to re-colonize northwestern Europe after the retreat of the glacial ice, (2) the increasingly leeched conditions of the area's soil in the warmer, moister Holocene climate, (3) the periodic burning and limited clearing of portions of the forest by Mesolithic peoples in early and middle Holocene, and (4) the large scale forest burning and permanent clearance first initiated by Neolithic peoples at the beginning of the late Holocene (Wright 1976: 583). The first five Blytt-Sernander pollen zones, including the Oldest Dryas, Bölling, and Older Dryas (Zones Ia, Ib, Ic), the Alleröd (Zone II), and the Younger Dryas (Zone III), date to the Late Glacial or terminal Pleistocene epoch from before 10,800 B.C. until about 8300 B.C. The vegetation changes that occurred in these five sequential zones indicates that the late Ice Age climate did not slip quietly and smoothly out of northwestern Europe. Instead it is clear that conditions whip-sawed in a rapid and erratic manner culminating in the retreat of the Fennoscandinavian, British and Alpine glacial complexes.

The earliest zone in the Blytt-Sernander system is Zone 1a: the Oldest Dryas which dates to before 10,800 B.C. Rigorous, full glacial climatic conditions prevailed in northwestern Europe during this time. This is reflected in the pollen profile by the presence of tundra vegetation and by the evidence of fine, wind-blown sediments. Key tundra plants species found in strata referable to this zone include: *Dryas octopetala* or mountain avens, dwarf willow (*Salix herbacea*), arctic willow (*Salix polaris*), dwarf birch (*Betula nana*), sea buckthorn (*Hippophaë rhamnoides*), and Öland rock rose (*Helianthemum oelandicum*). Based on the ranges and requirements of these plants in modern times, the summer temperatures during the Oldest Dryas are estimated to have averaged 8°to 9° Centigrade (J. Jensen 1982: 12).

A dramatic shift in late Pleistocene environmental conditions occurs in the subsequent Zone Ib, known as the Bölling. The Bölling dates to between 10,800 and 10,400 BC. Pollen spectra recovered from this zone indicates that northwestern Europe was generally characterized a open parkland vegetation at this time. The abrupt appearance of vegetation of this type suggests that the region experienced a brief rise in mean annual temperature that upset the delicate balance in arctic vegetation that had characterized the Oldest Dryas zone. Although the mild climate of the Bölling proved only temporary, it foreshadowed the conditions which were eventually to be established in northwestern Europe in the Holocene.

Like its beginning, the end of the Bölling zone is marked by abrupt changes both in the nature of deposition and pollen percentages in the peat bogs. The subsequent deposits of Zone Ic, the Older Dryas (10,400-10,200 B.C.), are characterized by fine, wind-blown sediments from which arboreal pollen is absent. Grass and sedge pollen dominant the pollen spectra of the Older Dryas suggesting that, for a brief 200-year interval at the end of the Pleistocene, harsh and cold tundra conditions returned to northwestern Europe.

Pollen recovered from Zone II, the Alleröd (10,200-9,400 BC), indicates that warm, interstadial conditions returned to northwestern Europe. Although tundra vegetation still was present around the margins of the retreating ice sheet in Scandinavia, the land south of this margin began to be reforested. Alleröd pollen spectra indicates that England and Denmark supported an open birch parkland characterized by: birch (*Betula pubescens*), quaking aspen (*Populis tremula*), rowan (*Sobus aucuparia*), and juniper (*Juniperus communis*). Pine woodlands appear to have dominated the territory south of the tundra in southern Germany, northern Poland and southern France. The mean annual temperature during this period was probably 5-6° Centigrade warmer than during Würm glacial maximum but still about 2.5° C. colder than the present.

By the end of the Alleröd and the beginning of the Younger Dryas zone in northwestern Europe, extinction had over taken such late Pleistocene megafauna as the woolly mammoth, woolly rhino, cave bear, European wild ass, a genus of goat (*Hemitragus bonali*) (Butzer 1971:483-484), and a species of giant deer known popularly as the "Irish elk" (Barnosky 1986). In addition, the spotted cave hyena, saiga antelope, Arctic lemming, the musk ox and a number of other animals species disappeared from mid-latitude Europe but continued to exist in eastern or northern Eurasia.

Most of these forms are linked to the steppe-tundra biotope so the rapid re-colonization of the tundra by boreal forests has often been cited as the cause of their extinction or regional disappearance. However, some characteristic tundra animals and steppe animals including reindeer, wolverine, European bison or wisent, and horse were able to survive in the new environmental conditions of northwestern Europe. At the same time, the open boreal forest vegetation had apparently attracted such forest animals northward as the common elk, beaver, brown bear (*Ursus arctos*), lynx, badger, and wild cat. Thus, it would appear that the animal life in northwestern Europe during the late Allerod had become increasingly varied and taken on a decidedly mixed and transitional character.

In Zone III, the Younger Dryas (9,400 B.C. to 8,300 B.C.) a marked and apparently world-wide climatic deterioration took place. In northwestern Europe, North America and elsewhere in the northern hemisphere continental glaciers began to re-advance (Mott, Grant, Stea, Occhietti 1986). Worldwide sea level which, in the preceding warmer conditions, had risen perhaps 40 meters or so above the Würm low dropped some ten meters at this time. The pollen of non-arboreal species once again came to dominate the northwestern European spectra suggesting the return of a tundra or a cold-tolerant forest-tundra biome across Britain and Denmark. Mean annual temperatures during the Younger Dryas were probably as low as those of the Older Dryas.

Zone IV, the Preboreal (8,300 to 7,500 B.C.), represents the first phase of the Holocene epoch. Pollen spectra from this zone suggest that environmental conditions became mild once again. During the Preboreal the tundra and tundra parkland that had expanded during the Younger Dryas was reclaimed once and for all by the forest. In Denmark and England the Preboreal vegetation was composed largely of light, open forest dominated by birch, aspen and pine. The birch and aspen had survived the previous cold conditions, pine arrived as a pioneer. The establishment of more-or-less continuous forest vegetation across the landscape of northwestern Europe totally altered the conditions of animal life. The once vast herds of reindeer and other game animals disappeared. Their place in the hunter's economy was taken by first by aurochs and elk and then, as the deciduous forest species began to arrive, by red and roe deer, wild boar (*Sus scrofa* L.), brown bear, wolf, badger, wildcat, pine marten, beaver, brown hare, and various species of voles (Jensen 1982: 15-16).

The British glacier disappeared altogether, the Alpine glacial complex was reduced to individual mountain glaciers, and the Fennoscandinavian glacier retreated northward and began to dissipate at a rate of about 300 meters per year. Vast quantities of water from their rapid melt-off were impounded in the Baltic region to form an huge inland fresh water lake. The ice plug holding this impoundment near the present coast of Danmark. When this plug was finally breached in 7912 B.C., it released the impounded water in a thunderous flood that carved a large river through Storebaelt in modern Denmark. The breaching, dated by varve counting and rounded to 8000 B.C., serves as a convenient end point for the Pleistocene epoch. At least in climatic terms, however, 8000 B.C. is a somewhat arbitrary boundary. The rigorous climate of the Pleistocene did not change abruptly at that point. As the vegetation change reflected in the Blytt-Sernander system indicates, the 2,500 years between the end of the Oldest Dryas around 10,800 B.C. and the onset of the Preboreal after 8300 B.C. were characterized by whip-saw reversals in climate from arctic to subarctic to arctic to temperate conditions in northwestern Europe.

At the very end of the Preboreal zone, sometime between about 7500 and 7000 B.C., a new foraging cultural tradition know as the **Maglemose** or **Maglemosian** emerged in southern Scandinavian and spread to adjacent regions of the European Plain at the northwestern end of the continent. This tradition, which was to persist for nearly 2,000 years, was characterized by distinctive microlithic stone repertoire as well as by flint axes, adzes and picks made either entirely by flaking or in combination with the new technique of stone grinding. Ground stone tools are made by roughing out a "blank" in the outline of tool either by flaking or "pecking" it with a hammerstone. The blank is then worked into desired shape by abrading against a grinding stone. Although their manufacture was laborious, modern experiments with them reveal that the smooth and regular shapes of groundstone tools markedly increases their cutting and chopping efficiency. The frequent occurence of axes and adzes in Maglemosian tool industries no doubt reflects the increased importance of wood and woodworking that followed on the return of the forests to northwestern Europe. Maglemosians also made spears harpoons, mattocks and other types of tools made of deer and elk bone and antler and made wide use of wood for dug-out canoes, axe handles, and spear shafts. The wooden bow and microlith-tipped arrow appears to have been the main weapons of the Maglemosians. Like their upper Paleolithic period predecessors, the Maglemosians made beads, pendants and other decorative objects of stone, antler and bone.

The bulk of the Maglemosian sites excavated to date have been small in size and located along the shores of inland lakes. These sites have produced abundant fish bones, fish hooks, basket traps and seine nets, leisters and other fishing equipment. Gron (1987: 319-320) suggests that such shoreline locations were primarily occupied in the summer. Although the Maglemosians exploited a wide range of plant and animal species present in the Postglacial forests, they appear to have placed particular emphasis on fishing and fowling and the hunting of large game such as the European elk, aurochs, red and roe deer.

Maglemosian culture in Great Britain is best represented at the celebrated site of **Star Carr** in the Vale of Pickering in Yorkshire, England. The site was located in boggy valley that, during the early Holocene, was occupied by a shallow lake. The site was discovered shortly after World War II by an alert amateur who recognized Mesolithic period tools in back dirt from a freshly dug drainage ditch. The site was brought to the attention of J. G. D. Clark at Cambridge who, at the head of a distinguished team of scientists from a number of disciplines, excavated Star Carr over a three year period from 1949 to 1951. Because the site had remained constantly wet in the anaerobic conditions of the bog since its formation, artifacts made of such perishable materials bone, antler, bark, and wood were recovered along with plant and animal ecofacts in a remarkable state of preservation. As a consequence of the high level of preservation and the correspondingly high degree of site resolution, Star Carr provides key insights into the subsistence economy of the early Mesolithic period. However, the wet conditions also complicated work at the site. Excavations eventually reached a depth below the ground water table. In order to keep working it became necessary to continually pumped out of the trenches and carefully shore up the sides of excavations with timber to avoid collapse.

The Mesolithic period occupation of the site was located beneath more than a meter of peat in two strata which together were some 15 to 40 centimeter thick. These strata lay atop a **solifluction layer,** that is, a deposit that had formed when masses of water-saturated sediment had slowly flowed from higher to lower ground in a mud stream or mud glacier. This deposit dates to the Younger Dryas and is further evidence of the arctic temperatures that returned to the Northern Hemisphere at the very end of the Pleistocene epoch.

Human occupation at Star Carr centered on what Clark calls "a loose platform of birch brushwood" that apparently stretched out into the boggy lake from its shore line. Within and around the platform were scattered glacial stones, wads of clay, animal antlers and bones, flint implements, and debris from flint tool making (Clark 1972: 20). This scattered debris covered an area of more than 200 square meters. The pattern of its distribution was carefully mapped in order to recover the outline of the site. Clark interpreted the stratigraphy as indicating that Star Carr had been occupied on at least two successive occasions, a conclusion confirmed by the slight typological differences found between the artifact assemblages recovered from the two strata.

The excellent conditions of preservation at the site allowed the recovery of bone and antler remains of 80 individual red deer, 33 roe deer, 11 elk, and nine aurochs. Bones of wild boar, beaver, red fox, marten, badger, hedgehog, and rabbit along with a number of aquatic birds were also found. Despite careful searching, no fish bones were recovered from the site (Clark 1972: 20; Wheeler 1978). In addition to these ecofacts, a host of tools made of perishable organic material was recovered including: barbed spear and harpoon points made of antler and bone, mattock heads, bâtons, pendants, and pins made of bone, and a large wooden canoe paddle fragment.

Stone tools were also well represented at the site. The bulk of the lithic assemblage was made up of microliths, burins, and scrapers. Pitts (1979: 36) makes a strong case for interpreting the Star Carr stone implement assemblage as animal hide and antler processing tools. Also recovered were flaked stone adzes, axes and gouges that were probably used in wood working. Large quantities of stone cores and chipping debris suggests that many of the finished artifacts were manufactured on site.

Clark concluded that the Star Carr had been a habitation site occupied on by a "micro-band" of hunter-gatherers who had (1) erected skin tents or temporary brush shelters on the birch platform and had (2) used it as a dry base within the bog upon which to work and (2) dropped debris from this work off or through the platform or had tossed into the surrounding water. Given the limited floor area of the platform, Clark concluded that it had been capable of accommodating no more than three or perhaps four families for a total of no more than twenty-five individuals.

Both shed and unshed antlers of roe deer, red deer and elk were recovered at the site in great abundance. Clark used the ratio of the antlers of the various species there to infer that Star Carr had been occupied "before the end of the year and into the following April" (Clark 1972: 23). Clark's argument for a winter use of Star Carr is ingenious. A specialist can quickly determine whether deer and elk antlers have been shed naturally or removed from the skull of the animal by a hunter. Clark found that (1) although some of the red deer antler recovered at Star Carr had been naturally shed, the bulk of them had been cut off the red deer's head, (2) roughly half of the elk antlers had been shed, half removed from the skull, and (3) all of the deer antlers had beem taken off slain animals. Roe deer shed their antlers in early October, elk before January, and red deer in April. Therefore, Clark concluded that the array of shed and unshed antlers found at Star Carr was most likely to have been produced by hunters using the site between late December and early April. He further suggested they made winter camp in the lowlands of the Vale of Pickering because the deer and elk, seeking shelter and browse during the harsh Boreal winter, were probably congregated in the lowland forests.

Deer and elk would have moved upland out of the Vale with the onset of early spring and summer as temperatures moderated and the vegetation at higher elevations began to green and Clark assumed that the Maglemosian hunters who occupied Star Carr had to respond to this seasonal shift with movement of their own. He noted that while the tool assemblage at Star Carr contained axes, adzes, microliths, flint cores, and burins, sites like Deep Carr in the surrounding hill country lacked axes and adzes and produced great numbers of microlithic arrow points. The contrast between the tool kits of lowland and upland Mesolithic sites in Britain is often taken to indicate cultural or ethnic boundaries between different occupants of the two regions. Clark rejects this interpretation, however. He suggests instead that the contrast reflects the difference between the activities carried on at upland and lowland sites by essentially the same people. He hypothesized that the winter residents of Star Carr hunted deer and elk, replenished their stocks of bone and antler, and took advantage of this brief sedentary period to fashioned barbed antler points and harpoons, bone mattocks, and other tools from their newly-acquired raw material. He concludes that these people then abandoned their encampment at Star Carr in the early spring and moved to the uplands on the heels of the migrating deer and elk where they spent the short Preboreal summer hunting and gathering as individual families or in all-male task groups. The absence of axes and other woodworking tools in the uplands sites reflects the use of this region primarily for summer hunting.

Unfortunately, Clark's interpretations of Star Carr are not without problems. First, if Star Carr was a winter camp, why weren't some structural remains and hearths found in it? Maglemosian sites in other parts of northwestern Europe have produced them and it seems unlikely that people would establish a winter camp in such a damp and boggy location without building fires and shelters. Second, while the pattern of shed and unshed antlers may reflect a midwinter occupancy, the unshed roe deer antlers recovered at the site point to a summer occupation, as do the bones of a crane and a possible white stork (Simmons, Dimbleby, Grigson 19–: 119) . A case can also be made that the pattern could equally well have been produced by the year round occupation of Star Carr, by repeated visits over many years, or by storing the antlers from year to year. We must recall, however, that to

> demonstrate seasonality one needs not only definite evidence that a site was occupied at a particular season, but, equally important, evidence that it was *not* occupied at other seasons (Simmons, Dimbleby, Grigson 19–: 119).

Finally, the artifactual and ecofactual remains at Star Carr appear to have resulted primarily from antler and skin working, stone tool manufacture and wood working. If Star Carr was a general habitation site, one would expect that the tools recovered from its vicinity would also contain evidence food processing, cooking, storage, and other base camp maintenance activities.

Zone V, the Boreal, (7,500 to 6,200 B.C.) witnesses the replacement of the birch-aspen-pine assemblage of the Preboreal by a forest dominated by hazel and pine after about 7,000 B.C. Hazel (*Corylus avellana*) arrives late from southern France but neither its appearance nor its disappearance indicates climatic change. Hazel is a pioneer species that thrives in open, disturbed habitats only to be crowded out by elm and lime in the Atlantic period (Larson 1982: 532-33) As the succession continued, new warm-loving species began to appear as well. These included: oak (*Quercus*), ash (*Fraxinus*), elm (*Ulmus*), and lime (*Tilia cordata*). Interestingly, the quaking aspen (*Populus tremula)*, a cold-adapted boreal species, is still present.

The vegetational changes that occur between Preboreal and Boreal were interpreted by Blytt-Sernander as representing the onset in northwestern European of the worldwide warm period known as the Hypsithermal (or Climatic Optimum). As noted above, the Hypsithermal is dated by Deevey and Flint (1957) to between 7000 B.C. and 300 B.C. In northwestern Europe, this would place it in the Boreal, Atlantic, Subboreal zones of the Blytt-Sernander model. However, paleoenvironmentalists are no longer altogether certain that the change in vegetation that mark the boundary between the Preboreal and Boreal pollen zones is due to climatic change. They prefer to seen at least some of this shift as due natural successional change in tree species that occurred as the Preboreal forests matured. In this view, changes in the Boreal pollen counts represent the replacement of pioneer species like birch, aspen, and pine that have high light requirements, short life spans and rapid growth cycles by climax species like oaks that are characterized by slow rates of reproduction but tenacious habits.

Despite the expanding diversity of the Boreal ecosystem, it was still not as productive from the standpoint of human subsistence as had been the steppe-tundra biome. However,

> from an ecological point of view, nature had evolved into a new stage of greater diversity and stability than before. Seasonal fluctuations in the availability of resources became less drastic. A broader spectrum of choice of existence now opened to the inhabitants of the land. Most probably the population gradually increased due in part to a lesser dependence on seasonal variations in the access to food, the presence of large game animals and the abundance of edible plant foods such as nuts, roots, bulbs, mushrooms, bark, fruits, plant juices and many more (Jensen 1982:17).

By the end of the Boreal zone, the elk had became locally extinct in Britain due to forest change perhaps aided by human predation (Simmons *et al*. 1981:116).

Worldwide sea level continued to rise during the Boreal and, by its close, England was separated from the continent. Yet, the former presence of Mesolithic peoples on the now inundated continental shelf of northwestern Europe was dramatically attested to by the recovery of an barbed antler harpoon head in the nets of a North Sea fisherman working the fish shoals of the shallow Dogger Banks 25 miles off the coast of England (Mellars 1974: 86-87).

By Zone VI, the Atlantic (6200-3300 B.C.), pollen evidence that most of northwestern Europe–with the exception of the Scottish highlands and a few other areas–pine species began to diminish in numbers as the region came to support a mixed forest dominated by oak and elm and a number of the tree species requiring warmer temperatures for optimum growth (Shotton 1977: 114). The appearance of this forest assemblage has generally been interpreted as evidence that the climate of northwestern Europe during the Atlantic zone was warmer, moister, and more "oceanic" than it is at present. However, here again, some paleobotonists regards the emergence of the oak-elm forests of the Atlantic to be due to the sequence of plant migration and succession rather than climatic change (Butzer 1971: 533). He considers the appearance of oak (*Quercus*), elm (*Ulmus*), alder (*Alnus*), lime (*Tilia*), ash (*Fraxinus*), holly (*Ilex*), ivy (*Hedera*), and other species in the pollen spectrum of the zone to be the result of their slow rates of migration from the nearest refuge location. Whatever its climatic implications, the Atlantic pollen zone represents the establishment of a stable forest at lower elevations in northwestern Europe. The hazel-pine forests of the Boreal was a pioneer assemblage that, in the long-run, could not compete with the climax assemblage that appeared near the end of the Boreal. This new assemblage of species forms what is known as the "mixed-oak forest" and the percentage of these species increases dramatically in northwestern European pollen profile after 6000 B.C.

The mixed-oak climax forest probably had a dense tree canopy and relatively sparse undergrowth and would thus have provided poor feed for big game. Jensen (1982:19) concludes that the interior of this forest was characterized by relatively low species diversity, low productivity year round, and poor habitat for red and roe deer and aurochs. Therefore, the forest interior during the Atlantic zone would not be ideal territory for hunting and gathering. However, high species diversity probably characterized certain microenvironment in and around the forest interior. Such species diversity might have been found in forest glades, inland lakes, bogs, streams, coasts, beaches, estuaries, islands, salt marshes, and dunes. In fact, these areas seem to have been the focus of human settlement and exploitation during the middle Holocene and especially in the Atlantic pollen zone. Vegetation followed a somewhat different developmental trajectory in the higher elevations of northwestern Europe: on the flatter terrain in these higher areas, deciduous forests commonly came to be replaced by blanket-peat-forming communities dominated by sphagnum moss (*Sphagnum eriophorum*). The expansion of blanket peat bogs in the highland ecosystems during the Atlantic pollen zone illustrates the complexity of gauging and interpreting ancient environmental change. These bogs may be due to natural causes such as increased groundwater due to climate changes or to the natural maturation of the soil under the new conditions of the Holocene. However, they may also have resulted from active human interference in the forest soil formation cycle through clearing and burning, and later, through cultivation. Very likely, individual bogs are the products unique combinations of natural and anthropogenic forces.

This change in forest morphology may account for the cultural changes evident in northwestern Europe at this time. Early in the Atlantic zone, the Maglemosian tradition gives way to its successor, the **Kongemose (ca. 5600-4500 B.C.)**. The distinction between the Maglemose and the Kongemose has largely been based on changes in the style of microliths and other artifacts. However, these stylistic changes may in fact be associated with more substantive alterations in subsistence and economic organization. In response to the optimal warmth and moisture of the Atlantic, the composition of the forests of northwestern Europe there had became increasingly deciduous in character. The wild harvest of nuts , berries, tubers and other plant foods in such forests would have far surpassed that of the earlier Boreal forests available to the Maglemosian peoples. This new forest would have provided the Kongemosians with a subsistence bounty they could hardily have ignored. The prominence of red and roe deer and wild pig bones in Kongemose middens probably also reflects the altered composition of the forests. Finally, riverine and coastal resources appear to have steadily risen in importance throughout the 1100 years that the tradition endured in England and southern Scandinavia.

The rise of the subsequent **Ertebølle (ca. 4500-3000 B.C.)** tradition marks the culmination of the Mesolithic period in northwestern Europe. The tradition takes its name from an immense shell site in Jutland in northern Danmark. In fact, the abundance of shell or "kitchen middens" like that of the Ertebølle site suggests that coastal resources had become central to subsistence at this time. This centrality may have come in response to the rising world wide sea level taking place at this time in the Middle Holocene. Nonetheless, in addition to the large quantities of mollusk shells (particularly those of oysters) that have been recovered in them, Ertebølle midden sites commonly produce the remains of a broad range of species. The variety represented indicates that the Ertebølle exploitation maritime and riverine resources, including sea mammals such as seals, in an effective and sophisticated manner.

The abundance of these midden remains indicates that their subsistence economy was highly productive. But shell middens are by no means the only type of site known for the Ertebølle. I n fact, sites of this tradition vary quite widely in character. This variation suggests that Ertebølle people commonly exploited the great microenvironmental diversity of their Atlantic zone habitat by establishing centrally-located base settlements and then using satellite camps for short term hunting or gathering of particular species.

The shell midden of **Morton Tayport B** near Fife in Scotland is a prime example of later Mesolithic subsistence. Radiocarbon dates from the site indicate that it was occupied sometime between ca. 4500 and 4000 B.C., squarely in the middle of the Atlantic vegetation zone. The site is as important to the study of economy in the Atlantic period as Star Carr was to the interface between the Preboreal and Boreal zones. The Morton Tayport B midden produced a remarkably large quantity of well-preserved faunal remains. The variety of species present in these remains indicate that the site's occupants exploited a broad-spectrum of inland and coastal habitats in combination. Fish bones from the site included cod, haddock, turbot, sturgeon, and various species of salmon indicate that these foragers fished both inshore and in deep water in the nearby estuary of the river Tay. Sea birds remains were probably those of falcons gathered from nests on the rocky coast. Forest ungulates, including aurochs, red and roe deer, indicate that these peoples hunted the inland forests. Elk was not present in the faunal inventory from the site suggesting that this species was already regionally extinct in the Atlantic. The bone counts indicate that the ungulates were apparently butchered on the site. The only curious feature of the faunal array from the site is the absence of marine mammals bones. However, this absence may indicate that the site was only seasonally, rather than continuously, occupied.

Sites like Morton Tayport B illustrate that by late Mesolithic times, broad-spectrum foraging in the Middle Holocene deciduous forests of northwestern Europe had reached its *apogée*. The degree of this subsistence and economic success is evident on two interrelated dimensions: (1) the longer-term, increasingly sedentary occupation of sites at this time and (2) the appearance of Ertebølle pottery within them. Sedentary living and pottery use are more generally associated with agricultural subsistence. As we shall see below, however, this not a necessary association and, in prehistory, the two traits have by no means always been found together. It appears more likely that the cause of sedentary living and the use of ceramics is not agriculture *per se*. Rather, sedentism and pottery appear to be correlated with high levels of subsistence productivity that allows food surplus to be concentrated and stored. Whether that surplus is the result of food collection or food production appears largely irrelevant.

In any case, techniques of ceramic making were acquired by the Ertebølle peoples sometime after ca. 3700 and 3100 B.C. By that time, agricultural peoples ultimately from the southeast of Europe had settled on the deltas of the Rhine and Meuse rivers in what is today the Netherlands. These people made pottery and, directly or indirectly, very likely conveyed the technology to the Ertebølle peoples (Gebauer 1995: 99). At about the same time, Mesolithic forest dwellers of adjacent European Russia begin making pottery also. It remains unclear whether this technology was independently introduced to both peoples from the same source, diffused from one to the other or, much less plausibly, was independently invented by one or both of them. Throughout the 4th millennium, Ertebølle people made large jars with pointed base as well as oval bowls thought to have been used as lamps. These vessels were made by means of the coil method. In this method potters first rolled out long snakes or coils of clay and then formed their pots by either: (1) laying a series of coil rings one atop the other or (2) drawing a single coil into a continuous spiral. In either case, adjacent coils must be joined to one another and this is generally done by smoothing them together. It is this smoothing process that sometimes gives the exterior surfaces of coiled pots a corrugated appearance. When smoothing is imperfectly done, pots exhibit lines of weakness between the coils and it is along such lines that the vessel is most prone to crack or break (Orton, Tyers and Vince 1993: 118).

The Mesolithic Wanes in Northwestern Europe During the 6,000 years of the early and middle Holocene epoch, the glacial and periglacial environments of northwestern Europe were transformed from open tundra and tundra parkland ecosystems to dense, stable, and biologically complex mixed oak forests. The Morton Tayport B site indicates that, during those 6,000 or more years, late Mesolithic peoples had learned how to adapt to the complexity and stability of that ecosystem. They had developed a stable and productive economy on food collecting on a base through the systematic and efficient exploitation of a remarkably wide range of plant and animal resources. Shortly after reaching its highwater mark in the Atlantic zone, however, the Mesolithic economy–and the system of social and cultural life built upon it–changed forever. Pollen percentages from strata dating to the end of Atlantic reflect a pronounced increase in the scale of forest burning and clearance heralds the arrival of agricultural peoples. This arrival signals the beginning of the end of the Mesolithic way of life in northwestern Europe.

During Zone VII, the Subboreal (3300 BC to AD 300), forests there came to be dominated by oak and beech, perhaps because the climate had become more "continental" instead of "oceanic" in character, that is, somewhat drier and cooler. More important still, the period witnessed the introduction of simple horticulture into northwestern Europe by Neolithic stage peoples. The earliest evidence of the practice of this new form of subsistence is found in the peat bog pollen profiles. It includes: (1) the increasing size and frequency of ash layers and other evidence of forest burning in the peat, (2) the rapid decline of elm (*Ulmus*) pollen, (3) the declining ratio of arboreal to non-arboral pollen reflecting an impoverishment of the forest and an increasing abundance of native weeds and scrub or secondary vegetation, and (4) the appearance in the profiles of the pollen of exotic plants–both domesticants and weeds derived from the continent, and ultimately, from the Mediterranean. Deforestation began in southern and central Europe after roughly 5000 B.C. Its onset after 3000 B.C. in the northwest reflects the slow but steady pace of expansion by agricultural peoples across Europe. This process will be discussed in the next chapter.

The date is arbitrary because the changing character of the vegetation reflected in the first five pollen zones of the Blytt-Sernander system indicates that the 3,000 years of between the Oldest Dryas and the Preboreal were characterized by rapid and dramatic oscillations of climate. They show that the end of the Pleistocene was not an abrupt event dating to 7912 B.C. but rather gradual one–sometimes characterized by whip-saw reversals of climate–that took place over 15,000 years from the Würm maxima ca. 18000 B.C. to the establishment of the modern sea level ca. 3000 B.C. (Butzer 1971:530).

Key Early Holocene Adaptations in Africa and Asia

The transformative effects of the Early Holocene

climate change were not limited to northern Europe, of course. Its impact was felt throughout the world. However, perhaps the most notable changes occurred in two key regions of the Old World: Saharan Africa and the Japanese Archipelago. We turn to a consideration of these two regions now as they seem to be emblematic of the Early Holocene epoch in the Old World beyond Europe.

Fishing on the Sahara: The Epi-Paleolithic of Northern Africa Greatest of the Earth's deserts, the Sahara has dominated northern Africa throughout historic times. Forming a mighty band of 20° of latitude in width that begins just inland from the Atlantic coast of West Africa, the Sahara stretches eastward nearly 6,000 kilometers to the Red Sea to span the continent. In the north, the boundary of the desert begins just inland from the fertile and relatively well-watered lands that fringe the Mediterranean Sea. From there it drops southward some 1,500 kilometers until it reaches the **Sahel**, or desert edge. The Sahel is the transition zone wherein desert gives way, first to savannah, and then to tropical rainforest. The harshness of the climate of the modern Sahara is legendary. It is a land of extreme aridity and wide fluctuations in temperature between night and day time. Buffeted by limitless sands carried endlessly before the wind, the ageing relief of the Sahara is in continuous surrender to the forces of erosion (Hugot 1981: 565). Such is the desert's hostility that, in recent times at least, permanent human settlement there clustered tightly and clung precariously to the few and precious oases found in its interior.

As arid as the Sahara is today, however, ample evidence attests to the fact that it has not always so. The Pleistocene climatic conditions that produced the cyclical advance and retreat of continental glaciation in the northern hemisphere had an impact on both the Sahara and Sahel. In general, interglacial periods and interstadial substages in the northern hemisphere were correlated with periods of greater effective moisture for vegetation, soil development and ground water recharge on the Sahara and Sahel. More effective moisture could result from increased precipitation, reduced evaporation or a combination of both factors (Haynes 1982: 633).

Conversely, the onset of continental glaciation in northern latitudes, or its resurgence at the end of an interstadial substage, generally signaled the return of drier conditions on the Sahara. Arid conditions reappeared there by 35,000 BP or earlier and reached a during the Last Glacial Maximum (LGM). From ca. 23,000 BP until the end of the Pleistocene, the region was characterized by **hyperaridity**, that is, by less than 1 millimeter of rainfall annually (Haynes 1982: 630). Such dryness is far greater than even that of the present in the region. At this time, massive dune fields formed in northern Africa both within the Sahara and well south of the current modern desert's boundaries.

Sand dunes, that is, active mounds of wind blown sand, are the characteristic land forms of sea coasts and deserts. The processes that form them on coasts need not concern us here. However, the presence of dunes in continental locations is generally due to long term conditions of aridity. Low effective moisture reduces vegetation and exposes the dry surface sediments to erosion by the wind. "Fossil" sand dunes are dunes are no longer actively forming and moving but have been stabilized by vegetation and have soils forming on their surfaces. If active sand dunes are sure signs of aridity, fossil dunes reflect the return of effective moisture to areas once arid. Vast lines of fossil sand dunes can be found stretching from the Atlantic side of Africa to the Sudan at its eastern end. These dunes formed under conditions profound dryness and their fossilized presence well south of the modern Sahara indicate how much larger the Sahara was during the LDG and Terminal Pleistocene than it is in the present day (van Zinderen Bakker and Maley 1979). Conditions of hyperaridity abated during the Holocene epoch. Dated sediment columns removed from playa lake beds in various parts of the Sahara suggest that the desert experienced at least three periods of increased rainfall during that epoch (Goudie 1977: 111; Faure 1981: 386).

This alternation between conditions of greater and lesser effective moisture caused the appearance and disappearance of thousands of small lakes in low lying regions and the expansion and contraction of savannah grassland vegetation across the Sahara. One of the most dramatic of these Holocene environmental changes has only recently been discovered in the extremely arid western part of northern Sudan. Today the region receives most 25 millimeters of rainfall per year and is entirely inhospitable. However, under increased effective moisture regime of the Early Holocene, bone-dry Wadi Howar was a

Holocene Shifts in Effective Moisture on the Sahara Desert

ca. 9000 to 6500 BC	> effective moisture
ca. 6500 to 5500 B.C.	< effective moisture
ca. 5500 to 2500 BC	> effective moisture
ca. 2500 to 1600 B.C.	< effective moisture
ca. 1600 to 500 BC	> effective moisture
ca. 500 BC to present	< effective moisture

400 kilometer flowing river that linked the northern Sudan to the Nile valley. At this time, the Wadi had numerous ground water outlets and freshwater lakes and supported a rich savanna fauna. Based on these data, the discoverers of this unknown tributary estimate that the southern edge of the eastern Sahara was some 500 kilometers further north than it is today (Pachur and Kröplin 1987). Such long term fluctuations in available moisture on the Sahara closely determined the nature and extent of human occupation by hunting and gathering peoples and later, as we shall see in the next chapter, proved particularly significant in the emergence–and later disappearance–of pastoral nomadism and perhaps agriculture there

Epi-Paleolithic Re-Occupation of the Sahara Archaeological sites dating to the Middle Stone Age are not uncommon on the Sahara suggesting that the relatively moist climatic conditions prevalent there roughly from the Last Interglacial stage until the onset of the LGM made the region attractive to early *Homo sapiens*. The Saharan expression of the Middle Stone Age is called the **Aterian** industrial tradition. Aterian assemblages are characterized by scrapers and a variety of other implements made on Levallois flakes. Some of these implements are carefully tanged and have been interpreted as projectile points that were formerly hafted on shafts. The Aterian tradition disappears from the Sahara before the onset of the hyper aridity associated with the Late Glacial Maximum period. Following the disappearance of the MSA from the Sahara, there is only scattered evidence of a subsequent Late Stone Age occupation there. This absence is no doubt due to the harsh climatic conditions that come to prevail there during the long millennia of the LGM. In fact, archaeologists often refer to the hyper arid climatic phase as simply the "**Hiatus**" because, while it endured, the Sahara appears either entirely devoid of human inhabitants or only very lightly occupied (Milburn 1984: 296). By forcing the abandonment–or virtual abandonment–of much of the Sahara, the LGM appears to have broken the continuous sequence of cultural development that had occurred there heretofore. When, during the Early Holocene, the regions was occupied once again, it was by people of a very different tradition whose cultural repertoires show little connection to the Aterian industry that, millennia before, preceded them there.

The moist conditions that returned to the Sahara at the very end of the Pleistocene ca. 9,000 B.C. persisted there until ca. 6,500 B.C. During this time, countless shallow lakes formed in the basins and low plains of the region. Dry remnants of these ancient lakes found on the modern surface of the desert are referred to as **playas**. When dated, their sediments within them are invariably found to be of Early Holocene age. Pollen from mesic or moisture-loving plants is commonly recovered from the sediment cores removed from these playas indicate that the greater moisture of the Early Holocene translated quickly into a vegetative re-colonization of the Sahara Desert. The return of this moisture and vegetation to the Sahara appears to have permitted the human reoccupation of the region. At first these re-occupation appears tentative and ephemeral. The few sites know to date to the 9th millennium B.C. are characterized by thin occupation layers and light scatters of debris found clustered around the banks of former playa lakes. Very likely these sites represent the ephemeral camps made by small groups of mobile hunter-gatherers who are Epi-Paleolithic in character. That is to say, they represent subsistence systems in which the hunting and gathering techniques developed in the Later Stone Age were adapted and applied to the new environmental conditions of the Holocene epoch. Epi-Paleolithic archaeological sites of this period, not a few of which are located well away from possible sources of permanent water, start to become more common as the Early Holocene wears on. At roughly the same time, a human presence is also reflected in the eastern Sahara at least by the appearance of Epi-Paleolithic age rock paintings and engravings in the so-called **Bubalus style**. The Bubalus style is characterized by naturalistic depictions of savannah wildlife like elephant, rhinoceros, hippopotamus, giraffe and other species that apparently throve on the Sahara during the Early Holocene but could not live there today nor during the Late Pleistocene period (Goudie 1977: 111; McIntosh and McIntosh 1983: 229). Still later, rock art painted by cattle raising peoples in the eastern Sahara depict cows, sheep and goats (Roset 1984).

The sites left by these Early Holocene occupants of the Sahara have been found to cluster around the vanished playas located there. This clustering lead John Sutton (1974, 1977) to labeled the life way that produced the sites the "Aquatic Civilization of Middle Africa" a name he shortened to simply the "Aquatic Culture" or, even less auspiciously, to the "**Aqualithic**." The content, as well as the placement, of Aqualithic sites convinced Sutton that they were the remains of "communities whose livelihood and outlook on life revolved around lakes and rivers." He saw the tradition developing in response to the increased rainfall and intensified monsoon conditions of the Early Holocene on the Sahara following the close of the Pleistocene. Their subsistence systems apparently combined fishing and shellfish gathering in the shallow lakes and water-courses with the harpooning of large aquatic animals like crocodile and hippopotamus. Assuredly, these latter activities are not for the fainthearted. Hunting of savannah ungulates and foraging for wild plants also seem to have figured into the Aqualithic subsistence mix.

Stone tool industries recovered at such sites are dominated by blades, often made by means of the punch blade technique. Ground stone bowls and the stone pestles–probably used in processing seeds and roots by grinding–are sometimes also found in these sites as well. Well-carved bone harpoons and fish hooks are also part of the tool kits at some Aqualithic sites. Most surprisingly, some later Aqualithic sites have produced ceramic sherds from round-bottom pots whose surfaces had been fully decorated with various wavy-line, or dotted, wavy-line, motifs. The earliest of these sherds have been recovered from archaeological contexts dating to between 7500 and 6500 B.C. , dates that place them among the oldest ceramics in the world. What is most striking about these early African ceramics is that they were invented and used by mobile, or at best, semi-sedentary foragers (Close 1995: 23). They therefore falsify the that common place idea that it was only among sedentary peoples that the development and manufacture of ceramics could tale place. Further, even in the absence of agriculture, the presence of ceramic cooking pots indicates that still another culinary revolution has begun in which such healthful and efficient foods as soups, porridge, and fish-stew (Sutton 1977).

At the Early Holocene peak of this tradition, Aqualithic sites appear distributed over a wide band of territory from Great Bend of the Niger river in West Africa eastward across the southern Sahara and Sahel to the upper Nile valley and from the upper Nile south to the Lake Turkana region in the Rift Valley of Kenya.. The distribution of these Aqualithic sites is strikingly similar to the distribution of modern speakers of the African languages of the Nilo-Saharan language family, perhaps suggesting a continuity of population between the that time and the present (Ehret and Posnansky 1982). But whether or not the bearers of Aqualithic culture were the ancestors of modern Nilo-Saharan speakers on the Sahara and Sahel today, it is Sutton's view that Aqualithic culture was based entirely on a particularly successful foraging subsistence. He attributes much of this success to their combining of hunting and gathering with the manufacture of ceramic pots. This combination enabled Aqualithic peoples to reap the nutritional benefits of the culinary revolution that the adoption of pottery invariably makes possible. So productive was this life way, Sutton asserts, that the simple presence of the Aqualithic tradition in Saharan Africa acted as a powerful impediment to the adoption and spread of Neolithic food production there during the Early Holocene.

Sutton's Aqualithic model is not without its critics, most of whom have focused on the question of the traditions reality on the ground. Some thinks the Aqualithic not a real tradition so much as an archaeological construct or reification. In this view, the vast geographic distribution of so-called Aqualithic sites makes it more likely that a great deal of historically unrelated archaeological materials have been lumped together into a single, artificial entity (Close 1995: 26). Others note that, even if the Aquatic style of subsistence was a unified practice across the Sahara as Sutton asserts, it may not have been unified and distinctive "way of life" or even a full time activity. In their view, it is more likely that aquatic foraging was but one part of a more complex, mixed subsistence system. For example, aquatic foraging might just as readily have been pursued only seasonally or occasionally by peoples otherwise engaged, in pastoralism say, or in manipulating or cultivating native African tubers or cereal grains (Smith 1980; Andah 1987; Robertshaw 1991).

Whatever the true nature of the Aqualithic, the onset of drier conditions on the Sahara after about 6,500 B.C. dry up the pools and brought the Epi-Paleolithic there to a close there. With the return of greater moisture after ca. 5,100 B.C., a new ceramic-using peoples with a mixed economy apparently based on foraging **and** the herding of domestic livestock claimed the area. Thus, the Middle Holocene witnesses the beginning of the Neolithic period on the Sahara and the eventual abandonment of aquatic foraging. We take up this subject in the next chapter.

Precocious Foragers: The Remarkable Mesolithic of Holocene Japan

No less than that of Pleistocene, the Holocene prehistory of Japan does not easily fit the standard conceptual schemes developed elsewhere in the Old World. Not only did the Mesolithic stage linger well into the Holocene on the archipelago, the standard order of the appearance of agriculture **before** the practice of sedentary living and the manufacture of ceramics was reversed there. The temperate climate and rich marine and terrestrial environments of the terminal Pleistocene and early Holocene Japan created an unparalleled abundance of wild resources on its islands and a robust and precocious foraging tradition called the **Jomon** developed there to exploit this environmental richness. Like the earlier Natufians a world away in the eastern Mediterranean, Jomon peoples lead hunting and gathering lives yet occupied largely **sedentary** settlements. But, while their lives as settled village foragers is not unprecedented, Jomon peoples are rendered extraordinary by their development and use of the earliest pottery on Earth.

Recent Climate on the Japanese Archipelago Hokkaido, Honshu, Shikoku and Kyushu, the four main islands of the Japanese archipelago, are located at roughly the same latitudes as North China and Manchuria. Due to their position on the rim of the North Pacific, these islands do not share the subhumid and semiarid climate prevalent on the continent. Annual rainfall in Japan is two to three times that of North China and is far more reliable and uniformly distributed throughout the year. Further, winters there, while cold, tend to be less severe in mot parts of the archipelago than on the adjacent mainland. Thus the climate of Japan is primarily temperate in character. Subtropical conditions are found only at the extreme southwestern end of the island of Kyushu and it is only at the extreme northern end of the archipelago on Hokkaido island that Japan's otherwise temperate climate gives way to subarctic conditions. The advantages of Japan's moderate rainfall and temperature are offset to some extent by three marked disadvantages, however. First, violent, nonperiodic weather disturbances are more numerous over archipelago than in similar latitudes in China (Trewartha 1981: 225-226). Second, the geological instability of the Japanese archipelago makes it subject to frequent earthquakes and volcanic activity. Finally, the Japanese islands are rugged, mountainous and possessed of only limited areas of low lying interior basins, plains and river valleys.

During the early Holocene, the wide elevation and latitudinal variation found in the interior of the Japanese archipelago came to support a rich mosaic of coniferous, deciduous and broad-leaf forest environments. Rising post-Pleistocene sea level meanwhile separated the Japan from China. This marine transgression lengthened the coastline, flooded coastal lowlands and lower river valleys, and expanded the estuaries and wetlands of the archipelago. At the same time, post-Pleistocene changes in circulation patterns in the North Pacific melded warm and cold ocean currents off the coast of Japan to create conditions of great richness for fish and other marine life. These factors conspired to form a natural environmental setting of arresting diversity and opulence.

Jomon Tradition (10,500-300 BC.) In a celebrated cave site located on the southern end of Kyushu called **Fukui rockshelter**, Japanese archaeologists recovered the earliest ceramic vessels known to science. Associated with these crude, hand built ceramic vessels was a lithic inventory that includes microliths as well as large, bifacial points and other materials of a typical terminal Paleolithic aspect. Radiocarbon determinations places the age of the archaeological context from which the pottery was recovered at between 12,400 and 12,700 BP. (Akazawa and Aikens 1986: 103-104). Ceramics of a similar age and context were also recovered in central Japan at the site of **Ishigoya cave**. The sherds from these two sites demonstrates that pottery, one of humankind's key technologies, first developed in Japan at the end of the Pleistocene.

Ceramics play a central role in the conception of the Jomon tradition as framed by Japanese prehistorians. This centrality is enshrined in the name that chose to designate it; *Jomon* means literally "rope" in Japanese and is a reference to the earlier potter's practice of decorating their vessels by pressing or rolling cords of rope or twine into the wet clay of their surfaces before firing them. This rope technique first comes into practice after about 8,000 B.C. during the Initial Jomon period. These early vessels are commonly conical with pointed bases, a shape that probably allowed them to be pressed in the ashes of a fire (Aikens 1995: 13-14). Bowls, globular cooking pots and other new shapes appear in the succeeding Jomon periods as do new techniques and motifs of surface treatment and decoration.

The Jomon Tradition in Japan

Dates	Period
10,700-7,500 BC	Incipient
7,500-5,000 BC	Initial
5,000-3,500 BC	Early
3,500-2,500/ 2,000 BC	Middle
2,500/2,000-1,000 BC	Late
1,000-300 BC	Final

Jomon ceramicists did not limit their efforts to making pottery vessels. They also produced great numbers of decorative anthropomorphic figurines of fired clay. Archaeologists presume were used for ritual and religious purposes. Over one-thousand fragmentary ceramic figurines were excavated at **Shakado**, for example, a habitation site located near the center of Honshu island. The Shakado site was successively occupied at least five times during the Early and Middle Jomon period. During its various occupations, numerous anthropomorphic figurines were manufactured only to be intentionally broken, apparently in the course of religious ceremonies (Yamagata. 1992). We can only speculate as to what such ritual breakage was meant to signify or effect.

A Jomon/Valdavia Connection? Whatever

the role of ceramics in the cultural life of Jomon peoples, not all archaeologists are convinced that the impact of its precocious invention was limited to Japan. Excavation by Betty Meggers, Clifford Evans and Emilio Estrada (1965) at the Valdivia site on the coast of Ecuador produced an early, sophisticated ceramic sequence that appeared there around 3000 B.C. apparently without local antecedents. To explain this abrupt appearance, the excavators evoked the mechanism of trans-Pacific diffusion and proposed that storm-tossed fishermen from the Middle Jomon culture of Japan introduced pottery making into the Americas. At the time of their discovery, the specimens from the Valdivia site were the oldest known yet ceramics recovered in the New World. This fact, coupled with their putative Jomon attributes and the location of Valdivia on the Pacific coast, strengthened the argument that the vessels were the product of diffusion despite the truly daunting nautical problems that confront drift voyagers in the Pacific (McEwan and Dickson 19–). Subsequently, pottery clearly unrelated to Jomon ware was recovered from earlier strata at the Valdivia site. More recently, excavations elsewhere have produced pottery dating to around 5000 and 6000 B.C. in the Brazilian Amazon and around 5000 and 4500 B.C. in northern Columbia. These finds do not eliminate the possibility that Jomon-style pottery making diffused from Japan to Ecuador. However, they dispose of the argument that ceramic technology was unknown in the Americas before 3000 B.C. and lend credence to the view that pottery developed independently in both the New and Old Worlds.

Jomon at Zenith Over much of prehistoric Japan, the extent to which people were able to live a more-or-less sedentary fashion increased markedly during the Early Jomon period (5,000 to 3,500 B.C). This trend is dramatically illustrated at the site of **Sugizawadai** on the coast of the Sea of Japan on the far northwestern end of Honshu island. Sugizawadai consists of a complex of four large structures arranged around a central plaza that is in turn surrounded by at least 100 storage pits and some 40 smaller buildings. The function of the four main buildings remains an enigma. One of them, the largest Jomon building yet uncovered, enclosed fully 220 square meters of interior space. This large building showed signs of having been rebuilt at least three times. During its final occupancy, six separate fire hearths were placed in the center of it on its long axis. The presence of these hearths makes this building reminiscent of the celebrated long house structures of the Iroquois. Iroquois long houses were occupied by clusters of related families that functioned as corporate landowning kinship groups. Each nuclear family within the cluster that occupied a communal Iroquois longhouse had its own fire hearth. These structural parallels may be spurious, of course, and the function of this enigmatic building, as well as those of the other three large buildings, remains unknown.

By its Middle period (ca. 3,500 and 2,000 BC.), Jomon culture reached the peak of it social complexity on the main island of Honshu. Perhaps nowhere is this complexity more evident than at the classic Middle Jomon sites of **Togari-ishi** and **Yosukeone** located near the center of Honshu. Working at Togari-ishi before World War II, archaeologists excavated 85 semi-subterranean residential pithouse structures. Each structure was built by first excavating a large hole to the depth of a meter or more beneath the ground surface. Large posts, probably two or meters high, were then erected inside the pits to serve as supports for a framework of smaller posts and branches. When covered with earth, this framework formed the pithouse roof. Excavations undertaken at the nearby site of Yosukeone immediately following the Second World War uncovered 27 additional pithouses build in this manner. The pithouses at this site and elsewhere represent a level of labor and capital rarely made by foragers. For the Japanese archaeologists, the sites of Togari-ishi and Yosukeone are emblematic of Jomon culture. Data from both of them continue to figure prominently in interpretations of Middle Jomon social and political organization. Although a great deal of information of all kinds was in fact recovered from these two sites, the most striking discovery was simply the conspicuous size and permanence of these Middle Jomon villages.

Whatever the nature of their social and political organization, at the end of the Middle Jomon period, inland villages in central Honshu like Togari-ishi and Yosukeone abruptly declined in number. Perhaps the scale of settlement and the density of population had exceeded the environmental carrying capacity at the center of the island. Alternatively, this settlement and population decline may simply reflect population movement in response to the rise of greater subsistence opportunities elsewhere. Following decline in central Honshu, a large increase in settlement occurred along the island's Pacific coast around the onset of the Late Jomon period (2,000-1,000 BC). During the Late and Final Jomon periods, the most common Jomon sites became the "**kitchen midden**" or large mound of discarded shells located along Honshu's Pacific coast. In a formal sense, Late Jomon shell middens are reminiscent of sites of the Ertebølle Culture of northwestern Europe discussed earlier. Unlike most European Mesolithic kitchen middens, however, Late Jomon shell mound sites also contain structural remains and, of course, pottery.

The settlement shift represented by the appearance of these Late Jomon shell middens reflects an increase in the importance of coastal and marine resources in Jomon subsistence. This increased importance may relate to some, as yet unrecognized, alteration in environmental conditions along the coast of Japan. Alternatively, such a new subsistence pose might have been due to innovations or improvements in ship building and deep sea fishing technologies. Whatever caused the increase in their frequency, shell mounds, continued to be formed and occupied throughout the Final Jomon period (1000 to 300 BC).

Burial practices at these shell mound sites appear to indicate that the organization of society increased in complexity during the Late Jomon period. For example, the large Late Jomon site of **Kashiwagi B** on Hokkaido island contained a series of large cemeteries enclosed by earth and stone embankments. Archaeologists found wide variation in the treatment of individual burials within these burial complexes. The forms of the graves, their position within or outside of the banked enclosures, and the richness of the offerings contained in the graves all varied widely (Ikawa-Smith and Kanjodori 1992). The high degree of variation within the burial populations of the Late Jomon is a good proxy measure of the degree of status and occupational differentiation within Jomon society as the tradition entered its final phase.

The waning years of Jomon tradition appears to have been a period of new economic, social and cultural challenges. That being the case, it is not surprising that Japanese archaeologists have found what they take to be evidence of widespread and intensified ritual activity during the Final Jomon period. Among this ritual evidence are the profusion of a new form of ceremonial figurine: a goggle-eyed anthropomorph that whose bulging eyes may represent the insight into the supernatural realm sought by the seer or the shaman. Does the increase of such new cultic forms indicate that cultural change and conflict in the Final Jomon period had stimulated attempts at religious and cultural revitalization?

What Was the Nature of Jomon Subsistence? From the outset, Jomon people were food collectors adapting to the extraordinary abundance and variety of wild resources accessible on the Japanese archipelago during the terminal Pleistocene and early Holocene. By the time the tradition had reached its peak in the late Holocene, Jomon society appears to have been characterized by a pronounced maritime subsistence system capable of supporting moderately large populations in sedentary village settlements. As noted, these are stunning and exceptional traits for hunters and gatherers to possess. But were the Jomon peoples simply collecting the foodstuffs upon which they depended? Some Japanese archaeologists are content with the notion that environmental richness is sufficient to account for the presence of such hallmarks of social complexity as pottery and large, sedentary village life in the Jomon. They point to the Northwest Coast of North America as evidence that, in the northern circum-Pacific region, the natural larder of ocean and shore was sufficiently abundant to support sedentary, "affluent foragers" able to live largely sedentary lives. In their view the success of Jomon foraging made agriculture unnecessary, even redundant (Akazawa 1986: 163). Scholars of this persuasion assume that the robust and efficient character of the Jomon food collecting economy actually prevented agriculture from developing indigenously in Japan. Agriculture, they say, was imported into Japan from the mainland of China via Korea and then only late in the Holocene epoch. After the arrival rice agriculture, the old forms of Jomon subsistence begin to enter a long period of economic transformation as the cusp of an advancing frontier of Neolithic stage rice agriculture expanded slowly throughout these islands.

Other scholars are not so sure. They suggest that the large storage pits, numerous grinding stones, digging sticks and stone hoes recovered from Jomon sites point to some other than passive foraging. They contend that it is more likely that, very early in their history, Jomon peoples practiced incipient cultivation of indigenous root, tuber and seed plants in combination with fishing, hunting and gatherings. They point to at least three classes of indirect evidence that such indigenous cultivation was practiced in combination with food collecting. First, there is clear indications that, as early as the terminal Pleistocene epoch, Jomon peoples were burning their forests. This repeated anthropogenic forest burning was no doubt largely intended to encourage the growth of edible wild plants but it may also have been connected with forest clearance for agriculture. Second, very limited quantities of seeds, pollen, phytoliths and macrofossils from buckwheat and millet as well as bottle gourd and other vegetables dating to the early and middle Holocene have been recovered in some Jomon sites (Anderson 1987: 276-277). Evidence of this has been recovered at the site of **Torihama** on Honshu island. Excellent organic preservation at this waterlogged site revealed that its Incipient and Early Jomon period residents lived in a sedentary village, used large dugout canoes and fashioned elaborately carved artifacts out of wood. More importantly, we know the Torihama people exploited a wide range of wild plants together with native species of gourds and beans, two species that, at least later, came be domesticated. Finally, beginning in the Middle Jomon period (3,600 to 2,500 B.C.), people in central Japan began to suffer from patterns of tooth decay like those of found among modern taro-eaters (Anderson 1987: 277; Turner 1979).

Are these data sufficient to support the conclusion that the Jomon peoples had begun to manipulate, control and domestic the indigenous wild plants upon which their food collection economy depended? Not in and of themselves. Was the sedentary existence of the Jomon people based partly or largely on agriculture? It is too soon to say. These questions, some of the most controversial and contentious faced by Japanese prehistorians, can only be resolved by additional excavations and further paleobotanical research.

Greater certainty surrounds our understanding of the appearance of rice cultivation in Japan. Current paleobotanical evidence indicates that rice cultivation: (1) did not arrive in Japan until the Late Jomon period (2,500 B.C. and 1,000 B.C.), (2) was not widely practiced outside of southern Japan until Final Jomon (1,000-300 B.C.) and (3) did not become a central element in subsistence over much of the archipelago until the subsequent **Yayoi period (300 B.C. to A.D. 300)**. If this date range for the spread of rice cultivation is correct, it indicates that Japanese subsistence practice was in marked contrast to that of Korea and China where rice cultivation became the dominate subsistence activity at a far earlier date. The retarded acceptance of rice cultivation on the Japanese archipelago is no doubt due in large measure to the ecological appropriateness and intrinsic strength of traditional Jomon subsistence strategies.

The cultural period that follows the Jomon in the Japanese archaeological sequence is known as the Yayoi. The name of the tradition is derived from its type site in Tokyo where its distinctive pottery was first recognized. It is not its pottery that renders the Yayoi watershed period in Japanese prehistory. Rather, it is the triumph of wet rice cultivation during the period and the revolutionary transformation of economic and social life on the archipelago that follows that triumph. The ultimate source of this new form of agriculture is clearly continental Asia. With the rise of the Yayoi, the indigenous Mesolithic cultures of Japan began to fade rapidly into oblivion.

Environment And Culture in the Early Holocene "Middle Passage"

As we have seen, world climate at the close of the Pleistocene epoch was markedly unstable. Over a period just under 2,000 years–brief in geological terms–temperature and moisture oscillated from very cold to warm to cold again before warming in the post-glacial millennia of the Early Holocene. Close behind this climatic whipsaw came massive waves of extinctions in the great steppe tundra megafauna of North Hemisphere in both the Old and New Worlds. Did the sharp, fast pace of these climatic oscillations cause the Post-Glacial extinctions? Surely they were a factor. But so too was the continued practice of profligate big game hunting with its game drives and attendant waste. Megafauna populations of the Late Glacial Maximum may have been able to withstand such hunting pressure, but the weakened and attenuated Terminal Pleistocene game in the ever-shrinking environmental refugia probably could not. Humankind, it would seem, plays the pivotal role in the extinctions at the end of the Ice Age.

Further, these environmental changes and megafauna extinctions intersected with rising human populations to create a "food-crisis" in the early millennia of the Holocene epoch. The solution to this crisis adopted worldwide appears to have been to broaden the human food niche through the systematic practice of more inclusive hunting and gathering. The practice of broad-spectrum foraging must often have included taming or "tending" of certain of the plants and animals being exploited. In most cases, such tending was isolated and exceptional and probably came to nothing. In some few instances, however, tending or taming of a wild population may have been undertaken on a scale sufficient to significantly altered the selective forces operating on it. That is, human selection began to replace natural selection among them in certain areas. In regions fortuitously endowed with genetically malleable plant or animal species, this replacement initiated the first incipient stirring of domestication. The first species to succumb and be brought into the human orbit in this fashion was the wolf. More, many more, species were to follow.

These climatic and subsistence events were accompanied–or in some cases preceded–by important changes in technology. Central among these were (1) the development of microlithic stone tools, (2) the use of stone grinding both in tool production and food processing, and (3) the invention of the bow-and-arrow. The bow-and-arrow appears to have also altered the selective forces operating on humans in many parts of the world. In a very few areas, these changes included the invention and use of ceramics. Discovery of that foragers made and used pottery in prehistory was most unexpected. Conventional archaeological wisdom has always considered the development of ceramics to have taken place **after** the rise of agriculture. However, recent archaeological excavations have conclusively demonstrated that the earliest ceramic vessels in fact appeared among

Mesolithic stage broad-spectrum hunter-gathers. The Terminal Pleistocene Jomon peoples of Japan were the first to make pottery vessels. Only two millennia later during the first millennium of the Holocene, Aqualithic peoples in the Sahara independently invented ceramics. In the next chapter we will see that Early and Middle Holocene foraging peoples in the Americas likewise developed or adopted ceramics.

Independent invention was not the only that foragers came by ceramic technology. Peoples belonging to the Ertebølle and other traditions in Europe successfully incorporated pottery into their foraging way of life after receiving them by diffusion from agricultural peoples farther south. Diffusion of this type further underscores the absence of a necessary connection between ceramics and agriculture. Ceramic-using foraging peoples all over the world appear to have had one thing in common: the centrality of marine or aquatic resources to their subsistence. All of them appear to have practiced some combination of fishing, strand lopping and aquatic or marine mammal hunting with their other food collection pursuits. Aquatic and marine resources tend to be predictable and abundant. Such resources are either infinitely renewable or outright inexhaustible. For these reasons, intensive food collecting in lacustrine or maritime environments sometimes fostered a degree of sedentary living that made the use of a fragile and bulky medium like pottery possible. Conversely, wild resources that are both highly predicable and inexhaustible are only rarely met by peoples foraging on the interior plains and forests of the world's continents. Small wonder then that ceramics do not appear first among terrestrial hunter-gatherers. But where ever it first appeared, pottery fostered an important new culinary revolution that conferred a pronounced selective advantage on peoples able to efficiently process their foods as nutritious soups and stews.

Finally, in accessing the culture change of the Mesolithic period in the Old World, it is important to note that these technical and subsistence developments were not concentrated in a single locus. Rather these often revolutionary changes appear earliest as widely scattered occurrences. Such dispersal suggests they were the product of independent inventions made in response to similar conditions and adaptive problems.

PART III:
PRELUDES TO THE PRESENT

Major Points Raised in Chapter Eleven:

1. The emergence of food production was the most fundamental transformation in human subsistence to occur in prehistory. This transformation apparently occurred independently in a number of different areas of the world after the close of the Pleistocene epoch. Possible causes for this transformation are examined in the first section of this chapter.

2. The second section deals with the origins of the southwestern Asian agricultural complex and with the domestication of wheat and barley, sheep, goats, pigs, and cattle. The rise of settled agricultural village life and the nature of early village economy and society is examined along with the evidence of a heightening of violence during the Neolithic period. The development of ceramics is discussed together with the uses of pottery in food processing and fermenting and its implication for human diet.

3. The idea that Neolithic culture spread from southwestern Asia in "the Wave of Advance" is examined in the second section as well. The role of population growth in the intensification of production near the end of the Neolithic period is given consideration. Productive intensification is seen in the development of irrigation agriculture, the rise of specialized, fully-nomadic pastoralism, and the "Secondary Products Revolution" in which the meat provided by animals becomes less important than the wool, milk, traction, and transportation they supply. A brief discussion of the domestication of the horse and the rise of wheeled vehicles ends the section.

4. The third section deals with agricultural complex that emerged in Africa south of the Saharan Desert. This complex combined cultivation of key European cultigens such as wheat, barley, sheep, goats, and cattle with native African domesticate plants such as finger millet, tef, noog, enseta, African rice, and possibly yams, oil palm, and sorghum. Domestication on the Ethiopian Plateau, the Great Bend of the Niger River, and the Sahel (the margin of the Sahara) and the Chad-Sudan region is discussed. The new conditions of disease and health created by the Neolithic in Africa are illustrated with an examination of the correlation between the distribution of malaria, sickle cell anemia, and shifting cultivation in West Africa and by the limits imposed on cattle-raising in many parts of Africa by the distribution of the Tsetse fly.

5. The fourth section deals with the East and Southeast Asian agricultural complexes. Carl Sauer's thesis that Asian agriculture began with the vegetative reproduction of plants is examined in light of the Hoabinhian occupation at Spirit Cave, Thailand. The origins of rice and millet cultivation is discussed and the role of this domestication in the rise of the Yangshao culture of China examined. The section ends with a discussion of the Jomon culture and the emergence of the Neolithic of Japan.

6. The fifth section deals with the rise of agriculture in four distinct regions of the New World. Maize or corn, the most important New World domesticant, appears earliest in the archaeological record of the central highlands of Mesoamerica. Parallel domestication took place in southeastern North America but was directed towards less productive wild species such as sheep sorrel and sunflower. The agricultural complex that emerged in the tropical lowlands of northern South America and southern Mesoamerica America centered on vegetatively-propagated root crops. This complex, which may have centered on a form of intensive, wetlands cultivation like the *chinampas* of modern highland Mexico, was sufficiently productive to support the first civilization in the New World, the Olmec, that emerged in the tropical lowlands by 1750 B.C. The most distinctive agricultural complex in the New World developed in the Andean highlands of western South America. While growing potatoes, maize, and other crops, peoples of Andean South America also domesticated the vicuna, guanaco, and the guinea pig.

7. In the final concluding section, eight generalization about the nature of early food production and consequences that the development occasioned are advanced..

CHAPTER 11

THE ORIGINS OF AGRICULTURE

Introduction

Was There a "Revolution"? The emergence of agriculture was the most fundamental transformation in human subsistence to occur in prehistory. This transformation apparently occurred independently in a number of different areas of the world after the close of the Pleistocene epoch. Describing the transformation of human subsistence from food collection to food production is easier than explaining why or how it came about in the first place. As with all prehistoric events and processes, complex events must be reconstructed using only their material traces. There is another impediment to our understanding of this transformation, however: the vast intellectual distance that separates us from the life of hunting and foraging. Our way of life is clearly supportable only by agriculture. We understand its advantages and cannot imagine life without them. The advantages of agriculture seem obvious, even self-evident to us and we presume that the replacement of food collection by food production offered a clear improvement in quality of life. However, the ancient people who took part in this transformation had no more idea of the future consequences of their acts than we do of ours. The obvious modern advantages of agriculture, being unimaginable at the time, could not have been influential in creating the transformation.

In the popular conception, agriculture is an "idea" that was "discovered" or "invented" by some unknown prehistoric genius who first observed that young animals could be tamed or that seeds planted in the spring resulted in fruit-bearing plants in the fall. This view presumes that most people are ignorant, uninventive, and prone to borrow someone else's solution to a problem rather than work it out on their own. I have christened it the "Ah-hah" or "Eureka" theory of the origins of agriculture.

However, the Ah-Hah theory neglects the abundant evidence that modern hunters and gatherers, at least, are not ignorant of the nature of plants or animals but, in Kent Flannery's words, are "sophisticated lay-botanist" with intimate knowledge and practical understanding of the working of the natural nature. After all, such knowledge is crucial to peoples who depend on wild plants and animals for the very existence. It seems unlikely that ancient hunters and gatherers were any more mystified by the general workings of the natural world than their modern descendants. Thus, explaining the unwillingness of hunter-gatherers to abandon food collection is as difficult as accounting for the emergence of food production.

But Why Should We Plant When There Are So Many Mongongo Nuts in the World? In 1651 the English philosopher Thomas Hobbes coined an eponymous phase to characterize life before the rise of the state; he said it was "solitary, poor, nasty, brutish, and short." Hobbes' phrase tends to stick in the mind and, until quite recently, most prehistorians took a Hobbesian view of the prehistoric hunting and gathering life. Consequently, the question "What prompted hunting peoples to take up agriculture?" was not difficult for them to answer. They thought it would have been obvious to prehistoric hunting-and-gathering people that agriculture offered them a stable food supply and a sedentary life; they presumed that prehistoric foragers adopted agriculture to free themselves from their Hobbesian world of violence and scarcity.

However, the current generation of scholars has found the question "why" to be more intractable. Logical objections have been raised against the idea that agriculture developed throughout the world simply because it appeared to prehistoric people to be a better idea. Rindos rejects such an explanation as "intentionalist" because it is based on the dubious assumption that the "ultimate benefits of agriculture for a society would have been observable in the early stages...(and that) a society aware of these benefits would act to take advantage of them" (Rindos 1984).

Further, field studies among hunting and gathering peoples have demonstrated that at least contemporary foragers do not lived in a Hobbesian world. The picture of foraging life among the *!Kung* or *Jul'hoansi* foragers of the Kalahari desert of southern Africa presented by Lee, DeVore and others, beginning in the late 1960s, persuaded many scholars that foraging peoples live in relative ease and are free of the unremitting toil that marks the life of the subsistence farmer (Lee 1965; Lee and DeVore 1968). According to Sahlins (1972), the *!Kung* or *Jul'hoansi* have created "the original affluent society" drawing only from the larder of nature. "Why should we plant," said one of Lee's *!Kung* or *Jul'hoansi* informants, "when there are so many Mongongo nuts in the world?" (Lee 1968) For such people, there is little inducement to farm. The advantages of food production that seem so obvious to us are not compelling to them. As a consequence of these logical and empirical approaches, the question, "What prompted hunting peoples to take up agriculture?" has, for the contemporary scholar, become "Why farm?"

Like the Hobbesian picture of hunter-gatherers, the view of food collectors as "affluent" may also be something of an exaggeration. Nonetheless, the fact remains that, despite the supposedly "obvious" advantages of food producing, ethnographically-known foraging peoples have generally resisted adopting agriculture altogether or have adopted only reluctantly. The Numa or Shoshonean-speaking peoples of the Great Basin of North America are a particularly well-documented case of such initial resistance. Throughout the early 19th century, these desert foraging peoples persistently rebuffed all efforts by Euro-Americans to induce them to abandon broad-spectrum food collecting in favor of settled farming. Late in the nineteenth century, however, subsistence agricultural practices indeed spread rapidly and widely among the Numa. The adoption of agriculture by the Numa did not occur because these peoples suddenly became aware of its advantages. After all, the Numa had lived adjacent to agriculturalists for centuries in both Pre-Columbian and historic times. Rather, the Numa took up agriculture only after the native plant ecology of the Great Basin–upon which their food collection subsistence depended–was profoundly modified by the arrival of large herds of domestic cattle and sheep in the company of Anglo settlers. The Numa's answer to the question "Why farm?" was simple: "Because we can no longer hunt and gather." The case of the Numa adoption of agriculture–with its antecedent anthropogenic environmental destruction–provides us with a model for explaining the permanent establishment of agriculture in the Old World.

As a general rule, foragers do not farm unless they have to, that is, unless new circumstances make their foraging adaptation impossible or difficult or impossible. Therefore the most promising strategy for explaining the transformations of some late Pleistocene food collecting subsistence systems into food producing ones is to examine the changing conditions antecedent to that transformation.

The rise of agriculture appears to have occurred after the close of the Pleistocene and to have been preceded worldwide by (1) climate change, (2) mammalian extinctions, (3) sea-level changes, and (4) human population increase. The impact of these events on human subsistence may have been profound. Mark Cohen (1977) argues that the population of prehistoric hunting and gathering societies grew throughout the Pleistocene but that the pressure of population on food resources became critical only at the end of the epoch. He suggests that this pressure caused the intensification of subsistence activities and triggered the domestication process.

However, the climatic and environmental changes must have dramatically lowered the human carrying capacity of much of the world during the terminal Pleistocene and early Holocene. The "push" of population pressure and the "pull" of environmental impoverishment appear to be plausible causes first, for the intensification of subsistence activities and second, development of agriculture out of this intensification. Of course, some subsistence strategies are more readily converted or into symbiosis and manipulation and control than others.

Reed (1986: 12), in an article provocatively titled "Wild Animals Ain't So Wild, Domesticating Them Isn't So Difficult," states that, contrary to popular myth, most wild animals are neither naturally fierce nor instinctively inclined to only fight or flee when confronted by humans. Instead, Reed suggests, most are really rather tame. In his opinion, the major impediment to the domestication of wild mammals during the upper Paleolithic period was not animal instinct but human hunting behavior. If Reed is correct, subsistence system that minimized hunting would have been the most able and most likely to take advantage of the natural tameness of many birds and mammals and initiate the domestication process. Conversely, systems centered on specialized big game hunting and game driving would have been the least likely context for early domestication.

Of course, the domestic dog represents an obvious exception to Reed's generalization. The earliest skeletal remains of domesticated dogs have been recovered from the archaeological sites of terminal Pleistocene hunting peoples in both the Old and New

Worlds. No doubt this is due to the compatibility of dogs with the hunting life and to the synergism that develops between hunters and their dogs. It no doubt also due to the fact that canines willingly act as scavengers and *coprophagers* (feces eaters) and would thus have been attracted to the vicinity of hunters' camps.

In any case, the taming of animals and the tending of plants certainly did not begin as a clear-eyed attempt at developing long-term solutions to "the Food Crisis." Rather, such practices were probably engaged in at many different times and places as means of hedging bets in the face of expanding human population and increasingly less productive and less predictable early Holocene environments. For example, taming and tending of animals and the cultivation of plants might have been initiated in order to

1. Reduce the risks arising from over reliance on increasingly unreliable wild foodstuffs,
2. Supplement wild food stuffs available primarily in one season with cultivated foodstuffs available in another,
3. Utilize the labor of marginal members of societies who were unable to participate fully in the primary food collection activities, or
4. Simply to add variety to a group's diet (Pryor 1986).

Whatever the explanation for the transformation of subsistence practice, once a people's cultivation practices began to include the systematic re-planting of the seeds they have gathered, they initiated a process of both "automatic" and intentional biological changes. That is, they begin to interpose human preference between the organisms and the selective forces of nature that formerly acted upon their populations and shaped their genetic makeup.

Under such human selection, the organism enters the first stages of domestication. Harlan, DeWet and Stemler are careful to stress that domestication is a process, not an event. According to them, "the process can be carried out over various ranges of time at different intensities. The process continues for every crop actively being cultivated and will continue for an indefinite period into the future. How far it can go, no one yet knows. If operated at a sufficient intensity for a sufficient period of time, dramatic changes in morphology and ecological adaptation can be expected" (Harlan, DeWet and Stemler 1976: 6).

Since domestication results from the genetic manipulation of a species by human activity, it is an on-going process that carries the species farther and farther from its wild ancestral parent stock as time goes on. Harlan, DeWet and Stemler (1976: 6) describe the results of this process for plants as follows:

> In the case of seed crops much of the manipulation is automatic and comes from planting harvested seed. As long as seed is only harvested and not planted, there is likely to be little or no genetic effect on wild populations. It is seldom possible to harvest as much as half the seed produced by wild populations and ample seed remains to maintain the stand. If there is any selection pressure at all, it will be in favor of such wild-type characteristics as shattering (spontaneous seed releases because of formation of abscission layers), seed dormancy, and indeterminate growth with maturation over a long period of time. It is the seed that escapes the harvester that will contribute to the next generation. As soon as man starts to plant what he has harvested, strong selection pressures are set up. Now there are two populations, one maintained by nature and one maintained by man, with selection going in opposite directions. Outside the realm of human activity, natural selection proceeds as before, favoring a wide dispersal of propagules. In the cultivator's field, however, all modifications that augment the 'harvestability' of plants have a *selective advantage* (Harlan, DeWet and Stemler 1976: 6-7).

A series of mutations begin to be favored in the cultivator's field that would normally be eliminated in the wild. Probably the first automatic changes occur in the seed dispersal mechanisms. Wild grasses and cereal grains are genetically highly plastic. Although wild plant species all have excellent mechanisms for seed dispersal, genetic mutations that retard natural seed dispersal occur continuously within their populations. Such mutations are fatal to progeny of the individual plants that carry them; natural selection acting on wild populations assures that only the genetic makeup of plants with efficient dispersal mechanisms will be transmitted to succeeding generations. Yet plants with retarded seed dispersal mechanisms are easier for humans to collect and would unconsciously be selected by seed gatherers like the Natufians.

As soon as the collectors began to re-seed the grain intentionally, a greater percentage of these mutations would become concentrated in the resulting grain populations than would be present in them naturally. As a consequence of human selection, the grain populations would be altered genetically over the generations. Further, these alternations would result in an increase in the productivity of the cereal grain plants per unit of labor invested in their collection.

Flannery (1973) suggested that the Natufians triggered a "genetic kick" within the wild cereal grain populations they were collecting by inadvertently favoring the mutations in the dispersal mechanisms of the wild wheat by replanting some of the seeds they collected· He suggests that the specific cause of the re-seeding that led to this "genetic kick" was the introduction of wild cereal grains into habitats outside their natural range. This would also have resulted in hybridization with local races of grasses, a further source of increased productivity. Selection favoring uniform rates of maturation is probably the next to occur in the population. In wild plants populations, seeds tend to appear over several week spans. This insures that at least some of the seeds will encounter favorable growing conditions in the environment.

However, domesticated cereal grains are generally harvested all at once. Therefore, seed production over long period is a disadvantageous as it is the plants which produce the most grain at a single time that make the greatest contribution to the next generation. For similar reasons, human seasonal harvesting also tends to reduce or eliminate prolonged seed dormancy in cereals. "Most wild grasses produce at least some seeds so dormant that they will not sprout under natural conditions for several years" (Harlan, DeWet and Stemler 1976: 8).

A trend toward larger seeds is also common, although not entirely universal in cereals. The increased size of domesticated seeds is no doubt partly due to conscious human selection of larger seeds for replanting. However, Harlan, DeWet and Stemler (1976: 6) state that "most of the change is probably due to seedling competition in the seed bed. The seedling that comes up first and grows the fastest is the most likely to contribute offspring to the next generation. Seedling vigor is closely related to seed size and energy storage in the endosperm. We also know that large seeds can emerge from deeper planting better than small seeds."

In addition to these "automatic" mutations and selections, a variety of others are imposed by the cultivators based on preference of taste, color, ease of processing, storage. As Harlan, DeWet and Stemler (1976: 8), point out:

> Some sorghums are boiled like rice, some cracked for porridge, some ground to make flour for bread or dumplings. Different cultivars are used for different purposes. A common procedure at harvest time is for the cultivator to go through his field and carefully select heads to be saved as stock seed for the next planting. The rest of the field is harvested for consumption. In this situation selection is total. The components of the population that contribute to the next generation are those chosen by the cultivator. The rest are eliminated from the gene pool. The fact that a cultivar makes good dumplings may confer no ecological advantage in the field, but the selection pressure may approach the absolute. It is this sort of human activity that has resulted in such a bewildering array of variation in most domesticated plants (Harlan, DeWet and Stemler 1976: 8).

As a general rule, the part of the plant most used by man is the part that is most modified by the process of human selection. For example, consider that the ears in corn, the tubers in yams, pods in peas, have little resemblance to the comparable organs in their wild ancestors.

Formation of the Major Agricultural Complexes The theory that domestication took place in a single limited area and then spread by diffusion throughout the rest of the world has been discredited. It now appears that plant and animal domestication occurred independently and repeatedly during the Holocene in numerous areas or "**hearths**" within six and perhaps seven large regions of both the New and the Old World. Part of the evidence for the existence of multiple independent regions of domestication is the fact that, during the Holocene epoch, a series of distinctive assemblages of domesticated plant and animal species, husbandry methods, food processing and storage techniques–as well as the associated weeds, commensal pests and diseases developed in a number of major geographic areas of the world. These assemblages retained their distinctive characters virtually until the onset of European global expansion in the 16th century. Throughout the remainder of this book, these distinctive geographical assemblages are called "**agricultural complexes**."

The agricultural complex that emerged in southwestern Asia at first centered on the cultivation of wheat and barley, lentils, bitter vetch, peas, chickpeas, and flax, and various other vegetables and spices. It also included the raising of sheep, goat, and cattle herding and the keeping of dogs (Harlan, DeWet, Stemler 1976:4; Zohary 1989: 358). Somewhat later, tree and vine fruits, pigs, horses, and donkeys were added to the array of domesticates. The techniques of animal husbandry appear to have emphasized the use of the animals primarily for meat. The raising of livestock as sources of "secondary" products such as milk, wool, or traction becomes as part of this complex only much latter. The array of food processing associated with this complex were derived in large measure from the grinding, pounding and grating techniques developed for processing wild plants by the incipient cultivators of the Epi paleolithic.

In Africa, native domesticates such as sorghum and millet and a host of other plants came to be combined with southwestern Asian cultivars such as sheep, coats and Near Eastern cattle (*Bos taurus*) that diffused across the Sahara. In contrast, the agricultural complex that developed more-or-less independently in East Asia came to center on the cultivation of yams, taro, millet, bamboo, wet and dry rice various fruit trees, a wide array of vegetables and such domesticated animals as dogs, pigs, chickens, water buffalo, and Zebu cattle (*Bos indicus*).

In Mesoamerica on the cultivation of maize, beans, amaranths, chili, beans, squash, cotton and a variety of other vegetables along with only minor domesticated animals as the Muscovy duck and the hairless dog. There were at least two other major centers of domestication in the New World: the southeastern United States, and lowland tropical South America. However, the agricultural complexes developed in these regions were profoundly influenced by diffusion there of the Mesoamerican complex.

Of course, we can never know with certainty how the these distinctive agricultural complexes developed in various parts of the world. However, it seems likely that, once cultivation began in the early Holocene, cultigens from contiguous regions were widely exchanged. Eurasia is a part of the world that was apparently blest with a large number of wild plant and animal species that were suitable for rapid domestication. The species of surrounding regions were also being brought under cultivation and such a situation must have laid the foundation for regional exchange. As Runnels and van Andel (1988) point, Old World food collecting societies had been part of trading networks through out the Upper Paleolithic period. These networks would have served as conduits for the exchange of cultigens.

Flannery suggests that such exchange would have had important evolutionary consequences. The movement of species out of their native ecozone into new ones would have changed the natural selective pressures operating on them. The demands and tastes of the new cultivators would have led them to place slightly different human selective pressures on the plant than had the previous cultivators. It is also necessary to remember that a good deal of variations exists among wild plants and animals as to: (1) suitability for domestication, (2) rate of genetic response or change to human selection, and (3) relative productivity.

For example, southwestern Asia was well endowed with wild cattle (*Bos primogenus*) and wild horses (*Equus*). Both of these animals had been hunted by humans since the Pleistocene. However, simple abundance and importance as a prey species does not account for the fact that these animals were domesticated. After all, the gazelle (*Gazella gazella*) was also abundant and widely hunted but it was never domesticated in prehistory, even by the those redoubtable incipient cultivators, the Natufians. Much later systematic attempts to domestic various species of gazelle, antelope and ibex undertaken by the ancient Egyptians during the Old Kingdom had no long term success (Zeuner 1963: 435). How can we explain this? It is possible that the gazelle was not domesticated in the Levant simply because Epi paleolithic peoples there chose to hunt the species intensively using mass game drives in which entire herds were slaughtered (Legge and Rowley-Conwy 1987). However, this does not explain the failure of the Egyptian efforts. The explanation appears to lie in that the natural characteristics of the gazelle. Although they are herd animals like the horses, cattle, sheep and goats, gazelles differ from those species in the amount of body contact with other members of their species they will tolerate. According to Walther, Mungall and Grau 1983: 35, 6), "gazelles and their relatives are not 'contact animals' but are rather 'distance animals' which keep a certain 'individual distance' between each other. In addition, the adult male gazelle has a well-developed sense of territoriality and vigorously strives to keep females within the boundaries they control at they same time expressing aggressive intolerance of incursions by other adult males. It seems probable that the wild ancestors of domesticated horses, cattle, sheep and goats were far less diffident about body contact and territoriality and were consequently easier for humans to herd and manipulate than gazelles or antelope.

The same logic may be extended to the New World. As we saw in Chapter 8 above, the wave of extinctions that swept the New World in the late Pleistocene and early Holocene eliminated a number of potentially domestica table animals including the horse and camel. The key herd animal that survived the extinctions were the pronghorn antelope (*Antilocapra americana*) and the buffalo (*Bison bison*). Pronghorns share most of the behavioral traits of their Old World cousins. American bison, for a variety of reasons, also have poor potential for domestication. Not surprisingly, both species remained wild into historic times. The absence of domesticated herd animals equivalent to the horse and cow in the Americas had a profound implication for the nature of New World civilization–a fact dramatically illustrated by the rapid military triumphs of mounted Spanish *Conquistadores* in Mexico and Peru in the 16th century.

Further, species that **were** suitable for domestication and human selective breeding did not change genetically at the same rate. For example, according to Harlan and Zohary, (1974: 1975, 1079), the cytogenetics, modern occurrence and ecological behavior of wild emmer wheat indicates that the most important difference between it and cultivated emmer is the more fragile *rachis* or dispersal mechanism of the wild races. That is to say, the wild and domestic wheat are separated by only a minor difference and would normally be included in the same species. This genetic proximity suggests that the domestication of wheat was both simple and rapid. Recent experiments in the measurement of domestication rates done by Hillman and Davies (1990) indicates that wild einkorn and emmer wheat and barley could have become completely domesticated in a period between 20 and 200 years.

In contrast, the domestication of maize appears to have involved far more complex genetic changes. Profound differences separate maize from the Mesoamerican grass species that was its parent. This suggests that domestication of maize took far longer than wheat, a hypothesis largely confirmed by the archaeological evidence from Mesoamerica and Peru. The earliest definite archaeological remains of maize were recovered in the Tehuacan valley. These tiny cobs, presumably from wild plants, are from archaeological contexts dating to ca. 5,000 B.C. The earliest definite evidence of fully domesticated corn does not appear in the Tehuacan sequence until ca. 3,500 B.C., some 1,500 years later. Thus it appears that wheat was domesticated at a far faster rate than corn and this relative difference had a profound impact on the subsequent pace of cultural evolution in the New World and Old World.

Finally, domesticated species differ from one another in productivity. Archaic peoples in both Mesoamerica and the southeastern portion of North America were culturally "pre-adapted" to agriculture and both groups succeed in domesticating local species. The domestication of native plants in both areas apparently took place more-or-less independently (Smith 1989). However, a higher population density and a more complex level of socio political organization were achieved sooner in the former than in the latter area and, eventually, Mesoamerican domesticates came to dominate the agriculture of the Southeast as well. What accounts for this difference? Simply, the fact that the domesticates from the two areas were by no means of equal productivity. While the Archaic peoples of Mesoamerica domesticated amaranths (*Amaranthus*) and such familiar species as corn, beans, and squash, the peoples of southeastern North America had only the sunflower (*Helianthus annuus*), a native North American squash (*Cucurbita pepo* spp. *ovifera*), (*Cucurbita texana*), chenopod (*Chenopodium berlandieri*), and the distinctly unpromising sumpweed or marshelder (*Iva annua*) (Smith 1989: 1567-1568).

Corn produces more foodstuffs per unit of cultivated land than any of the native southeastern crops. Its nutritional quality is further enhanced by boiling dried cereals in a .5 percent lime solution before grinding into flour (Katz, Hediger and Vallcroy 1974: 768). In addition, the nutritional disadvantages of corn can be eliminated or at least reduced by growing it with beans. Beans fix nitrogen in the soil in which they are grown and, when eaten, supply many of the nutrients missing in maize.

In summary, as the Holocene worn on, those species that proved to be complementary to one another came to be organized into areal agricultural systems complexes. In this way–by exchange, by trial and error, and natural and human selection–the various combinations of domesticates began to be organized into agricultural complexes. Such complexes came to include complementary plant and animal domesticates and methods of husbandry. In a related manner, each complex also developed distinctive cuisines, food processing and storing technologies. These complexes were probably first organized in the regions considered by Vavilov to be primary centers of domestication. However, it does not follow from this early organizational success that domestication took place only in those few centers or even began earliest in them.

In any event, after their successful development and integration, the agricultural complexes began to diffuse and spread outward from their primary regions into surrounding territory. This process of expansion and adaptation of a population to new circumstances is called "adaptive radiation." Two engines apparently drove the adaptive radiation of early agricultural socio-cultural systems: trade and population growth. As Runnels and van Ardel (1988) point out, the dispersal of crops and cropping techniques would probably have taken place within the existing "market channels" or trade networks of the ancient world.

In addition, agricultural systems would have expanded into surrounding territory because they were demographically unstable. That is, because agriculturalists are unable or unwilling to control the growth of their populations in order to remain below the carrying capacity of their environment (Rindos 1980: 752). Population expansion would have forced these early agricultural peoples to intrude into comparatively lightly populated territory occupied by hunting and gathering peoples. Agriculturalists would generally come to dominate and replace food collectors in such settings, due to their greater population growth and the alterations of the natural environment that followed from their forest burning, land clearance and introduction of domesticated animals.

Expansion also occurred into regions settled by peoples engaged in subsistence centered around indigenous cultigens. If the agricultural complex possessed by arriving peoples was superior to the indigenous cultigens and cultivation practices, the foreign complex would, overtime, supercede the local one. However, in the process of this supersession, attractive indigenous crops were retained and integrated into the emigrant complex. As noted above, this happened when the Mesoamerican agricultural complex diffused to southeastern North America. Sunflower and a native North American form of squash continued to be cultivated in the Southeast following the arrival of the new complex and sunflower has become fairly important modern specialty crops. However, maize and beans edged out marshelder or sumpweed and chenopod as domesticates and reduced them to the status of garden pests.

A parallel process of expansion and incorporation appears to have taken place on the margins of the cultivators fields among so called "weed" species. As Crosby points out:

> 'Weed' is not a scientific term in the sense of species, genus, or family, and its popular definitions are protean; so we must pause to define it. In modern botanical usage, the world refers to any plant that spreads rapidly and out competes others on disturbed soil. Before the advent of agriculture, there were relatively few of these plants representing any given species; they were the 'pioneers of secondary successions or colonizers,' specializing in the occupation of ground stripped of plants by landslides, floods, fires, and so forth (Crosby 1986: 149).

However, land clearance for agriculture dramatically expanded the habitat of such pioneers and turned them into unintended cultivars dependent upon human environmental impact and subject to human manipulation. Therefore, it is not surprisingly that new varieties of weeds were produced as the by-products of cultivation. These new varieties emerged as hybrids when wild and domesticated races of cereals and other plants were grown together in the same fields (Harlan, DeSweet, and Stemler 1976: 8-9).

In Southwestern Asia early agriculture produced a new array of cereal plants consisting of cultivated cereal varieties in manmade habitats, wild cereals growing in primary niches, and weedy sibling forms of the same cereal plant growing in and on the margins of the cultivated fields (Zohary 1989: 359-360).

Some weed species were ultimately domesticated. Both rye and oats appear to have entered the repertoire of Old World domesticates in this manner when the southwestern Asian agricultural complex diffused into the interior of Europe. Others continued on as pests to bedevil the cultivator ever after.

As the newly formed agricultural complexes diffused, they took their weed species with them to the new territories. The best evidence for the arrival of agricultural peoples in northern Europe is the appearance of such characteristic Mediterranean weeds as ribwort plantain (*Plantago lanceolata*), stinging nettle (*Urtica dioica*), and various kinds of sorrels (*Rumex* spp.) (Roberts 1989: 120). Much later, the spread of European agriculture that began in historic times introduced Old World weeds into the New World, Australia, and the islands of the Pacific with devastating effect (Crosby 1986).

A Fortunate Few The peoples of those favored areas of the world where the key agricultural complexes developed experienced sustained population growth and geographic expansion. As a result of the accidents of history and ecology, these people made a disproportionately large contribution to both the genetic and the linguistic makeup of the world's population. If Colin Renfrew is correct, the distribution of the Indo-European family of languages may be a prime example of this phenomenon. Since the 18th century, students of language have been aware that diverse peoples inhabiting the Old World from Western Europe across Anatolia and Iran to the Indian subcontinent as far as western China speak basically similar languages. These modern tongues are all part of the Indo-European language family and are all apparently derived from the same mother tongue.

Philologists, that is, students of historical linguistics, have interpreted these language similarities as evidence of vast ancient migrations of Indo-European speakers out of the steppe lands of Central Asia. In recent times, archaeological data has been enlisted in the reconstructions of these linguistic migrations. For example, the appearance of the distinctive "Beaker" style of pottery drinking vessel and in a new form of burial mound or *kurgan* in Western Europe after ca. 2,500 B.C. is commonly taken as evidence of the intrusion of members of the Indo-European language family into Europe during the Bronze Age. Similarly, the axe-welding charioteers who conquered the Indus Valley between ca. 2,000 and 1,000 B.C. are assumed to have spoken some form of Indo-European and to have replaced the non-Indo-European peoples who resided there previously (Basham 1954: 28-30).

Historical linguists have generally estimated that these migrations took place comparatively late in prehistory, that is, during the last three to five thousands years. However, Colin Renfrew (1987) has recently argued on the basis of both linguistic data and archaeology that primitive forms of the Indo-European languages were spoken across Europe thousands of years earlier than had hitherto been supposed. In fact, it is his view that it was Indo-European speaking peoples who first developed the southwestern Asian agricultural complex and then introduced this form of agriculture into Europe and eastern Eurasia.

Renfrew's book ignited a fire storm among historical linguists and archaeologists and the issue remains a long way from decided. However, if he is correct, the vast distribution of Indo-European languages in the Old World may ultimately have been due to the fortuitous, early Holocene placement of the ancient speakers of Indo-European in the hearth of the European agricultural complex.

The Origins of the Southwestern Asian Agricultural Complex

Southwestern Asia–often called the Near East or the Middle East–is that area occupied today by the modern states of Turkey, Iran, Afghanistan, Iraq, Saudi Arabia and the smaller states of the Persian Gulf and the Arabian Peninsula (see Map 1). It also includes the Levant, the land fringing the eastern end of the Mediterranean and occupied in modern times by Syria, Lebanon, Jordan, and Israel. Although in strict physical geographic terms Egypt is part of North Africa, its history and culture bind it inextricably to southwestern Asia.

Seven major physiographic features dominate the landscape of the Southwestern Asia:

1. The coastal plains of the Mediterranean Sea and the Persian Gulf.
2. The alluvial flood plains of the Tigris and Euphrates rivers of Mesopotamia and the Nile valley of Egypt.
3. The rugged mountains, plateaus, valleys, and piedmont of the Pontus Mountains along the Black Sea in northern Anatolia, the Taurus Mountains in southern Anatolia, and the Zagros Mountains of Iran and Iraq. These mountains, that stretch in a broad and more-or-less continuous arc from the western edge of Anatolia to the Persian Gulf, form the northern boundary of southwestern Asia.
4. The semi-arid steppe lands and basins of Anatolian Plateau in central and western Turkey.
5. The semi-arid to arid steppe lands and deserts of Syria and the Arabian peninsula.
6. The semi-arid highlands of the Levant between the Mediterranean coast and the great rift valley.
7. The great rift valley containing the Dead Sea and the Jordan river.

Holocene Climate of Southwestern Asia The agricultural complex that emerged during the Holocene in southwestern Asia centered on the cultivation of barley and wheat, lentils, bitter vetch, peas, chickpeas, and flax, and various other vegetables and spices. It also included the raising of sheep, goats, and cattle and the keeping of dogs. Somewhat later, tree and vine fruits, pigs, horses, camels, and donkeys were added to the array of domesticates (Harlan, DeWet, Stemler 1976:4; Zohary 1989: 358). At least three main hearths of primary domestication have been identified in Southwestern Asia:

1. The eastern Mediterranean coast of the Levant from the middle Euphrates river drainage south to the Dead Sea in what is today Syria, Lebanon and Israel was probably the region where the earliest domestication of the dog, possibly the pig, einkorn and emmer wheat, barley, lentils and certain vegetables took place.
2. The foothills and intermontane valleys of the Zagros and Taurus mountains of southern Iran, northern Iraq and eastern Turkey was very likely were goats, sheep and perhaps pigs were first domesticated.
3. The Anatolian Plateau, western Anatolia and Greece for cattle (*Bos taurus*).

In addition, domestication of a number of important species took place later in southwestern Asia beyond the primary areas. These secondary domestications were probably stimulated by the earlier, primary ones. Two of the most important secondary hearths of domestication include:

1. The alluvial flood plain of the Tigris and Euphrates in Mesopotamia where the wild ass or onager and the camel may have been domesticated, and
2. The plains of southern Russia along the coast and north of the Black Sea between the Ukraine and Turkestan just beyond the northern boundary of southwestern Asia that may have been the location for the earliest domestication of the horse.

Origins of Wheat and Barley Cultivation The agricultural complex that emerged in the area that extends south from the middle drainage of the Euphrates River along the coast and inland hills of Lebanon, Israel, and Jordan as far as the Dead Sea (Bar-Josef 1989: 632). This complex, which appears sometime between 8300 and 7000 B.C., is the first definite food producing subsistence system known in prehistory. It centered earliest on the cultivation of wheat, barley, pulses and certain other vegetables. Sheep, goats, and pigs together with additional plant species were added to it in the Levant sometime before 7000 B.C. Wheat and barley are the common names given to the annual cereal grains of the genuses *Triticum* and *Hordeum* respectively. Both are grasses (*Gramineae*) and were apparently the first of this family to be domesticated. Although today barley is raised primarily as a speciality crop, wheat is the second most important food crop in the world. Only rice surpasses its contribution to the world's modern diet.

Archaeological evidence from the Levant and the Zagros mountains indicates that emmer and einkorn wheat were the earliest wheat species to be intentionally cultivated. Like all domesticated wheat species, cultivated emmer and einkorn are very closely related to their wild relatives and would normally be included in the same species. The most important difference between it and cultivated emmer is the more fragile rachis of the wild races (Harlan, and Zohary 1966: 1075, 1079). That is, domesticated emmer is simply wild emmer that has lost its seed dispersal mechanism through human selection. The thick rachis of domesticated emmer wheat hold the seed on the stalk; to remove it the wheat must be threshed.

As we saw in Chapter 10, by the end of the Pleistocene, large stands of wild wheat and barley were established over much of southwestern Asia. Wild emmer probably occurred as it does today as a large-seeded and robust race in the upper Jordan Valley of Israel and as a minor component among other wild cereals in Turkey, Iran, Iraq, and the southern Russia. Wild einkorn occurs naturally in massive stands in parts of the Zagros highlands of Iraq, Iran, and Turkey today and no doubt did so in prehistory as well. Wild barley was also part of the prehistoric natural grass cover of the open-woodlands of the Zagros highlands of Iran and Iraq. However, as it is less cold tolerant than einkorn wheat, it grew at slightly lower elevations. Outside of this Zagros, wild barley spread as weed throughout much of the eastern Mediterranean region (Harlan and Zohary 1966).

According to (Harlan 1967: 197-201), natural stands of wild wheat in Turkey produce seeds that are far higher in protein than modern hybrid domestic wheat. Experimental harvesting of these natural stands indicate that wild wheat seeds would have been easy to harvest and process. So why bother to plant? Of course, gathering the seeds of wild wheat and barley was precisely the strategy adopted by the Natufians, the complex foraging peoples who occupied the Levant during the Epi-Paleolithic period. The emergence of the "complex foraging" adaptation or "incipient agriculture" among the Natufians was discussed in the preceding chapter. Here we briefly return to the Levantine archaeological record to determine how the Natufians adaptation was transformed.

The archaeological record of the Epi-Paleolithic indicates that the rise of agriculture in the heartland of southwestern Asia can not be treated as a unitary phenomenon. Rather, it consisted of two distinct stages: (1) an earlier period of incipient agriculture which Kislev (1984) terms the "agro-technical revolution," and (2) the actual "domestication revolution," that took place later. The Epi-Paleolithic period transformation of simple foraging into "complex" foraging centered on wild cereal collecting by peoples like the Natufians of the Levant and the Karim Sharirians in the Zagros foothills of Iraq was an agro-technical revolution.

As we saw in Chapter 10, temperature appear to have been elevated worldwide after 13,000 B.P. and, while interpretations of the paleoenvironment of the Levant vary, the current consensus is that the climate of the region became was characterized by more effective moisture and by cooler, milder temperatures than at present (Byrd 1989: 171). These new climatic conditions had particularly important consequences in the Levant: they prompted the colonization of the uplands by Mediterranean woodlands vegetation and markedly expanded the distribution of wild cereal grasses and nut tree species.

According to Henry (1989: 227), the subsistence strategies and patterns at the base of the Natufian tradition were developed at this time to take advantage of the expansion and increased concentration of these grasses and nuts. These new "complex foraging" subsistence strategies included the elaboration of existing techniques and tools for grinding, pounding and grating wild plants, the construction of permanent facilities for storing food, the occupation of semi-sedentary or even fully sedentary settlements, the accumulation of wealth, and finally the fashioning of fledgling hierarchies of political and social inequality.

Recall that the existing zooarchaeological and archaeobotanical data from the Levant contains **no** morphological evidence for the domestication of plants or herd animals by the Natufians (Byrd, 1989: 159). None-the-less, the productivity of their subsistence allowed them to "experience, if at somewhat tenuous levels, all the traditional and fateful consequence of agriculture, while not growing plant food by cultivation and still relying on gathering natural products" (Gould 1990: 26). Thus, the Natufians foraging life way came to share a more ominous characteristic of fully agricultural people: the loss of the natural fertility controls practiced by mobile hunter-gatherers.

Mobile foraging peoples are motivated to keep their populations low because they exert little control over the amount of wild food stuffs available and their lifestyle makes storing food difficult. At the same time females find it necessary to maintain a proper spacing between the birth of their offspring as carrying more than one child is extremely difficult. Consequently, hunter-gatherers consciously limit the size of their populations by practicing infanticide, avoiding intercourse for long periods after child birth, even inducing miscarriages by various means. Foraging women also unconsciously limit their fertility by nursing their babies for extended periods of time after birth and thereby postponing the start of their ovulation.

In contrast, women remaining in the same locality throughout the year are less inclined to practice conscious population control since, not being burdened with carrying their children as often, they can space births closer together (Sussman 1972). Sedentary women may also simply be more fertile than nomadic women. There is strong evidence that a critical minimum of fat in the body is necessary for ovulation and sedentary women may tend to have more fatty tissue in their bodies than the more mobile foragers (Howell 1979; Frisch 1980). Finally, sedentary living must, as a rule, offer better prenatal conditions for pregnant women than the more mobile life of the broad-spectrum foragers making it more likely that pregnancies will go to term. Whatever the explanation, the Natufians let the population genie out of the bottle and, in doing so, "crossed the physiological, demographic, social, and even ideological divides that separate foragers and food-producers" (Henry 1989: 227).

As a consequence, by late Natufian times both the size of individual communities had become larger and the number and geographic distribution of such sites had increased over earlier times. Both of these trends reflect an increase in population in the late Natufian (Henry 1989: 218). Unchecked population growth is a danger for foragers, whether complex or simple. Since they do not control the production of the foodstuffs upon which they depend, they cannot increase it as need. The Natufians paid the price for crossing the population Rubicon when the Levantine environment changed once again in the worldwide warming trend of the early Holocene. In the new warmer, drier conditions of the early post-Pleistocene Levant, the distribution of the Mediterranean woodlands began to shrink. In the face of this shrinkage, the Natufian system began to unravel. The fundamental instability of complex foraging adaptation had become apparent.

The new environmental circumstances presented them with a fateful choice: return to simple foraging at reduced levels of population and cultural complexity, or increase the output of wild species by controlling and manipulating. Clearly, some of these peoples took the first course and resumed the hunting and gathering in small and mobile bands. Others among the late Natufians–or perhaps their immediate successors–took the second course (Henry 1989: 219). In doing so, they ushered in the "domestication revolution" in the Levant. The ancient relations of humankind to the natural world were altered forever. Prehistorians can not be certain when the domestication revolution actually began in southwestern Asia. However, the initial phases of it were over before 7,000 B.C. as virtually all of the plant species of the early southwestern Asian agricultural complex have been recovered from archaeological sites dating to before that time (Harlan 1977: 364).

Sheep and Goat Domestication Ungulates are hoofed mammals. Herbivorous in diet, the ungulates are adapted for running and social life in herds. First as prey animals, later as domesticates, no mammals have figured more importantly in the prehistory and history of humankind. There are two orders of hoofed mammals: *Artiodactyla*, the even-toed ungulates and *Perissodactyla*, the odd-toed ungulates. Pigs, hippos, camels, and ruminants (sheep, goats, cattle, deer, giraffes, antelopes) belong the first order; horses, tapirs and rhinoceros to the second. Ruminants, members of the suborder *Pecora*, have proven to be particularly important domesticated animals because their complex stomach and digestive system enables them to eat efficiently process cellulose, a ubiquitous substance found in leaves, grasses, and woody plants, but indigestible by humans.

Current archaeological evidence indicates that goats and sheep–referred to collectively as caprines–were the first of the ruminants to be domesticated and among the first animals of any kind to enter the human orbit. Only the dog preceded them. There are some obvious biological reasons for the early domestication of caprines and, according to Reed (1969), wild goats and sheep in some sense were "pre-adapted" for life with humans. As noted, their capacity to digest substances that humans find uneatable means that do not compete for the same food in a landscape. In addition, they are easy to tame when captured as young and their "neural and endocrine apparatus suits them for sociability" (Reed 1969: 367).

However, these are traits shared by all ruminants, antelope and deer no less than sheep and goats. Why weren't those other species domesticated as well? The answer perhaps is that, unlike gazelles and a number of other ruminants, wild goats and sheep do not defend territories. Further, wild goats and sheep herds are structured into dominance hierarchies under the control of a single, dominant "alpha" male. Consequently, when their young are raised in captivity, they come to regard the human herdsman as their leader. These two attributes allow caprines to be herded together in compact groups of mixed sex under the control of a herdsman who acts *de facto* as the alpha male. Conversely, the absence of social structures based on dominance in gazelle and deer herds combined with the instinct of their bucks to defend territories prevents those ruminant species from being herded and inhibits their domestication.

Clutton-Brock (1981: 56) speculates that humans could have exploited these caprine pre-adaptations to domestication either by taming and rearing young animals or by habituating a wild herd to human presence and then inducing them to stay in a favored situation such as near a supply of water. Recent experiments with European elk (*Alces alces*) indicate that wild ruminants can be "attached" to a particular location through regular feeding and then semi-domesticated. Wilkenson (1972) reports that after a period of continuous feeding, elk became so tame that they could be milked regularly. No attempt was made to restrain the elk or prevent them from interbreeding with fully wild animals. As a result, wild elk were lured to the feeding ground by the presence of the tame elk and the population was continually renewed.

Establishing control over breeding–not attempted in the elk experiment–would have been the logical next step on the path towards fully domesticating these animals. Of course, herding these elk even after they were domesticated would probably have proved to be impossible. The western Asian wild ancestors of domesticated sheep and goats were foothill and mountain-dwelling animals. They were not part of the forest and grasslands biomes of lower elevation, although their domesticated descendants are found concentrated in these regions today. The mountain and foothill biomes to which these wild caprines were adapted were places where the pressure of predation is not as great as the need to find sufficient food. Therefore, in their native mountain habitats, wild goats and sheep occupy relatively small home ranges, are not territorial, and will graze and browse on a great variety of grasses, herbs, and woody plants (Clason, and Clutton-Brock 1982: 146; Schaller 1977).

Wild caprines do not flourish on the plains and forests of lower elevations because just the opposite selective pressures operate there. That is, food is abundant, but the pressure of predation on the herbivores who live there is very great. Natural selection favors wild ungulate species that, like the gazelle, are able to remain permanently on guard against predators and to elude them through great speed. Caprines can survive in such lowland environments only if they are protected both from predators and the competition of native ungulates. Once freed of these natural selective pressures by human protection, however, the voracious browsing and grazing habits and the catholic food tastes of the caprines suits them perfectly to such surroundings.

The existing archaeological evidence suggests that the earliest domestication of wild sheep and goats occurred independent of the domestication of cereal grain. This process seems to have first taken place in the Taurus Mountains of eastern Anatolia and Zagros Mountains of western Iran. Caprines may have been herded there as early as 8000 B.C. (Hole 1984: 54, 1989: 97). As it is extremely difficult to distinguish the skeletal remains of wild sheep from those of wild goats, it has not been determined archaeologically which caprine species was actually first to be domesticated.

The earliest remains of possibly domesticated caprines in southwestern Asia have been recovered at the site of **Zawi Chemi Shanidar** in the Zagros foothills of northern Iraq in contexts radiocarbon dated to ca. 9000 B.C. A number of other sites in the Taurus-Zagros region have also produced caprine bones from the period between roughly 9000 and 7000 B.C. (Hole 1984: 54). In contrast, bones of domesticated caprines do not appear in the early Neolithic archaeological record in the Levant until after ca. 7000 B.C. Before the rise of pastoralism there, the cereal-growing peoples appear to have relied on gazelle hunting to supply them with meat. Between ca. 7000 B.C. and 6000 B.C., domestic goats and sheep became the principal source of meat and raw materials for the early farmers of the region (Clutton Brock 1981: 51). The dispersal of the domesticated caprines after this time appears to have been rapid. Domestic sheep or goats (the identity is uncertain) have been recovered at the cave of Haua Ftea in Cyrenaica in North Africa in archaeological contexts dating to before 4800 B.C. and perhaps as early as 6400 B.C. (Reed 1969: 372).

Domesticating the Pig The domestic pig is descendant from the wild boar, *Sus scrofa*, a species common throughout Europe, eastern Eurasia, and North Africa but absent south of the Sahara, in Australia and the Pacific Islands and in the New World. Although primarily a broad-leafed, deciduous forest dweller, the wild boar was apparently capable of adapting to a wide variety of Holocene environments in the Old World. Adult wild boars are dangerous animals. When separated from their parents, however, wild piglets are easy to tame.

One of your authors (Dickson) became aware of this some years ago in the course of archaeological field work in a mountainous rural region of Costa Rica. There he encountered a half grown peccary, a New World relative of *Sus scrofa*, living as a pet in the house of a farm family. The peccary was friendly and craved as much attention as a puppy. The owner told me he had captured it after killing its mother during a hunt and that, although it was being raised as a pet, it was destined to be eaten when it reached proper size!

Surely such domestications must have happened countless times in prehistory and it is not surprising therefore that the pig was apparently domesticated independently in both southwestern and southeastern Asia early in the Holocene. Recognizing the signs of that domestication in skeletal remains, however, is not often easy. According to Clutton-Brock (1981: 72), compared to wild pigs, domestic pigs are generally smaller and have skulls with shorten frontal regions. She used these criteria in her analysis of the pig bones recovered at the site of Jericho in Israel, and concluded that domestication of wild pigs–or at least human interference with and management of wild populations–had begun by 7000 B.C. in the Levant.

Pig domestication may have occurred even earlier in the Taurus and Zagros foothills. Small quantities of pig remains have also been recovered from about the same time period at the early village sites of **Jarmo** in the Zagros foothills of Iraq and at **Argissa-Magula**he bones of domesticated goats and sheep have been recovered from about the same time period. It thus appears that in southwestern Asia, the domestication of the pig began more or less simultaneously with that of goat and sheep but that species remained far less important there economically.

When it was introduced into forested Europe by early Neolithic settlers, however, the domesticated pig flourished. There the pig moved from a minor element in the Neolithic farm yard to a central focus of subsistence. Although some pigs were probably penned near the settlements and fed household scraps, the bulk of the Neolithic pigs were surely raised by means of open range husbandry. That is, they were marked in some way and then turned loose to graze in the forests. They may have been herded and closely-watched by day and then penned at night or they may simply have been allowed to roam in the vicinity and collected when needed. Under these circumstances, domestic sows routinely bred with wild boars.

Open range or free forest grazing was also a common form of husbandry for sheep, goats, cattle and horses in parts of the Old World until well into historic times. It was widely practiced in the frontier regions of the New World, especially in the American South. The following description of free forest pig raising in east Texas before the Depression conveys the flavor of this informal farming system.

> Every spring when Mr. L. was a little boy, too little to carry a gun, he'd ride out the bottom with his two hog dogs and a tow sack-full of corn split at the top so he could hang it over the back of his saddle. When the dogs treed a bunch of hogs, they'd circle the herd, steadily backing. The commotion triggered protective instincts in all the hogs anywhere around close, and they'd come to defend the herd. Once the hogs were bunched up, the child would call off the dogs and shell and ear of corn. As the hogs ate, he would count those with his family's ear mark and report their range to his father. The bottom was dotted with community pens built on the old Caddo (Indian) mounds where folks would single out fattening hogs and mark their choates. Once a couple of wild boars got in with the Bottom stock, real tush hogs, big old red piney woods rooters. The men got together and hunted them for weeks without success. Hogs are smart. For years the boy's stirrup leathers carried a tusk mark where he jerked up his leg just in time. Hogs can't swim. When the river got up, men would meet on the riverbank and cut a log raft. Then they'd pole through the flood

> waters and find the surviving hogs clinging to hummocks or downed trees, then edge the raft over slowing until the hog's instincts took over and it jumped aboard. Then they'd ease down to someplace like Harris Mounds, where the sight of high ground would entice the cargo to safety (Fisher 1990: 10A).

Cattle Domestication The placid European domestic cows of modern times, *Bos taurus*, are descendants of *Bos primigenius* the wild cow or aurochs. The last aurochsen were driven into extinction in the 17th century in Poland (Kurtin 1968: 189), so we can not directly observe their behavior as we can that of the wild relatives of the domestic pig, goat, sheep and dog. However, feral cattle–domestic cattle that have reverted to the wild–provide some insight into the behavior of the aurochs. If the ancestors of the domestic cow behaved as the did the feral Spanish longhorn cattle of 18th and 19th century Texas and California they were formidable indeed.

Frank Dobie (1941: 21) collected accounts of the feral longhorns and concludes that the bulls were exceedingly fierce, quick, willing to attack without provocation, and hard to kill. According to Dobie, contemporary observers considered feral longhorns "more dangerous to footmen than grizzly bears." If anything, the Aurochsen of late Pleistocene Europe were just as tough. If they were, it is fair to say that Aurochsen, some of whom stood as tall as two meters at the withers, were among the fiercest of creatures encountered by peoples living in the Old World during the early Holocene epoch. In light of this we are left to wonder what possessed our Neolithic forebearers to attempt to domesticate such formidable animals.

Issac (1962), following the 19th century scholar, Eduard Hahn, suggests that only religious motivation would have been sufficiently compelling to induce people to undertake such a task. Isaac proposes that bull sacrifice was an important element in early Neolithic religious life and suggests that bulls might have been obtained for sacrifice by driving herds of aurochsen into large enclosures or natural traps and then extracting the appropriate individual males. The remainder of the herd might have been kept in the enclosure and allowed to breed as a source of additional sacrificial animals. Overtime those animals might also have been exploited for their meat and hides.

It is true that by late prehistoric and early historic times, the bull was an important element in the religions of the peoples on Crete, in Egypt and Mesopotamia and elsewhere in the eastern Mediterranean. Examples of the veneration and sacrifice of the bull can also be found in our own religious traditions. *Exodus* 32: 1-32 describes the worship of the image of a "Golden Calf" erected in Moses' absence by the Israelites fleeing Egypt. *Leviticus* 1: 2-9 states the rules governing the sacrifice of a "bull without blemish" as a burnt offering to God.

Yet can we presume that bull worship and interest began in the early Neolithic? In fact, evidence of the worship and ritual treatment of cattle as early as 6000 B.C. has come to light at the site of **Çatal Hüyük** in Turkey. A specially decorated shrine room at that site was found to contain a number of horn cores of *Bos primigenius* set into a clay bench. The room also contained human figurines, fertility symbols including modeled human breasts and a painted mural depicting a great black bull surrounded by little human figures who appear to be trying to catch it. Although these data lend support to Isaac's thesis, we will never know with certainty whether the domestication of cattle had a religious motivation.

Whatever the motive, cattle appear to have been brought into the repertoire of the southwestern Asian domesticated animals much later than dogs, goats, sheep and pigs. Although their domestication may have begun by 7000 B.C., The earliest definite archaeological evidence of domestic cattle date to ca. 6500 B.C. at the site of **Argissa-Maghula** in Greek Thessaly and to ca. 6100 B.C. at the site of **Nea Nicomedia** in Macedonia (Reed 1969: 372).

The Rise of Sedentary Agricultural Village Life Villages are "small clusters of dwellings situated in surroundings otherwise empty of housing though economically important for subsistence" (Leeds 1976). Note that this definition excludes mention of the kind of subsistence practiced in the surrounding empty area by the residents of such small clusters of dwellings. There is good reason for such an omission. Although it is often asserted that some kind of agriculture is a necessary prerequisite for the establishment of settled village life, the archaeological and ethnographic record contains a number of important exceptions.

In environments where key resources are particularly abundant and spatially concentrated, settled village life can sometimes be supported by a foraging economy. A number of such environmental settings can be found along the coastal rim of the northern Pacific ocean. The early Holocene **Jomon** peoples of coastal of Japan established settlements that appear to have been occupied for all or most of the year based on hunting-gathering-strand loping-and-fishing. On the Northwest Coast of North America, Native American food collecting peoples like the Tlingit and Haida had, by late prehistoric times, established settlements every bit as impressive as those of the Jomon. "Although in former times these foragers practiced no agriculture, they were yet able to live in large, richly furnished wooden houses in socially-stratified, more-or-less permanent villages. But of course, the environments they inhabited were also truly exceptional. The Northwest Coast is rich almost beyond measure. Along that remarkable coast, key wild resources such as salmon, sea mammals, whales, mollusks and a host for forest plants and animals are geographically concentrated, extraordinarily abundant and highly predictable in their in their occurrence. "A somewhat lower level of sedentary living was also practiced by the intensive acorn harvesting Native American people like the Maidu of central California (Basgall 1987).

On the other hand, a high degree of residential mobility has also been found to be practiced by certain full time agricultural peoples. The Rarámuri peoples who inhabit the Rejogochi valley region in the rugged mountain country of the Sierra Madre Occidental in the northern Mexican state of Chihuahua are such a people. The Rarámuri:

> engage in four types of residential mobility. First, during the growing season households move to residences located at fields outside the Rejogochi valley, living there from several days to several weeks at a time. Second, some households move to spend the cold, wet months of the year in rock-shelters or winter houses. Third, entire households sometimes relocate temporarily to work for wages outside the Rejogochi community. Fourth, during the winter and spring, households also spend several days and sometimes weeks at a time in the pueblo center of Basíhuarc, organizing and celebrating religious ceremonies at the local church (Hard and Merrill 1992: 605).

As these examples illustrate, the presumed link between sedentism and food production must be decoupled. In the absence of other kinds of evidence, the simple presence of permanent villages can not be taken as proof of the practice of agriculture. Conversely, the **absence** of such settlements does not establish that subsistence was based solely on hunting and gathering.

An prime example of this decoupling can be found on the Levantine coast of the Mediterranean. There, the Natufians and other complex foraging peoples were able to live a settled existence while practicing cereal collection and gazelle hunting. However, Flannery (1973) notes that Natufian villages were laid out along very different lines than those built by the fully agricultural peoples who followed them in the Levant. According to Flannery, the earliest settled villages to appear in the archaeological record of southwestern Asia are all "circular hut compounds." This type of settlement appear after 9000 B.C. among the Natufians and lasts into the subsequent Pre-Pottery Neolithic period until the sometime between ca. 7000 B.C. At first these compounds were supported entirely by foraging. Almost without exception, only the bones of wild animal species have been recovered from them. Early compounds have produced wild cereals, later ones cereals in the earliest stages of domestication.

The circular hut compound apparently emerged as a new settlement type out of the base camp settlement pattern practiced by hunting and gathering peoples in the region during the late Pleistocene epoch. Compounds contain between ten and fifty circular huts arranged in a circle around a central courtyard. The huts have stone foundations and most of them enclose interior floor space sufficient to house only one person, although some larger huts exist and may have been used by two to three people. Granaries, storage pits and other storage facilities are usually situated outside the huts in the open courtyard, a placement that leads Flannery to conclude that the foodstuffs stored in them were shared by all of the residents in the compound.

Archaeological debris recovered from the courtyards indicates that these open areas were also the locations of many craft and processing activities. Circular compounds similar to the archaeological examples from southwestern Asia continue to be used at present in Africa. Based on what is known of manner in which contemporary African compounds are used, Flannery hypothesizes that the ancient circular hut compounds were very likely occupied by "groups of related polygynous males, each accompanied by one to three females and their children." In his view, these polygynous extended households were the primary units of food production, food sharing and storage among these early Holocene settled foragers.

Beginning around 7000 B.C., villages of rectangular houses began to replace the circular hut compound villages in southwestern Asia. This change was clearly connected with the great subsistence shift from food collection to food production of the early Holocene. While the floral and faunal remains recovered from the circular hut compound were generally of wild species or species in the early stages of domestication, those recovered from the rectangular house villages were overwhelmingly domesticated. Villages of this pattern generally occupy between one and four hectares; a rare few are even larger.

Individual houses generally enclosed between 25

and 35 meters of roofed, interior space in the main room. However, the rectangular shape of these houses made it easy to add additional rooms as the number of residents living in it increased or to wall up unused rooms as the numbers contracted. In addition, many houses have unroofed patios or courtyards attached to them. In contrast to the circular huts of earlier times, the rectangular houses appear sufficiently large to accommodate nuclear families. The tools and debris recovered from them suggests that a wide range of craft and processing activities–both those of men and of women–were carried on within or immediately adjacent to these houses.

Perhaps the most important contrast between these villages and the circular hut compounds was the fact that each house had storage facilities located inside it rather than outside in some open, public space. Flannery interprets the placement of the storage facilities within the individual houses and the presence of craft and processing debris within the walled courtyards to mean that in these ancient villages:

1. It was not obligatory to share foodstuffs stored in the house with all members of the village, and
2. Craft and processing work space in the courtyard-and by implication the output of this work–was also not shared.

After 7000 BC the number of rectangular house agricultural villages in southwestern Asia grows steadily. The size of these villages–and the scale of the construction in some of them–also grows over time. Thus, it appears that the emergence of rectangular house villages was followed by a dramatic increase in population in the area. The community pattern of these ancient rectangular house villages seems immediately familiar to anyone who has visited the rural parts of southwestern Asia. Tiny farming hamlets essentially duplicating this pattern are still found scattered throughout the area. In these modern southwestern Asian farming village, the individual household is the basic unit of production, storage, and sharing. Based on the analogy of these historic villages, Flannery (1973: 39-40) argues that their replacement of the circular compound community pattern in the archaeological record signals the emergence of the single family household as the primary productive unit in the economic life of the area and the corresponding replacement of the earlier, communally-organized polygynous extended families in that role.

Thus Flannery postulates that the new form of village community pattern reflects the emergence of a new form of ownership and property rights which in turn resulted in a pronounced intensification in the production of foodstuffs. Although hunter gatherers produce a lower gross output of foodstuffs than do cultivators, they also spend far fewer hours in the food quest. Since the higher output of farmers is achieved larger through an increased work effort on their part, foragers' output of food energy **per unit of labor** is higher than that of agriculturalists. Farmers have more but must work harder to get it (Service 1979: 12). What persuades them to make such a trade off?

Flannery concludes that the narrowing of communal property rights radically altered the incentive structure. If communal ownership and sharing in fact characterized economic life in the foragers' circular hut compound–and the placement of the storage facilities in the public courtyard suggests they did–then little incentive existed for individual members to increase their productive output. In contrast, when freed of the obligation of sharing with all members of the village, individual households had an incentive to increase production since their members profited directly from such increases. These new incentives had a dynamic impact on agricultural production.

But the key result of the transformation was not simply the high levels of economic surplus that village households began to produce. According to North and Thomas (1977: 230), the new concept of property rights also created powerful incentives to acquire and apply new knowledge to economic activity. In their view, this jural change represents the most significant element in the dramatic growth Neolithic technological innovations as well as production. The relative decline in communal sharing with the rise of the agricultural village was correlated with a greater acceptance of the social inequality that inevitably accompanies increasing economic surplus (Lenski and Lenski 1970: 176).

Yet, we know that in modern peasant villages, accumulation by individuals and households is seldom unrestrained. Instead, it is commonly offset or checked by the operation of what anthropologists call "pan-village sodalities." As the name implies, sodalities are communal religious or social organizations meant to strengthen solidarity, unity, and interdependence within the village. Sodalities extract what Annis (1987) calls a "cultural tax" from all villagers in return for these community services. This cultural tax is commonly paid by each household in the form of labor or food surplus. The sodality them redistributes the food surplus throughout the village as welfare or as public feasts during religious festivals and organizes the labor service to undertake public works.

Of course, such cultural taxes may simply

function as social leveling mechanisms that squander collected resources in order to keep all households at roughly the same economic level. Alternatively, these taxes may be used to provide essential infra structural support to village economic production and growth or make its military and political survival possible. The percentage of income extracted by sodalities and the extent to which the proceeds of such extractions were well or poorly used no doubt varied from village to village.

Village Economy and Society Flannery's analysis nicely illustrates the utility of using our knowledge of historic agricultural villages to interpret the archaeological record of these early Neolithic rectangular house sites. By analogy with modern examples, the economies and societies of this early villages of the Neolithic in southwestern Asia probably had a number of other common attributes. Their infrastructure was still characterized by:

1. Simple technology, which, overtime comes to include the manufacture of ceramic and ground stone artifacts,
2. A presumably "extensive" cultivation and herding system appropriate to their technological simplicity, abundant land resources, and relative labor scarcity,
3. A subsistence system generally capable of producing a moderate surplus of food energy but at a relatively low ratio of output of food energy per unit of labor compared to foragers,
4. Diets to which domestic plants contributed a greater percentage of the calories than either hunted or domesticated animals but which nonetheless, reflect a continued dependence upon wild food resources,
5. Te systematic disregard of wild food resources whose distribution and seasonal availability places their collection in conflict with the schedule of agricultural activity, and
6. An emphasis on storage and accumulation.

Given the foregoing infrastructure, the demographic structure of these Neolithic farming villages was probably characterized by:

1. Densities of population per square kilometer generally higher than those achieved by hunting and gathering societies,
2. High rates of population growth through time,
3. A maximum population size in any given village limited by the amount of food stuffs produced, stored, or collected during the season of **minimum** availability, and
4. Endemic and epidemic diseases which began to become a problem as human population and settlements increased in size, concentration, and degree of sedentariness.

Larger, denser population sizes and relatively high rates of growth among contemporary–and therefore by extension, Neolithic–subsistence farmer may be in part a reflection of the more localized nature of resources critical to the farmer and to the greater sedentariness of farming life. However, the large populations of cultivators is mainly due to conscious procreative decisions made by peoples aware of:

1. The significant economic contribution of children's labor to their subsistence,
2. The comparative ease of bearing and tending children in a sedentary settlements,
3. The larger demand for labor in subsistence agricultural production, and
4. The relaxation of the necessity of limiting population made possible by the fact that food production and food storage can be increased as needed. After all, it is rational to have large families when the cost of feeding each new member is offset by the contribution that additional person will eventually make to food production.

The social organization of early Neolithic farming villages was probably characterized by small, largely autonomous villages generally more sedentary in residence pattern, more permanent in membership, and more complex in organization than the band organization of foragers.

1. Kinship probably remained the chief principle governing social life but was supplemented to a growing degree by a sense of community based on ties other than kinship.
2. Corporate kin groups were probably the primary land-holding institutions that both control access to joint resources and coordinate the labor of group members,
3. An ideology of descent which stresses maintenance of land ownership through many generations, with continued participation of decreased ancestors in the affairs of the descent group" (Flannery 1973: 28),
4. Economic distribution and exchange based on reciprocity within the household and extended family and redistribution of some

surplus within the village,
5. The appearance of full time specializations beyond the sexual division of labor, and
6. The appearance ascribed statuses and roles beyond those of age and sex.

Early Neolithic villages were probably largely autonomous entities governed by headmen and councils of elder male family heads. Only much later were they incorporated into political units larger than the village. Compared to foraging bands, the early Neolithic villages were probably characterized by more permanent membership and more complex political organization. However, on at least one important political dimension, early Neolithic villages seem to have experienced a reduction in social complexity relative to foraging groups: their social formations may not have been integrated into extensive, extra-band regional networks. At least in the early millennia of the Neolithic, the wide ranging regional trading, marriage, and political and religious exchange fields appear to be truncated with the emergence of self-sufficient, endogamous (in-marrying), inward-looking agricultural villages.

For example, Leone (1968) uses archaeological data from the American Southwest to test the hypothesis that an increasing dependence on agriculture leads to increasing social distance between the minimal economic units needed to make agriculture a successful economic base. As both of his variable co-varied positively, he concluded that, as dependence on agriculture increased during the Neolithic period, villages became more endogamous, that is more prone to require that their residents find marriage partners within the village.

Neolithic Spiral of Violence The increasing isolation and self-sufficiency of these villages may be reflected on another dimension: the intensification of feuding and raiding beginning in the early Neolithic. Bryony Orme contrasts the life ways of food producers with that of foragers on one final dimension: the presence of feuding and raiding among farmers. Orme states that:

> most differences between hunters and farmers are a matter of degree, and there is no sharp dividing line between the two means of subsistence. However, the last advantage to be considered here occurs so regularly amongst people who farm, and so rarely amongst people who live by hunting and gathering, that it could almost be used as a subsistence indicator. Yet its significance in this context has not really been explored, either by ethnographers or by prehistorians. The advantage in question is warfare: farmers fight whilst hunters do not. Hunters-gatherers rarely have sufficient people in one place for long enough to achieve organized warfare, even if they sometimes manage a battle or a raid, whereas farmers are constantly on the warpath or in the war-canoe. From the point of view of most readers, warfare will hardly seem to be an advantage but, judging by numerous ethnographies...farmers relish their battles as strongly as a football match and in much the same spirit (1977: 46).

The evidence of this intensification of violence leads Lenski and Lenski (1970: 176) to comment on the ethical consequences of the horticultural revolution and to note that the technological and structural progress of the Neolithic does not imply ethical progress: "it is one of the great ironies of evolution that progress in technology and social structure is often linked with ethical regress. Horticultural societies provide several striking examples. Some of the most shocking, by the standards of our own culture, are the increases in headhunting, cannibalism, human sacrifice, and slavery, all more common in the technologically progressive horticultural societies than in the more primitive hunting and gathering groups."

A unique illustration of the intensification of violence in southwestern Asia during the early Neolithic and emerging need for defense against it is found at Tell-es-Sultan in Israel, the site of the Biblical town of Jericho. A tell is a large mound that contains the accumulated remains of human occupation. Tell-es-Sultan or ancient Jericho is located on an semi-arid plain in the Jordan River valley on the left bank of the river northwest of it's junction with the Dead Sea. Near the base of the tell is a permanently flowing spring, the Ain es Sultan or the Fountain of Elisha. This spring creates an oasis in the midst of a harsh and semi-arid countryside and, from earliest times, it constituted the primary reason for settlement at this spot.

The site was first occupied after 9000 B.C. by Natufian peoples who hunted gazelle, used stone sickles to exploited the great stands of wild emmer wheat that grew in this part of the Jordan Valley, and built an enigmatic walled enclosure above the spring that may have been a shrine. Kathleen Kenyon, the excavator of the site, relates the "shrine" to the presence of the spring on the grounds that people tend to worship the mysterious sources of the things they depend upon in the material world (Kenyon 1979:25).

The Natufian occupation at Jericho was buried beneath 3.06 meters (13 feet) of successive deposits containing an innumerable succession of living floors marked with slight rises thought to be the bases of circular huts built of perishable materials. Although the excavator called this occupation the "Proto-Neolithic" and considered it to represent the transition from food collecting to food production at the site, no evidence of domesticated animals or plants were recovered from it. Presumably the economy of Jericho during this time was still based on the collecting of wild food stuffs in a setting sufficiently productive to allow sedentary living without agriculture.

Around 8000 B.C. more permanent structures were built in a dense cluster on the site. These structures are round to oval, beehive-shaped houses 4 to 6 meters in diameter. They have stone-lined, semi-subterranean foundations that support superstructures built of plano-convex or hog-back mud bricks set in mud mortar. They appear to be arranged in a classic–albeit an extraordinarily large–circular hut compound. This occupation, first known as the "hog-back brick phase," was later christened the Pre-pottery Neolithic A (PPNA) or "Sultanian." The PPNA in the Levant begins around 8300 B.C. and lasts until ca. 7300 B.C.

It is followed by the Pre-pottery Neolithic B (PPN B) that begins ca. 7300 B.C. and lasts until about 6000 B.C. (Bar-Josef 1989: 632). The PPN A occupation at Jericho has produced evidence of domesticated plants in the form of carbonized seeds of emmer wheat, two-row barley, fig trees, pulses, and lentils. However, this occupation produced no certain evidence of domesticated animals. Meat at the site was apparently obtained exclusively by hunting of wild gazelle, cattle, goats, and boar (Kenyon 1979: 27).

Kenyon (1979:28) makes two speculations about the Pre-pottery Neolithic A occupation at Jericho based on the site's area during this time. First, she suggests that the PPN A occupation at Jericho covered 10 acres (4.05 hectares) and probably could have contained a population of 2,000 people or more. Second, Kenyon concludes that population densities of this kind would have been sustainable only by the practice of some form of irrigation agriculture drawing upon the water from the spring.

Both of these views have become deeply embedded in general literature. However, Bar-Josef (1989: 634-635) estimates that the PPN A occupation only covered about two acres at the site. Further, he notes that, like a number of other PPN A and PPN B sites, Jericho is located on the edge of an alluvial fan. An alluvial fan is a cone shaped land form found along the edges of mountain fronts in semi-arid and arid regions. Such land forms consists of sediments that have been carried down the mountain side by the periodic flow of stream floods or sheet floods. They are called fans because, as these sediments move downslope, they spread outward from the vertex of the cone to form a broad, fan-shaped base.

Bar-Josef contends that PPN A and PPN B sites were located near alluvial fans because the water table tends to be high in them and their soil are continually being fertilized by silt deposited with each periodic flood. It is his view that cultivation on these fans could be carried on without irrigation. In fact, the problem in such settings is not insufficient water but rather its surfeit during periods of flood run-off–a point we will return to in a moment. Further, the fact that both domesticated cereals and nitrogen-fixing pulses were being grown at Jericho and contemporary sites raises the possibility that PPN A farmers had already learned how to maintain soil fertility in their fields by rotating cereal crops with legumes (Moore 1985: 24). Therefore, it is highly probable that the population at Jericho during the PPN A was not as large as was originally thought to be and that the agricultural technology used there was sufficiently productive to support that population without recourse to irrigation.

However, the most spectacular–and unexpected–discovery at Jericho was the sophisticated system of defensive architecture found to have surrounded the PPNA circular hut village. If the Proto-Neolithic settlements at Jericho had walls or other defensive works around them, they were made of perishable materials and have not survived in the archaeological record. Sometime shortly after 8000 B.C., the PPNA peoples began the construction of a free standing stone wall nearly 2 meters thick. This wall eventually encircled the entire village–a circumference of approximately one half mile. There appears to have been at least three main building phases and, during the second one, the wall was fronted by a large ditch 8.25 meters (27 feet) wide cut into bedrock to a depth of 2.75 meters (9 feet).

What was purpose of this wall? Bar-Josef (1986) asserts that it was built as a kind of levee designed to protect the compound from flood water from the nearby *wadi*. That the wall could have served as a flood control facility is undeniable. However, it also is clear that this was at best a secondary purpose for its construction. This is because on the inside of the face of the wall the builders anchored a large, round tower massively built of stones set in mud mortar. The tower survives to a height of 7.62 meters (25 feet) and was originally higher still; presumably the wall was built to this height as well (Kenyon 1979: 26-27). Such a tower would be most useful in defending the site against men, not water.

The top of the tower was reached through a door

at it base of the tower inside the wall. The door led to a stone passage with a staircase of more than 20 steps that leads up through the center of the tower to its roof. The excavators say of this passage, "It was a great feat of mason's craftsmanship to trim these slabs to their rectangular shape and hammer-dress them to a smooth finish with the tools available, stone axes, adzes and hammer-stones. It was a still more remarkable architectural feat so to build the roof and consolidate the superstructure as to enable them to carry the overlying weight of at least 6 meters of solidly packed stone, and to survive for some nine thousand years in spite of innumerable earthquakes" (Burleigh, Cornwall, Kurth, Röhrer-Ertl 1981: 19).

Although modest by the standards of later construction in the southwestern Asia, the PPNA defenses at Jericho were built on a scale that was utterly without precedent in the archaeological record of this early time. The nature and scale of the wall, ditch and tower complex at Jericho tells us that true warfare–that is, organized violence carried on for extended periods of time–was becoming a part of the human experience in southwestern Asia as early as 8000 B.C.

Surely the construction and maintenance of defensive systems of this kind demand sustained communal cooperation. Perhaps this work was undertaken by the village warrior sodality. However, it is possible that the wall reflects the emergence of some form of authoritative leadership at Jericho since, in later millennia, true warfare is the hallmark of chiefdoms and state forms of political organization. The defensive wall and tower complex at Jericho also indicates that the PPNA community could produce sufficient surplus of food stuffs to enable them to divert labor from subsistence tasks in order to build it. Of course, it may have been such economic surplus that had attracted and tempted Jericho's neighbors and made the defenses necessary in the first place.

The tower must have been a particularly startling presence when it first arose on the Neolithic landscape–a skyscraping human artifice against a horizon broken otherwise only by natural eminences. No doubt the gloomy keep loomed large in the imaginations of the time. And why not? Its battlements foretell the tragic fate of humankind in the millennia to come: alarms, tumult, wars and rumors of war. But what enemies, what threats, what terrors, what guilt inspired the Neolithic peoples of Jericho to build such a remarkable defensive complex? The absence of written records makes this question difficult to answer and there are only fragmentary clues in the archaeological record.

The PPNA level at Jericho has produced appreciable quantities of exotic materials derived from distant locations: marine shells from the Mediterranean coast and the Black Sea, amulets and beads of greenstone and, most notably, obsidian determined to be derived from a region of Anatolia some 500 miles distant (Kenyon 1979: 29; Bar-Josef 1989: 636). This exotic material is convincing evidence that the early Neolithic residents of Jericho were involved in long distance trade. The Dead Sea region is rich in salt, sulphur, bitumen and other useful minerals. Perhaps it was these materials that were being exchanged for the exotic goods.

Although we don't know how the trade was organized, it seems likely that Jericho's size and architectural precocity resulted in part from her participation in it. Geographically the city is positioned to dominate the major east-west trade route across the Jordan Valley and to control the entrance to the hill country of western Palestine. Very likely both economics and geography necessitated the construction of the imposing PPNA defensive works. All of this remains speculative, however. Both the intentions and identities of Jericho's rivals–and of her trading partners–are lost forever. The site appears to have been abandoned by PPNA peoples by 7000 B.C. or earlier.

After a short hiatus, Jericho was re-occupied sometime after 6800 B.C. in the Pre-pottery Neolithic B period by what appear to be a new people. At Jericho and elsewhere in the Levant the PPNB period is marked by the appearance of Flannery's "rectangular house villages" built with "cigar-shaped" bricks and characterized by floors plastered with chalk burned in kilns or in open fires. Typically these houses were cellular in pattern and built around unroofed interior court yards where cooking and other crafts were performed.

The nature of subsistence has changed also. While PPNA peoples in the Levant appear to have relied on gazelle hunting to supply their meat, in the PPNB, "the domestic goat and sheep had become the principal source of meat and raw materials for the early farmers of the region and cattle and pigs were probably also undergoing the process of domestication at this time" (Clutton Brock 1981: 51). The PPNB occupation at Jericho grew steadily during the period but interestingly, it remained **unwalled** during the first 10 successive construction levels. By the time a wall was built, accumulated debris had raised the living surface of the mound to nearly 24 feet (7.32 meters) above the original ground surface. During this first part of the PPNB occupation, rectangular houses streched from the top of the mound down its sloping sides. Clearly the residents during this time saw no reason to huddle behind a stone barricade. Perhaps no one threatend them or perhaps someone did but the villages felt that the size and residential density of their village defense enough–and perhaps they were right.

Yet they changed their mind rather abruptly. Midway through the PPNB occupation, a stone wall was built around the village. Perhaps significantly, this construction took place at the same time as another, particularly gristly, event: the mass burial of forty or more individuals beneath the floor of a single house (Kenyon 1979: 35). The simultaneous burial of this many individuals is clearly out of the ordinary. It is strong evidence of a disaster: a fire, a plague, ethnic rioting, the violent suppression of a rebellion, an enemy raid or massacre.

Further, the skulls had been removed from most of the skeletons in the mass grave. Kenyon (1979: 35) considers this decapitation to been done for ritual purposes after the death of the individuals. This because other parts of the PPNB occupation at Jericho produced "plaster skulls," that is, human crania the lower part of which "had been covered with plaster, modeled into the likeness of human features. Each head has a most individual character, and one cannot escape the impression that one is looking at real portraits. The eyes are inset in shells. In the case of six of the heads, the eyes are made of ordinary bivalve shells, with a vertical slit between two sections giving the appearance of the pupil. The seventh head has cowrie shell, and the horizontal opening of the shells give him a distinctly sleepy expression. The molding of the feature, mouth, nose, ears, and eyelids is fine and delicate, giving the impression that the people whose portraits we here see had small and well-fashioned features. The top of the skull is always left uncovered, though in one case the skull is painted with broad bands of dark paint, perhaps to represent a head dress. One curious fact is that in only instance is the moulded over the upper teeth, and the herds have therefore a somewhat squat appearance" (Kenyon 1979: 35-35).

Similar plaster skulls have been found at **Tell Ramad** in Syria and the site of Beisamoun in Israel and headless skeletons found buried at the site of **'Ain Ghazel** in Jordan may indicate that decapitation and head decoration was practiced at that site too. It is Kenyon's view that these plaster skulls were remains of deceased relatives kept by their descendants as momentos or even worshiped as ancestral deities. However, it is more plausible to consider the plaster skulls "trophy heads" taken in warfare or headhunting raids and the headless skeletons buried at 'Ain Ghazel and Jericho to be the victims of such raids.

If this interpretation is correct, the construction of the wall in the PPNB period may have been prompted by the same events that had left more than forty people dead. We shall never be certain but whatever the immediate cause of the decision, the building of the wall around Jericho suggests a renewal of warfare and a rising level of violence occurred in the PPNB.

Making Pottery Clay is a fine-grained soil or sediment that is highly plastic when wet. The plasticity of wet clay allows it to be readily formed into a desired shape by modeling or hand building, by successively coiling clay "worms" atop one another, by forming clay into molds or by shaping it on a wheel. Whatever methods are used, when clay is heated to temperatures in excess of about 475° Centigrade and allowed to cool, it hardens into a porous, earthenware that retains its shape permanently and does not disintegrate in water. When temperatures are raised to higher levels and heating prolonged, the pottery that results is a harder, less porous stoneware or porcelain.

But heating drives the water out of the clay and causes it to shrink and crack. To prevent the shrinkage of the clay from cracking or breaking the vessels during firing, it is generally necessary to knead a **"temper"** of some coarse, non-plastic material into the wet pottery clay before using it. Potters use a wide range of materials as temper–sand, volcanic ash, crushed rocks and minerals like limestone, quartz, feldspar or mica, ground pottery sherds, even organic materials such as fiber, grass or crushed mollusk shell.

While the technology of the pottery making is simple, its creative possibilities are infinite. The skilled potter can vary and refine his handiwork virtually without limit. This creative potential has meant that ceramic styles and vessel types have appeared and disappeared at a rapid pace throughout prehistory. The speed at which pottery styles change combined with the great fragility of ceramic vessels are two reasons why archaeologist love the craft so much. Once broken, a fragile ceramic vessel becomes a mass of well-nigh indestructible potsherds. Incorporated into archaeological context, these potsherds become temporal markers and cultural "fossils" that record the passage of time and the pace of culture change.

Pottery making is a craft most commonly found in food producing societies. In fact, the last generation of prehistorians generally assumed that the making of ceramics, the growing of crops, and settled village life appeared in prehistory virtually simultaneously. Greater understanding of the archaeological record has revealed this to be false. It is now clear that the earliest ceramics in human prehistory were made after 12,000 B.C. by Mesolithic hunting, fishing, shellfish and plant gathering peoples of the Jomon tradition living on the coast of Japan. As noted in Chapter 10, the richness of their environment allowed the Jomon peoples to achieve a semi-sedentary or perhaps fully sedentary existence based on food collecting, rather than agriculture.

The earliest ceramics in North America were also made and used in a similar coastal setting among semi-sedentary, strand lopping peoples living on the on the southeast coast of Florida perhaps as early as 2500 B.C.(?) The discovery of these early fiber tempered wares further undermined the old generalization that ceramics were only found with agriculture. It now appears that the necessary precondition for ceramics is simply a semi- or fully sedentary lifestyle. Yet some degree of sedentism is a necessary precondition for ceramic use, it is not sufficient as cause in and of itself. Archaeological work in Palestine has revealed that settled village life there began more than two millennia before the development of ceramics.

However, the connection between ceramics and agriculture was considered so important by archaeologists working in the Levant, following the discovery of the late emergence of the potter's craft, the culture traditions dating to the period between ca. 8,300 B.C. and 6,000 B.C. in the Levant was henceforth referred to by the rather unfortunate term of "Pre-pottery Neolithic."

Little is known of the process leading to the invention of ceramic technology near the end of the Pre-pottery Neolithic B. Pottery making is relative simple and it is therefore unlikely that clear archaeological evidence of the various steps in its development will ever be recovered. Nonetheless, various scenarios have been suggested.

Amiran (1965: 240) finds a clue to the origin of ceramics in the resemblance between pottery making and the preparation of bread and porridge. Although the aims of these two crafts are different, the technical activities–skills, methods, procedures, routines–that people engage in to accomplish them are similar. Bakers and potters both prepare their raw materials by grinding, combining ingredients, mixing them with water, kneading, shaping or molding them, and firing them in ovens. Could the long practice of bread and porridge making have led to the rise of the potter's craft?

Amirian (1965: 242) also speculates that ceramics may alternatively have emerged sometime in the Pre-pottery Neolithic as a by-product of the threshing activity, the use of mud in house construction and the construction of clay-lined basins in house floors. Straw, chaff, and husks are the inevitable by-product of any type of threshing activity and the archaeological evidence indicates these substances were found in abundance on Pre-Pottery Neolithic sites. Clay mixed with straw, chaff, and husks became the primary construction material used in Pre-pottery Neolithic houses: the walls were made of mud bricks, and the floors and roofs were lined with straw-tempered mud plaster. Depressions or basins were built into the floors and courtyards of many of these houses. The sides of these basis were lined with thick layers of straw-tempered clay. Although these basins are located near ovens and hearths in what were probably the food preparation areas of the hours, their exact purpose is uncertain. Amiran suggests they may have been storage bins, general purpose household receptacles, even "stone boiling ovens" in which water and food were placed and then brought to a boil with red hot stones heated in nearby campfires. Amiran speculates that the PPNB peoples were inspired to experiment with intentional pottery making after repeatedly observing that the accidental firing of their floor basins transformed the straw-tempered clay that lined them into ceramics.

Whatever the cause of the craft's development, after 6000 B.C. pottery making was added to agriculture, settled village life, and polished stone tools to complete the full Neolithic technological repertoire in southwestern Asia (Mellaart 1975). The practical applications of ceramics are great indeed and, whenever ceramic containers first become integrated into a prehistoric economy, it soon becomes the most radical and important of Neolithic crafts (Mumford 1967: 139). Although ceramic vessels are more fragile than containers made of stone, skin, or basketry, they are more versatile than all of these and are "cheaper" in that it takes far less labor to make them.

The great plasticity of ceramics allowed the ready manufacture of receptacles of a boundless variety. Soon after the development of the craft in southwestern Asia, a host of vessels of different sizes, shapes, and functions began to appear including: bowls, pitchers, chamber pots, chimney pots, feeding troughs, braziers, lamps, beehives, dovecotes, churns, and sling stones. Lacking stone outcrops on the southern Mesopotamian flood plain, the early 'Ubaidian peoples who settled there made their hoes and sickles of a ceramics heated to temperatures high enough to harden and vitrify the clay. Of course a variety of less immediately practical objects such as gaming pieces–and much later, smoking pipes–were also made of ceramics there as were numerous aesthetic objects where the emphasis is on the object itself rather than on its use such as figurines, beads, and jewelry (Matson 1965: 212-213).

But the most important advantage of pottery is the fact that food storage and preparation can be more efficiently done in clay pots than in leather, stone or basketry containers. Pots can be made into moisture-proof, leak-proof, vermin-proof vessels for storing or transporting water, oil, wine, beer, grain and other essential commodities. Cooking pots may be placed directly on a fire and thus allow food to simmer for long periods of time without constant attention. This feature would have been of particular utility during periods of peak daily activity in the fields (Glassow 1972: 297). This is because a primary advantage of pottery is the ease with which its users can soak and leach foodstuffs. Soaking and leaching are important food processing strategies used among other things to soften or hydrate plant tissue, aid detoxification, and precipitate starch from plant foods (Stahl 1989: 175-176).

Not the Yeast of Our Worries Unbeknownst to the early farmers, by manipulating plants and animals in the visible world, they were silently "domesticating" organisms in the unseen microbial" realm as well. Many of the microbes domesticated at this time were benign or even positively useful. Yeast, for example, makes fermentation possible. Because the systematic practice of fermentation is impractical, even impossible, in the absence of sealable pottery jars, "the domestication of yeast must have been an unintended consequence of the development and use of ceramics. Fermentation processes dependent upon various forms of yeast include pickling, cheese-making, bread-making, culturing and all the various other processes involved in producing vinegars, sauces (especially fish sauces), lactic acid fermentations of various vegetables like sauerkraut and so on. Fermentation is an important means of preserving food, improving its flavor–or hiding its defects–with sauces and relishes, removing toxic components, utilizing food wastes, and supplying essential secondary food constituents (Stanton 1969: 463).

Of course let us not forget the fermentation of wine and beer. Early wine appears to have been made from honey and palm sap, as well as dates, figs, raisins, apples, pomegranates, and numerous other fruits. Beers, on the other hand, were fermented chiefly from wheat and barley. Beer brewing begins with the preparation of malt. To obtain malt, grain kernels are forced to germinate either by soaking them in water for a prolonged period or by sprinkling them with water and allowing them to stand in a place protected from direct sunlight. Germinating grain kernels produce **diastase**, an enzyme that converts part of the grains's starch into maltose or malt-sugar. After the grain begins to sprout, it is ground into flour and made into malt cakes. Malt cakes can either be eaten as bread or used in the brewing process. Malt sugar aids the fermentation process and improves the flavor of beer (Forbes 1954: 278). But beer has another advantage as well. The primary defect in cereal porridge is its lack vitamins of the B complex. Happily, most beers are rich in B complex vitamins and thus form an important supplement to the diet of cereal agriculturalists (Linton 1955: 94). But of course, human beings, unlike most other species, must effect adaptation on the mental plane as well as the material one. Beer and wine help here too. After all, as A. E. Houseman long ago observed, "Malt does more than Milton can/To justify God's ways to man."

Given its utility, it is not surprising that evidence of the Neolithic fermentation practices are widespread. This is particularly true in southwestern Asia where such evidence comes from many sources, linguistic as well as archaeological. One example of this linguistic evidence of Neolithic period fermentation will have to suffice. Recall that the earliest written language in the world is Sumerian, an ancient tongue spoken on the vast alluvial plain of southern Mesopotamia region know as Sumer by at least 3500 B.C.
The Sumerian language is well understood by scholars able to read cuneiform script. By studying cuneiform writings, these scholars have discovered linguistic evidence that the Sumerians were apparently not the first people to inhabit Sumer. The Sumerian language contains a number of non-Sumerian loan words. Very likely the Sumerians learned and borrowed these words from the early Neolithic language (or languages) spoken in Southern Mesopotamia when they arrived there. Non-Sumerian loan words or survivals are especially evident in the Sumerian agricultural and beer-brewing vocabularies. The words for wheat, malt, mash and emmer are all adoptions (Hallo 1971: 19). This suggests that the Neolithic settlers who preceded the Sumerian on the southern alluvium had a well-developed brewing tradition and that the Sumerians adopted their practices–or at least the words for them–when they arrived in Sumer.

But whatever the language of these ancient brewers, one can not help but be impressed by the Neolithic achievement in harnessing the microbes that make fermentation possible. The control and practical understanding of the ferments is especially remarkable in that it took place in a pre-scientific context and must have been done by trial and error. In addition to mastering fermentation, Neolithic peoples began the domestication of a whole new range of organisms, the so-called lower plants such as mushrooms, fungi, and algae were also. Such domesticated yeasts, bacteria and lower plants offer three advantages to the cultivator: short production cycles, modest space needs and low water requirements. At the same time, the relative simplicity of their husbandry requirements allows the character of the final product to be precisely controlled (Stanton 1969: 468).

"Unfortunately, the silent domestication process had a more sinister aspect as well. As humans began to live more intimately with animals, this interaction altered the transmission patterns of the pathogens that afflicted them and stimulated the evolution of new ones. Overtime humans came to share the diseases of the creatures they were exploiting. For example, in the Old World, cattle pathogens evolved into measles, tuberculosis, and smallpox while diseases of pigs, and perhaps also of ducks and dogs, became the human scourges of influenza and pertussis. Many others were to follow (Diamond 1997: 207). At the same time, the curtailment of mobility that followed the replacement of hunting and gathering by food production brought people into more sustained contact with human and animal waste, materials that provide ideal environments for microbial growth and transmission.

But this brings us once more to beer! We can be certain that early Neolithic villagers were not overly careful about the disposal of the human and animal wastes that accumulated around their settlements. In the absence of a clear recognition of the relationship between fecal matter and disease, little in the way of systematic effort was made to prevent the contamination of food and water by sewage. Diseases spread by fecal contamination include diarrhoeal infections (which are particularly dangerous for small children), typhoid, and cholera. Intestinal parasites such as worms and protozoa are also readily spread in this manner. To a large extent, these diseases can be avoided by doing what every wise traveler in a Third World country does. That is, by not drinking the water and making do with beer and wine instead! Alcoholic ferments were thus an important cultural adaption to the new conditions of disease and health created by the silent domestication processes active in early settled village life. More, much more, will be said of the role of silent domestication processes of disease domestication in later chapters.

Neolithic Wave of Advance The Neolithic lifeway began to expand out of southwestern Asia after 7,000 B.C. Traditional explanations of this expansion have often seen it as the result of large-scale migrations of peoples or of intentional colonization efforts. A different sort of explanation, the so-called Neolithic "Wave of Advance" model, has been proposed by Albert Ammerman and L. L. Cavalli-Sforza (1979). These authors see nothing intentional in the Neolithic expansion. Instead, they consider it to have been a slow, steady, unintended transformation that occurred as a consequence of random settlement movement caused by the interaction of population growth and environmental carrying capacity.

Their model is based on the assumption that pronounced population increases followed the establishment of food production and on the observation that population growth is logistic in form, that is, populations initially grows at an exponential rate but begins to slow as the human carrying capacity of a region is approached. Populations experiencing exponential growth double every eighteen years. A region can absorb additional people until its human carrying capacity is reached. After that, only a portion of each successive generation can be supported on the land where they were born. Others have to move elsewhere to find vacant land to cultivate. For purposes of calculation, the authors specified the carrying capacity of farming regions during the Neolithic at five people per square kilometer.

In addition, fields under cultivation decline in fertility over time. The authors assume that, under Neolithic conditions, fertility could be restored only by periodically abandoning the fields and allowing them to lie fallow for extended periods. Ammerman and Cavalli-Sforza calculated that the foregoing conditions would create continuous and random settlement movement within a region. The average movement of settlement per generation was specified by the authors as eighteen kilometers.

When this problem was simulated by computer, it was found that the combination of logistic population growth and random settlement movement would create a continuous "wave" of population and settlement that would expand radially outward from the original region at the rate of one kilometer per year. Ammerman and Cavalli-Sforza compared the results of their simulation model with the archaeological record of the earliest appearance of the Neolithic life way in various parts of the Old World. In their opinion, its prediction of a slow but steady rate for the spread of agriculture fit available archaeological evidence of Neolithic expansion from southwestern Asia throughout Europe and eastern Eurasia.

In any case, Neolithic wave of advance reached southern Europe shortly after ca. 5,500 B.C. By this time, Neolithic culture was characterized by a large and complementary suite of domesticated plants and animals, settled villages, pottery making, and stone tool technology based on grinding. The culture complex spread along the shores of the Mediterranean and was established in Greece before 5,000 B.C. Sheep and goats, and to a lesser extent, pigs and cattle, were the primary animal domesticates; emmer and eincorn wheat the chief plant crops (Murray 1970: 110). By 4000 B.C., early farmers of the Linear Pottery culture had carried the Neolithic way of life up the Danube into Hungary and down the Rhine into the

Netherlands and France. Peoples of this tradition arrived in Spain about this time as well. Cattle were the preferred livestock of the Linear Pottery peoples and, in addition to emmer and einkorn wheat, they cultivated bread wheat, hard wheat, club wheat, barley, millet, flax and peas (Murray 1970: 111).

The southwestern Asia Neolithic also spread south to the Nile Valley and this arrival seems to have stimulated the transition to full agrarian life in Egypt after about 5550 B.C. After that time, the complex spread upstream along the Nile to the Ethiopian Plateau. The complex was also carried eastwards to Arabia, the Yemen and on to the Indus River valley and north India by 3,000 B.C. By 1,500 B.C., southwestern Asian domesticated plants were being cultivated in China (Harlan *et al.*, 1976: 4). Recall that Colin Renfrew hypothesizes that the spread of Indo-European languages from Anatolia across three continents came with the adoption of agriculture and the gradual but inevitable population growth that then took place as larger areas of land were settled.

In front of the advancing wave The archaeological record of the Neolithic wave of advance leaves some important question unanswered. Did the expansion of the agricultural peoples into these new areas simply sweep away and genetically swamp the foraging peoples there before them? Or, did their arrival stimulate local hunter-gatherers to experiment with domestication of their own wild plant and animal species? Or, more interesting still, had the indigenous peoples already independently domesticated local species **prior** to the arrival of the southwestern Asia cultigens?

In Europe, these is is clear archaeological evidence that the dog was domesticated by European Mesolithic peoples but only equivocal evidence of the independent domestication of certain other species there prior to the arrival of Neolithic peoples has been recovered there. For example, the bones of small cattle, pigs, and sheep have been reported from a number of early Mesolithic sites in northwestern Europe and the diminutive size of these specimens has sometimes been taken as evidence that they were domesticated. Murray (1971: 110), however, concludes that, at least in the case of the sheep, the remains are those of a small, **wild** animal varieties that had adapted to the environmental changes that occurred between the Final Pleistocene and Holocene periods.

Evidence that peoples belonging to terminal Mesolithic cultural traditions like the Ertebolle and Ellerbek in Europe may have engaged in limited agricultural activity has also been reported. However, these peoples were already beginning to have contact with Neolithic peoples moving northward and this contact may have stimulated the foraging peoples to experiment with the domestication of local populations of aurochsen and boars (Murray 1970: 110).

We will probably never be certain of the nature of the relations established between Neolithic emigrants and the indigenous foraging peoples who found themselves in front of their advancing wave. However, as the dismal history of European colonial expansion in the New World and Australia clearly indicates, long term contact with farming peoples is usually fatal to the foraging way of life. It appears that the food producing and food collecting subsistence modes are simply incompatible with one another. The practical contrasts between farming and foraging social formations may be summarized in a few dichotomies:

1. The infrastructures of foraging and farming systems contrast in the amount of land they require. In the same environment, agriculturalists use less land per capita, but use it more intensively; foragers use more land per capita, but use it less intensively.
2. The demographic structure of foragers and farmers contrast in size and rate of growth. Foragers do not control the production of foodstuffs upon which they depend, farmers do. Therefore, while hunter-and-gatherers must adapt to their environments by limiting their population size to fit their production, subsistence farmers adapt by expanding their production to fit their population.
3. Foragers and farmers have contrasting impacts on their environments. In order to buffer themselves against long and short term variations in the availability of any given species, broad-spectrum hunter-gathers tend to exploit a wide range of wild plants and animals within their environments. In order to buffer themselves against variations in production of domestic plants and animals, agriculturalists tend to increase production and rely on storage. Presented with the same problem, foragers tend to preserve and utilize the natural diversity of their environment, while agriculturalists tend reduce the diversity of the natural habitats they colonize and occupy.

Foraging systems are designed to take advantage of environmental diversity and foragers adapt to nature by mimicking it. Agricultural systems are designed to simplify nature and farmers adapt to nature by over riding its checks. By clearing land, draining swamps, eliminating predators, removing wild plants and animals that compete with domesticated species, and hunting and gathering in the vicinity of permanent settlements, agriculturalists control and transform the natural habitats they colonize and occupy. Such simplification, particularly when it takes place in tandem with rapidly expanding human populations, creates environments that are inhospitable to broad-spectrum hunting-and-gathering.

Neolithic Production Intensifies When agricultural economists use the term "intensification" they are referring to increases in agricultural productivity that results from increased work hours and effort and a greater capital investment (Pryor and Mauer 1982: 34). Capital investment can take many forms including use of fertilizers (in pre-industrial economies this means animal dung, human wastes, organic composts and green manures), the greater use of machinery and draft animals, the draining, irrigating, terracing or fencing of land, and the more frequent cropping of a given area. In addition, intensification often involves such arduous practices as growing seedling plants in forcing beds and then systematically weeding, pruning and tending them once they have been transplanted in the fields.

In practice, intensification means moving away from extensive or "shifting" cultivation towards constant or "permanent" use of the plot season after season. In shifting cultivation systems, plots are cropped for a few years until their fertility declines and then abandoned and left fallow until fertility is restored. In permanent fields, soil nutrients must be restored by adding fertilizers or by growing complimentary crops by rotation on the plots or both. Intensive cultivation systems typically produce far greater quantities of food than extensive ones, but only at the cost of far more labor per unit of output. All things remaining equal, most subsistence farmers living in a pre-market economy chose to work only enough to meet their subsistence requirements. What would induce such people to chose the hard toil of intensive agriculture?

According to the agricultural economist, Ester Boserup (1965), they do not chose to do so. In her view, subsistence farmers move to intensify their cultivation only when they are forced to do so by the pressure of an expanding population on their food supply. In extensive agriculture, lands are planted in crops for shorter periods than they are left fallow (Conklin 1954). If the system is to remain viable, there must be always be empty land for farmers to cultivate once they have abandoned the fields they are currently farming. As a growing farming population begins to fill its empty land, the length of time any given plot may be left fallow in the rotation cycle begins to be shortened. When all the empty land has been filled and no land is available for rotation, the system breaks down. Henceforth, fields must be continuously cropped and their fertility restored by means other than fallowing. At the same time, agricultural production per unit of land must be increased to feed the enlarged population. These two ends can only be met by substituting greater labor, skill and capital for land in the production equation. That is, by intensifying agriculture.

The archaeological record clearly indicates that the Neolithic period witnessed a remarkable expansion human population. During the Paleolithic period populations are estimated to have grown at an average rate of only 0.01 percent per year. This rate apparently dropped even lower to about 0.003 percent per year during the Mesolithic period. However, population growth rates apparently accelerated during the Neolithic period to a staggering 0.1 percent per year (Hassan 1981; Carneiro and Hulse 1966). Correspondingly, the archaeological record indicates that after ca. 6,000 B.C. the hitherto extensive Neolithic agricultural adaptation began to be intensified in a number of ways. Key elements in this intensification were the development of:

1. Irrigation agriculture,
2. Specialized nomadic pastoralism,
3. An increased emphasis on secondary products of animals such as milk, wool, traction, park transport and riding rather than simply their meat, and,
4. Plow agriculture

Rise of Irrigation Agriculture Prior to the advent of irrigation, cereal cultivation was dependent upon naturally available water. Consequently, farming was found in two settings:

1. The banks of flowing streams and springs, alluvial fans, swamp margins and other locations where water was perennially available or the water table was sufficiently high to ameliorate fluctuations in rainfall, and
2. Those regions where annual rainfall was sufficiently high and properly timed to permit "dry" or rainfall farming. In modern southwestern Asia, these latter conditions are generally not found outside regions receiving less than 12 inch of annual rainfall (Oates and Oates 1976: 111).

Settlement was effectively closed to early Neolithic agriculture in lands where these conditions were not met. The development of irrigation, which seems to have begun early in the 6th millennium B.C., changed this situation dramatically. This effects of this development are particularly evident in the watershed and flood plain of the Tigris and Euphrates rivers. The watersheds of these two rivers are located in the Taurus and Zagros mountains in modern Turkey and Iran. Upon leaving these mountains, the rivers enter a vast alluvial plain characterized by low relief and semi-arid to arid climate. This region was called Mesopotamia (literally "the land between the rivers") by the Greeks and today is largely occupied by the modern nation of Iraq.

Once on the plains, the two rivers flow generally southeast following parallel courses south of the Taurus and Zagros mountains and north of the Arabian desert. The city of Baghdad, near the center of the plain, marks the dividing line between the northern and southern section of Mesopotamia. Rainfall or dry farming can be practiced in the Taurus and Zagros foothills and there is sufficient moisture for farming in the marshy delta region on the southeast end of Mesopotamia where the two rivers entered the Persian Gulf.

Conversely, low rainfall over most of Mesopotamia proper makes dry farming on the alluvium impossible. Sustained cultivation there requires irrigation in one form or another. Between the foothills, where rainfall agriculture can be practiced, and the flood plain, where it cannot, lies an extensive marginal belt where dry farming is possible in wet years and impossible in dry ones (Oates and Oates 1976: 111). As long as sufficient well-watered land remained available, the precariousness of the rainfall in these transitional zones made them unattractive to early Neolithic dry farmers. When Neolithic populations growth expanded the need to penetrate this marginal belt, the most successful settlers would have been those able to supplement dry farming with water management.

Shortly after 6000 B.C., the **Hassuna** cultural tradition emerged at the edge of the Zagros foothills in a transitional region known as the Assyrian steppes. The Assyrian steppes are above and northwest of the flood plain. The type site, **Tell Hassuna**, is located near the modern city of Mosul or Al Mawsil in Iraq. Hassuna sites are all located within regions receiving 8 inch of rainfall annually in modern times. Although average annual rainfall would be minimally adequate for cereal dry farming, variations from year to year in this region would make it a risky enterprise at best. Of course, annual rainfall there during the sixth millennium may have been slightly higher than today. In any case, no direct archaeological evidence that the Hassuna peoples used irrigation has come to light. However, colonization of the Assyrian steppes appears unlikely without at least some rudimentary form of water management to guarantee a crop in years of poor rainfall. If they did exist, no doubt these works were highly localized and used merely an adjunct to farming practices dependent mainly on rainfall.

With the emergence of the **Samarran tradition** by 5500 B.C., Neolithic settlement spread even farther downstream on the Tigris. Samarran sites are located between Mosul and Baghdad on the fringes of the northern Mesopotamian alluvial plain in areas that were on or beyond the 8-inch rainfall isohyet in settings clearly too arid for dry farming. Archaeological evidence of water management has been discovered at the Samarran sites of **Choga Mami** and **Tell es-Sawwan**. At Choga Mami, the Samarrans practiced a rudimentary form of irrigation called "water-spreading" This method involved constructing shallow channels across braided streams on a great alluvial fan at the base of the Zagros foothills (Sherratt 1984: 98; Oates and Oates 1976). Interestingly, like Jericho, both Choga Mami and Tell es-Sawwan were walled.

Similar technical developments were taking place at about the same time in Khuzistan on the northeast edge of the flood plain in the Susiana region (Adams 1962: 112-113) and on the **Deh Luran plain** (Hole, Flannery, and Neeley 1969: 355-356). Prior to about 5000 B.C., villages in these regions clustered in the northern part of the upper plain where rainfall is highest. After about 5000 B.C. the number of villages increased and they were clustered much farther south around fossil water courses in areas of markedly decreased rainfall. Thus, the archaeological record of the 6th millennium B.C. contains evidence of a general movement downward from the well-watered foothills into regions where lower rainfall make reliance on dry farming hazardous or even impossible.

Settlement on the rich soils of the semi-arid alluvial plain of southern Mesopotamia finally began–primarily along the Euphrates river–sometime shortly before 5000 B.C. Almost 2,000 years after their establishment in the nearby Levant and Zagros and Taurus foothills, agricultural peoples had finally colonized the region. This earliest flood plain tradition is called the **'Ubaidian** after the type site of **Al 'Ubaid** on the Euphrates in Iraq. The appearance of U'baidian settlements in southern Mesopotamia indicates that the technical and organizational problems presented to irrigation agriculturalists by the seasonal flood cycle of the Euphrates River had been partly overcome at this early date.

We know very little of the nature of irrigation practiced in these early millennia. However, on the basis of extensive site survey on the lower Euphrates Valley, Adams (1974:2-4) concludes that at the time of initial settlement, the river was characterized by a braided pattern of small and intertwined stream branches rather than by a continuous channel. According to him, such small, braided channels could be controlled by small-scale, localized irrigation systems. Agriculture in these early millennia could proceed without the use of large, centralized canal systems (see also Adams 1981; Buringh 1957:39-40).

Adams and Nissen's later work revealed that both the human settlement pattern and irrigation system that emerged in early times on the lower Euphrates drainage closely paralleled the natural drainage system of the river. They note that generally the: "inhabitants of the region lived in discontinuous enclaves, indicating that population were low in relation to the potentially arable area. Water rather than land was the critical determinant, and so the network of permanent settlements closely adjoined a largely natural system of meandering, bifurcating, and rejoining distributaries. To be sure, these natural watercourses were periodically diked, straightened, and artificially maintained, and they certainly supplied numerous artificial canal off take leading into nearby fields (Adams and Nissen 1972:86).

The clustering of early settlements near the lower Euphrates also suggests that water was not drawn inland over great distances by means of large canals. This conclusion is strengthened by Wright's (1981:324) observation that along the southern margins of Sumer, surface finds of 'Ubaidian clay sickles tend to be restricted to areas of about 5 kilometers or less around Late and Terminal 'Ubaid period sites. Since these clay sickles certainly mark the former locations of the 'Ubaid period fields Wright views their limited distribution around the sites as an indication that branch canals from the Euphrates at that time were but three to five kilometers long. Thus it would appear that there was little effort made to override the river and transform it into a man-made system at this time. Instead, the early "natural flow" irrigation practiced on the southern alluvium probably mimicked the region's natural drainage patterns. During the 'Ubaid period the raw number of small, isolated settlement on the flood plain appears to have increased steadily. The archaeological record of settlement change indicates that both population and agricultural productivity in southern Mesopotamia expanded dramatically between the sixth and fourth millennia (Oates 1972:303).

In sum, during the 6th millennium B.C., population increase and general movement down from the foothills towards the alluvial plain took place in the Tigris and Euphrates watershed and flood plain. This movement appears to reflect the development of techniques of irrigation. The earliest use of irrigation apparently took place in zones transitional between foothills and flood plain where the precariousness of annual rainfall encouraged the development of water management as a supplement to dry farming. Near the end of the 6th Millennium, the refinement in irrigation technology was sufficient to allow settlement on the Tigris and Euphrates flood plain itself where low rainfall made reliance on dry farming impossible. This trajectory of agricultural intensification and population expansion led eventually to the rise of the state in Mesopotamia–a process considered in the next chapter.

Specialized, Fully-Nomadic Pastoralism Peoples who raise grazing animals like sheep, goats, cattle, camels, and horses must supply them with two key elements: water and pasturage. In some regions, the distribution and availability of these key comodities changes dramatically with the seasons. Where this is the case, herding peoples have often responded by developing patterns of movement that may be classed as either pastoral nomadism or transhumance.

Herding peoples who practice pastoral transhumance maintain year-round villages. A segment of the population–often the majority–remains in the village to tend the fields and gardens while another segment accompanies the herds to seasonal pastures at higher elevations or in ecological zones located elsewhere. In contrast, fully nomadic pastoralists are people of shifting settlement who subsist primarily by herding domesticated animals. Seeking pasturage and water for their herds, such peoples move with their animals and all their worldly goods from one place to another with the changing seasons. Although milk, yogurt, meat, cheese, blood and other products of the herds figure largely in their diets, most pastoral peoples consume grain, vegetables and other agricultural foodstuffs as well.

It may be possible to live entirely by some combination of herding and hunting and gathering. Hole (1989: 97) suggests that, in the early stages of animal domestication in southwestern Asia, this in fact was the nature of subsistence. However, the fully nomadic pastoralist peoples encountered in history and ethnology have all been dependent upon neighboring farming peoples for agricultural supplies. Thus, while transhumant pastoralism is a mixed system of plant and animal raising, fully nomadic pastoralism is one-half of a division of labor between settled farmers and nomadic herders. In such a symbiotic arrangement, farmers and herders each practice a specialized form of food production and depend upon trade with the other for products they do not raise themselves.

Lees and Bates (1974) hypothesize that during the Neolithic period in southwestern Asia, the growth of plant cropping systems based on irrigation fostered the development of fully-nomadic pastoralism as a complimentary specialization. The argue that irrigation agriculture and nomadic pastoralism are parallel intensifications in food production that grew of generalized rainfall farming and animal husbandry in response to the increasing pressure of populations on agricultural land. Early irrigation farmers probably both raised cereal grains and practice animal husbandry on the lower Mesopotamian plains. Very likely they practiced some form of seasonal transhumance. However, Lees and Bates hypothesize that the expansion of irrigation systems down stream on the Euphrates would have had two immediate effects on their agricultural practice:

1. Greater and more contant labor would have been demanded of farmers in the form of flood control and canal construction and maintenance, and
2. The new land continuously being brought under cultivation was located farther and farther from the best summer grazing lands located in the upland zones.

Lees and Bates (1974: 190) conclude that the increasingly conflicting labor demands of field and herd would have made it harder and harder for the same individuals both to herd and to practice irrgation farming efficiently. At some point, local economic groups would have begun to abandon transhumance and become specialized as either farmers or herders. Nomadic pastoralism may also have been developed as a hedge against the eractic fluctuations in the environment of the Mesopotamian flood plain. While the sedentary farmer must endure floods or plagues of locust, faced with such circumstances, the nomadic herdsmen simply moves on. Alternatively, the nomad starves in years of low rainfall and poor pasturage but the irrigation ditches still carry water to the crops.

Thus, nomadic pastoralism and irrigation agricultural have contrasting strengths and weaknesses. Where social and political boundaries between the practioners of each were permeable and interpenetrating, the two kind of subsistence may simply have been separate sectors of a single economic system. Gilbert (1983: 110) suggests that in a unitary system of this kind, "the response to subsistence failure in one economic sector might be greater reliance upon another, that could remain viable under the same circumstances. Thus, pastoralism and agriculture could have been inseparable complements in certain areas, perhaps from quite early times." In such a partnership, he suggests "herding as a productive specialization might have expanded and contracted with the surrounding social climate" (Gilbert 1983: 115).

However, we must be careful to avoid overstating the case for the compatibility of "the desert and sown." The mobility and warlike nature of nomadic pastoralists–especially following the domestication of the horse and camel–meant that their trade relations with settled villagers easily and often gave way to raiding as the balance of force between nomad and farmer shifted back and forth throughout history. As Lees and Bates (1974: 191) note, "the very existence...of politically discrete and potentially predatory nomadic pastoral populations raises the 'overhead' costs of agriculture in contioguous regions, and thereby limits production" (Lees and Bates 1974: 191). One reason for the remarkable precocity and early success of the Old Kingdom of Egypt may have been that, unlike the plains of southern Mesopotamia, the harsh North African deserts surrounding the Nile valley effectively insulated farmers there from attacking nomads.

Utter Uses for Cows A second phase of agricultural intensification appears to have been initiated in Southwestern Asia and Europe after 4000 B.C. Before that time, Neolithic herdsmen raised domestic livestock primarily for their meat. After ca. 4000 B.C, agricultural peoples began to systematically exploit the "secondary products" of these species including: milk, wool, traction, pack transport, and riding. Andrew Sherratt (1983, 1984) has called this profound change in Neolithic economic life the **Secondary Products Revolution**.

Milk Although it often comes as a surprise to white Americans, most of the world's people do not drink cow or goat milk after infancy. In fact, many of these people find that consuming even small quantities of milk makes them heartily sick. In large measure this is because, as adults, they lack the emzyme lactase used by the body to break down milk sugar or lactose. The absence of this emzyme and the consequent inability to digest milk is very high among Asia peoples, American Indians, Melanesians, native Australians, and most of the native peoples of Africa with the exception of pastoralists like the Maasai, Fulani and Tusi. Conversely, the presence of lactase is very high among the pastoral Bedouin peoples of southwestern Asia (but not the non-pastoral peoples of the region) and among Europeans and Euro-Americans.

Since milk is such a ready source of vitamins and proteins, the ability to digest lactose is of great selective advantage. Yet we can not be certain when and among what people this mutation first became widespread. It need not have occurred as soon as livestock were domesticated. Lactose is broken down when milk is cultured into yogurt, cheese and other milk products, milk could have been used to produce these products long before it was directly consumed.

Sherratt suggests that selective pressure favoring milk-drinking became increasingly important with the northwards spread of agriculture in Europe. These pressures may have been from two sources. First, lactose appears to behave biochemically like vitamin D, a vitamin essential for the absorption of of calcium by the body. Due to their lower exposure to sunlight, the bodies of peoples in northern latitudes make less vitamin D than peoples further south. Overtime, natural selction may have favored lactose tolerant individuals over intolerant ones in such climates because of their greater calcium absorption.

Second, Sherratt (1983: 95) argues that the density of the primary forest in Europe during the early Neolithic probably limited grazing and made it difficult to raise large herds of livestock for meat. In his view, the expansion of cattle raising in Neolithic Europe did not begin until forest clearance was well under way and intensive, plow agriculture was being practiced. Before that time, Neolithic peoples able to milk their small herds of cattle would have extracted more benefit from them than those who merely kept them for meat. The greater efficiency of dairying under these circumstances would have been a further advantage that fell to lactose tolerant people.

Wool Wild sheep have stiff and hairy outer coats that cover woolly undercoats. While the outer coats cover the animal year round, the undercoat grows only in the winter. Wild sheep–as well as certain primitive domesticated breeds–naturally shed this soft undercoat each spring. In contrast, domesticated sheep bred for their wool have lost the hairy outer coat. Their fleece consists entirely of "undercoat" which they cannot shed. The fleece of most domesticated sheep must be sheared from their bodies each spring by the shepherd.

Since the outer coats of wild sheep and goats are not particularly desirable, the animals were probably domesticated as a source of meat, fat, skin and horn rather than of textile fibers. However, when sheep shed naturally in the spring, the undercoat falls away in dense mats. These mats can be gathered or plucked from the animal and made into felt or twisted into crude yarns and woven into textiles. The breeding of woolly sheep presumably resulted simply from the preserving individual animals who retained this undercoat on their bodies long enough for the shepherd to collect it. Animals with longer, softer fibers in their undercoats would also have been favored.

Wool weaving probably developed out of basketry and the weaving of plant fibers since there techniques are easily transferable and creative experimentation is very much a part of the weaving process. Once weaving with wool yarn began, it became closely connected with sheep breeding. According to Ryder (198-:?), "the evidence shows that the development of the fleece cannot be understood without reference to the rise of textile crafts, technologies that developed in step with biological changes in the coat. In some cases innovations in technology led to selective breeding for particular coat characteristics. In other cases the influence flowed the opposite way, as particular features of the sheep's coast inspired new ways to make use of it." In any case, weaving textiles out of wool rather than fashioning a garment out of sheep skin conserves the flock rather than diminishes it. The systematic use of such a secondary product no doubt resulted from the same impulse that led herdsmen to obtain nutrition from their animals' milk rather than their meat.

Two interesting bench marks suggest that Neolithic peoples began selective sheep breeding for wool between roughly 6000 and 5000 B.C. Neolithic peoples introduced domestic sheep to the islands of Corsica and Sardinia shortly after 6000 B.C. Some of these sheep escaped and founded a race of "wild" or feral sheep known as Mouflon. Although the coat of the Mouflon is not identical to that of wild sheep, it exhibts little evidence of the fleece of later domestic breeds (Ryder 198-: 112). However, selection for fleece apparently had already begun by this time in greater Mesopotamia. This is attested to by the recovery of a statue of a woolly sheep dating to early in the 6th Millennium from the site of **Tepe Sarab** in Iran (Clutton-Brock 1981: 57).

Animal traction and plow agriculture The simplest and most basic work perfomred by livestock among cereal farming peoples is the treading on the newly harvested grain to separate it from the chaff. Little equipment is needed to use animals to winnow grain in this way. A horse, donkey or ox need only be driven in circles around and around on the threshing floor.

To make use of the strength of an animal, not just its weight, it is necessary to harness it in some manner. The horn yoke, in which ropes are attached to a pole lashed to the horns of one or more animals, was probably the earliest form of harness for cattle. Donkeys can be harnessed with a simple yoke fitted over their necks just in front of the shoulder blades.

The most significant use of this traction in prehistory followed from the harnessing of draft animals to the plow. The plow is a devise for cutting and breaking the soil that is pulled by a draft animal. It is used to control weeds, prepare fields for planting or to cover seeds broadcast on the field surface. Plowing is also an effective means of aerating the earth and bringing subsoil nutrients and moisture to the surface. There are basically two types of plows: the scratch plow or **ard** and the moldboard plow. The scratch plow is the simplest form and was apparently the first to be invented. Both scratch and moldboard plows slice the soil, but the latter type is designed to turn the soil over as well (Pryor 1985: 728).

The earliest plows were made entirely of wood, although the pointed plowshare was probably hardened in the fire. With the development of metallurgy, plows were provided first with bronze, later with iron shares. Plows made largely of wood are unlikely to survive once they have been discarded and they rarely are recovered from the sorts of special microenvironment–dry caves, tombs, bogs and so forth–where organic decay is held in abeyance. As a consequence, three types of evidence have been used in reconstructing the prehistory of the plow: plow marks preserved in paleosols buried under earth mounds, models or figurines, and artistic representations. Using these data, Sherratt (1983: 91-92, 98) concludes that the development of the plow and the use of livestock for traction probably occurred in greater Mesopotamia sometime between ca. 6000 and 4000 B.C.

Sherratt (1984: 98) considers the Hassuna peoples on the Assyrian steppes as likely inventors as the preconditions for the development of traction cultivation using scratch plows and oxen were present among them at an early date. Soil preparation would have been necessary for cereal-growing on the steppes and cattle were at that time becoming widespread on lowland sites while the Hassuna tradition flourished. Sherratt is also persuaded that the instrument was probably a basic element in the intensive irrigation cultivation systems that later made farming possible on the southern alluvium. Initial agricultural settlement on the alluvium began before 5000 B.C. and, if Sherratt is correct, the plow is present there from the beginning. Unfortunately, the first definite evidence–artistic representations of yoked oxen pulling plows–do not appear on the southern Mesopotamian alluvium until after 4000 B.C. At the other end of the Eurasian land mass, plow marks dating to as early as ca. 3700 B.C. have been reported from Danmark. The presence of the plow in far northwest Europe at this date suggests its arrival in the more southerly parts of Europe before the beginning of the 4th millennium.

Prior to the invention of the plow, farmers depended upon their own muscles augmented by simple tools like hoes, digging sticks, mattocks, or shovels. As a consequence, farming throughout most of the Neolithic period was limited to shallow cultivation in those setting where vegetation could be cleared with fire and soils broken by minimally amplified human muscle power. Typically, cultivation systems centered on the hoe or digging stick are extensive in nature and characterized by low levels of food output production. By harnessing animal power to the plow, late Neolithic farmers were able to expand both their net agricultural output and the geographic distribution of their subsistence system by (1) farming more acreage per household than formerly possible, and (2) ringing soils too intractable for hoe agriculture under cultivation.

Wheeled Vehicles The earliest archaeological evidence of wheeled vehicles in prehistory date to shortly before 3000 B.C. in Eurasia. Although the credit for their invention is often assigned to the Sumerian peoples on the southern Mesopotamian flood plain, recently archaeological work in the Soviet Union suggests that the region north of the Zagros and Taurus mountains between the Black and the Caspian may in fact deserve primacy (Piggot 1968). It is worth noting that the potter's wheel comes into use at roughly this time in southwestern Asia. Since both of these devices make use of rotary motion around a shaft, their development may have been related.

A number of key technical difficulty had to be overcome in order to build efficient wheeled vehicles: a means of lubrication had to be developed in order to reduce the friction between the wheel and the axle, the wheel had to be equipped with a bearing to allow it to spin freely, and a proper means of harnessing animals intended to pull such vehicles had to be devised. That these and other technical problems were solved to a tolerable degree by the beginning of the 3rd Millennium B.C. is reflected in the relatively common appearance after that time of the four-wheeled wooden wagons in the archaeological record. These early wagons, pulled first by oxen and later also by donkeys, were heavy, difficult to move over rough terrain, and no doubt awkward to steer in the best of circumstances. Nonetheless, with their development provided farmers with an ideal device "to convey bulky foodstuffs from the fields to the settlements, and to carry farmyard manure in the opposite direction" (Childe 1954: 209).

Of course, humankind could never–can never–leave it at that. It appears instead that we must take whatever comes to hand and fashion a weapon out of it–be it a stone, a branch or a computer chip. The millennia that followed the invention of the stolid farmer's cart witnessed the appearance of the spoke-wheeled, horse-drawn war chariot, a military device that proved to be the fighter plane, the tank and the H-bomb of its age.

Domestication of Horses and Camels Sometime after 3000 B.C., the horse was added to the repertoire of domesticates in the Old World. The earliest domestication appears to have been accomplished by peoples of the Tripolye culture located on the plains of southern Russia along the coast and north of the Black Sea between the Ukraine and Turkestan. Domesticated horse appear to have been used first as a source of meat and milk; their use as beasts of burden, draft animals and mounts comes much later.

Cattle were probably the earliest domestic animals used for pack transport and traction. No doubt the performance of both tasks was markedly enhanced by the domestication of the horse, donkey and later the camel. Horses were being used to pull wheeled carts in Eurasia by 2600 B.C. In order to use the horse in this way, however, it is necessary to develop a new kind of harness. Neck yokes cannot be used on horses since these animals hold their necks erect and would be choked by such a device. Instead, horses harnesses must be positioned to fit across the horse's chest. Once developed, such harnesses placed the formidable traction power of the horse at the disposal of humankind. Direct evidence that horses they were mounted and ridden is not present in the archaeological record until ca. 2000 B.C. (Renfrew 1987: 138). Riding it thus appears, was the last secondary use of livestock to be developed.

From 2000 B.C. onwards, "classic" pastoral nomadism–mounted riders driving cattle, sheep, goats, and camels and moving their belonging by pack animals–emerged in southwestern Asia. After this time, the domesticated horse spread rapidly over Europe, eastern Eurasia and North Africa. The most profound long term effect of the domestication of the horse and camel were military. Mounted warriors could move quickly to strike or flee and a mounted bowman was a particularly dangerous opponent. Yet, the effectiveness of light cavalry against well drilled or well fortified infantry eventually proved to be limited. Therefore, the most decisive military use of the horse in ancient times came from its traction power.

The perfection of the two-wheeled, horse-drawn war chariot around 1800 B.C. brought a radical change in the nature of warfare in Southwestern Asia. Mobility, firepower, carnage and terror on the battlefield were raised to new levels by teams of galloping horses drawing fast, light, sturdy chariots containing both archer and driver. The population best able to take advantage of the possibilities of chariot warfare were steppe dwellers, whose way of life assured an easy access to horses. Accordingly, waves of barbarian conquerors equipped with chariots overran all the civilized lands of the Middle East between 1800 and 1500 B.C. (McNeill 1982: 9-10).

On the other hand, the military effectiveness of cavalry–mounted warriors as opposed to charioteers–was not markedly enhanced until much later with the invention of the foot stirrup in China in the fifth century A.D. This device made it possible for a cavalryman to use the shock of his horse's charge to drive his lance against an enemy without himself being thrown from the saddle by the force of the blow. The stirrup diffused rapidly and reached Europe shortly after A.D. 700. In the subsequent 8th century, widespread use of this simple device effected a profound revolution in European military technology and organization. Infantry was eclipsed and, for the next five hundred years, the heavily armored, mounted knight and his levée dominated the battlefields of medieval Europe. The Age of Chivalry had begun..

Origins of the African Agricultural Complex

Little is known of the origins of the agricultural complexes of Sub-Saharan Africa. When European first began exploring the Africa, they found people who simultaneously raised the familiar southwestern Asian domesticated animal species such as sheep, goats, cattle, camels and horses along with a variety of domesticated plants of native African origins hitherto unknown in Europe. Did this unusual amalgamation mean that Sub-Saharan Africa was a "primary" or independent center of species domestication that only later received domesticated species of southwestern Asian origin? Or, alternatively, was it a "secondary" center where the manipulation and eventual domestication of native plants was stimulated by the arrival of agricultural peoples from the north or the east across the Sahara? Unfortunately, archaeological knowledge is insufficient at present to resolve the question and the nature of the steps in the rise of agriculture in Africa remain an enigma.

Physiography of Africa The second largest continent on Earth, Africa has been joined to Eurasia throughout the Cenozoic. The long term intimacy of this contact lead the geographer, Sir John Mackinder to refer collectively to Africa and Eurasia as the "world island." The geography of the African continent is dominated by four major physiographic features:

1. **The African craton**, the massive, single tectonic plate of great age that forms the continent of Africa, has been little deformed for a long time. Viewed in its entirety, the craton constitutes a huge elevated plateau with escarpments that rise abruptly from the coastal edges. The escarpment appears least hospitable on the West Africa side of the continent. There one encounters

> an unindented coastline with few natural harbors, where sandbars conceal the entrances to rivers and high surf pounds the beaches. A long strip of the Guinea and Sierra Leone coast is still marked 'all approaches dangerous' on Admiralty maps. There were no inland seas or navigable rivers flowing across the breadths of the continent. (de Gramont 1976: 35).

The most common landforms in the interior of the craton are flat, rolling elevated plains and plateaus that geomorphologists call **erosion** or **planation surfaces.** The highest plateaus are found in the Ethiopian massifs, and in eastern and southern Africa. Lower planation surfaces characterize West Africa. The stair step topography of these elevated plains and plateaus create countless waterfalls and rapids in the rivers that drain them. Needless-to-say, such topography impedes movement across Africa by rendering long stretches of its rivers unnavigable and making overland travel there difficult (Hance 1975: 44-45). The Sahara Desert is the most spectacular of these elevated planar surfaces. The desert stretches without interruption across North Africa from the Atlantic coast to the India Ocean. In general, the elevation of the African craton rises as one moves from the north and the west to the south and the east.

2. **The Congo Basin** is located in the central interior of the African craton. It is a vast, shallow depression encircled by ranges of low hills. The uplands that ring the eastern edge of this basin form the watershed of the headwaters for the Congo River. The Congo, the fifth longest river in the world, stretches from the highland savannah country of the central African interior to the Atlantic coast. Other great rivers include the Niger, the Senegal and the Gambia in central and western Africa, the Nile that drains the East African highlands into the Mediterranean and the Tana that follows from those same highlands into the Indian Ocean on Africa's east coast. Of all these rivers, the Nile has figured most significantly in human history. No less than the Nile, the valleys of the tropical Africa have rich alluvial soils and abundant water for their irrigation. However, the natural impediments to human exploitation of these river valleys are also manifold. While these great rivers are the habitat of such fearsome creatures as crocodiles and hippopotamuses, even more dangerous to humans are the various parasites and insect born diseases that thrive in the waterborne environments. At the same time, the
3. **The Rift Valley** is the dominant landform of East Africa. It is a complex of steep sided, structural trenches or *grabens*, many of which contain deep lakes lacking outlets to the sea. The complex extends from the Dead Sea in Palestine to Mozambique in East Africa. The crustal movement that created this complex has also engendered attendant volcanic and tectonic activity throughout its existence.
4. **Mountain ranges of chiefly of volcanic, rather than structural or tectonic, origin** Volcanic activity accompanying the formation of the Rift Valley system in the late Cenozoic era released large outpourings of fluid basalt. This volcanic activity created much of the highlands of East Africa. During the Plio-Pleistocene, volcanic eruptions there created the highest mountain peaks in Africa, Mounts Kilimanjaro and Kenya. By contrast, mountain ranges created through folding, faulting or uplift are few in number due to the stability of the cration The Ruwenzoris mountains–the fabled "Mountains of the Moon"–in Uganda and Zaire East Africa constitute one such structural range, the Atlas mountains of extreme northwestern Africa another.

Holocene Climates of Africa The equator cuts across the middle of Africa. As a consequence, the Earth's second largest continent, is also its warmest. Mean annual temperatures are high over most of the continent and about 90 percent of the land area may be classified as tropical, a higher percentage than any other continent (Hance 1975: 46). Nonetheless, the modern African climate is far from uniform and, on a continental scale, is characterized by four major climatic zones:

1. the Equatorial,
2. the Sudanic,
3. the Desertic and
4. the Mediterranean.

These four zones extend across the continent on both sides of the equator as mirror opposite of one another. Equatorial Africa is characterized by a broad zone of tropical climate and, across its western and central regions, by extensive equatorial forest. Temperatures are high and rainfall, averaging in excess of 78 inches annually, is evenly distributed throughout the year. The Sudanic climate zone is located north and south of the equatorial zone the Equatorial zone. The northern sudanic zone is known as the **Sahel**. Here modern annual rainfall levels decline along a sharp gradient and natural vegetation and agricultural practice change in response. Rainfall in the Sudanic zones is concentrated in the summer and average rainfall, which grades from a high of 78 inches to less than 10 inches annually, varies erratically from year to year and from place to place. Natural vegetation in the Sudanic zone typically ranges from humid savannah woodlands to dry, open grassland reflecting the semi-arid conditions and extended winter dry season. Extensive deserts–the Sahara and the Kalahari, respectively–are located both north and south of the Sudanic zones. Modern rainfall on the Sahara has averaged less than ten inches annually, temperatures are high and vegetation sparse. Mediterranean climate zones are located south of the Kalahari desert on the southernmost tip of Africa and north of the Sahara along the Mediterranean coast of the continent. Mediterranean climates are characterized by rainfall that ranges from 14 to 35 inches annually that comes almost entirely in winter. Summers are hot, dry, and sunny (Goudie 1984: 87-88).

As noted in Chapter 10, the climate of late Pleistocene Africa appears to have been cooler and drier than the present time. Consequently, during the late Pleistocene the Sudanic and Desertic zones expanded markedly and the great African forests retreated into limited refugia to be replaced by open savannah forests and savannah grasslands. This arid phase ended near the close of the Pleistocene about 9000 B.C.. The early Holocene was characterized by a wet phase that affected much of Africa. Between ca.. 9000 B.C. and 6500 B.C., intensified monsoon circulation patterns appear to have brought more moisture to inland Africa and eastern Eurasia. During this time, the Sahara and a number of other desert regions on those landmasses were relatively well-watered savannah grasslands dotted with lakes and watercourse. Wet conditions diminished briefly between ca. 6500 and 5100 B.C. but returned again from ca. 5100 to 2200 B.C. After 2200 B.C., however, worldwide temperatures dropped permanently and created the climatic configurations of the modern world. Apparently the intensity of the monsoons diminished and these conditions resulted in increasing aridity in many parts of Africa and eastern Eurasia. The modern climatic configuration in Africa developed over the last 2,500 years in the late Holocene of the.

Environmental Impediments to Agriculture in Africa The climatic zonation of Africa in combination with its landform and soils geography had a profound impact on the nature of the agricultural complexes that developed on the continent and the rate of its diffusion. Consequently, African agricultural complexes contrast with those of southwestern Asia and much of the rest of the Old World in a number of important ways.

First, African soils are, as a rule less suitable for agriculture than those of the other major areas of the world.

> Usually the best agricultural soils in the tropics are 'new' soils, those of the large river deltas, in areas of recent volcanic activity, and on not too steep mountain slopes that are continually rejuvenated through the action of mild erosion. In Africa, these soils are concentrated in a very few narrow river basins and in some of the volcanic highlands in East Africa, Cameroon, Rwanda, and Burundi. Much of Africa is very old geologically, and the parent material is acidic or granitic. The topography is mostly flat or undulating, and the derived soils are old, highly weathered and, consequently, low in nutrients (Popenoe 1986: 169).

It is easy to understand why South America, with its chain of volcanic mountains along the length of the continent, or Southeast Asia, with its vast river deltas, should have more fertile soils than Africa (Popenoe 1986: 169).

Second, the Sudanic regions of Africa receive the bulk of their annual precipitation between May and October and experience a severe dry season during the winter months. Wheat and barley are unsuitable crops in the summer rainfall areas and thus indigenous plant species such as pearl millet, finger millet, and sorghum must be grown. Third, the plough did not diffuse into Africa until the nineteenth century. Consequently, traditional African agriculture, based on the hoe and digging stick, was very labor intensive in the past and much of it remains so today. The late arrival of the plow may have been due to the geographic isolation of Africa from much of the rest of the Old World. Yet, this isolation did not prevent a host of eastern Eurasian food plants and animals from diffusing into Africa. Alternatively, the absence of the plow may relate to the difficulty of raising cattle for traction in many parts of Africa due to the presence of the tsetse fly.

Fourth, due to the tropical conditions that prevail on the continent, the silent domestication of disease pathogens seems to have played a particularly significant role in human adaptation in Africa from the Neolithic onwards. The distribution of the hereditary disease known as **sickle cell anemia** in Africa is a particularly telling example of the relationship between cultivation practices and the spread of disease. Individuals who are homozygous for sickle cell anemia die in childhood. However, heterozygous individuals have a limited immunity to malaria. Such individuals have an adaptive advantage in regions where malaria is endemic since they are less likely to contract the disease than individuals who do not carry sickle cell anemia.

Since there is a close correlation between the distributions of malaria and of sickle cell anemia in West Africa, it is apparent that the genetic disorder persists in human population there due to natural selection. However, there is more to it than that. Malaria is transmitted to humans by mosquitos. Mosquito eggs and larva do best in sunny pools of standing water. Since these conditions are seldom met in the deep forest of the Equatorial zone, mosquitos are rare in such habitats. When cultivators clear the forest to create their fields however, they open up the forest canopy and inadvertently increase the amount of standing water in sunlit locations. Thus, the expansion of farming into the tropical forests enlarged the habitat of malaria-carrying mosquitos, expanded the distribution of malaria, and created conditions favoring the persistence of sickle cell anemia in human populations in West Africa (Livingstone 1958).

The dreaded "sleeping sickness" or **trypanosomiasis** is still another tropical African disease the distribution of which is closely related to human agricultural practice in (Deshler 1960: 541-542; Lambrecht 1964). Trypanosomes are single-celled parasites that live in the blood streams of vertebrates and are transmitted from host to host by many species of **tsetse fly** (*Glossina*) and other species of bitting insects. The primary African hosts for trypanosomes include buffalo, elephants, giraffes, wart hogs and various species of antelope. These wild game animals are immune to trypanosomiasis, and goats, sheep and some strains of native West African cattle are resistant to it. However, when insect vectors transmit the protozoa either to Zebu cattle, to pigs, or to humans, the disease is usually fatal. The various species of tsetse fly are found in many lowland environments ranging from the immediate vicinity of permanent water to the edge of semi-arid savannahs.

> The landscape epidemiology of trypanosomiasis is the savanna. In East Africa it is the dry savanna woodland which is colonized by tsetse vectors of the *Glossina morsitans* group. In West Africa it is the more humid savanna woodland through which run streams and rivers bordered by outliers of high forest. These riparian thickets are the habitat to which the chief vector, .*Glossina palpalis*, if confined. The concentration of the fly along the water courses where there is so much human activity–water collecting, washing, fishing, etc.–ensures a high degree of vector-host contact (Desowitz 1980: 470).

However, tsetse flies are generally absent from the uplands and it is in that zone that African pastoralists concentrated their adaptation. The distribution of the tsetse fly has determined–or at least severely limited–the distribution of human settlement. In East Africa, livestock trypanosomiasis infests much of the best potential grazing country. Deshler (1960: 541) estimates that more than half of the total area of Tanzania, a third of Uganda, and a quarter of Kenya are those closed to pastoralists by the presence of the disease and its vectors.

> ...the tsetse fly, not the lion, that is the true King of the Beasts in Africa. Surely, any animal that can deny more than 15 million square kilometers to livestock production and settlement must take pride of place. This is an area that could potentially provide 125 million head of cattle to the protein starved continent. Animal trypanosomiasis has forced herdsmen to concentrate on the limited amount of fly-free pasturage, which has led to overgrazing and attendant soil erosion. On the other hand, for those less in sympathy with human needs, the tsetse fly and the trypanosome can be viewed as the most stalwart guardians of African savanna ecosystems and their magnificent fauna (Desowitz 1980: 469).

Actually, the modern distribution of trypanosomiasis may be more constricted than in the past. Rainfall was higher over much of Africa during the Early Holocene and consequently the tsetse fly belt may have extended further north into the Sahara and prevented early Neolithic pastoral peoples from moving southwards into Sub-Saharan Africa until conditions became drier during the Middle Holocene (Shaw 1977, Clutton-Brock 1989: 205). However, the Neolithic pastoralists who arrived in northern Kenya near the end of the Middle Holocene may have been impeded in their movements further south by the presence of trypanosomiasis. Of course, numerous other animal borne diseases besides trypanosomiasis afflict domestic cattle in contemporary times in East Africa. Malignant catarrhal fever, endemic in wildebeests and deadly to livestock, is just one example. The abundance of epizootic diseases like these must have slowed the original spread of the Neolithic pastoralism in East Africa much as they complicated the introduction of European cattle and ranching there during the Colonial period (Gifford-Gonzalez 1998:193; Huxley 1935:30-31).

Twilight of the Later Stone Age The technology of the Stone Age lingered in Africa well into the Holocene "epoch. It closed only when, late in the Holocene", stone-using peoples were contacted, transformed or displaced by iron-using peoples. In the "Early Holocene", "well before this Iron Age transformation, a cultural tradition called the **Aqualithic** emerged Sahara and Sahel region Africa from a Later Stone Age technological base. As we saw in Chapter 10, the greater moisture of the Early Holocene climate allowed Aqualithic peoples "to build a successful subsistence adaptation centered on fishing and aquatic mammal hunting. At its peak during Early Holocene, sites referable to this tradition were distributed over a vast region from the Great Bend of the Niger river in West Africa eastward across the southern Sahara and Sahel to the upper Nile valley and from the upper Nile south to the Lake Turkana region in the Rift Valley of Kenya. The return of drier conditions across North Africa between ca. 6,500 BC and 5,500 B.C. lead to a marked contraction in the distribution of aquatic-based subsistence in northern Africa. However, this arid episode ended ca. 5,500 B.C. and mean annual rainfall levels rose again in northern and central Africa between ca. 5,500 and 2,500 B.C. With the return of more favorable environmental conditions, peoples practicing the old Aqualithic way of life were able to expand and reclaim some of the territory they had abandoned earlier. However, they were unable to reestablish the lifeway over the entirety of its former geographic range. By this time the foragers of the Aqualithic tradition were no longer alone in central and eastern Africa and were therefore not the sole beneficiaries of the return of more equitable conditions to the Sahara. Peoples possessed of a new, more dynamic way of life had arrived in the region and the Aqualithic peoples now had competition.

Early Pastoralism in Northern Africa The new and competing, life way that appeared in Sub-Saharan Africa at this time was based on the raising of livestock, particularly sheep, goats and cattle. Cattle husbandry and the presence of ceramics among these people has lead prehistorians to classify these people as "**Pastoral Neolithic.**" This is a broad and not altogether helpful use of terms. Strictly speaking, the phrase "Neolithic culture" refers to the traditions of Eurasia characterized by the manufacture of pottery and ground stone tools and by the practice of settled village agriculture. Although, the so-called Pastoral Neolithic peoples of Africa possessed domesticated livestock, they seem to have lived nomadic, rather than sedentary, lives. More importantly, there is no clear evidence that Pastoral Neolithic peoples raised domesticated plants. Instead they seemed to have combined stock raising with hunting and gathering in a unique subsistence system that has been called "carnivorous pastoralism."

Both the origin and nature of this Pastoral Neolithic life way is obscure. First, as no wild ancestors of domesticated goats and sheep have been found in Africa, it is widely concluded that these animals were introduced on to the Sahara. Very likely, the ultimate ancestral source of these African sheep and goats was southwestern Asia. On the other hand, the auroch or *Bos primogenus*, the wild ancestor of domestic cattle was present in Pleistocene Africa. African aurochsen, therefore, could have been the source of the domestic cattle of the Pastoral Neolithic tradition. But were they? Studies of the molecular genetics of contemporary cattle have revealed a deep phylogenetic division exists between Zebu (*Bos indicus*) cattle and those of the taurine or (*Bos taurus*) line. The genetic difference between these two division is expressed phenotypically in the hump of Zebu cattle and the absence of a hump in cattle of taurine origin. The genetic distance between the Zebu and taurine cattle indicate that they followed different paths to domestication: the Zebu presumably derived from wild cattle populations in south Asia, the taurine from a distinct population in southwestern Asia and perhaps still another one in Africa.

Humped, humpless and crossbred cattle are all common in contemporary Africa. However, DNA studies of these modern animals suggests that the original or indigenous breeds of African cattle were of humpless *Bos taurus* stock like those southwestern Asia. The molecular genetics of the humped Zebu cattle in modern Africa suggests that this type was introduced there only late in prehistory, perhaps no earlier than about 1,000 B.C. Zebu and taurine cattle are entirely interfertile, of course, and the two strains have been cross-bred with one another ever since (Bradley, Loftus, Cunningham and MacHugh 1998: 84). Simply because the earliest African cattle are descended from ancestral *Bos taurus* stock does not prove that the ancient African pastoral life way based on cattle raising was derived from southwestern Asia. The genetic distance between modern African taurines and those of western Eurasia is large enough to prevent us from rejecting the hypothesis that African and western Eurasian cattle were **independently** domesticated from ancestral aurochsen populations that were largely isolated from one another on each of the two continents (Bradley, Loftus, Cunningham and MacHugh 1998: 85).

Unfortunately, these genetic studies still leave unresolved the question of when the presumed independent African domestication of wild cattle took place. The fragmentary remains of large grazing animals have been recovered from archaeological contexts dating to between 9,500 and 8,000 B.P. in the Nabta Playa and Bir Kiselba regions of the eastern Sahara are often cited as proof of the early date for the beginning of this process. Unfortunately, the poor state of the preservation that these remains are in prevents investigators from being certain that the bones are those of cattle rather than buffalo. However, if subsequent analysis proves them in fact to be *Bos*, Wendorf and Schild (1994), their excavators, assert that they then **must** have been domesticated. In their view, this is because the Nabta Playa and Bir Kiselba regions would simply have been too dry during the Early Holocene for cattle to survive unaided by attentive humans. While this may be so, more certain evidence of the actual presence of cattle will be required if the eastern Sahara is to be accepted as early independent center of African cattle domestication.

Far from the eastern Sahara along Africa's Mediterranean fringe domestic cattle are definitely present after ca. 4500 B.C. at the site of Capeletti in Algeria. The Capeletti remains, which appear to be those of *Bos taurus*, were found in association with the bones of sheep and goats (Roubet 1978). It is possible that these cattle were independently domesticated from North African populations of *Bos primigenus* but, given the late date of the occupation at Capeletti, these animals were just as likely to have descended from stock introduced into the area from southwestern Asia. Whatever the source of the aurochsen ancestry, by 4,000 B.C. domesticated cattle had begun to spread widely and rapidly across Africa north of the Equatorial rainforests. This expansion was no doubt due in large part to the return of moist climatic conditions on the Sahara from ca. 5500 to 2500 B.C. in the middle Holocene. Based on the representations at the celebrated rock art sites at **Tassili** in the central Sahara, the cows of this southward expansion were of long horned, humpless *Bos taurus* stock (Grigson 1991).

Near the end of the Middle Holocene after ca. 2,500 B.C., the great desertification of the Sahara began in earnest and continued for nearly 1,000 years. At the height of this dry cycle, the great playa lakes and wadis that had dotted the Sahara dried up and the savannah plants and animals that had thrived there in since the Pleistocene were banished. In the wake of this dessication, cattle herding peoples were either forced southwards into the Sahel and Sub-Saharan Africa by increasing aridity or simply expanded in that direction following in the wake of the retreating tsetse fly belt (Clutton-Brock 1989: 205).

Pastoral Neolithic South of the Sahara The dessication of the Sahara appears to be the engine that pushed pastoral nomads southwards in tropical West Africa. Archaeologists are uncertain as to exactly when such northern African cattle herders first arrived in the equatorial region there. Direct archaeological evidence of their presence there is fragmentary at best. Nonetheless, compelling **indirect** evidence that cattle have been present in tropical West Africa for a long time is provided by the small local breed of cow known as N'Dama. Unlike almost all other kinds of domestic cattle, the small, humpless N'Dema have a natural immunity to tsetse-born trypanosomiasis infection (Epstein and Mason 1984: 21). Exactly how long it took for such immunity to evolve in the N'Dama breeding population is unknown. We can be certain, however, that some considerable time was necessary.

The earliest commonly accepted archaeological evidence of domesticated animal herding in East Africa comes from sites in the Ileret region on the shore of Lake Turkana in the northwestern Kenya. The dates of cattle remains from these sites indicate that pastoralism was established there sometime after ca. 3,000 B.C. (Robbins 1972; Barthelme 1985; Marshal *et al.*1984). After this time, the Pastoral Neolithic tradition appears to spread slowly out of the Lake Turkana region to the south and west. The remarkable rockshelter site of Enkapune Ya Mune in the eastern Rift Valley of southern Kenya has produced pottery that has been elsewhere associated with a pastoralist adaptation in contests dated top ca. 2,910 B.C. (Ambrose 1998:381). This date is well within the range of the earliest introduction of pastoralism in Kenya. However, faunal remains associated with this pottery at the site do not include the bones of any domesticated animals (Marean 1992). In fact, analysis of the bone remains at the site indicate that domesticated animals were not raised by people at Enkapune Ya Mune until ca. 2000 B.C. The data f rom Enkapune Ya Mune, at least, suggests that Pastoral Neolithic pottery was present in southwestern Kenya among peoples who possessed no domesticated animals but continued to subsist by foraging. How can this apparent time lag be explained?

On one hand, the presence of Neolithic pots without Neolithic cattle at Enkapune Ya Mune may indicate that the presence of animal diseases in East Africa, particularly trypanosomiasis and malignant catarrhal fever, slowed the spread of pastoralism from the north across the region (Gifford Gonzalez 1998). On the other hand, the absence of domesticated animals at Enkapune Ya Mune may not demonstrate that pastoralists were not present elsewhere in the Kenya Rift Valley. Prior to the arrival of the pastoralists, much of East Africa appears to have supported a small groups of hunting and gathering peoples who centered their adaptation on mountain vegetation zones adjacent to the highland plateaus. Because early pastoralists would have most closely adapted to the open grasslands, they may not have come into direct competition with the indigenous foraging peoples. It seems very likely that the two peoples would have established symbiotic relationships based on the exchange of complimentary products. The foraging peoples of Enkapune Ya Mune may have been but one part of an economic system based on the exchange of the wild resources for the milk, blood, meat or hides of domesticated cattle. Perhaps it was such exchange, no doubt part of subsistence risk reduction strategy practiced by both peoples, that brought the Pastoral Neolithic pottery to Enkapune Ya Mune long before domesticated animals were raised there (Ambrose 1984: 232-233).

It should be noted, however, that not all specialists accept the notion of an early arrival of pastoralism into East Africa. Robertshaw (1995) rejects the earliest radiocarbon dates reported for pastoralism there. He asserts further that when such subsistence did arrive in East Africa, it was not practiced as an self-sufficient form of subsistence. Based on his work at **Gogo Falls** in western Kenya, Robertshaw concludes that the establishment of pastoral subsistence in Kenya **followed** the rise of settled agriculture. Therefore, as in the Near East, pastoralism in East Africa was a subsistence specialization "peripheral" to and dependent upon settled village agriculture (Robertshaw 1991; Robertshaw and Collett 1983).

In any case, sometime after ca. 1,000 B.C., humped Zebu cattle from the Indian subcontinent begin to appear in the East African archaeological record (Bradley, Loftus, Cunningham and MacHugh 1998: 85). The infusion of such stock must have lead to marked improvements in the Pastoral Neolithic herds. Modern Zebu cattle are generally more resistant to nutrient stress brought on by drought than taurines and need less water and forage to survive and gain weight more quickly when the environmental stress is over. Further, their crossing with local breeds may also have resulted in more disease-resistant cattle capable of producing greater quantities of milk (Clutton-Brock 1989: 200-201; Marshall 1989: 236-237).

It is during drought conditions that the adaptive value of domestic cattle becomes most obvious. The human body needs fat to survive and, in its absence, suffers severe metabolic problems. Food and water stress, such as is experienced during droughts, sharply reduces the body fat of most mammals. Sustained drought conditions can render their meat of little nutritional value to humans: "even when the amount of lean meat is unlimited a person can starve to death" (Smith 1986: 200). Like other mammals, cattle lose body fat in the face of food and water stress. However, they retain a higher percentage than most African game animals (Speth and Spielmann 1983: 10). At the same time, cattle provide a ready source of fats and proteins in the form milk. In addition, African pastoralist also extract blood from their animals without killing them. Cattle blood is a basic ingredient in the diets of these people. As the North African climate became drier during the late Holocene, peoples in North and East Africa probably began to turn to pastoralism to take advantage of the greater fat retention of domestic cattle and their ability to provide valuable secondary products (Smith 1986: 200). In this regard it is important to note that the economy that developed in Africa at this time was its apparent combination of herding with hunting and gathering. The subsistence of the African Pastoral Neolithic was, at least at first, a kind of pastoral foraging or "carnivorous pastoralism." The practice of herding without a parallel reliance on the cultivation of domesticated plants renders the Pastoral Neolithic unique. However, as time went on, these early pastoralists may also have grown some cereal grain crops as a further hedge on their adaptive bets.

To be successful in these regions of Africa, however, dry lands pastoralism must include three key elements: population mobility, a vast territory for movements, and the maintenance of sizable livestock holdings. Rainfall and vegetation is highly localized in this area and becomes more so during droughts. Faced with drought conditions herdsmen must be prepared to move their livestock over long distances to find rangelands with forage and water. In the course of such movement herdsmen sustain considerable animal losses from stress and predators. However, if their herds are sufficiently large to begin with, enough animals survive to allow the herd to be rebuilt at the conclusion of the drought (Little 1989: 239).

Herders and Hunters in Southern Africa Much of the vast territory of southern Africa continued to be occupied by hunters and gatherers until about 2,000 years ago. After that time, this began to change for ever. Beginning in the first millennium B.C, the bones of cattle and sheep and the scattered remains of pottery begin to appear in the archaeological record of the region as far south as the Cape of Good Hope. It is generally presumed that both the herd animals and the pottery were brought into southern Africa by the ancestors of the cattle-herding peoples whom the Dutch encountered when, in the 17th century, they began their colonization of the Cape. These indigenous peoples, whom the Dutch referred to as **Hottentots**, lived in large village encampments or *kraal* and spoke languages that contain distinctive "click" consonant sounds characteristic of the **Khoisan** or **Khoikhoi** language family. Most scholars still accept the view that Khoi speaking peoples introduced both pottery and fat-tailed sheep and cattle into southern Africa in the course of their migration there from the region of Botswana or Zimbabwe to the north, during the first millennium (Nurse, Weiner and Jenkins 1985). However, on both linguistic and archaeological grounds a strong case can be made that the Khoi did not arrive in the south until at least the second millennium B.C. If this is correct, the first pottery and cattle must have reached the Cape by the process of diffusion from group to group of indigenous hunting people (Sadir 1998).

Whenever the Khoi people did arrive, they came with tools that were still predominately made of stone. Nonetheless, these agro-pastoralist were generally able to replace the hunting populations in the areas where chose to settle. The penetration of southern Africa by Khoi pastoralists was followed in the first millennium AD by the arrival of a second great wave of agro-pastoral peoples. These new peoples possessed iron metallurgy and were part of the great expansion and dispersal of Bantu-speaking peoples that initiated the onset of the Iron Age throughout much of Sub-Saharan Africa (Phillipson 1984: 272-280; Bousman 1998). Recent archaeological work in southern Africa indicates that such peoples had established themselves in the Kalahari desert by A.D. 500. By A.D. 800 they were combining cattle raising with grain agriculture. They also were engaging in complex long distance trade with the peoples surrounding the Kalahari (Denbow and Wilmsen 1986; Denbow 1986).

Although the expansion and spread of agro-pastoral peoples occurred at the expense of indigenous hunting and gathering peoples, foraging nonetheless continued to be practiced as the primary subsistence mode in areas of southern Africa that, like the Kalahari desert, were marginal to or unsuited for agriculture. European settlers called such foraging peoples "Bushman" and often actively sought to displace or remove them. While these European efforts further constricted the distribution of hunter-gathers in southern Africa, they did not result in their total elimination. In fact, such peoples persisted into the modern era in souther Africa. Their survival of South African hunting peoples presented anthropologists with a unique opportunity to study the foraging mode of subsistence first hand. In turn, such first hand ethnographic work has in turn been used by archaeologists has a source of ideas and insights into the nature of life among the prehistoric hunter-gatherers known only from the archaeological record. The use of such first hand accounts of living peoples to interpret the archaeological record is known as the **ethno-archaeological approach**. Sometimes the process is reversed, however. Anthropologists and historians of culture often find the archaeological record provides a useful check on or expansion of the ethnographic record of the people they are studying. In working backwards from the ethnographic present into the past using archaeology such scholars are employing what is called the **direct historical method**.

The ethnographic accounts of peoples known variously in the past as the *San*, the *!Kung*, the !Kung Bushmen or the *Jul'hoansi*, are the most widely used by archaeologists seeking to employ the ethno-archaeological method to interpret the material remains of prehistoric foragers. The *Jul'hoansi* have been intensively studied and filmed by a number of very talented ethnographers over the last 50 years. As a consequence, the *Jul'hoansi* are among the most fully studied hunting and gathering peoples known to science. It was this intensive scientific scrutiny revealed the ability of the *Jul'hoansi* to thrive in the harsh conditions of the Kalahari Desert helped to dispel dismal Hobbesian view of the foraging life as "solitary, poor, nasty, brutish, and short." The study of *Jul'hoansi* art and mythology has also provided deep insights into the interpretation of the rock art of the Later Stone Age found throughout southern Africa.

This research also made it clear that the more powerful and numerous agricultural peoples–white and black–who surround the *Jul'hoansi* presented very great political and economic obstacles to their survival as a people. Of course, it is not just anthropologists and archaeologists who find the *Jul'hoansi* intrinsically interesting. Much of the general public hold this admirable people in high regard and, for better or worse, have made them into a kind of New Age cultural icon.

The Rise of Farming South of the Sahara

Cattle were not the only domesticated species to

appear south of the Sahara during the Middle and Later Holocene epoch. A host of indigenous African plant species also began to be drawn into the web of human manipulation and control. Unfortunately, little direct archaeological evidence of this domestication process has been recovered in Africa. Nonetheless, botanical, genetic and geographical studies suggest that native plants were domesticated in at least four main hearths in Sub-Saharan Africa:

1. the Great Bend of the Niger river,
2. the Sahara proper or its margins in the Sahel,
3. the savannah grasslands of Sudan-Chad region of northeast Africa, and .
4. the Ethiopian plateau,

The Great Bend of the Niger River The Niger river rises on the Fouta Djallon plateau in Equatorial forest zone of Guinea not far from the Atlantic coast of West Africa. It then flows inland to the northeast where it leaves the forests of the Equatorial zone and crosses into the savannah country of Mali. In central Mali, the Niger spreads out across a broad and treeless plain to form a vast, swampy "inland delta" of braided streams, twisting channels and interconnected lakes. Below the legendary city of Timbuktu in Mali, the Niger gradually begins to turn, first to the east, then to the southeast until it forms a great bend. This broad turn carries the Niger back towards the Atlantic where it reenters the Equatorial forests. On the coast of Nigeria, the Niger forms a huge delta where, after a traverse of more than 4,000 kilometers, it flows into the ocean at the Gulf of Guinea. Direct archaeological evidence of domestication west and northwest Africa is very limited (McIntosh and McIntosh 1979). Nonetheless, on biological and ecological grounds a strong case can be made that the savannah country in the vicinity of the Great Bend of the Niger appears was the setting for the domestication of some of the continent's most significant and widely used cultigens including: **African rice** (*Oryza glaberrina*), **African yams** (*Dioscorea* sp.) and **oil palm** (*Elaeis guineensis*).

African rice was among the most important native cereal on the continent until the introduction of Asia varieties of rice in historic times. The wild progenitor of African rice was almost certainly a savanna annual species that grew around water holes and playas that fill with water during the rainy season and dried out after its close. Only much later was rice brought into the humid, forested regions of African where it was raised as a wet season crop (Harlan, DeSmet, Stemler 1976: 5; Harlan 1992: 65). African yams are raised for their edible underground tubers. The yam tuber is the plant's storage organ that annually produces a vine during the rainy season. Although the vine dies away at the end of the growing season, the tuber endures. It is the tuber that allows the plant to survive the dry months before the rains return in the following growing season. Although wild yams are flowering plants that reproduce sexually and form seeds, domesticated yams must be propagated vegetatively. Most commonly this simply means that small pieces are cut from stored yam tubers and planted in mounds of earth. After thousands of years of vegetative reproduction by humans, domestic yams have largely lost their capacity to reproduce sexually. Many varieties of domestic yams are thus completely dependent upon human intervention for their continued reproduction (Alexander and Coursey 1969: 406-407, 412). The first Europeans to visit West Africa in the 15th Century found yams to be the primary staple crop supporting the dense populations living in the region. Both the importance of yams there during historic times and the fact that the great rain forests of the apparently resisted penetration by agricultural peoples until quite late in prehistory led Alexander and Coursey (1969: 420-421) to suggest that Neolithic settlement of the Equatorial zone did not begin until after the yam had been domesticated. It is their opinion that yam-collecting hunting and gathering peoples at the margins of the Equatorial forest were stimulated to experiment with yam domestication by contact with cereal agriculturalists arriving on the southern margins of the Sudanic zone. On the basis of linguistic and geographical evidence, George Peter Murdock credits the Mande-speaking peoples of the middle Niger with the domestication of these indigenous plants (1959). This identification has been widely criticized but insufficient archaeological evidence exists to decide the issue.

Finally, although botanists have identified the savannah county of the Niger Bend as the most likely region for the first domestication of the oil palm, the earliest evidence of its cultivation has been found in the forested tropical interior of West Africa on Lake Bosumtwi in Ghana. Pollen cores extracted at this site record a marked increase in oil palm pollen ca., 1,500 to 1,000 BC. This increase very likely reflects the beginning of oil palm cultivation at this site (Stahl 1993). If the oil palm was domesticated in the savannah country to the north, its domestication there must have occurred there substantially earlier than 3,500 years before the present. Oil palms tree are prized for their oily and fibrous fruit. Fat pressed from this fruit produces a viscous liquid that ferments rapidly and easily into a sweet palm wine greatly favored throughout much of traditional West Africa. Additionally, oil palm wood and fronds are put to many craft and building uses in Africa.

The Lower Sahara Proper and its Margins in the Sahel The vast plains region of the Sudanic zone stretches across North Africa for nearly 6,500 kilometers from Senegal and Mauritania on the west to the Ethiopian Plateau on the east. While a region this huge exhibits great biotic diversity, the Sudanic zone is nonetheless dominated by desert, semi-desert and savannah grasslands or grassy steppes that fringe the great Sahara and Libyan deserts to the north. The western portion of the Sudanic zone between Mauritania and Chad is known as the Sahel. Based on ecological and botanical criteria, the lower margins of the Sahara and the northern reaches of the Sahel was very likely the region where the wild savannah ancestors of such species as **cotton** (*Gossypium herbaceum*) and **watermelon** (*Citrulus lanatus*) and **sesame** (*Sesamamum indicum*) and numerous other minor crops were first domesticated. In addition, the important African cereal grain **pearl millet** or **bullrush millet** (*Pennisetum typhoideum*) is also likely to have originally been brought into cultivation in this region since it is the most drought-resistant of all African hot-season millets and cereals (Harlan 1989: 340).

The Savannah Grasslands of Sudan-Chad Region of Northeast Africa. The eastern end of the Sudanic zone is today occupied by the modern nation-states of Chad and the Republic of the Sudan. In prehistory, this region was almost certainly the center for the domestication of **sorghum** (*Sorghum vulgare*), Africa's greatest contribution to world agriculture. Sorghum is a coarse annual cereal grass with an extensive root system that renders it extremely resistant to drought. When fully matured, the plant resembles corn in appearance except that its seeds emerge on flowering shoots called panicles rather than on ears.

While there is no archaeological evidence of sorghum's prehistoric domestication in Chad and Sudan, a race of wild sorghum that is fully fertile with the domesticated forms still grows in abundance on savanna grasslands of these two nations (Harlan 1989: 337). The most primitive of sorghum known today is the bicolor race. Bicolor was probably the first sorghum to be domesticated, it is still widely cultivated. The Guinea sorghum races of West Africa, the kafir races of South Africa, and *caudatums* sorghum grown from Chad to eastern Sudan and Uganda all appear to have developed out of bicolor race as local adaptations. Bicolor apparently reached in India by 2000 B.C. where the distinctively Indian sorghum race called *durras* was developed *Durras* sorghum was in turn introduced into Africa from India probably in Islamic times (Harlan 1989: 336-337). Innumerable varieties of sorghum are cultivated throughout the modern world. The plant's capacity for growth in hot, dry settings, however, renders sorghum an especially important human food and animal forage in semi-arid regions of the Earth.

The Ethiopian Plateau The highlands of Ethiopia cover about two-thirds of land area of that northeastern African nation. Rugged and heavily eroded, the highlands average over 2,000 meters in elevation and contain a number of peaks that are well above 4,000 meters in height. The Great Rift Valley runs from northeast to southwest through the middle of Ethiopia and this great tectonic structure divides the highlands into two separate massifs. The Ethiopian or Amhara plateau, the larger and higher of these massif, is located on the west side of the Rift. The smaller massif on the eastern side is further subdivided into the Somali and the Ogaden plateaus. The Ethiopian plateau slopes west to the Nile Basin and forms the watershed for the Blue Nile, a major tributary of the Nile river proper. Climate is closely related to elevation in the highlands and, over most of the Ethiopian plateau, is equitable, temperate and characterized by minimal annual changes in rainfall and temperature (Parker 1995: 108).

Like the continent's three other hearths, little archaeological evidence of early plant domestication has been recovered on the Ethiopian plateau. Based on his analysis of plant morphology and native husbandry practices, however, the great Russian plant geographer N. I. Vavilov concluded that the region was most likely the setting for the domestication of such uniquely Abyssinian crops as **tef** (*Eragrosstis tef*), **noog** (*Guizotia abysinica*), **ensete** (*Ensente ventricosa*), **khat** or **murrá** (*Catha edulis*), **coffee** (*Coffea arabica*), a host of medicinal plants, and **finger millet** (*Eleusine coracana*). Millet is the common name given to numerous species of domesticated cereal grasses that are major cereal grain crops in modern Africa, Russia, southern Europe, China, and India where it is commonly eaten as a porridge or ground into flour. Various species of millet are grown in the United States but here they are used chiefly for hay and animal forage. Tef is also a cereal grain used for both human consumption and animal fodder. Tef stalks are also used as straw. Ensete is a variety of banana grown not for its fruit but for the starchy fibers of its edible stalk. Noog is a flowering plant that produces seeds from which an edible cooking oil is made.

Of these crops, only coffee, the stimulant khat and finger millet have had much impact on agriculture outside of northeastern Africa. Further, despite the unique character of the Ethiopian Complex, by the historical period, agriculture on the Ethiopian plateau was based largely on higher-yielding cultigens like barley, lentils, flax, wheat and cabbage whose origins were originally in southwestern Asia. Unfortunately, archaeology offers little insight into the dates of the diffusion of these exotic domesticates on to the plateau. However, the indigenous Ethiopian constellation of plants had almost certainly been independently domesticated before the arrival of these foreign plants. After all, it seems highly unlikely "that wheat and barley growers would bother to domesticate tef and finger millet, or that farmers with a complete crop complex would take up ensete growing" Harlan (1992: 68).

The Eastern Eurasian Agricultural Complexes

The eastern portion of the Eurasia continent occupies one-third of the Earth's surface and contains nearly two-thirds of its contemporary human population. Physically, its vast surface area is dominated by three major landform types:

1. **A parallel series of high, rugged mountains**. The mountainous roots of the continent stretch in a continuous line across eastern Eurasia from the Pontus Mountains on the Black Sea in Turkey to the **Tsingling Mountains** that divides China into northern and southern parts. This double line alternately merges to form a solid nucleus and diverges to enclose high plateaus or intermontane basins (Cressey (1951: 16). The Himalayas of South Asia are the mightiest members of this long chain of mountain ranges.
2. **The high plateau of Mongolia**. Located north of the continuous line of mountains, the Mongolian high plateau is a vast, rolling, grassy steppe broken within at a number of points by high isolated mountain ranges. Characterized by hot summers, cold winters and low rainfall, the high plateau constitutes the easternmost segment of the great Eurasian plain that stretches back across the middle of the Eurasian continent to the doorway of Europe in Hungary and the Black Sea region.
3. **Numerous mighty river systems** These great rivers radiate outward from the interior of eastern Eurasia to flow into the oceans that ring the continent. The Lena, Indigirka and the Kolyma flow into the Arctic Ocean. The Amur, the Hwang or Yellow, the Yangtze and the Mekong flow into the Pacific Ocean. The Granges, Brahmaputra, Indus and numerous lesser streams flow into the Indian Ocean. The fertile soils on the alluvial plains of their great valleys have made those on the southern and central parts of the continent into the chief centers of agriculture and population in eastern Eurasia..

These three great landform types form the key attributes of eastern Eurasia's environments. Correspondingly, mountains, steppes and river valleys are the prime determinants of the distribution of human population on this portion of the continent in recent centuries. During the early Holocene, they demarcated and determined the location of the two great hearths of agricultural origin in eastern Eurasia:

1. Tropical southeastern Eurasia
2. Northeastern Eurasia

Holocene Climates of Eastern Eurasia As we saw in Chapter 8, the modern climate zones of eastern Eurasia run roughly parallel across the continent. These zones grade from Arctic, Subarctic, to a Midlatitude climates as one moves from north to south across the continent. This Midlatitude zone includes vast stretches of semi-arid to arid terrain in the central and western parts of the continent and a humid temperate zone in central China and Japan. The humid sub-tropical and tropical portions of the continent known as "monsoon Asia" is found in the south and southeast. The present-day weather pattern in this region is dominated by pronounced seasonal changes in atmospheric circulation known as the **monsoons**. Winds from the Indian Ocean blow for about six months from the northeast and then reverse during the next six months to blow from the southwest. Monsoon circulation is caused by temperature differences between the warm eastern Eurasian and African landmasses and the cool surrounding oceans. This thermal gradient creates low pressure over land and high pressure over the ocean which in turn produces an inflow of moist air and heavy rainfall to South and East Asia and West Africa (COHMAP members 1988: 1049). Rainfall in those areas is thus markedly seasonal in its distribution and highly variability within the year and from region to region. When it arrives, it is characterized by heavy downpours (Trewartha 1981: 207).

The monsoon pattern is most pronounced in South Asia or the Indian sub-continent. Winter begins there in January when the onset of the northeast monsoon reverses the winds and brings cool, dry weather. By March temperatures begin to warm appreciably. The highest temperatures of the year are reached in the dry spring months prior to the onset of the rains in June. The southwest monsoon brings the bulk of the annual rain to the subcontinent. The onset of the rains is marked by turbulent storms and high daily temperatures. The rains are concentrated between June and September and are followed by a drier, post-monsoon period that lasts until December. The south coast along the Indian ocean and northeastern regions of the sub-continent are characterized by high rainfall and humidity. Conversely, the monsoon rains do not reach northwestern and interior Eurasia and North Africa and these regions are hot and dry. In fact, nearly one-half of the Indian subcontinent may be classed as arid to semi-arid, with the most rainfall deficient lands concentrated in the **Thar desert** at the northwest end of the sub-continent (Trewartha 1981: 192).

Current paleoenvironmental and archaeological evidence indicate that this aridity is a middle to late Holocene development and that between ca. 10,000 B.C. and 4,000 B.C., the Thar region, like the Sahara and Arabian deserts, were relatively well-watered. These and other data indicate that, during the early Holocene, there was an intensified monsoon pattern in this part of the world. This intensified pattern probably resulted from the higher annual temperatures of the so-called "thermal maximum" of the early Holocene. During this time, North Africa and Eurasia were between 2 and 4 degrees warmer than they are at present. Since the contrast in temperatures between the land and the surrounding oceans were greater, early Holocene monsoons were stronger than they are at present and greater rainfall was carried farther inland to the Thar, Sahara and Arabian deserts as well as other parts of South and East Asia (COHMAP members 1988: 1049-1050).

The changes in rainfall, temperature, and the monsoon circulation patterns that occurred during the early Holocene seemed to have less impact on tropical southeast Asia than did the rise in world wide sea level that occurred at the same time. Lowered worldwide sea level during the late Pleistocene had exposed vast tracts of the continental shelf and joined insular southeastern Asia to the mainland. The severance of this connection during the early Holocene, "the mainland of Southeast Asia was reduced from 1.7 to 1 million square kilometers, and the length of the coastline increased by a factor of 1.8 giving a more than 3-fold increase in the ratio of coast to land" (Glover 1980: 152). The littoral zone in the tropical supports a particularly rich biota and this richness no doubt offset the net loss of land in the area.

Early Domestication in Southeast Asia The initial development of eastern Eurasian agriculture appears to have occurred independently of that of southwestern Asia. However, isolation of these areas from one another was never complete and, over time, domesticated species were exchanged between them (Glover 1977: 156; Chang 1970: 182). By late prehistoric times, three major agricultural complexes were established in east and southeast Asia.

In tropical and subtropical southeast Asia, cultivation centered on vegetatively-reproduced root crops like **yams** (*Discorea* sp.), and **taro** (*Colocasia* and *Alocasia* sp.) but including bamboo, fruit tree species like bananas, and cereal seed crops such as rice, millet and **Job's tears** (*Coix lacryma jobi*). In the agricultural complex established in south China seed crops such as dry and wet rice and a wide range of vegetables form the focus of cultivation. In the cooler and more arid regions of northern China, millet is the primary cereal crop. Dogs, pigs, Asiatic or zebu cattle (*Bos indicus*) with their characteristic hump, **water buffalo** (*Bubalus bubalis*), ducks and chicken are the primary domesticated animals in all three complexes. Three main regions of primary domestication have been posited in East Asia:

1. The deciduous forest zone of northern southeast Asia (including modern Viet Nam, Laos, Thailand, Burma, southern China, western Malaysia),
2. The lower Yangtze river valley of central and southeastern China, and
3. The Huang Ho or Yellow river valley and the inland loess plains of northern China.

Domestication in eastern Eurasia is commonly assumed to have begun earliest in tropical southeast. As noted above, rising worldwide sea level inundated vast tracts of southeastern Asia and the demands of these new conditions no doubt created adaptive stresses. Thus, early Holocene in eastern Eurasia came to be characterized by a dramatic increase in the size of one of the world's richest natural ecosystems: the tropical littoral zone. The celebrated American cultural geographer Carl Sauer notes that, in the lowland tropical deltas, the diverse ecological zones as the tropical forests, freshwater lakes and rivers, tidal estuaries and the ocean are all found in close proximity to one another.

In an influential book published in 1952 entitled *Agricultural Origins and Dispersals*, Sauer suggests the environmental richness of the tropical littoral of Southeast Asia presented a unique array of subsistence opportunities to the Mesolithic stage foraging peoples who came to inhabit it in the early Holocene. He reasons that fishing peoples living along these deltas would have been able to develop a semi-sedentary or even sedentary settlement pattern based on food collecting. Further, he thinks that the requirements of fishing would have led them to gather plants whose fibers could be processed into fishing or nets or from which fish poisons, drugs, or medicines could be obtained. Perhaps only secondarily would they have sought plants for food.

In any case, Sauer concludes that the gathering of plants eventually led these fisherfolk to experiment with the **vegetative propagation** of these useful wild plants. Manipulation and experimentation with tuberous species like wild yams and taro led eventually to the domestication of these plants. In this way, Sauer sees fishing peoples living in the delta areas of the great rivers of tropical Southeast Asia as the initiators of the world wide process of domestication (Sauer 1952: 27).

Sauer never considered his theory to be testable by archaeological means. When it was published in the early 1950s, little archeology had been undertaken in southeastern Asia and virtually none of that work had been done in tropical lowland riverine zones. Sauer was not convinced that archeological work there would meet with much success anyway. In his view, the rising sea level of the early Holocene would have destroyed all the evidence (Sauer 1952: 23). Further, the hot and humid climate of Southeast Asia creates poor conditions of preservation of plant macrofossils and pollen and renders the recovery of direct evidence of plant domestication an unusual event.

Therefore, considerable interest was generated in the archaeological community in the late 1960s when Charles Gorman, the excavator of Spirit, Banyan Valley, and Tham Pa caves in northwestern Thailand, reported the recovery of a limited quantity of plant materials–including shell fragments, seeds, nuts, and plant parts of possibly domesticated plants–from deposits containing stone tools of the **Hoabinhian**, the Southeast Asian Mesolithic industrial tradition that dates to the Pleistocene-Holocene boundary mentioned earlier. The Hoabinhian cultural tradition has been reported over a vast area of the Southeast Asian tropics from Burma as far east as Viet Nam and from south China to northern Sumatra, an area roughly the size of Europe (Glover 1977: 148). The tradition appears in this area at least as early 17,000 B.P. and lasts until ca. 8,500 to 7,000 B.P. or even later in some areas. The culture is characterized by implements flaked with rather primitive technique, unifacial split river cobbles were used to make discs, short axes, and almond-shaped tools of uncertain function and hammerstones and grinding stones were common tools. In the final Hoabinhian, a crude, cord marked pottery appears (Glover 1977: 145).

The subsistence base of the Hoabinhian initially appears to have been broad-spectrum hunting and gathering. Over time the presumably uniform ecological adaptation of Hoabinhian was modified to include a series of new elements perhaps including early domesticated plants (Gorman 1971). **Spirit Cave** is the most important Hoabinhian site excavated to date. It is located high on limestone cliff in the deciduous forest zone of northwestern Thailand near Thai-Burma border. Although the site lies in fine hunting country, it is perched on a steep slope far away from the nearest running water (Higham 1983: 138).

It was excavated in the late 1960s and early 1970s by Chester Gorman. Gorman described five strata in the cave deposits and assigned these layers to two cultural levels: the "Hoabinhian" of Level I, dating to between 12,000 B.C. and 7,000 B.C., and an "Advanced Hoabinhian" in Level II dating to between 7,000 B.C. and 5,600 B.C. The strata assigned to Level I produced plant remains of some ten separate genera of three different types:

1. Trees or perennial plants which grow wild today in the surrounding primary or secondary forests, including butternut, almond, candlenut, and betelnut,
2.. Definite annual and perennial plants represented in the region today as cultigens, including pepper, cucumber, bottle gourd, and Chinese water chestnut,
3. Possible annual legume crop plants, including some poorly preserved specimens tentatively identified as either common bean (*Phaseolus*) or broad bean (*Vicia)* and pea (*Pisum*) and a native American and African palm (*Raphia*).

The recovery of this plant material in the context of the Hoabinhian, dated to between 12,000 and 5600 B.C. This early date, the presumed cultivated or domesticated status of some of the plant species, and the presence of seeds originally identified as possible members of either the common bean or broad bean led some scholars to conclude that southeastern Asia experienced an agricultural revolution earlier than either southwestern Asia or the New World, and that Sauer's theory of agricultural origins had been partially vindicated.

However, the tentative identifications of the presence of peas, a palm species, and beans has been widely disputed. As Harlan and DeWet (1973:52) point out, the identifications of two cool adopted Mediterranean species, *Pisum* and *Raphia*, and an American bean species in a tropical cave are doubly suspect and Yen (1977), who made the identifications, concedes that they are not certain but notes that modern representatives of these species are domesticated and can be found growing in the vicinity of the cave in the present day.

A series of Hoabinhian cave sites in west central Thailand have recently been found to contain a different array of plants, which also appear wild but have modern descendants that are cultivated today in Southeast Asia for use as folk medicines (Pyramarn 1989). Thus, whether or not these plants were tended or merely exploited in the wild during early Holocene times, their presence in the caves very likely reflects the first stages of the domestication process in the region and suggests that Hoabinhian hunter-gatherers were incipient agriculturalists.

Around 6,800-6,000 B.C. at Spirit Cave, Gorman reports evidence of an apparent site unit intrusion. At that time the Hoabinhian stone tool industry was enriched by the addition of cord marked ceramics, quadrangular adzes, polished slate knives. The slate knives are identified as rice cutting tools by Gorman who considers the entire assemblages as indicates the arrival of alien rice farmers in the cave's vicinity by 7th millennium B.C.

The occupation of Spirit Cave ends around 5,500 B.C. However, evidence that broad-spectrum hunting and gathering continued to be practiced in the area after this time was recovered by Gorman at nearby **Banyan Valley Cave**. Further, a way of life not unlike the Hoabinhian persists in the forested mountain zones in Southeast Asia until sometime between A.D. 1,000 to 1,500 or later (Glover 1980: 171) and forest hunting and gathering adaptation was continued to be practiced by the isolated peoples in Southeast Asia.

The plant remains and other evidence from the Hoabinhian cave sites in Thailand lead to two conclusions about Sauer's theory of agricultural origins. First, if Thai specimens in fact indicate that the Hoabinhians were in the first stages of plant domestication, the absence of any evidence of vegetatively-reproduced plants within their assemblages is particularly damaging to Sauer's theory. Gorman (1977) even asserts that, based on this evidence, we must abandon the notion that vegetative reproduction was the first form of plant domestication practiced in Southeast Asia. Second, the Spirit Cave plant data does not oblige us to reject Sauer's view that the lowland tropical river deltas of Southeast Asia were natural locations for the experiments with plant cultivation. Spirit Cave, located high on a steep slope for away from the nearest running water in the mountainous interior of Thailand, is in a far different geographical setting than the one posited by Sauer (Higham 1983: 138). In fact, Sauer may have been correct in asserting that his theory cannot be tested archaeologically and other kinds of data must be used. With that in mind, let us turn to the botanical reconstructions of the domestication of rice.

Origins of Rice Cultivation Rice is the common name given to grasses of the genus *Oryza*. At least twenty wild species of rice have been distinguished (T. T. Chang 1989: 408) but only a single domesticated one: *Oryza sativa*. In addition, taxonomists recognize two main sub-species or races of rice: the short-grained *O. japonica* and the long-grained *O. indica* (Crawford and Shen 1998: 863). These two cultivated rice subspecies are perhaps the most important domesticates in human economic and demographic history. On this account, we would therefore expect that botanists and archaeologist respectively would have closely (1) mapped the distribution of wild rice, and (2) studied the archaeological evidence for the origins of domesticated rice. It is rather surprising, therefore, to discover that the distribution of modern wild rice is poorly known and, until quite recently, the date and place of origin of domesticated rice very poorly understood.

However, recent archaeological work has begun to fill the large gaps in our knowledge of the origins of rice and the picture that is emerging is a surprising one. Once thought to have been brought into the human orbit first in Southeast Asia relatively late in prehistory during the Middle Holocene, domesticated rice has been recovered from Chinese contexts dating to the Early Holocene epoch. Plant parts known as **phytoliths** apparently from rice plants of a far earlier date may indicate that wild rice tending began in the Terminal Pleistocene in East Asia. Clearly, we are witnessing a profound transformation in our knowledge of the domestication process in Asia.

Nonetheless, our knowledge of the ecology of wild rice ecology in in pre-agricultural times remains sketchy. Harland maintains that at the time

> the deltas of the great rivers of southeastern Asia were probably vast seas of wild rice. The mouths of the Ganges river in Bangladesh, the delta of the Irrawady river in Burma, the Chao Phraya river in Thailand, the Mekong river in Viet Nam, and other lesser river deltas were probably once covered with rice of a floating type adapted to deep water and the rapid rise and fall of the rivers during the monsoon season. During high water, seed heads probably

> covered huge areas, sticking up above the water. When the water receded, tangled masses of stems must have covered the ground. These floating rices were mostly perennial but were not as strongly rhizomatous as the related African perennial rice. The stems elongate very rapidly as water rises, and when the water recedes, the tangled mass tends to root at the nodes (Harlan 1977: 370).

Thus, at first glance, the great river deltas of tropical Southeast Asia appear to provide the ideal setting for the domestication of wild rice. In fact, they may have been just the opposite. If Harlan's reconstruction is correct, these great deltas constituted what Binford calls an "optimum habitat zone," that is, a natural ecology so rich in wild foodstuffs as to offer foraging peoples no inducement to try to manipulate wild species artificially. The floating rice was probably both abundant and easy to harvest from canoes or rafts. If someone had suggested cultivation to the early Holocene foraging peoples exploiting this rich floating resource, their reply might have paraphrased Lee's *!Kung* or *Jul'hoansi* informant: "Why should we plant when there is so much wild rice in the world?" Even if there were inducements to tend the floating rice, however, natural conditions in the deltas would have mitigated against human control. According to Harlan (1977: 370-371):

> the floating rices are generally poor seeders and the environment almost impossible for the initiation of agriculture. Without control of water, no planting or selection is really possible. These areas eventually became the rice bowls of Asia, but not until dikes and canals could be built. The environment of the delta is not the place to begin domestication of rice.

For this reason, Harlan suggests that the arid savannahs of Southeast Asia were far more suitable locations for the early manipulation of rice than the tropical lowland river deltas. In monsoon eastern Eurasia, the savannah are characterized by wet and dry seasons. Shallow natural depressions on the savannah fill with water during the rainy seasons and dry up after the rains abate. Annual, rather than perennial, varieties of wild rice are well adapted to such seasonal alterations and, during the early Holocene, these shallow natural savannah depressions would no doubt have made ideal rainy season habitats (Harlan 1977: 370-371). Manipulating and controlling annual wild rice in such settings would have been far simpler than controlling perennial rice in the lowland tropical deltas.

The domestication process would have been initiated when wild rice foragers simply began re-seeding the annual rice by broadcasting seeds across the surface of natural depressions just before the onset of the rainy season. T. T. Chang (1989: 410) suggests that this process was probably gradually refined by planting the seeds in holes with a dibble stick or by seeding them in rows. Transplanting growing plants to paddies would have come much later and have been part of a land-use sequence in which shifting or swidden agriculture gave way to the establishment of permanent fields and then to fields with low earth and stone dams around them to retain water. The earliest permanent fields would have depended on rainfall; artificial irrigation would have developed much later. Chang consider this entire developmental progression to have been driven by increasing human population density. He thinks this domestication process took place somewhere in "a broad belt that extended from the southern foothills of the Himalayas, across Upper Burma, northern Thailand, and Laos, to north Vietnam, and southwest and south China."

Chang suggests rice domestication took place there between 15,000 and 10,000 years ago at many separate times and places within this vast area, but he readily concedes that the archaeological record from this vast area is far to incomplete to allow us to be certain (Chang 1989: 410). At present, the earliest evidence of domesticated rice comes from northern Thailand. Deposits excavated by Gorman at Banyan Valley Cave, a small cave site near Spirit Cave in Thailand, produced quantities of rice seeds and hulls (*Oryza*) in layers that also contained pottery. The rice, recovered from layers dating to 5,500 B.C., may have been obtained by the cave residents through trade with nearby rice farmers but its size and morphological characteristics suggest to Yen (1977: 588) and others that these specimens may be of wild rice gathered by the cave's residents. Of course, it is possible that they were "tending" this wild rice and were thus incipient cultivators.

Organic preservation is generally poor in Southeast Asia and this has complicated the search for archaeological evidence of plant cultivation. However, evidence of plants have been recovered from the valley of the Mekong river in Thailand at the sites of Non **Nok Tha** and **Ban Chiang**. Rice chaff was mixed into ceramic clay as a temper and the imprint of the rice was preserved on the surface of some of the pottery dating to shortly after ca. 4,500

B.C. Unfortunately, whether or not this rice was domesticated has not been determined. In South Asia, evidence of rice cultivation has been reported from the site of **Koldihawa** on the flood plain of the Ganges river in India ca. 4,500 B.C. also but, for a number of reasons, this date may be suspect (Jacobsen 1979: 483).

China's Emerging Record of Rice Domestication Until very recently, the earliest definite archaeological evidence of rice cultivation from China was recovered the site of **Hermudu** in the lower valley of the Yangtze River. The site, located in central southeastern China near the coast of the China Sea, dates to some time between ca. 5,500 and 4,500 B.C and contains the remains of a large, well established farming village consisting of wooden buildings erected on pilings along the edge of a lake. The inhabitants were skilled in woodworking, pottery making, weaving and ivory carving. In addition to domestic rice, they grew bottle gourds, water chestnuts and sour jujube fruit trees and raised dogs, pigs, and water buffalo. They also hunted a wide range of wild animals and gathered acorns and shellfish (K. C. Chang 1981: 152-154). Zhimin (1989: 645) states that some form of slash-and-burn or swidden agriculture would have been the appropriate method of cultivation given the dense forest growth of the Yangtze valley. In any case, the high degree of cultural and economic development reflected at Hermudu clearly indicates that domesticated rice farming had begun well before occupation of the site. But how much earlier?

The answer may lie beyond the Tsingling Mountains in China's northern interior. Rice grains recently recovered from deposits at the sites of **Pengtoushan** and **Jiahu** in Yangtze valley in south central China suggest that the process was well underway between 6000 and 7000 B.C., at least a thousand years before the occupation of Hermudu (Crawford and Shen 1998: 862). Even earlier dates have been proposed by the Sino-American archaeological team excavating **Diaotonghuan Cave** in the Dayuan Basin just below the Yangtze river in south central China. In their view, domesticated rice began to be grown in the Dayuan region between 7000 and 8000 B.C.(Zhao 1998: 895). The chief evidence of the presence of domesticated rice at Diaotonghuan cited by its excavators are plant **phytoliths**.

The word phytolith means literally "plant stone" but in the emerging science of phytolith analysis, the term is used in a much more restrictive sense. There it refers to the microscopic silica deposits that form in the cellular structures of many kinds of plants. Silica is taken up from groundwater and distributed within the plant. In the process. the silica comes to take on the distinctive shape of the plant cells which it fills, consequently, phytoliths from plants of different genus and species vary in size and shape. Phytoliths are common in soils, sediments and archaeological deposits. This is because the normal processes of decay, burning or processing detaches large quantities of them from the plants that produced them. Once deposited, phytoliths resist decay for long periods of time. It is therefore possible to extract phytoliths from deposits and identify the genus and, sometimes, the species, of the plants that produced them. Phytoliths thus provide archaeologists with an important–although as yet little used–new tool for identifying changing patterns of plant growth and use (Piperno 1988: 11, 40).

Phytolith analysis promises to be particularly useful in recognizing the appearance of domesticated plants in the archaeological record. This is because, the process of human selection that has operated on domesticated plants has, in general, resulted in plants that have larger fruits and seeds than their wild ancestors. As a consequence, the cells of many domesticated plants are larger also and so too are the phytoliths than form within them. The Chinese phytolith analyst, Zhijun Zhao, has used this principle to develop accurate measurements that allow him to distinguish phytoliths produced by wild rice plants from those from domesticated ones. Using this method in the analysis of the phytoliths from the deposits at Diaotonghuan cave, Zhao asserts that phytoliths from domesticated rice plants appear in a layer dated to between 11,000 and 7,000 B.C. at the site (Zhao 1998). These phytoliths are the only evidence of domestication in the layer, however. Animal bones and cultural material from the layer indicate that Diaotonghuan cave's occupants made their living chiefly by hunting and gathering during that time. These data suggest that, if rice was under cultivation ast this time, it was apparently raised as a supplemental food. In fact, it is not until after 5000 B.C. that evidence of rice is sufficiently abundant in the cave deposits to suggest that it had become a staple food of the people who lived there. Nonetheless, if Zhao's analysis is correct, the rice phytoliths from Diaotonghuan constitute the oldest evidence of domesticated rice yet recovered. What's more, they suggest that rice cultivation is among the oldest form of agriculture in the world. Of course, confirmation of these results at other sites must be obtained before these early dates are widely accepted.

Origins of Millet Cultivation As one moves from the humid, well-watered regions of coastal and tropical China, annual rainfall declines sharply and the climate becomes subhumid, then semiarid. Precipitation in northern China and Manchuria is lowest in the winter months, increases in frequency in the spring, and reaches its highest levels in the summer (Trewartha 1981: 220-221). Not surprisingly, therefore, the agricultural complex or sub-complex that emerged in North China contrasts markedly with the one that emerged in the south. There, dispite the apparent early appearance of rice, the the domesticated plant that came to be central to early agricultural subsistence was the small grain known as millet.

Recall from our earlier discussion of African agriculture that millet is the common name for a number of species of drought-resistant domesticated grasses used as human cereal grain, animal forage, and hay. Millet is grown throughout the world but is especially popular China, India, Africa and the Soviet Union. Although finger millet was probably domesticated in Africa, foxtail (*Setaria italica*), broomcorn, pearl and most other important millet species appear to have been independently domesticated in north China by peoples living along the upper and middle reaches of the Huang Ho or Yellow river.

The Huang Ho rises in the mountains of northern China, cuts eastward, then southweard in a huge bend, then eastwards again through the great loess deposits of north central China. It leaves the bend to form a vast deltaic plain and enters the Yellow Sea between China and Korea. During the sixth millennium B.C. at least three recognizably distinct early Neolithic cultural traditions emerged in the upper and middle reaches of the Huang Ho drainage: the **Peiligang**, the **Cishan** and the **Dadiwan** cultures. These precocious early Neolithic traditions were based on the raising of millet and pigs. Zhimin (1989: 645) suggests that the dry, savannah grassland environment of the Huang Ho valley necessitated the periodic fallowing or abandonment of fields to allow them to recover their fertility.

Farmers lived in semi-subterranean pithouses and made ceramic pots, bowls, tripods in a coarse but serviceable red ware. Carbonized grains of millet, probably foxtail millet, have been recovered from their storage pits and pottery. The importance of cereal grains in their subsistence is further reflected in the large number of grinding stones, stone sickles, and saddle-shaped querns on their sites. Animal bones recovered indicated that pigs and dogs were their only domesticated animals and th+at, as in the south, hunting and fishing remained important subsistence activities. As K.C. Chang (1981: 152) put it "there is little question that millet farming and pig raising, the twin pillars of ancient civilization of North China, are now established as dating at least from the sixth millennium B.C., almost in sight of the very beginning of the postglacial period." By approximately 5,000 B.C., these three distinct early Neolithic life ways had coalesced into a single distinctive regional Neolithic tradition known as the **Yangshao** culture.

Although it was centered in the upper and middle reaches of the Huang Ho, the impact of Yangshao culture extended south as far as the middle drainage of the Yangtze. The Yangshao cultural tradition dates to between about 5,000 and 3,000 B.C. Yangshao peoples lived in large, mud walled villages on the terraces of the Huang Ho near their fields and made ceramic vessels in a strikingly original style. Over time enlarged their agricultural repertoire to include limited quantities of cattle, sheep and goats, and the first definite insect domesticant, the silkworm (K.C. Chang 1977: 95). Although fragmentary horse remains have been recovered in a Yangshao archaeological context, it is not certain if these animals were under domestication during the period (Ben-Shun 1989: 105).

The Yangshao culture is followed and replaced by the Longshan cultural tradition (3,000 to 2,000 B.C.) over much of north China. During this period, The domesticated horse is definitely added to the agricultural repertoire during this time, as are sheep and goats, Chinese artichokes, soybeans and numerous fruit tree species. The roots of Chinese village life and civilization are clearly visible in the Longshan. After 2,000 B.C., the emergent Shang dynasty ushers in Chinese history.

Introduction of Rice to Japan If substantiated, the new evidence of plant cultivation and domestication in the early Holocene in the Yangtze river valley will force a revision in our understanding of Chinese prehistory. What effect will these controversial new dates have on our understanding of the rise of food production in Japan?

As we noted in Chapter 10, a number of Japanese archaeologists have made a strong inferential case that, prior to the introduction of rice cultivation into Japan, peoples of the Jomon tradition tended, manipulated and incipiently controlled certain plants native to the archipelago. Some of these scholars further contend that Jomon sedentism was based partly or largely on native agriculture. While their data are suggestive, they have yet to demonstrate the indigenous domestication of indigenous wild plants in Japan. The entire question, the most controversial and contentious one faced by Japanese prehistorians, can only be resolved by additional excavations and further paleobotanical research.

Of course, the question become even more complicated if it is true that rice was under cultivation in North China during the early Holocene. Until quite recently, most archaeological and paleobotanical evidence from Japan suggested that rice cultivation: (1) did not arrive in archipelago until the Late Jomon period (2,500 B.C. and 1,000 B.C.), (2) was not widely practiced outside of southern Japan until Final Jomon (1,000-300 B.C.) and (3) did not become a central element in subsistence over much of the archipelago until the subsequent **Yayoi period (300 B.C. to A.D. 300)**. If this date range for the spread of rice cultivation is correct, it indicates that Japanese subsistence practice was in marked contrast to that of Korea and China where rice cultivation became the dominate subsistence activity at a far earlier date. If this contrast is correct, how can we explain the retarded acceptance of rice cultivation on the Japanese archipelago. Was it due to the ecological appropriateness and intrinsic strength of traditional Jomon subsistence strategies as some prehistorians have suggested? Or have we simply missed the evidence that rice was present there much earlier than we have supposed? Only further archaeological research can resolve the issue.

Yayoi Tradition (300 BC-AD 300) The cultural period that follows the Jomon in the Japanese archaeological sequence is known as the Yayoi. The name of the tradition is derived from its type site in Tokyo where its distinctive pottery was first recognized. It is not its pottery that renders the Yayoi watershed period in Japanese prehistory. Rather, it is the triumph of wet rice cultivation during the period and the revolutionary transformation of economic and social life on the archipelago that follows that triumph. The ultimate source of this new form of agriculture is clearly continental Asia and is not accidental that the earliest known Yayoi sites are located at the far western end of the archipelago on Kyushu island nearest the Korean coast across the Sea of Japan. Not surprisingly, culture of the Early Yayoi period culture in the west shows marked similarities with contemporary cultures in Korea and western China whereas Yayoi sites found in eastern Japan possess a more insular character and have more traits in common with the Jomon.

The Yayoi Tradition in Japan

Dates	**Period**
300 BC-100 BC	Early
100 BC-AD 100	Middle
AD 100-AD 300	Late

The exceptional preservation at the waterlogged site of **Itazuke** on the northwest side of Kyushi island has provided archaeologists with an rare window on this subsistence transformation. Itazuke appears to have been occupied at the very cusp of the transition between the Jomon and the Yayoi periods at a point on the archipelago closest to the Korean pennisula. The recovery there of the remains of paddy-field ridges, drainage ditches, distinctive *ishibocho* stone reaping knives, wooden farming tools and carbonized rice grains is therefore particularly significant. The evidence from Itazuke indicates that wet rice agriculture was being practiced in southern Japan on a fully organized and sophisticated basis by at least 300 BC. (Aikens and Higuchi 1982: 199-205). Yet despite the elaborate nature of its rice husbandry, the Early Yayoi agricultural complex apparently altogether lacks domesticated animals besides the dog. Barn yard staples like domesticated cattle and horses do not appear in the Japanese archaeological record until the Late Yayoi period; pigs and chickens only in the subsequent **Kofun period** (A.D. 300 to 600) (Anderson 1987: 277-278).

Anthropometry, the science of human skeleton and body measurement, has also been brought to bear on the question of Yayoi period subsistence. Careful measurement of burial population reveals that the shapes of crania changed between the Jomon and the Yayoi period. The new cranial shapes have been interpreted as reflecting either the arrival of a new racial groups–presumably agricultural peoples from mainland eastern Eurasia–or the effects of a pronounced change in diet. In either case, this cranial change is commonly cited as further strong, if indirect, evidence that the transition from food collecting to food producing took place in Japan no earlier than the late Holocene epoch.

In sum, two conclusions about the rise of rice agriculture in Japan seem secure: first, that such husbandry was imported into Japan from China across the Korea peninsula and second, that its rise to prominence occurred only during the Final Jomon and Yayoi periods. While accepting these two conclusions, archaeologist still differ on the significance of this this change. As we have seen some scholars contends that, when rice cultivation first appeared in Japan during the Late Jomon, it was not a revolutionary innovation. Rather, they suggests such husbandry was simply grafted onto an existing systems of indigenous agriculture that had been centered on such native eastern Eurasian root crops like wild taro and yams and on the incidental cultivation of imported seed crops mentioned earlier including buckwheat, millet, beans and gourds (Koyama 1978). While we known these plants were present in the Jomon, the extent of their cultivation and dietary importance remains unknown. However, we can be sure that, whether mere graft or profound foreign innovation, the practice of paddy rice agriculture transformed Japanese subsistence forever. The adoption and spread of rice husbandry did not simply change diet and nutrition on the archipelago. The new foodstuff was accompanied by a suite of other culture traits ultimately derived from continental Asia. Weaving using looms was part of this suite, as were new styles of architecture, tomb construction. New pottery forms also appear in the Yayoi including distinctive pedestaled bowls with geometric decorations. Jars with long necks also become common as do a variety of basins and pots with wide mouths and deep interiors are part of these Yayoi cultural suite. Somewhat later, bronze and iron metallurgy was added to the mix.

While the rise and spread of the Yayoi marks a break with Japan's Jomon past, a marked continuity between the prehistoric Yayoi culture (and for that matter, the later Kofun period) and that of historic Japan is clearly evident. This continuity is nicely reflected in **Shinto**, the traditional religion of Japan. Numerous Shinto symbolic and sculptural motifs–and, one imagines, the concepts that lay behind them–can be traced back to the Yayoi and Kofun periods.

Yayoi Economy and Society In Holocene Japan, no less than elsewhere at this time, Neolithic peoples began to place increasing reliance on stored resources. This greater reliance is clearly reflected at the **Toro** site, a Middle to Late Yayoi village was built on a natural levee in the delta of the Abe River on the Pacific coast of Honshu island. The Toro village is a kind of Japanese Pompeii. Some time shortly after the beginning of the Late Yayoi period it was completely buried by flooding on the Abe river. What was no doubt a catastrophe for the Toro villagers has proven to be a great boon for archaeology. Flood deposits covered and preserved the architecture and much else at the site. Excavations have so far revealed that the village consisted of a dense cluster of at least twelve pithouses surrounded by large granaries with raised floors. Stored food remains reveal that village subsistence was based on both the cultivation of wet or paddy rice and the dry farming of millet, melons and gourds. Peach stones were also recovered in the excavations. While these stones may be evidence that the villages had domesticated peach orchards, they may just as readily indicate that the villagers gathered wild fruit in the surrounding forests. In addition to the stored food remains, the Toro site contained the remains of skillfully carved wooden farming tools including plow parts and a host of every day wooden objects like bowls, spoons and swords. The fact that many of these artifacts were still more-or-less in the place where they were last used by the inhabitants makes Toro a particularly vivid window into the life and material culture of a rice farming village in the Late Yayoi period (Aikens and Higuchi 1982: 226-239).

Sites from the Yayoi period have also provided important evidence that pronounced social stratification was emerging in Japan at this time. Perhaps the clearest evidence of this differentiation is the appearance of elite burial mounds. Japanese archaeologists are not fully in agreement as to whether the design of these mounds is of indigenous origin or reflects cultural diffusion from Korea or China. In either case, the Late Yayoi site of **Tatetsuki** in south central Honshu island contains one of the most notable of these mounds.

Numerous other mounds of widely varying size and elaboration have been found scattered throughout the archipelago. Despite these variations, collectively these mounds indicate that by the Late Yayoi period, special treatment was being afforded to a very limited number of high status individuals. Such special treatment is, in and of itself, powerful evidence that a pronounced degree of social ranking had emerged on the Japanese archipelago by the Late Yayoi period. It seems likely that the individuals buried in these mounds were the leading members of locally-important kin groups who dominated social and economic life in the immediately surrounding area. Such mounds are thus the material remains of what was probably a host of small, independent, kin-based polities there. As time went on, some of the larger ones may been transformed into chiefdoms capable of exercising power and control over whole regions rather than of merely local districts. Even larger and more elaborate mounded tombs appeared during the succeeding Kofun period suggesting that, once started, the trend towards greater centralization of political power continued apace.

Archaeologists have also found increasing evidence of competition, warfare and violence during the period of mound construction. The Late Yayoi people had a penchant for building their habitation sites in defensible, upland locations and then surrounding them with earthen embankments and moats. This is clear, if mute, testimony that competition during the period often turned violent. The multiple villages excavated at the site of

Yoshinogari on Kyushu island is a particularly good example of this trend. Not only are the Yoshinogari villages carefully position on the local landscape to maximize defensive potential, their builders enhanced the strength of the placements by encircling entire settlements with two moats. As might be expected, mounds built within the moated area contained burials of apparently high status individuals of well-developed social hierarchy (Hudson and Barnes 1991). Taken in aggregate, the defensive preparation and social stratification evident at Yoshinogari and other Late Yoyai sites in western Japan indicates that, by the end of the period, the processes lead to the formation of chiefdoms and states were well under way. These processes continued at an accelerating pace throughout the subsequent **Kofun period (AD 300-700)**. At the same time, contact between Kofun Japan and the surrounding higher cultures in Korea and China intensified through trade and diffusion. By the end of the Kofun period, culture change due both to internal processes of institution building and external contact and diffusion created the Ritsuryo state, the first social formation to unite virtually the whole of the Japanese archipelago. The basic elements of historic Japanese culture and society were now in place.

Origins of Agriculture in the Americas

Although the question has been disputed bitterly and often, most scholars agree that agriculture initially developed independently in the New World (but cf., Carter 1988; Jett 1983). With the possible exception of the bottle gourd and linted diploid form of cotton (Pickersgill 1989: 435) for which no American wild ancestors have been found, "native American farming was based on indigenous plants and animals, and on a technology which owes nothing to contacts with Europe or Asia" (Bray 1977: 225). This independence is reflected in the paucity of domesticated animals in Native American agriculture and the fact that no domesticated animals of Old World ancestry are know to have been present in the New World prior to 1492 (but cf., Carter 1988; Johannessen 1986).

Native fauna in the New World was greatly impoverished by the massive extinctions that occurred at the end of the Pleistocene. Particularly important in this regard was the extinction of the camelids in North America and horses, mastodons and mammoths throughout the New World. Native American agriculturalists did their best with the wild species that remained and domesticated the dog, the Moscovy duck, the turkey, the Guinea pig, the llama, the vicuna and several other species. Nonetheless, as beasts of burden and meat animals these New World domesticates are simply not equivalent to the goats, sheep, pigs, cattle, horses and camels that formed the back bone of mixed husbandry of Eurasia and Africa. New World peoples therefore lacked the dietary, transport, labor and military advantages that animals afforded the people of the Old World. This weakness in their agricultural systems began glaringly apparent following the early 16th century arrival of mounted Spanish Conquistadores in Mexico and Peru.

However, the dearth of animal domesticates in the New World also had a very positive impact on human health and longevity in Pre-Columbian times. New World peoples developed fewer human/animal diseases precisely because they had far less frequent and intimate contact with domesticated animals. As a consequence, Pre-Columbian native peoples lived in relatively disease-free environments. Ironically, after 1492, this circumstance was also destined to turn into a profound disadvantage. Lacking immunity to Old World illnesses, Native American peoples in truly terrifying numbers shortly after the arrival of Europeans and Africans on their shores. Yet, with the possible exception of syphilis, the New World had no similarly virulent diseases for which Old World peoples had no immunity. This imbalance in pathogens and immunity initiated the great population change of the last 500 years and had profound consequences for the Native Peoples of the Americas.

Physiography of the Americas The geography of North and South America is dominated by four major physiographic features:

1. **Great north-south trending mountain ranges** At the western end of the New World, a virtually unbroken chain of mountainous terrain extends south from Alaska and the Canadian arctic through the Pacific Coast ranges of North America into the tropical highlands of Mesoamerica. From there it continues across Isthmus of Panama to join the Andean highlands of western South America that extend virtually to the southern tip of the continent at Tierra del Fuego.
2. **Expansive semi-arid steppe land and arid deserts** The rain shadow of the north-south trending mountain ranges create large regions of low rainfall. Such regions include the intermontaine deserts and the Great Plains in North America and the vast Pampas that stretch from the Andes foothills eastward to the Atlantic between the Gran Chaco to the Rio Colorado in South America.

3. **Vast expanses of forest land** In North America temperate forests or forest-prairie begin in the Mississippi drainage and stretch eastward to the Atlantic ocean. Tropical forests dominate central and southern Mesoamerica and northern South America.
4. **Immense dendritic river systems draining the interiors of both continents** In North America, the Mississippi-Missouri-Ohio system drains the Rockies and the steppe lands on the west and the prairie and forests on the east. In South America, the Amazon and Orinoco rivers drains the Andean highlands of western South America and the great tropical forest of the lowlands at the northern end of the continent.

New World Agricultural Hearths Archaeological and ethnographic research indicates that there were four main geographic areas of primary domestication in the New World. By late prehistoric times, four related agricultural subcomplexes were established in and beyond each of these areas:

1. Central and southeastern North America,
2. The tropical highlands and dry interior basins of Mesoamerica,
3. The tropical lowlands of Mesoamerica and northern South America, and
4. The Andean highlands of western South America.

The crops and cropping techniques that emerged in these four areas closely reflects the character of the climatic and vegetation that emerged in them during the Holocene epoch.

The cool, dry climatic conditions of late Pleistocene altered dramatically in North America around 10,000 years ago. It was at this time that the modern North American plant assemblages began to take shape. As elsewhere in the world during the generally warmer and moister conditions of the early Holocene resulted in the circumscription of the deserts and tundra regions and the expansion of the grasslands and forests (Roberts 1989: 92). These warmer conditions prevailed in North America as elsewhere on Earth until around 2000 B.C. when mean annual temperatures began to decline (COHMAP 1988: 1048).

The climate and vegetation that characterizes the natural regions of the New World at the present-day were essentially established during the late Holocene. By that time, permanent land ice had largely become restricted to Greenland. Arctic and sub-arctic climatic zones that stretch from east to west across vast interior plains of Canada to form two broad parallel bands of transcontinental climate. These zones are respectively characterized by tundra and spruce-dominated boral forest. In contrast, the climatic zonation of far southern Canada and the continental United States is more complex. Instead of parallel bands, the zones form a checkerboard of often sharply contrasting climatic regimes.

This checkerboard pattern is partly a reflection of the physiography of the continental United States. For example, the climate of the west coast of North America is humid to subtropical in character. However, the mountains of the high cordillera that run roughly parallel to the west coast greatly restrict the penetration of marine influences into the interior of the continent. Consequently, there is no gradual transition from humid maritime to humid continental climates in western North America as is the case in Eurasia (Trewartha 1981: 281). Instead, semi-arid and arid conditions appear abruptly on the eastern side of the coastal mountains and dominate the Great Basin of the intermontane region and steppes of the Great Plains beyond the Front Range of the Rockies.

Rainfall increases steadily as one moves eastward across the Great Plains towards the Mississippi valley and humid temperate conditions characterize the central lowlands of the Midwest and plateaus and mountains of the north eastern United States. Vegetationally this climatic transition is marked by the change from short-grass steppe to tall grass prairie to broad-leaved deciduous forests containing oaks. Until about 2000 B.C., the boundary between the forests and the prairie in the Midwest was much farther east than it is at present time. This indicates **drier** conditions prevailed at midcontinent during the early and middle Holocene (Roberts 1989: 90).

Central and Southeastern North America The central lowlands of North Americas of North America stretch from the western edge of the Appalachian mountains to the Mississippi river drainage and from the northern shores of the Great Lakes south to the Ozark Plateau. It is a region of low altitude and slight relief. Its climate is continental in character with sharp seasonal differences. The growing season in the central lowlands is hot and rainy, its winters cold and stormy. This climate, in combination with its deep, rich loess soils and flat terrain, make the central lowlands of North America the most productive agricultural region in the modern world (Hunt 1974: 345). The Great Lakes and the major rivers drain the central lowlands in three main directions: northward to Hudson Bay, northeastward through the Great Lakes and the St. Lawrence River, and southward to the Mississippi River (Hunt 1974: 330-353).

Rainfall declines as one moves across the region from east to west. From a high of about 50 inches or more in eastern Tennessee, mean annual precipitation declines gradually to about 40 inches in the Ohio country, 20 inches near the 100th meridian, and then drops to about 15 inches or less on the semiarid Great Plains the Missouri River. This rainfall cline is reflected in the distribution of vegetation across the region. As one moves from east to west, broad-leaved deciduous oak forests gradually give way to tall grass prairie and thence, to short-grass steppe. Until about 2000 B.C., the boundary between the forests and the prairie in the central lowlands was much farther east than it is at present time. This indicates **drier** conditions prevailed at midcontinent during the early and middle Holocene (Roberts 1989: 90). Interestingly, the earliest evidence of domesticated **sunflower** (*Helianthus annuus* var. *macrocarpus* Ckll.) has been dated to 4,200 BP (Crites 1993; Rieseberg and Seller, 1990), near the end of these drier climatic conditions. Perhaps these conditions may have fostered mutualistic relations between humans and the wild ancestors of this edible plants that lead to the development of plant husbandry in the region.

Southeastern North America Some of the highest levels of annual rainfall and humidity in North America are found in the Southeast, that corner of the continent that streches north from the Gulf and Atlantic coastal plain to the Ohio river valley on the north and west across the Mississippi to the Ozark Plateau. The modern climate of the Southeast ranges from temperate to humid and subhumid subtropic characterized by mild winters, hot summers, abundant, year round rainfall, and a long growing season. Yet, exceptions to this generally benign climatic pattern in the form of prolonged droughts and severe winters are not unknown in historic times (Butzer 1971: 69-70). The modern climate of the southeast supports a potential climax vegetation of great biotic richness and variety. Vegetation ranges from spruce-fir assemblages in the higher elevations of the Appalachians, through upland forests of pine and mixed hardwoods in the plateau provinces, to the piney woodlands of the coast plains. By the middle Holocene, Archaic peoples living in the abundant environments of the Southeast had begun to manipulate and cultivate indigenous plant species such as sunflower, marshelder, native squash and chenopod (Smith 1989). Southeastern domestication occurred contemporary with–but apparently independent of–the domestication of corn and other tropical species taking place further south in Mesoamerica.

By 1000 B.C., cultigens domesticated in Mesoamerica began to arrive in many parts of North America and spread widely over much of the continent. In the Southeast, corn began to replace local cultigens in the native subsistence. By late Prehistoric times, a distinctive agricultural complex had come to characterize the temperate regions of North America. Farming peoples there combined corn, beans, amaranths, squash and other Mesoamerican plants with species like sunflower and chenopod that had been domesticated independently in North America. Thus, the North American agricultural complex was the product of both independent domestication in native plants in the southeastern United States and of the diffusion of Mesoamerican agricultural complex into North America.

The northern limits of maize cultivation became the limits of the North American agricultural complex in general. This limit was reached at roughly 47.5° longitude in what is today southern Canada. For the most part, land south of this line averages at least 120 frost free days during the growing season. Yarnell (1964) reports finding no early historic reports or archaeological evidence of maize growing north of this line despite the fact that some Native American corn varieties could mature in 70 days or less.

Domesticated animals were a decidedly minor element in the native North American economy. For the most part, only the dog and the turkey were significant. Although dogs were commonly eaten, turkeys were raised chiefly for their feathers. On first encounter with the Spanish explorer Coronado and his men, Pueblo Indians villagers in the American Southwest were shocked when their well-armed and formidable "guests" slaughtered and ate at a sitting the turkeys they had for years been carefully nurturing!

Mesoamerica The region known as Mesoamerica includes the modern state of Mexico and the Central American republics. These states are located on the Isthmus of Panama, "share a general focus on the Gulf of Mexico and the Carribean Sea and an intermediate location between North America and South America" (West and Augelli 1989: 3). The great mountainous spine of North America continues south across area creating the fundamental Mesoamerican physiographic and cultural distinction between highlands and lowlands.

Four major natural regions characterize Mesoamerica:

1. Subtropical dry lands
2. Tropical highlands,
3. Tropical dry basins and valleys, and

4. Tropical lowlands.

Existing archaeological evidence suggests that the path to agriculture was slightly different in each region.

Sub-Tropical Dry Lands The subtropical dry land extends from the southwestern North America into northern Mexico. During the late Pleistocene and early Holocene, much of these lands appear to have been moderately well-watered. However, by the middle Holocene conditions began to become more like the present (Wright 1987: 486). Today, the area is characterized by high annual temperature, low, eratic average annual rainfall concentrated in summer monsoons and low humidity. Cactus and other xerophytic vegetation tend to dominate plant assemblages at lower elevation.

These arid conditions did not prove to be a great impediment to native agriculture. Mesoamerican domesticated plants were being cultivated there after 1000 B.C. By late Prehistoric times, elaborate systems of irrigation were being used by the **Hohokam** peoples on the drainage of the Gila and Salt rivers of southeastern Arizona and by the **Anasazi** peoples on the Colorado Plateau.

Tropical Highlands The tropical highlands are located in the rugged mountain spine that runs north to south through Mesoamerica. The highlands begin at elevations above about 4,000 feet and are distinguished from the drylands by their moderate, mainly summer, precipitation and cool to cold temperatures throughout the year. Although badly over-forested in modern times, the tropical formerly supported temperate oak-conifer forests at the middle elevations, tropical cloud forests on high, exposed slopes, and open woods and grasses in valleys (West 1964: 370-371). The tropical highlands has been the most significant natural region in Mesoamerica from prehistory to the present. Not only were the tropical highlands a primary center for early domestication in the New World, they were a locus for early state development, came to support the highest population densities and elaborate cultural traditions in the Americas. Below approximately 4000 feet in elevation, the highlands begin to give way to the tropical lowlands. The geographic division between the tropical highlands and lowlands has had a profound impact on human culture and subsistence throughout Mesoamerican prehistory and history.

Tropical dry basins and valleys Within the lowland tropics of Mesoamerica are dry interior tropical basins and dry valleys where the rain-shadow of the surrounding mountains limits annual precipitation to between 20 to 40 inches (500-1000 mm). As a consequence of their special climatic conditions, vegetation in such basins and valleys range from steppe-grasslands to fully-arid, deserts dominated by thorny, scrub and cactus vegetation. Dry caves and rockshelters are common in some of these valleys and from the rockshelters of the semi-arid **Tehuacan Valley**, archaeologists have recovered the earliest evidence of agriculture in the New World (West 1964: 381).

The agricultural complex that emerged in the tropical highlands and interior dry basins of central Mesoamerica centered on the cultivation of beans, squash, amaranths, chili, tomatoes, vanilla and other vegetables. A number of cactuses were also part of this complex as were such fruit trees as avocado, pineapple, sapodilla, guava, papaya, and cacao. Cotton and sisal were raised for fibers, tobacco and other plants as drugs, medicines and hallucinogens. Despite the great richness and diversity of its domesticated species, the Mesoamerican agricultural complex was centered on the cultivation of one crop in particular: maize or corn. Maize is the common name for *Zea mays*, a domesticated form of New World grass that is today the world's third most important food crop after rice and wheat.

The nature of the domestication process is a particularly well understood in this region for two reasons. First, the abundant dry rockshelters in the arid and semi-arid valleys of highland Mexico provide excellent micro-environments for the preservation of organic remains. Second, over the last forty years, archaeological field work has been undertaken in the region by some of the most able scientists in modern archaeology. In Chapter Five the excavation and interpretation of one particularly notable highland Mexican rockshelter site, the **Coxcatlan Cave** in the Tehuacan Valley was discussed in detail. This site was excavated as part of a long term archaeological and botanical research project that sought nothing less than the explanation of the origins of Mesoamerican corn agriculture.

The nearly four meters of stratified deposits excavated at Coxcatlan contained an unparallel record of the ecological and cultural change that took place in the Tehuacan Valley during the Holocene epoch. Among other things, the deposits produced fragments of the earliest maize ever recovered. The primitive nature of this maize suggests that it was in the first stages of domestication. The Coxcatlan record and other data indicate that wild maize had begun to be domesticated by highland **Preceramic** (or **Archaic**) peoples sometime after 5,000 B.C. but that process of social and cultural change that followed from that domestication was a slow one. It was not until sometime between ca. 4,000 to 3,500 B.C. that maize and other cultivated plants had begun to make a significant contribution to the diet in the highlands.

During this time, corn became a fully dependent upon humankind. Like all wild grass species, the wild ancestor of corn must have had an efficient mechanisms for dispersing its ripe seeds. Human selections over countless generations has entirely removed that mechanism from its modern descendants, however. Corn seeds adhere tightly to their cob and are wrapped in a "shuck" or thick sheath of leaves which must be removed in order to free the kernels for planting. For this reason, modern corn can longer re-seed itself. Lacking an efficient mechanism of seed dispersal, corn is completely dependent upon humankind for its survival.

At some point in this process, maize began to diffuse from its original highland Mesoamerican hearth. Other areas of the New World were profoundly influenced by the diffusion maize and other crops and cropping techniques from Mesoamerica after this time. When Europeans arrived in the New World, they found native peoples raising maize from the mouth of the St. Lawrence in Canada and as far south as central Chile.

It is easy to understand why this diffusion occurred, After all, as George Beadle (1980: 112) notes, maize "is the most efficient of all cereals in converting solar energy, carbon dioxide, water and mineral nutrients into foodstuff." It is furthermore one of the most environmentally-adaptable domesticated plants. Yet, for all its efficiency, productivity and adaptability, maize has an important drawback: it is deficient in such essential elements of human nutrition as the amino acids tryptophan and lysine, the metals iron and zinc, as well as calcium, vitamin C, and niacin. Diets deficient in these key elements have been linked to pellagra and other deadly diseases.

Since maize does not supply these nutrients, to avoid pellagra and achieve a balanced diet, Native American agricultural peoples had to obtain them from other sources. Perhaps the best source would be protein rich animal flesh. Animal protein can of course be obtained both by hunting wild game and raising domestic stock. However, in sharp contrast to its huge repertoire of cultivated plants, the Mesoamerican agricultural complex included only four species of domesticated animals: the Muscovy duck, the turkey, the stingless bee and the hairless dog. Although dogs and Muscovy ducks were regularly eaten, they are inefficient meat producers. Of course, turkeys are eatable, but in the New World they were raised chiefly for their feathers.

Compared to the Old World, this was a tiny array of domesticated animals and this had important dietary implications for the native peoples in the New World: it meant that, while Old World agricultural peoples had relatively easy access to protein in the form of meat, milk, and blood, New World peoples were obliged at first to obtain the animal protein they needed to supplement their high corn diet primarily by hunting. However, overtime, natural selection favored two additional solutions:

1. Joint cultivation of beans with corn, and
2. the treatment of maize with slaked lime or wood ash in preparing it for consumption.

Beans and corn are complementary crops: the amino acids lacking in one are abundant in the other and vice versa (Kaplan 1965). In addition, the traditional Native American food preparation technique of soaking raw maize kernels in a lime solution prior to grinding it frees up existing lysine and tryptophan, improves the balance between the ammino acids leucine and isoleucine, and adds key vitamins and minerals to the resulting maize flour (Katz, Hediger and Valleroy 1974).

When maize agriculture diffused northwards into North America, the peoples who adopted it compensated for its dietary deficiencies both by growing large quantities of common beans and by alkali maize processing using either slacked lime or wood ashes to improve the amino acid balance. After ca. A.D. 800 peoples of the Mississippian cultural tradition of southeastern North America cooked maize in shell tempered ceramic vessels which may have made a similar contribution (Osborn 1988). Alkali-processing of corn, a high rate of bean consumption, and the ingestion of protein chiefly from wild game produced a sufficiently balanced diet as long as populations remained relatively low. However, when densely-populated cities and states arose in Mesoamerica in later times, hunting could no longer supply protein in sufficient quantity to any but the elite–and the fatal flaw in the Mesoamerican agricultural complex became evident.

Tropical Lowlands of Mesoamerica and Northern South America Tropical lowlands constitute as much as one-third of the continental portion of Mesoamerica and South America. Lowlands begin on the lower mountain slopes and continue down the coastal plain to sea level. Tropical lowlands in Mesoamerica are found along the wide coastal plain on the Gulf of Mexico and the Carribean and on the narrow coastal plain along the Pacific.

In South America tropical lowlands characterize the Orinoco plain and the Amazon plain that extends from the eastern foothills of the Andes in the interior of the continent all the way to the Atlantic ocean. Today much of the potential climax vegetation of the tropical lowlands is extremely humid, full-canopied rainforest where annual temperatures are high and rainfall is heavy throughout the year. The establishment of these rainforests in the lowlands took place relatively recently during the Holocene. Cool, dry climatic conditions during the late Pleistocene times appear to have limited rainforest vegetation to a but a few refugia within the lowlands (COHMAP 1988: 1049). Elsewhere in the lowlands, rainfall regimes characterized by winter drought support tropical savannas. Finally, in east, central South America, the rain shadow of the Andes creates the vast, rolling savannah grassland known as the Pampas on the Paraná-Paraguay Plain.

The agricultural complex that emerged in the tropical lowlands centered on vegetatively-propagated root crops including such tropical species as sweet and bitter manioc or cassava, sweet potato, American Yam, arrowroot and several other species. Manioc are starchy and high in carbohydrates but relatively low in protein. Fortunately, the tropical lowland rivers are rich in fish, amphibians, and aquatic reptiles. Such wild species are high in protein and, since they are natural dietary complement to the root crops, remained important elements in the subsistence system of lowland farming peoples.

As noted earlier, Carl Sauer's view that root and tuber crops were domesticated earliest on the alluvial flood plains of the great rivers of the tropics has received wide acceptance, although his view that vegeculture was also the earliest form of agriculture altogether has remained much disputed. Sauer's view received strong theoretical support in the writings of the archaeologists Donald Lathrap (1977). Lathap and his students took the view that the domestication of key root and tuber crops. was well underway by 6,000 B.C. in the New World tropical woodlands of northern or perhaps northwestern South America. Unfortunately, little direct archaeological evidence of the process had been recovered rom the tropics. At the time of his writing, the earliest archaeological remains of primitive, but nonetheless domesticated, maize from the Tehuacan Valley in Mexico has been dated to around 5,000 B.C. Although, the dates of the beginning of agriculture in the tropical lowlands still remain indispute, the picture has recently changed dramatically with the recent recovery of: (1) evidence of possible anthropogenic burning in the tropical forest after ca. 4000 B.C and (2) phytoliths of what appear to be domesticated squash from contexts dating to ca. 8000 B.C.

Anthropogenic forest burning has been suggested as an explanation for the charcoal been found throughout soils of the forests of the upper Rio Negro region of Columbia and Venezuela. This charcoal dates to between ca. 4000 B.C. and the present and indicates that fires in this portion of the Amazon Basin became widespread and frequent by the end of the middle Holocene. Probably the majority of these fires occurred naturally during extremely dry period. However, some of it may have resulted from intentional fires set by people clearing the land for agriculture (Saldarriaga and West 1986). If so, this charcoal offers a clue as to the date of the earliest onset of cultivation in the lowlands. By ca. 4,000 B.C. cultivated Mesoamerican maize appears to have diffused into the lowland tropics of the Amazon in northern South America (Bush, Dolores, and Collinvaux 1989) and were being grown in concert with locally domesticated lowland root crops such as bitter and sweet manioc.

Even more provocative is the recovery of phytoliths of what appears to be domesticated squash (probably *Cuburbita moschata*) at the Vegas Site 80 located in the tropical forest that borders the southwestern coast of Ecuador (Piperno and Pearsall 1998) . Here again as in the Chinese case discussed earlier, the new technique of phytolith analysis appears to produce results that refute the conventional wisdom. If these identifications (and their dates) are correct, they force us to push estimates for the emergence of agriculture back earlier and earlier in the Holocene epoch in the New World as well as the Old.

Olmec and Chavin That the new subsistence system that emerged from this combination was apparently a powerful and productive one is evidenced that New World civilizations–the **Olmec** culture in Mesoamerica, the **Chavin** culture in Peru–emerged earliest in the tropical lowlands. Further, recent archaeological work has revealed that, by late Prehistoric times, New World peoples developed a unique form of intensive cultivation in the permanently-wet, seasonally-flooded, or poorly-drained regions of the lowlands generally ignored by modern agriculture. This system involves the construction of mounds or ridges of swamp mud, decaying plant matter, human feces, kitchen wastes, and other organic composts in the wetlands.

Aerial photographs reveal that the Carribean coastal lowlands in Central and South America, the Orinoco and Amazon basins, the tropical savannahs of Bolivia and other regions in the lowland tropics contain vast networks of mounds, ridges and causeways apparently built as raised agricultural fields and terraces (Parsons and Denevan 1967). Since the raised beds over most of the tropics have been abandoned since shortly after the Conquest, it is difficult to determine how they were farmed or how productive they were.

Fortunately, an apparent descendant form of this intensive cultivation system exists today in the famous "floating gardens" or ***chinampas*** in the Lake Chalco-Xochimilco area near Mexico City. The so-called floating gardens are really raised agricultural earth mounds built into the wetlands there. The construction of these mounds enable farmers to control drainage in waterlogged environments while at the same time making use of the natural moisture and continually fertilizing the soil with compost. In this way, modern Lake Chalco-Xochimilco *chinampaneros* produce about 2.5 times the crop yield of the *milperos* who raise corn extensively in rain-fed fields (Parson 1976: 243; Dickson 1980: 705).

Of course, the Lake Chalco-Xochimilco region is located in the highland zone. Its is probably fair to assume that raised fields in the warmer climates of the lowlands were even more productive. Agricultural ridge fields system have also been reported from locations as diverse as the high Andes, even the North American Midwest (Parsons and Denevan 1967: 97; Gallagher 1989). Clearly, this unique New World form of intensive agricultural was remarkably flexible.

Andean Highlands of Western South America Although rugged, elevated topography characterizes both the Brazilian and Guiana Highlands, the most dramatic and extensive mountains in South America are found in the Andean system on the western side of the continent. The Andes system consists of three distinct mountain ranges that extend some 4,400 miles from the Caribbean Sea south to Tierra del Fuego at the tip of South America where the mountains again plunge into the sea. The Andean system and its associated plateaus, more than 500 miles across in central Bolivia, shrinks to a width of no greater than 20 miles in Chile. While the average height of the mountain is about 13,000 feet, numerous volcanic peaks exceed altitudes of 20,000 or more feet (Calrson 1954: 45).

As a consequence of this dramatic relief, western South America a region of extreme rainfall variation. Both rainless deserts and rainforests can be found along the west coast of South America and environmental zonation changes with elevation in the Andes. Rainfall and snow tend to increase with altitude, temperatures tend to decrease. Consequently, when one ascends the flanks of mountains in the Andes, one moves by stages form an equatorial forest in the lowlands, to the ***altiplano*** with its montane forest and still-higher ***puna*** steppe tundra and grassland. On the higher peaks above the *puna*, one encounters a virtual polar desert.

This zonation has profoundly affected human adaptation in the Andean highlands. Because of the extreme verticality of the Andes, these markedly different ecological zones are separated from one another by very short overland distances. According to John Murra (1985a, 1985b) such ecological variety and proximity encouraged the development of political systems organized as ***Archipièlagos Verticales***. An archipelago is a group of islands. In Murra's judgement, the separation of the vertically-ascending Andean lifezones rendered them "island-like" in character. At the same time, the distinct nature of the resources in each zone made them highly complementarity to one another.

For historical and ecological reasons, the high *puna* above 13,000 feet (4000 meters), became the most productive zone from the standpoint of human subsistence. Murra contends that, as a consequence, the ethnic group centered in the high *puna*, could simultaneously control the other geographically-dispersed ecological tiers. The most notable Andean polity organized as a "vertical archipelago" was the late prehistoric Inca empire.

The agro-pastoral subsistence system developed by the native peoples of the Andean highlands that was the most distinctive agricultural complex in the New World. In the warm mountain valleys of middle elevation, such plants as arracacha, achira, yam bean, Lima bean, and South American species of amaranth and chili pepper came under human control. At higher elevations, the cold tolerant South American species of chenopod and, more importantly, the potato (*Solanum tuberosum*) became the crops of choice (Pickersgill and Heizer 1978; Hawkes 1989: 482-491). Andean peoples also domesticated that ubiquitous source of modern drugs and medicine, the coca plant (*Erythroxylum novogranatense*) from the wild coca species (*Erythroxylum coca*) (Pickersgill 1989: 436).

Between approximately 5000 B.C. and 3000 B.C., early agricultural peoples in Andean South America received key domesticates from Mesoamerica–including maize, squash, avocado, and probably tomato–and made them staples in their diets. Presumably, this southward diffusion took place along the Isthmus of Panama. Interestingly, the South American potato did not diffusion northwards into Mesoamerica in PreColumbian times. Likewise, the distinct species of cotton, chili peppers and amaranths that appear to have developed independently in South American never were taken northward (Pickersgill and Heizer 1978: 136).

Finally, the highlands South America supported

two wild, camel-like animals, the vicuna and the guanaco, not found in North or Central America. The domesticated llama and the alpaca were derived from one or both of these wild species. Native South Americans raised the llama as a beast of burden, the alpaca for its wool and both for their meat. Citing both archaeological and modern distributional data, (Browman 1989: 265) argues that intensive human use of these camelids began shortly after 7,000 B.C in the Andes and that this use resulted in their domestication by 4,000 B.C. The combination of these animal and plant domesticates was particularly complimentary in the higher elevations of the Andes where the potato became the focus of cultivation and llama and alpaca dung became an essential element in fertilizing the potato crop (Bray 1977: 225). Meanwhile, at lower elevations, a different complex based on seed crops like corn, beans, squash, amaranths, chili and tropical forest root crops was favored. As a further source of meat, Native South Americans domesticated the wild ancestors of the small rodent known erroneously as the guinea pig.

Agriculture Was Revolutionary After All

Available resources impose an impassable limit on demographic growth (Livi-Bacci 1997: 35) . This limit is called the environmental carrying capacity. Throughout the Pleistocene hominid populations grew to reach the carrying capacity of the area they inhabited and then budded off to colonize new territory. In this manner, much of the Earth's surface appears populated by the terminal Pleistocene. At the close of the epoch, the capacity of the planet to support human hunting and gathering population appears largely to have been reached apparently initiating what Mark Cohen (1977) has called a "food crisis" in world prehistory. In the face of this crisis, Early Holocene foraging peoples began first to **broaden** the range of wild plant and animal species they exploited and, later, to intensively and incessantly **manipulate** certain of these wild species to their advantage.

As we have seen, the Earth's plant and animal resources did not prove to be static. Human manipulation may have lead to more productive hunting and gathering in some regions and to outright ecological damage in others. However, in a few key regions of the Earth, wild species capable of favorable genetic change began to respond to such manipulation. In the limited regions of the world so favorably endowed, genetically malleable wild species began to become domesticated. In time, subsistence systems in these select regions were transformed from food collection into food production. This early transformation appears to have taken place independently in at least six, and possibly seven, regions of the Earth. As we have seen, the earliest earths of agriculture include: southwestern Asia, southeastern Asia, sub-tropical and tropical West Africa, the North China plain in the Old World; Mesoamerica, the Andean Highlands of Western South America and perhaps also southeastern North America in the New. In these regions, the cap on human demographic growth reached at the end of the Pleistocene was lifted. The expansion of the human carrying capacity stimulated a sharp increase in population throughout the Holocene epoch. As we have seen in this chapter, the archaeological record allows us to make some notable generalization about the truly **revolutionary** consequences that such development occasioned.

Early food production did not occur uniformly throughout the world. This is largely because species suitable for domestication and susceptible to human manipulation are not found everywhere. The independence of the domestication process in the six or seven primary regions in Eurasia, Africa and the New World combined with the contrastive wild species available for manipulation on each of these continents, resulted in distinctive agricultural complexes (and sub-complexes) and cuisine within each of them by about the Middle Holocene.

The greatest contrasts existed between the agricultural complexes that developed in the Old and New Worlds. Old and New World agricultural complexes sharply diverged from one another both in the species chosen for domestication and in the techniques used in their husbandry. The most notable difference was the relative importance of animals in each system. Animal domestication played a primary role in the Old World transformation to food production while in the New World animals was relatively insignificant. As a consequence, Old World peoples came to have far greater access to protein in the form of meat, milk and eggs and to animal traction for transport, farm labor and warfare. This and other contrasts proved to have profound consequences in the later shared history of the two regions.

A"silent" or unconscious processes of microbe domestication occurred more-or-less in tandem with the conscious domestication of plants and animals. Unbeknownst to these early farmers, their manipulation of the plants and animals was altering relations between themselves and microbial realm as well. They were "domesticating" unseen organisms as well those that stood visible before them in their fields and pens. Many of the microbes domesticated at this time were benign or positively useful. Yeast, for example, makes fermentation possible and numerous others could be added to this list. However, as humans began to live more intimately with animals, their interaction altered the transmission patterns of the pathogens that afflicted these animals and stimulated the evolution of new ones. Overtime humans came to share the diseases of the creatures they were exploiting. In the Old World, for example, cattle pathogens evolved into measles, tuberculosis, and smallpox while certain pig diseases, and perhaps also those of ducks and dogs, became the human scourges of influenza and pertussis. Many others were to follow (Diamond 1997: 207). At the same time, the curtailment of mobility that followed the replacement of hunting and gathering by food production brought people into more sustained contact with garbage, offal and fecal matter, materials that provide ideal environments for microbial growth and transmission. Of course, New World peoples domesticated far fewer animal species than those living in the Old World. They therefore lacked the dietary, transport, labor and military advantages that animals afforded people in the Old World. However, they also developed far fewer human/animal diseases than did Old World peoples as a consequence. This advantage was quickly nullified after 1492 when the arrival of Europeans and Africans introduced Old World diseases into the New World. Native Americans, lacking immunity to these pathogens, died in huge and terrifying numbers to begin the great population changes that took place in the Americas over the last 500 years.

Early food production was strongly correlated with the appearance of sedentary village life. In most regions of the Earth, wild resources important in human subsistence are neither uniformly available in space nor constantly available throughout the year. It is primarily for these reasons that most foragers practice mobile subsistence strategies. Since they exercise a great degree of control over the location and abundance of the foodstuffs on which they depend, Agriculturalists are generally therefore able to live in more-or-less sedentary fashion and to store portions of their crops. Exception to this rule of mobile foragers, settled agriculturalists exist, of course. Key examples include the Jomon of Japan, the Natufians of the Levant and the Native American peoples of the Northwest Coast of North America. The Northwest Coast peoples were perhaps the most successful sedentary foragers but, of course, they inhabited were also truly exceptional in the richness, abundance and predicably of the wild resources they exploited. When these unusual conditions pertain, people need not farm. An insightful *!Kung* or *Jul'hoansi* forager on the Kalahari desert of southern Africa reputed answered an ethnographer's question as to why they didn't practice agriculture with the counter question: "Why should we plant when there are many *Mongongo* nuts in the world? A Haida or Tlingit Indian from the Northwest Coast might have responded similarly. Few other foragers on Earth have been so lucky.

The social, economic and political organization of sedentary agricultural societies is markedly different than in societies based on hunting and gathering. Again, with the limited exception of fortunate foragers like those on the Northwest Coast, the populations of food producers are generally larger than those of food collectors. This difference appears to be due to a complex of causes. Particularly important factors are (1) the relative ease of producing and storing a surplus of foodstuff in agricultural subsistence systems, (2) the improved conditions for bearing and raising children created by greater sedentism and (3) the positive contribution that additional children make to agricultural subsistence all favored the growth of population during the Neolithic period. Neolithic populations were also generally concentrated in larger, more concentrated and more permanent settlements than those of foraging bands at a comparable level of technological complexity. Overall, such differences suggest that Neolithic villages were more likely to have had more permanent membership and more complex forms of social and political organization than foragers. These early villages were probably largely autonomous bodies governed by headmen and councils of elders. At the same time, their economic production and distribution was probably organized and conducted along households and extended family lines. No doubt pan-village mechanisms for redistributing some surplus **within** the villages existed as well. However, the interconnecting–often vast–trading, marriage and interactive networks that tend to connect foraging bands with one another were largely absent. The trend towards self-sufficiency likely lead early Neolithic villages into increasing isolation and social distance from their neighbors.

Once developed, systems of food production spread quickly and easily into adjacent regions. Agriculture subsistence represents an ecological **simplification** of human subsistence. That is, it narrows human dependence to a limited range of plant and animal species whose numbers and production are largely controlled by the farmers themselves. The various arrays of domesticated species are integrated, flexible and generalized subsistence solutions that work in a wide range of environments.

Agriculturalists newly arrived in a territory are less obliged to learn the ins-and-outs of their new ecological setting than foragers would be. This is because they and their animals are able to change it much of the setting to meet their own demands. By clearing, burning and later by plowing and irrigating, they override over ride local constraints and create new, relatively homogenous environment settings. The various agricultural complexes spread outward from the primary regions of their formation in steadily radiating waves. These relentless Neolithic waves-of-advance created a favored the spread of the genes and the languages of those peoples whose remote ancestors had been fortunate enough to participate in the original domestication of plants and animals.

Agricultural subsistence, especially when combined with domesticated animals, has far greater impact on the natural environment than does hunting and gathering. By late prehistoric times, few environments in the world had not felt the impact of humankind. Nonetheless, all impact is not of the same magnitude. The impact of hunters and gatherers on a given landscape is comparatively modest. However, as agricultural systems spread into new regions, they transform and simplify their natural ecology. By radically changing the natural conditions of a region and by increasing the size of the human population supported within it, agricultural colonists make it difficult or impossible for hunting and gathering subsistence to remain viable there. Finally, agricultural peoples and their animals introduced diseases of various kinds into the regions which they colonized.

The rise of agriculture resulted in an increase in the scale of inter-societal violence and warfare. Although the tower, the mass grave at the site and the plastered trophy heads at the site of Jericho are among the earliest and most dramatic signs of heightened violence in the Neolithic, ample evidence exists elsewhere. Yet why was the Neolithic so violent? No doubt the peoples of the earlier times were plagued with local disputes and enmities that lead to violence and killing. Nonetheless, there are at least three primary reasons why violence appears to have taken place on a larger, more intensive and more deadly scale during the Neolithic than during earlier prehistory. First, Neolithic populations generally appear to have been larger and their settlements more concentrated and more sedentary than those of earlier periods. Population size and density translates directly into violence; more people in a region mean more chances for scrapes and ill will to develop. The presence of more people means greater likelihood that damage will be done once trouble starts. Second, unlike most foraging economies, those of the Neolithic societies accumulated and stored food surplus. As a consequence, when warfare and feuding occurred, it could have been carried on for longer periods of time and involved more participants than heretofore. Finally, the growth of Neolithic populations no doubt lead to local resources scarcity which in turn fostered competition over land and water between neighboring farming groups. When this competition took place in districts surrounded by open land, the effects of this competition would have been ameliorated by out-migration. On the other hand, when such scarcity and competition forced farmers to emigrate, it would have fostered competition and hostility with the foragers who occupying the land beyond the farming frontier. Of course, when the environment was a circumscribed one, this option would have been unavailable.

Population growth lead to the intensification of agricultural production which in turn lead to further populations growth and, ultimately, to the development of greater levels of societal complexity. Later in the Neolithic, population growth appears to have stimulated the practice of more intensive forms of cultivation and husbandry including irrigation agriculture and its twin, fully nomadic pastoralism. As "the Neolithic wears to a close," continued population growth leads to the emergence of more and more complex forms of social organization. In large part this is because there is a strong relationship between the population size of a society and the complexity of its social organization (Carneiro 1967). As societies increase in size, they must develop certain forms and organizations which they previously did not need or could not support.

Conclusions

The domestication of plants and animals that began in the Holocene epoch eventually broke the resource constraints that had limited the size of human populations throughout earlier prehistory. The transformation of human subsistence from food collection to increasing reliance on food production initiated a period of greater sedentary settlement and accelerating population growth. Agriculture spread rapidly during the Middle and Late Holocene epoch and a Neolithic pattern of settled village life was established throughout much of the world. The rise and spread of spread of Neolithic-stage agriculture also lead to marked changes in the patterns of disease, environmental destruction, agricultural intensification. violence and warfare.

When such intensification and warfare occurred in "circumscribed" environments, that is, in areas whose cultivatable land was sharply bound and limited by natural environmental features like mountains, seas or deserts, it resulted in the evolution of still complex forms of human organization and effort: **civilizations, states** and **cities** (Carneiro 1970).

REFERENCES CITED

Abbate, Ernesto, A. Aibianelli, A. Azzaroli, M Benvenuti, B. Tesfamariam, P. Bruni, Nicola Cipriani, R. J. Clarke, G. Ficcarelli, R. Macchiarvelli, G. Napoleone, M. Papini, L. Rook, M. Sagri, T.M. ecle, D. Torre, and I. Villa. 1998. A One-million-year-old *Homo* Cranium from the Danakil (Afar) Depression of Eritrea. *Nature*. 393: 458-460.

Abrams, M.H. 1961. *A Glossary of Literary Terms*. Holt, Rinehart and Winston, New York.

Adams, Dawn A. 1997. Famous Felids. *Science* 276: 909-910.

Adams, Robert McC. 1962. Agriculture and Urban Life in Early Southwestern Iran. *Science* 136: 109-122.

Adams, Robert McC. 1974. Mesopotamian Irrigation Agriculture. In *Irrigation's Impact on Society*, edited by T. E. Downing and Mcguire Gibson, pp. 1-6. University of Arizona Press, Tucson.

Adams, Robert McC. 1981. *Heartland of Cities: Surveys of Ancient Settlement and Land Use on the Central Floodplain of the Euphrates*. University of Chicago Press, Chicago.

Adams, Robert McC. And H. J. Nissen. 1972. *The Uruk Countryside: The Natural Setting of Urban Societies*. University of Chicago Press, Chicago.

Adovasio, James M. and Ronald C. Carlisle. 1988. The Meadowcroft Rockshelter. *Science* 239: 713-714.

Adovasio, James M., J. D. Gunn, J. Donahue, R. Stuckenrath, J. Guilday, and K. Lord. 1980. Yes Virginia, It Really Is That Old: A Reply to Haynes and Mead. *American Antiquity* 45: 588-595.

Agenbroad, Larry D. 1984. New World Mammoth Distribution. In *Pleistocene Extinctions*, edited by Paul S. Martin and Richard Klein, pp. 90-108. Tucson: University of Arizona.

Agrawal, D. P. 1982. The Archaeology of India. *Scandinavian Institute of Asian Studies Monograph Series* 46.

Aguirre, Emillano. 1997. Human Evolution in the Plio-Pleistocene Interval. In *The Pleistocene Boundary and the Beginning of the Quaternary*. Edited by John A. Van Couvering. Pp. 129-137.Cambridge University Press, New York.

Aikens, C. Melvin. 1995. First in the World: The Jomon Pottery of Early Japan. In *The Emergence of Pottery: Technology and Innovation in Ancient Societies*. Edited by William K. Barnett and John W. Hoopes, Pp. 11-21. Smithsonian Institution Press, Washington, D.C."

Aikens, C. Melvin and Takayasu Higuchi. 1982. *Prehistory of Japan*. Academic Press, New York.

Aikens, C. Melvin, Kenneth M. Ames, and David Sanger. 1986. Affluent Collectors at the Edges of Eurasia and North America: Some Comparisons of Observations on the Evolution of Society among North-Temperate Coastal Hunter-gatherers. In *Prehistoric Hunter-gatherers in Japan: New Research Methods*, edited by Takeru Akazawa and C. Melvin Aikens, pp. 3-26. University of Tokyo Press, Tokyo.

Aikens, C.Melvin and Takayasu Higuchi. 1982. *Prehistory of Japan*. New York: Academic Press

Aitken, M. J. 1990. *Science-Based Dating in Archaeology*. Longman. New York.

Akazawa, Takeru and C. Melvin Aikens, editors. 1986. In *Prehistoric Hunter-gatherers in Japan: New Research Methods*. University of Tokyo Press, Tokyo.

Akazawa, Takeru. 1986. Hunter-Gatherer Adaptations and the Transition to Food Production Japan. In *Hunters in Transition: Mesolithic Societies of Temperate Eurasia and Their Transition to Farming*, edited by Marek Zvelebil, pp. 151-163. Cambridge University Press, Cambridge.

Albritton, Claude C., Jr. 1980. *Abyss of Time*. Freeman, Cooper and Company, San Francisco.

Alexander, J. and D. G. Coursey. 1969. The Origins of Yam Cultivation. In *The Domestication and Exploitation of Plants and Animals*, edited by Peter J. Ucko and G. W. Dimbleby, pp. 405-441. Gerald Duckworth, London.

Allen, J. Chris Gooden, Rhys Jones, and J. Peter White. 1988. Pleistocene Dates for the Human Occupation of New Ireland, Northern Melanesia. *Nature* 331: 707-709.

Alvarez, Luis W., Walter Alvarez, Frank Asaro, and Helen V. Michel. 1980. Extraterrestrial Cause for the Cretaceous-Tertiary Extinction *Science* 208: 1095.

Alvarez, Walter, Philippe Claeys, and Susan W. Kieffer. 1995. Emplacement of Cretaceous-Tertiary Boundary Shocked Quartz from Chicxulub Crater. *Science* 269: 930-935.

Ambrose, Stanley H. 1984. The Introduction of Pastoral Adaptations to the Highlands of East Africa. In *From Hunters to Farmers: the Causes of Consequences of Food Production in Africa*, edited by J. Desmond Clark and Steven A. Brandt, pp. 212-239. University of California Press, Berkeley.

Ambrose, Stanley H. 1998. Chronology of the Later Stone Age and Food Production in East Africa. *Journal of Archaeological Science* 25: 377-392.
Amirian, Ruth. 1965. The Beginings of Pottery-Making in the Near East. In *Ceramics and Man*, edited by Frederick R. Matson, pp. 240-247. Aldine Publishing Company, Chicago.
Ammerman, Albert J. and L. L. Cavalli-Sforza. 1979. The Wave of Advance Model for the Spread of Agriculture in Europe. In *Transformations: Mathematical Approaches to Culture Change*, edited by Colin Renfrew and Kenneth L. Cooke, pp. 275-293. Academic Press, New York.
Andah, R. W. 1987. Early Food Producing Societies and Antecedents in Middle Africa. *West African Journal of Archaeology* 17:129-170.
Anderson, Atholl. 1987. Recent Developments in Japanese Prehistory: a Review. *Antiquity* 61: 270-281.
Anderson, E. 1984. Who's Who in the Pleistocene: A Mammalian Bestiary. In *Quaternary Extinctions: a Prehistoric Revolution*, edited by Paul S. Martin and Richard G. Klein. pp. 40-89. University of Arizona Press, Tucson.
Anderson, P. 1974. *Passages from Antiquity to Feudalism*. New Left Books, London.
Anikovich, Mikhail. 1992. Early Upper Paleolithic Industries of Eastern Europe. *Journal of World Prehistory* 6(2): 205-445.
Annis, Sheldon. 1987. *God and Production in a Guatemalan Town*. University of Texas Press, Austin.
Anonymous. 1997. Early Music. *Science* 276: 205.
Anonymous 19–. The End of the Ices Ages? *New Scientist* 1669: 49-52.
ApSimon, A. M. 1980. The Last Neanderthal in France? *Nature* 287: 271-272.
Artamonov, M.I. 1965. Frozen Tombs of the Scythians. *Scientific American* 212(5): 100-109.
Avery. Graham, Kathryn Cruz-Uribe, Paul Goldberg, Frederick K. Grine, Richard G. Klein, Michael J. Lenardi, Curtis W. Marean, W. Jack Rink, Henry P. Schwarcz, Anne I. Thackeray, Michael L. Wilson. 1997. The 1992-1993 Excavations at the Die Kelders Middle and Later Stone Age Cave Site, South Africa. *Journal of Field Archaeology* 24: 263-291.
Axelrod, Daniel I. 1985. Rise of the Grassland Biome, Central North America. *The Botanical Review* 51(2): 164-201.
Bada, Jeffrey L. 1985. Aspartic Acid Racemization Ages of California Paleo-indian Skeletons. *American Antiquity* 50(3): 645-647.
Bahn, Paul G. 1977. Seasonal Migration in Southwest France During the late Glacial Period. *Journal of Archaeological Science* 4: 245-257.
Bahn, Paul G. and Claude Couraud. 1984. Azilian Pebbles: An Unsolved Mystery. *Endeavour* 8(4): 156-58.
Bailey, Robert C., G. Head, M. Jenike, B. Owen, R. Rechtman, and E. Zechenter. 1989. Hunting and Gathering in Tropical Rain Forest: Is It Possible? *American Anthropologist* 91(1): 59-82.
Bailey, Wendy J. 1993. Hominoid Trichotomy: A Molecular Overview. *Evolutionary Anthropology* 2(3): 100-108.
Baillie, M. G. L. 1982. *Tree-Ring Dating and Archaeology*. University of Chicago Press, Chicago.
Baker, V. R., G. Benito, and A. N. Rudoy. 1993. Paleohydrology of Late Pleistocene Superflooding, Altay Mountains, Siberia. *Science* 259: 348-350.
Bannister, Bryant and William J. Robinson. 1975. Tree-Ring Dating in Archaeology. *World Archaeology* 7(2): 210-225.
Bannister, Bryant. 1969. Dendrochronology. In *Science in Archaeology. Second Edition*, edited by Don Brothwell and Eric Higgs, pp. 191-205. Praeger Publishers, Inc., New York.
Bar-Josef, Ofer. 1980. Prehistory of the Levant. *Annual Review of Anthropology* 9: 101-133.
Bar-Josef, Ofer. 1986. The Walls of Jericho: An Alternative Interpretation. *Current Anthropology* 27: 157-162.
Bar-Josef, Ofer. 1989. Early Farming Communities in the Jordan Valley. In *Foraging and Farming: The Evolution of Plant Exploitation*, edited by D. R. Harris and G. C. Hillman, pp. 632-642. Unwin Hyman, London.
Bar-Josef, O., M. Arnold, N. Mercier, A. Belfer-Cohen, P. Oldberg, R. Housley, H. Laville, L. Meignen, J.C. Yogel, and B. Vandermeersch. (1996). Dating of Upper Paleolithic Layers in Kebara Cave, Mt. Carmel. *Journal of Archaeological Science* 23: 297-306.
Barker, Philip. 1982. *Techniques of Archaeological Excavation*. Universe, New York.
Barnosky, Anthony D. 1986. 'Big Game' Extinction Caused by Late Pleistocene Climatic Change: Irish Elk (*Megaloceros Giganteus*) in Ireland. *Quaternary Research* 25: 125-135).
Barry, R. G. 19??. Late Pleistocene Climatology. In *The Quaternary in North America, Volume 2: The Pleistocene*, edited by H.T. Wright, pp. 390-407. Minneapolis: University of Minnesota Press.
Barry, Roger G. 1992. Climate-Ice Interactions. *Encyclopedia of Earth System Science, Volume* 1:517-523.
Barthelme, J. W. 1985. Fisher-hunters and Neolithic Pastoralists in East Turkana, Kenya. *Bar International Series* 254.
Basgall, M. E. 1987. Resource Intensification Among Hunter-Gatherers: Acorn Economies in Prehistoric California. *Research in Economic Anthropology* 9: 21-52.
Basham, A.1. 1954. *The Wonder That Was India*. The Grove Press, New York.
Beadle, George W. 1980. The Ancestry of Corn. *Scientific American* 242: 112-119.
Beaumont, Peter B., Hertha de Villiers and John C. Vogel. 1978. Modern Man in Sub-Saharan Africa Prior to 49,000 Years B.P.: Review and Evaluation with Particular Reference to Border Cave. *South African Journal of Science* 74: 409-419.

Bellwood, Peter. 1997. *Prehistory of the Indo-Malaysian Archipelago*. University of Hawai'i Press, Honolulu.

Ben-shun, Chow. 1989. The Domestic Horse of the Pre-ch'in Period in China. In *The Walking Larder: Patterns of Domestication, Pastoralism, and Predation*, edited by J. Clutton-Brock, pp. 105-107. Unwin Hyman, London.

Bender, B. 1978. Gatherer-Hunter to Farmer: A Social Perspective. *World Archaeology* 10: 204-222.

Bender, B. 1981. Gatherer-Hunter Intensification. In Economic Archaeology, edited by J. A. Sheridan and G. N. Bailey. *Bar International Series* 96: 149-157.

Bengtson, Stefan 1991. Oddballs from the Cambrian Start to Get Even. *Nature* 351: 184–185.

Bennett, C. L. 1979. Radiocarbon Dating with Accelerators. *American Scientist* 67(4): 450-457.

Berger, T. D. and Eric Trinkaus. 1995. Patterns of Trauma Among the Neanderthals. *Journal of Archaeological Science* 22: 841-852.

Berggren, W. A. and John A. Van Couvering (Editors). 1984. *Catastrophes and Earth History: The New Uniformitarianism.* Princeton University Press, Princeton, New Jersey.

Berggren, W. A., L. H. Burckle, M. B. Cita, H. B. S. Cooke, B. M. Funnell, S. Gartner, J. D. Hays, J. P. Kennett, N. D. Opdyke, L. Pastouret, N. J. Shackleton, and Y. Takayanagi. 1980. Towards a Quaternary Time Scale. *Quaternary Research* 13: 277-302.

Berner, Robert A. 1997. The Rise of Plants and Their Effect on Weathering and Atmospheric CO2. *Science* 276: 544-546).

Binford, Lewis R. 1972. *An Archaeological Perspective*. Seminar Press, New York.

Binford, Lewis R. 1973. Interassemblage Variability–The Mousterian and the "Functional" Argument. In *The Explanation of Culture Change* edited by Colin Renfrew, pp. 227-254. Duckworth, Liverpool.

Binford, Lewis R. 1981. *Bones: Ancient Men and Modern Myths*. Academic Press, New York.

Binford, Lewis R. 1983. *In Pursuit of the Past*. Thames and Hudson, Inc., New York.

Binford, Lewis R. 1984. *Faunal Remains from Klasies River Mouth*. Academic Press, New York.

Binford, Lewis R. 1985. Human Ancestors: Changing Views of Their Behavior. *Journal of Anthropological Archaeology* 4:292-327.

Binford, Lewis R. 1986. Reply to Singer and Wymer. *Current Anthropology* 27: 57-62.

Binford, Lewis R. and Sally R. Binford 1966. A Preliminary Analysis of Functional Variability in the Mousterian of Levallois. *American Anthropologist* 68(2): 238-296.

Binford, Sally R. and Lewis R. Binford. 1969. Stone Tools and Human Behavior. *Scientific American* 220: 70-83.

Binford, Lewis R. and Chuan Kun Ho. 1985. Taphonomy at a Distance: Zhoukoudian, "The Cave Home of Beijing Man?" *Current Anthropology* 26(4): 413-442.

Binford, Sally. 1968. Variability and Change in the Near Eastern Mousterian of Levallois Facies. In *New Perspectives in Archeology*, edited by Lewis and Sally Binford, pp 49-60. Aldine, Chicago.

Bird, Junius B. 1969. A Comparison of South Chilean and Ecuadorian 'Fishtail' Projectile Points. *Kroeber Anthropological Society Papers* 40: 52-71.

Bischoff, James L., Kenneth Ludwig, José Francisco Garcia. E. Carbonell and Manola Vaquero, Thomas W. Stafford, Jr., and A. J. T. Tull. 1994. Dating of the Basal Aurignacian Sandwich at Abric Romani (Catalunya, Spain) by Radiocarbon and Uranium-Series. *Journal of Archaeological Science* 21: 541-551.

Blackwell, B, H.P. Schwarcz and A. Debenath. 1983. Absolute dating of Hominids and Paleolithic Artefacts of the Cave of La Chaise de Vouthon (Charente), France. *Journal of Archaeological Science* 10: 493-513.

Blanc, Alberto C. 1961. Some Evidence for the Ideologies of Early Man. In The Social Life of Early Man edited by Sherwood L. Washburn. *Viking Fund Publications in Anthropology* 31: 119-136.

Blanc, Alberto. 1961. Some Evidence for the Ideologies of Early Man. In The Social Life of Early Man. Edited by Sherwood L. Washburn, *Viking Fund Publications in Anthropology* 31:119-136.

Blanton, Richard E. 1988. Review of *Death, Sex, and Fertility* by Marvin Harris and Eric B. Ross. *American Anthropologist* 90(2): 439-440.

Blumenberg, Bennett. 1983. The Evolution of the Advanced Hominid Brain. *Current Anthropology* 24(5):489-523.

Blumenschine, R. J. 1986. Carcass Consumption Sequences and the Archaeological Distinction of Scavenging and Hunting. *lournal of Human Evolution* 15:639-659.

Boardman 1985. Comparison of Soils in Midwestern United States and Western Europe with the Interglacial Record. *Quaternary Research* 23: 62-75.

Boëda, E., J. Connan, D. Dessort *et al.* 1996. *Nature* 380: 336-338.

Boggs, Sam Jr. 1987. *Principles of Sedimentology and Stratigraphy*. Merrill Publishing Company, Columbus.

Bonnefille, R. and J.C. Roeland and J. Guiot. 1990. Temperature and rainfall Estimates for the Past 40,000 Years in Equatorial Africa. *Nature* 346: 347-349.

Boorstin, Daniel J. 1983. *The Discoverers*. Random House, New York.

Bordaz, Jacques. 1970. *Tools of the Old and New Stone Age*. The Natural History Press, Garden City.

Bordaz, Jacques. 1970. *Tools of the Old and New Stone Age*. The Natural History Press, Garden City.

Bordaz, Jacques. 1970. *Tools of the Old and New Stone Age*. The Natural History Press, Garden City.

Bordes, Francois and C. Thibault. 1977. Thoughts on the Initial Adaptation of Hominids to European Glacial Climates. *Quaternary Research* 8:115-127.

Bordes, François and Denise de Sonneville-Bordes. 1970. The Significance of Variability in Paleolithic

Assemblages. *World Archaeology* 2(1): 61-73.
Bordes, François. 1968. *The Old Stone Age*. McGraw-Hill Book Company, New York.
Bordes, François. 1972. *A Tale of Two Caves*. Harper and Row, Publishers, New York.
Bordes, François. 1973. On the Chronology and Contemporaneity of Different Paleolithic Cultures in France. In *The Explanation of Culture Change*, edited by Colin Renfrew, pp. 217-226. Duckworth, Liverpool.
Boserup, Esther. 1965. *The Conditions of Agricultural Growth: The Economics of Agrarian Change Under Population Pressure*. Aldine Publishing Company, Chicago.
Bousman, C. Britt. 1998. The Chronological Evidence for the Introduction of Domestic Stock into Southern Africa. *African Archaeological Review* 15(2): 133-150.
Bowler, Peter J. 1984. *Evolution: the History of an Idea*. University of California Press, Berkeley.
Bowman, Sheridan. 1990. *Radiocarbon Dating*. University of California Press, Los Angeles.
Brace, C. Loring. 1962. Cultural Factors in the Evolution of the Human Dentition. In *Culture and the Evolution of Man*. Edited by Ashley Montagu, pp. 343-354. Oxford University Press, New York.
Brace, C. Loring. 1979. Krapina, 'Classic' Neanderthals and the Evolution of the European Face. *Journal of Human Evolution* 8:527-550.
Bradley, Daniel G. Ronan T. Loftus, Patrick Cunningham and David E. MacHugh. 1998". Genetics and Domestic Cattle Origins. *Evolutionary Anthropology* 6(3): 79-86.
Bradley, J. P., R. P. Harvey, and H.Y. McSween, Jr. 1997. No "Nanofossils" in Martian Meteorite. *Nature* 390: 454.
Bradley, R.S. 1985. *Quaternary Paleoclimatology*. Allen and Unwin, Boston.
Braidwood, Robert J. 1958. Near Eastern Prehistory. *Science* 127: 1419-1430.
Brain, Charles K. 1981. *The Hunters or the Hunted? An Introduction to African Cave Taphonomy*. The University of Chicago Press, Chicago.
Bray, Warwick. 1977. From Foraging to Farming in Early Mexico. In *Hunter-Gatherers and First Farmers Beyond Europe: an Archaeological Survey*, J. V. S. Megaw, pp. 225-250. Leicester University Press, Leicester.
Bricker, Harvey M. 1976. Upper Paleolithic Archaeology. *Annual Review of Anthropology* 5: 133-148.
Britten, Roy J. 1986. Rates of DNA Sequence Evolution Differ Between Taxonomic Groups. *Science* 231: 1393-1398.
Broecker, W.S. and J. van Donk. 1970. Insolation Changes, Ice Volumes and O-18 Record in Deep-Sea Cores. *Reviews of Geophysics and Space Physics* 8: 169-198.
Browman, David L. 1989. Origins and Development of Andean Pastoralism an Overview of the past 6,000 Years. In *The Walking Larder: Patterns of Domestication, Pastoralism, and Predation*, edited by J. Clutton-brock, pp. 256-268. Unwin Hyman, London.
Bryan, A.L. 1986. Paleoamerican Prehistory as Seen from South America. In *New Evidence for the Pleistocene Peopling of the Americas*, edited by A. L. Bryan, pp. 1-14. Orono: Center for the Study of Early Man, University of Maine, Orono.
Budinger, F.E. 1983. Evidence for Pleistocene Man in America: the Calico Early Man Site. *California Geology* 36: 75-82.
Burch, Ernest S., Jr. 1972. The Caribou/Wild Reindeer as a Human Resource. *American Antiquity* 37: 339-368.
Burckle, Lloyd H. 1995. Current Issues in Pliocene Paleoclimatology. In *Paleoclimate and Evolution with Emphasis on Human Origins*. Edited by Elisabeth S. Vrba, Geoge H. Denton, Timothy C. Partridge, and Lloyd H. Burckle, pp. 3-7. Yale University Press, New Haven and London.
Butzer, Karl. 1976. Pleistocene Climates. In Ecology of the Pleistocene: A Symposium. Edited by R.C. West and W.G. Haag, pp. 27-44. *Geoscience and Man* XIII.
Buringh, P. 1957. Living Conditions in the Lower Mesopotamian Plain in Ancient Times. *Sumer* 13: 30-46.
Burke, Ann C. and Alan Feduccia 1997. Developmental Patterns and the Identification of Homologies in the Avian Hand. *Science* 278: 666-668.
Burrus, Thomas L. and Herbert J. Spiegel. 1976. *Earth in Crisis; an Introduction in the Earth Sciences*. The C.V. Mosby Company, St. Louis.
Butzer, Karl. 1971. *Environment and Archeology: An Ecological Approach to Prehistory. Second Edition.* Aldine-Atherton, Chicago.
Butzer, Karl. 1978. Climate Patterns in an Un-glaciated Continent. *Geographical Magazine* 51: 201-208.
Butzer, Karl. 1981. Cave Sediments, Upper Pleistocene Stratigraphy and Mousterian Facies in Cantabrian Spain. *Journal of Archaeological Science* 8: 133-183.
Butzer, Karl. 1982. Geomorphology and Sediment Stratigraphy. In *The Middle Stone Age at Klasies River Mouth in South Africa*, edited by Ronald Singer and John Wymer. Chicago: The University of Chicago Press.
Butzer, Karl. 1984a. Late Quaternary Environments in South Africa. In *Late Cenozoic Palaeoclimates of the Southern Hemisphere*, edited by J. C. Vogel. Rotterdam: A.A. Balkema.
Butzer, Karl. 1984b. Archaeology and Quaternary Environment in the Interior of Southern Africa. In *Southern African Prehistory and Paleoenvironments*, edited by Richard G. Klein, pp. 1-64. Rotterdam: A.A. Balkema.
Butzer, Karl, G. J. Fock, L. Scott, and R. Stuckenrath. 1980. Comment on: Red Ochre and Human Evolution: A Case for Discussion by E. E. Wreschner. *Current Anthropology* 21: 635.
Butzer, Karl W. and Leslie G. Freeman. 1986. Series Editors' Forward. In *Cities of Clay; The*

Geoarchaeology of Tells by A.M.Rosen, pp. xii-xv. University of Chicago Press, Chicago.

Butzer, Karl and Glynn Isaac, editors. 1975. *After the Australopithecines: Stratigraphy, Ecology, and Culture Change in the Middle Pleistocene*. Aldine, Chicago.

Byrd, Brian. 1989. The Natufian: Settlement Variability and Economic Adaptations in the Levant at the End of the Pleistocene. *Journal of World Prehistory* 3(2): 159-197.

Cameron. W.F. 1969. *Toward a History of Geology*. M.I.T. Press, Cambridge.

Campbell, Bernard G. 1982. *Humankind Emerging*. (Third ed.). Little, Brown and Company, Boston.

Campbell, I. H., G. K. Czamanske, V.A. Fedorenko, R. I. Hill, V. Stepanov. 1992. Synchronism of the Siberian Traps and the Permian-Triassic Boundary. *Science* 258: 1760-1763.

Carbonell, E., J.M. Bermúdez de Castro, J.L. Arsuaga, J. C. Díez, A. Rosas, G. Cuenca-Bescós, R. Sala, M. Mosquera, X.P. Rodríguez. 1995. Lower Pleistocene Hominids and Artifacts from Atapuerca-TD6 (Spain). *Science* 269: 826832.

Carlisle, R. C. and James M. Adovasio, editors. 1982. *Meadowcroft Rockshelter: Collected Papers on the Archaeology of Meadowcraft Rockshelter and the Cross Creek Drainage* Department of Anthropology, University of Pittsburgh, Pittsburgh.

Carneiro, Robert L. 1967. On the Relationship Between Size of Population and Complexity of Social Organization. *Southwestern Journal of Anthropology* 23(3): 234-243.

Carneiro, Robert L. 1969. The Measurement of Cultural Development in the Ancient Near East and in Anglo-Saxon England. *Transactions of the New York Academy of Sciences* II, 31: 1013-1023.

Carneiro, Robert L. 1970. A Theory of the Origin of the State. *Science* 169: 733-738.

Carneiro, Robert and Daisy Hulse. 1966. On Determining the Probable Rate of Population Growth During the Neolithic. *American Anthropologist* 68: 179-181.

Carter, George F. 1980. *Earlier Than You Think*. College Station: Texas A&M University Press

Carter, George F. 1988. Cultural Historical Diffusion. In *The Transfer and Transformation of Ideas and Material Culture*, edited by Peter J. Hugill and D. Bruce Dickson, pp. 3-22. Texas A&M University Press, College Station.

Cartmill, Matt. 1974. Rethinking Primate Origins. *Science* 184:436-443.

Cartmill, Matt, David Pilbeam, and Glynn Isaac. 1986. One Hundred Years of Paleoanthropology. *American Scientist* 74: 410-420.

Cerling, Thure E., J.M. Harris, Bruce J. MacFadden, Meave G. Leakey, Jay Quade, Vera Eisenmann, and James R. Ehleringer. 1997. Global Vegetation Change Through the Miocene/Pliocene Boundary. *Nature*

Chang, K. C. 1977. *The Archaeology of Ancient China*. (3rd Edition) Yale University Press, New Haven.

Chang, K. C. 1981. In Search of China's Beginnings: New Light on an Old Civilization. *American Scientist* 69: 148-160.

Chang, Kwang-Chih. 1977. *The Archaeology of Ancient China*. Yale University Press, New Haven.

Chang, T. T. 1989. Domestication and the Spread of the Cultivated Rices. In *Foraging and Farming: The Evolution of Plant Exploitation*, D. R. Harris and G. C. Hillman, pp. 408-417. Unwin Hyman, London.

Chase, Philip G. and Harold Dibble. 1987. Middle Paleolithic Symbolism: a Review of Current Evidence and Interpretations. *Journal of Anthropological Archaeology* 6: 263-296.

Chase, Philip G. and April Nowell. 1998. Taphonomy of a Suggested Middle Paleolithic Bone Flute Form Slovenia. *Current Anthropology* 39(4): 549-553.

Cheney, D.R., R. Seyfarth, and B. Smut. 1986. Social Relationships and Social Cognition in Nonhuman Primates. *Science* 234:1361-1366.

Childe, V. Gordon. 1954. Rotary Motion. In *A History of Technology, Volume I: From Early Times to the Fall of Ancient Empires*, edited by Charles Singer, E. J. Holmyard, and A. R. Hall, pp. 187-237. Oxford University Press, London.

Claeys, Philippe, Jean-George Casier and S. V. Margolis 1992. Microtektites and Mass Extinctions: Evidence for a Late Devonian Asteroid Impact. *Science* 257: 1102-1107.

Clark, George A., Marc A. Kelley, John M. Grange, and M. Cassandra Hill. 1987. The Evolution of Mycobacterial Disease in Human Populations: a Reevaluation. *Current Anthropology* 28: 45-62.

Clark, Grahame. 1977. *World Prehistory in New Perspective*. Cambridge University Press, New York.

Clark, J. D. 1974. Africa in Prehistory: Peripheral or Paramount? *Man* 10: 175-198.

Clark, J. G. D. 1971. *World Prehistory: A New Outline*. Cambridge: Cambridge University Press

Clark, J. G. D. 1972. *Star Carr: A Case Study in Bioarchaeology*. Addison Wesley, Reading, Mass.

Clark, J. G. D. 1980. *The Mesolithic Prelude*. University of Edinburgh Press, Edinburgh.

Clark, J.G.D. 1975. *The Earlier Stone Age Settlement of Scandinavia*. Cambridge University Press, Cambridge.

Clark, J. G. D. and Stuart Piggott. 1965. Prehistoric Societies. Alfred A. Knopf, New York.

Clark, LeGros. 1959. *The Antecedents of Man*. Harper Torchbooks, New York.

Clark, LeGros. 1964. *The Fossil Evidence for Human Evolution*. (Second ed.). The University of Chicago Press, Chicago.

Clark, R. M. 1975. A Calibration Curve for Radiocarbon Dates. *Antiquity* 49: 251-266.

Clark, T. H. and C. W. Stearn. 1960. *The Geological Evolution of North America*. The Ronald Press Company, New York.

Clarke, David. 1976. Mesolithic Europe: The Economic Basis. In *Problems in Social and Economic Archaeology*, edited by G.G. Sieveking, I.H. Longworth, and K.E. Wilson, pp 449-467. London:

Duckworth.
Clarke, David L. 1978. *Analytical Archaeology. Second Edition*. Columbia University Press.
Clason, A. T. and Juliet Clutton-Brock. 1982. The Impact of Domestic Animals on the Vegetation During the First Phases of Animal Husbandry in the Mediterranean and Near East. In Paleoclimates, Paleoenvironments, and Human Communities in the Eastern Mediterranean Region in Later Prehistory, edited by J. L. Bintliff and W. Van Zeist. *Bar International Series* 133: 146.
CLIMAP Project Members. 1976. The Surface of the Ice-Age Earth. *Science* 191: 1131-1137.
Close, Angela E. 1995. Few and Far Between. In *The Emergence of Pottery: Technology and Innovation in Ancient Societies*. Edited by William K. Barnett and John W. Hoopes, Pp. 23-37. Smithsonian Institution Press, Washington, D.C.
Cloud, Preston. 1988. *Oasis in Space*. W.W. Norton and Company, New York.
Clutton-Brock, J. 1989. Cattle in Ancient North Africa. In *The Walking Larder: Patterns of Domestication, Pastoralism and Predation*. Edited by J. Clutton-Brock. Pp. 200-206. Unwin Hyman, London.
Clutton-Brock, Juliet. 1981. *Domesticated Animals from Early Times*. University of Texas Press, Austin.
Clutton-Brock, Juliet. 1989. Cattle in Ancient North Africa. In *The Walking Larder: Patterns of Domestication, Pastoralism, and Predation*, pp. 200-206. Unwin Hyman, London.
Cochrane, Eric. 1973. *Florence in the Forgotten Centuries*. The University of Chicago Press, Chicago.
Cohen, Mark N. 1977. *The Food Crises in Prehistory*. Yale University Press, New Haven.
Cohen, Mark. 1977. *The Food Crisis in Prehistory*. Yale University press, New Haven."
COHMAP members. 1988. Major Climatic Changes of the Last 18,000 Years: Observations and Model Simulations. Science 241: 1043-52.
Colinvaux, Paul A. 1989. The past and Future Amazon. *Scientific American* 260(5): 102-108.
Collier, Stephen. 1993. Sexual Dimorphism in Relations to Big-Game Hunting and Economy in Modern Human Populations. *American Journal of Physical Anthropology* 91: 485-504.
Collins, Desmond and Rufus Onians. 1978. The Origins of Art. *Art History* I(i): 1-25.
Columbia University Press. 1975. *The New Columbia Encyclopedia*. Columbia University Press, New York.
Conkey, Margaret W. 1981. A Century of Paleolithic Cave Art. *Archaeology* 34(4): 20-28.
Conklin, H. C. 1954. An Ethnoecological Approach to Shifting Cultivation. *Transactions of the New York Academy of Sciences* (Series 2) 17(2): 133-142.
Conroy, Glen C., G.W. Weber, H. Seidler, P. V. Tobias, A. Kane, B. Brunsden. 1998. Endocranial Capacity in an Early Hominid Cranium from Sterkfontein, South Africa. *Science* 280: 1730-1731.
Coon, Carleton S. 1976. *The Hunting Peoples*. Harmondsworth, England: Penguin Books, Ltd.
Coon, Carleton S. 1976. *The Hunting People*. Penguin Books, New York.
Copleston, F. C. 1955. *Aquinas*. Penguin Books, Baltimore.
Cotter, John L. 1981. The Upper Paleolithic–however it Got Here, It's Here (Can the Middle Paleolithic Be Far Behind). *American Antiquity* 46(4): 926-928.
Cracraft, Joel. 1988. Early Evolution of Birds. *Nature* 331:389-390.
Crawford, Gary W. and Chen Shen. 1998. The Origins of Rice Agriculture: Recent Progress in East Asia. *Antiquity* 72: 856-866.
Cressey, George B. 1951. *Asia's Lands and Peoples: a Geography of One-third the Earth and Two-thirds its People*. Mcgraw-Hill Book Company, Inc, New York.
Cressey, George B. 1955. *Land of the 500 Million: a Geography of China*. McGraw-Hill Book Company, Inc., New York.
Cressey, George B. 1963. *Asia's Lands and Peoples: A Geography of One-third of the Earth and Two-thirds of its People*. (3rd edition) McGraw-Hill Book Company, Inc., New York.
Crites, G. D. 1993. Domesticated Sunflower in Fifth Millennium B.P. Temporal Context: New Evidence from Middle Tennessee. *American Antiquity* 58: 146-148.
Cronin, J. E., N. T. Boas, C. B. Stringer, and Y. Rak 1981. Tempo and Mode in Hominid Evolution. *Nature* 292:113-121.
Crosby, Alfred W. 1986. *Ecological Imperialism*. Cambridge University Press, New York.
Crowley, Thomas J. and Gerald R. North. 1991. *Paleoclimatology*. Oxford University Press, New York.
Da Vis, Richard S. 1998. The Enisei River of Central Siberia in the Late Pleistocene. *Journal of Archaeological Research* 6(2): 169-194.
Dahl-Jensen, D., K. Mosegaard, N. Gundestrup, G. D. Clow, S. J. Johnsen, A.W. Hansen, and N. Balling. 1998. Past Temperatures Directly from the Greenland Ice Sheet. *Science* 282: 268-271.
Dalrymple, G.B., and M.A. Lanphere. 1969. *Potassium-Argon Dating: Principles, Techniques, and Applications to Geochronology*. W.H. Freeman, San Francisco.
Damon, P. E., C. W. Ferguson, A. Long, and E. J. Wallick. 1974. Dendrochronlogic Calibration of the Radiocarbon Time Scale. *American Antiquity* 39: 350-66.
Damon, P.E. 1970. Radiocarbon as an Example of the Unity of Science. In: *Radiocarbon Variations in Absolute Chronology*, edited by I.U. Olsson, pp. 571-593. Wiley, New York.
Daniel, Glyn. 1960. *The Idea of Prehistory*. Penguin Books, Baltimore.
Daniel, Glyn. 1967. *The Origins and Growth of Archaeology*. Galahad Books, New York.
Daniel, Glyn. 1971. From Worsaae to Childe: The Models of Prehistory. *Proceedings of the Prehistoric*

Society 27: 140-153.

Danmuth, John E. and Rhodes W. Fairbridge. 1970. Equatorial Atlantic Deep-sea Arkosic Sands and Ice-age Aridity in Tropical South America. *Geological Society of America Bulletin* 81: 189-206.

Danmuth, John E. and Rhodes W. Fairbridge 1970; Fairbridge, Rhodes W. Effects of Holocene Climatic Change on Some Tropical Geomorphic Processes. *Quaternary Research* 6: 529-556.

David A. Breternitz. 1966. An Appraisal of Tree-Ring Dated Pottery in the Southwest. *Anthropological Papers of the University of Arizona* 10.

Davidson, I. 1974. Radiocarbon Dates for the Spanish Solutrean. *Antiquity* 48: 63-35.

Davies, Geoffrey F. 1992. Plates and Plumes: Dynamos of the Earth's Mantle. *Science* 257:493-494.

De Gramont, Sanche. 1976. *The Strong Brown God: The Story of the Niger River*. Houghton Mifflin Company, Boston.

Deacon. 1984. Later Stone Age people in southern Africa," In *Southern African Prehistory and Paleoenvironments*, edited by Richard G. Klein, pp. 221-328. Rotterdam: A. A. Balkema.

Dean, Jeffrey S. 1970. Aspects of Tsegi Phase Social Organization: A Trial Reconstruction. In *Reconstructing Prehistoric Pueblo Societies*, edited by W. A. Longacre, pp. 140-174. University of New Mexico Press, Albuquerque.

Deetz, James and Edwin S. Dethlefsen. 1967. Death's Head, Cherub, Urn and Willow. *Natural History* 76(3): 29-37.

Deetz, James. 1967. *Invitation to Archaeology*. The Natural History Press, Garden City, New York.

Deevey, F. S. and R. F. Flint. 1957. Post-glacial Hypsithermal Interval. *Science* 125: 182-184.

Delcourt and Delcourt 1985. Quaternary Palynology and Vegetational History of the Southeastern United States. In *Pollen Records of Late-Quaternary North American Sediments*, edited by V.M. Bryant and R.G. Holloway, pp. 1-37.

Delcourt, H.R. and P.A. Delcourt. 1985. Quaternary Palynology and Vegetational History of the Southeastern United States. In *Pollen Records of Late-Quaternary North American Sediments,* edited by Vaughn M. Bryant and Richard G. Holloway, pp. 1-37. American Association of Stratigraphic Palynologists Foundation, Dallas, Texas.

Ehret, C. and M. Posnansky. 1982. *The Archaeological and Linguistic Reconstruction of African History*. University of California Press. Berkeley, California.

Delpech, F. and H. Laville. 1988. Climatologie et chronologie de la grotte Vaufrey. Confrontation des hypothèses et implications. In La Grotte Voufrey. Paléoenvironnement-Chronologie-Activités Humaines, edited by J.-P. Rigaud. *Mémoires de la Société Préhistorique Française* XIX: 381-388.

Delporte, H. 1976. Les Sepultures Mousteriennes de la Ferraisse. In Les Sepulturas Neandertaliennes, pp. 8-11. *IX Congres UISPP Pretirage*. Nice: CNRS.

DeLumley, Henri. 1969. A Paleolithic Camp at Nice. *Scientific American* 220:42-50.

deMenocal, Peter B. 1995. Plio-Pleistocene African Climate. *Science* 270: 53-59.

deMenocal, Peter B. and Jan Bloemendal. 1995. Plio-Pleistocene Climatic Variability in Subtropical Africa and the Paleoenvironment of Hominid Evolution: A Combined Data-Model Approach,. Pp. 262-288. In *Paleoclimate and Evolution, with Emphasis on Human Origins*. Edited by Elisabeth S.Vrba, George H. Denton, Timothy Partridge and LIoyd H. Burckle. Yale University Press, New Haven.

Denbow, James R. 1986. A New Look at the Later Prehistory of the Kalahari. *Journal of African History* 27: 3-28.

Denbow, James R. and Edwin N. Wilmsen. 1986. Advent and Course of Pastoralism in the Kalahari. *Science* 1509-1515.

Dennel, R. W., H.M. Rendell, and R. Hailwood. 1988. "Late Pliocene Artefacts from Northern Pakistan," *Current Anthropology* 29(3): 495-498

Dennell, Robin. 1983. *European Economic Prehistory: A New Approach*. Academic Press, New York.

Denton, G. H. and S. C. Porter. 1967. Neoglaciation. *Scientific American* 222: 101-110.

Deraniyagala, P. E. P. 1955. *Some Extinct Elephants, Their Relatives and Two Living Species*. Ceylon National Museum Publications.

Deshler, Watler. 1960. Livestock Trypanosomiasis and Human Settlement in Northeastern Uganda. *The Geographical Review* 50: 541-554.

Desowitz, Roberts S. 1980. Epidemiological-ecological Interactions in Savanna Environments. In *Human Ecology in Savanna Environments*, edited by David R. Harris, pp. 457-477. Academic Press, London.

Diamond, Jared. 1997. *Guns, Germs, and Steel: The Fates of Human Societies*. W. W. Norton and Company, New York.

Dibble, David S. and Dessamae Lorrain. 1971. Bonfire Shelter: a Stratified Bison Kill Site, Val Verde County, Texas. Texas Memorial Museum. Miscellaneous Papers No. 1.

Dibble, Harold L. 1987. The Interpretation of Middle Paleolithic Scraper Morphology. *American Antiquity* 52(1): 109-117.

Dickson, D. Bruce. 1980. Ancient Agriculture and Population at Tikal, Guatemala: an Application of Linear Programming to the Simulation of an Archaeological Problem. *American Antiquity* 45(4): 697-712.

Dickson, D. Bruce. 1985. The *Atlatl* Assessed: a Review of Recent Anthropological Approaches to Prehistoric North American Weaponry. *Bulletin of the Texas Archaeological Society* 56: 1-38

Dickson, D. Bruce. 1988. Anthropological Utopias and Geographical Epidemics: Competing Models of Social Change and the Problem of the Origins of Agriculture. In *The Transfer and Transformation of Ideas and Material Culture*, edited by Peter J. Hugill and D. Bruce Dickson, pp. 45-72. Texas A&M University Press, College Station.

Dickson, D. Bruce. 1990. *The Dawn of Belief: Religion in the Upper Paleolithic of Southwestern Europe*.

University of Arizona Press, Tucson.
Dickson, D. Bruce. 1991. Origins and Dispersals: Carl O. Sauer's Impact on Anthropological Explanations of the Development of Food Production. In *The Legacy of Intellect: Carl O. Sauer and the Berkeley School of Geography*, edited by Martin S. Kenzer. University of California Press, Berkeley.
Dillehay, Tom D. 1989. *Monte Verde: A Late Pleistocene Settlement in Chile: Paleoenvironment and Site Context*. Washington: Smithsonian Institution Press.
Dillehay, Tom D. 1997. *Monte Verde: A Late Pleistocene Settlement in Chile: The Archaeological Context and Interpretation*. Washington: Smithsonian Institution Press.
Dillehay, Tom D. and Michael B. Collins. 1988. Early Cultural Evidence from Monte Verde in Chile. *Nature* 332: 150-152.
Dillehay, Tom D., Q. M. Pino, E. M. Davis, S. Valrela, and R. Casamiquela. 1982. Monte Verde: Radiocarbon Dates from an Early-Man Site in South-Central Chile. *Journal of Field Archaeology* 9: 547-550.
Dixon, R. A. 1941. *Economic Institution and Cultural Change*. McGraw-Hill Book Company, Inc., New York.
Dobie, Frank. 1941. *The Longhorns*. Little, Brown and Company, Boston.
Dobzhansky, T. 1937. *Genetics and the Origin of Species*. Columbia University Press, New York.
Dole, Gertrude E. 1973. Foundations of Contemporary Evolutionism. In *Main Currents in Cultural Anthropology*, edited by Raoul Naroll and Frada Naroll. Appleton-Century-Crofts, New York.
Dorozynski, Alexander. 1994. A Family Tree of European Bears. *Science* 263: 175.
Dunnell, Robert C. 1971. *Systematics in Prehistory*. The Free Press, New York.
Dury, G. F. 1965. Theoretical Implications of Underfit Streams. *U. S. Geological Survey Professional Paper* 452.
Dury, G. F. 1977. Underfit Streams: Retrospect, Perspect and Prospect. In *River Channel Changes*, edited by K. J. Gregory, Pp. 281-293.
Dyson, Freeman J. 1988. *Infinite in All Directions*. Harper and Row, Publishers, New York.
Edwards, Stephen W. 1978. Nonutilitarian Activities in the Lower Paleolithic: a Look at Two Kinds of Evidence. *Current Anthropology* 19: 135-37.
Edwards, R. Lawrence, J. H. Chen, T.-L. Ku and G. J. Wasserburg. 1987. Precise Timing of the Last Interglacial Period from Mass Spectrometric Determination of Thorium-230 in Corals. *Science* 236: 15471553.
Eicher, D.L., A.L. McAlester, and M.L. Rottman. 1984. *The History of the Earth's Crust*. Prentice Hall, Inc., New York.
El-Najjar, Mahmoud Y. 1979. Human Treponematosis and Tuberculosis: Evidence from the New World. *American Journal of Physical Anthropology* 51: 599-618.
Elster, Jon. 1983. *Explaining Technical Change*. Cambridge University Press, Cambridge.
Epstein, H. and I. L. Mason. 1984. Cattle. In Evolution of Domesticated Animals. Pp. 6-27. Edited by I. L. Mason. Longman. London.
Erwin, D. H. 1993. *The Great Paleozoic Crisis*. Columbia University Press. New York.
Fairbridge, Rhodes W. 1976. Effects of Holocene Climatic Change on Some Tropical Geomorphic Processes. *Quaternary Research* 6: 529-556.
Falguères, Christophe, Jean-Jacques Bahain and Hassane Saleki. 1997. U-Series and ESR Dating of Teeth from Acheulian and Mousterian Levels at La Micoque (Dordogne, France). Journal of Archaeological Science 24: 537-545.
Falk, D. 1980. Language, Handedness and Primate Brains: Did the Australopithecines Sign? *American Anthropologist* 82(1): 72-78.
Farrand, William R. 1975. Sediment Analysis of a Prehistoric Rockshelter: The Abri Pataud. *Quaternary Research* 5: 1-26.
Farrand, William. 1985. Rockshelter and Cave Sediments. In *Archaeological Sediments in Context*, edited by Julie K. Stein and William Farrand, pp. 21-39. Center for the Study of Early Man. Institute for Quaternary Studies, University of Maine at Orono.
Farrand, William R. 1992. Eustatic Curve. *Science* 256: 1225.
Fenton, C. L., and M. A. Fenton. 1952. *Giants of Geology*. Doubleday and Company Inc, New York.
Ferguson, C.W. 1968. Bristlecone Pine: Science and Esthetics. *Science* 159: 839-845.
Figgins, James D. 1927. The Antiquity of Man in America. *Natural History* 27(3): 229-239.
Fisher, Don. 1990. Stories Missed from a Place Too Fierce to Be Anything but Free. *Bryan/College Station Eagle*, Sunday, April 1, 1990: 10a.
Fladmark, K.R. 1979. Routes: Alternate Migration Corridors for Early Man in North America. *American Antiquity* 44: 55-?
Flannery, Kent V. 1968. Systems Theory and Early Mesoamerica, In *Anthropological Archaeology in the Americas* edited by Betty J. Meggers, pp. 67-87. Anthropological Society of Washington, D.C., Washington, D.C.
Flannery, Kent V. 1969. Origins and Ecological Effects of Early Domestication in Iran and the Near East. In *The Domestication and Exploitation of Plants and Animals*, edited by Peter J. Ucko and G. W. Dimbleby. Aldine Publishing Company, Chicago.
Flannery, Kent V. 1973. The Origins of the Village as a Settlement Type in Mesoamerica and the Near East: A Comparative Study. In *Man, Settlement and Urbanism*, edited by Peter J. Ucko, Ruth Tringham and G. W. Dimbleby, pp. 23-53. Gerald Duckworth and Company, Ltd, London.
Fleagle, J.G., T.M. Bowen, J.D. Obradovich, and E.L. Simons. 1986. Age of the Earliest African Anthropoids. *Science* 234:1247-1249.
Flint, Richard. 1971. *Glacial and Quaternary Geology*. Wiley, New York.

Forbes, R. J. 1954. Chemical, Culinary, and Cosmetic Arts. In *A History of Technology, Volume I: From Early Times to the Fall of Ancient Empires*, edited by Charles Singer, E. J. Holmyard, and A. R. Hall, pp. 238-298, Oxford University Press, London.

Forsten, Ann. 1988. Middle Pleistocene Replacement of Stenonid Horses by Caballoid Horses. *Palaeogeography, Palaeoclimatology, Palaeoecology* 65: 23-33.

Fowler, Melvin and Richard MacNeish. 1972. Excavations in the Coxcatlan Locality in the Alluvial Slopes. In *Prehistory of the Tehuacan Valley. Volume 5. Excavations and Reconnaissance,* R. S. MacNeish, M. L. Fowler, A. Garcia Cook, F. A. Peterson, A. Nelken-Terner, and J. A. Neely, pp. 219-340. University of Texas Press, Austin.

Fox, Robert B. 1970. The Tabon Caves. *Monographs of the National Museum* 1. Manilla.

Fox Robert B. 1978. The Philippine Paleolithic," In *Views of the Past; Essays in Old World Prehistory and Paleoanthropology*, edited by Leslie G. Freeman, pp. 59-85. The Hague: Mouton Publishers.

Frayer, David W. 1978. Evolution of the Dentition in Upper Paleolithic and Mesolithic Europe. *University of Kansas, Publications in Anthropology* 10.

Freeman, Derek. 1974. The Evolutionary Theories of Charles Darwin and Herbert Spencer. *Current Anthropology* 15(3): 211.

Freeman, Leslie G. 1966. The Nature of the Mousterian Facies in Cantabrian Spain. *American Anthropologist* 68(2): 230-237.

Freeman, Leslie G. 1975. By Their Works You Shall Know Them: Cultural Developments in the Paleolithic. In *Hominisation and Verhalten*. Edited by Gottfied Kurth and I. Eibl-Eibesfeldt, pp. 234-261. Gustav Fischer Verlag, Stuttgart.

Freeman, Leslie G. 1978. Mousterian Worked Bone from Cueva Morin (Santander, Spain): A Preliminary Description. In *Views of the Past*, edited by L. G. Freeman, pp. 29-51. Mouton Publishers, The Hague.

Freeman, Leslie G. 1980. The Development of Human Cultures. In *The Cambridge Encyclopedia of Archaeology*, edited by A. Sherratt. Crown Publishers, New York.

Freeman, Leslie G. and J. Gonzalez Echegaray. 1981. El Juyo: A 14,000-Year-Old Sanctuary from Northern Spain. *History of Religions* 21(1): 1-19.

Frisch, Rose E. 1980. Fatness, Puberty, and Fertility. *Natural History* 10: 16-27.

Frison, George. 1974. *The Casper Site: A Hell Gap Bison Kill on the High Plains*. New York: Academic Press.

Frison, George. 1984. The Carter/Kerr-Mcgee Paleo-Indian Site: Cultural Resource Management and Archaeological Research. *American Antiquity* 49(2): 288-314.

Frison, George. 1978. *Prehistoric Hunters of the High Plains*. Academic Press, New York.

Frison, George C., R. L. Andrews, J. M. Adovasio, R. C. Carlisle, and Robert Edgar. 1986. A Late Paleo-Indian Animal Trapping Net from Northern Wyoming," *American Antiquity* 51(2): 352-361.

Frison, George C. and George M. Zeimans. 1980. Bone Projectile Points: An Addition to the Folsom Cultural Complex. *American Antiquity* 45(2): 231-237.

Fritts, H.C., G. Robert Lofgren, and Geoffrey A. Gordon. 1981. Past Climate Reconstructed From Tree Rings. In *Climate and History*, edited by R.I. Rutberg and T.K. Rabb, pp. 193-213, Princeton University Press, Princeton, New Jersey.

Gallagher, James P. 1989. Agricultural Intensification and Ridged-Field Cultivation in the Prehistoric Upper Midwest of North America. In *Foraging and Farming: The Evolution of Plant Exploitation*, edited by D. R. Harris and G. C. Hillman, Pp. 572-584. Unwin Hyman, London.

Gamble, Clive. 1981. Interaction and Alliance in Paleolithic Society. *Man* 17(1): 92-107.

Gargett, Robert. 1989. Grave Shortcomings; the Evidence for Neanderthal Burial. *Current Anthropology* 30(2): 157-190.

Gasse, F., M. Arnold, J. C. Fontes, M. Fort, E. Gibert, A. Huc, Li Bingyan, Li Yuanfang, Liu Qing, F. Mélières, E. Van Campo, Wang Fubao and Zhang Qingsong. 1991. A 13,000-year Climate Record from Western Tibet. *Nature* 353: 742-745.

Gebauer, Anne Birgitte. 1995. Pottery production and the Introduction of Agriculture in Southern Scandinavia. In *The Emergence of Pottery: Technology and Innovation in Ancient Societies,* edited by William K, Barnett and John W. Hoopes, pp. 99-112. Smithsonian Institution Press, Washington, D.C.

Geist, V. 1981. Neanderthal the Hunter. *Natural History* 90: 29-33.

Gibbon, Ann. 1998. Old, Old Skull Has a New Look. *Science* 280: 1525.

Gilbert, Allan S. 1983. On the Origins of Specialized Nomadic Pastoralism in Western Iran. *World Archaeology* 15(1): 105-119.

Gillespie, R. J., A. J. T. Jull, and T. H. Zabel. 1984. Radiocarbon Measurement by Accelerator Mass Spectrometry: An Early Selection of Dates. *Archaeometry* 26(1): 15-20.

Gilson, Etienne. 1936. *The Spirit of Mediaeval Philosophy*. C. Scribner's Sons, New York.

Gladkih, Mikhail I, Ninelj L. Kornietz, and Olga Soffer. 1984. Mammoth-Bone Dwellings on the Russian Plain. *Scientific American* 251(5): 164-175.

Glassow, Michael A. 1972. Changes in the Adaptations of Southwestern Basketmakers: A Systems Perspective. In *Contemporary Archaeology: a Guide to Theory and Contributions*, edited by Mark P. Leone, pp 289-302. Southern Illinois University Press, Carbondale and Edwardsville.

Glover, Ian C. 1977. The Hoabinhian: Hunter-gatherers or Early Agriculturalists in South-east Asia? In *Hunter-Gatherers and First Farmers Beyond Europe: an Archaeological Survey*, edited by J. V. S. Megaw, pp. 145-166. Leicester University Press, Leicester.

Glover, Ian C. 1980. Agricultural Origins in East Asia. *The Cambridge Encyclopedia of Archaeology*, Pp.

152-161. Cambridge University Press, Cambridge.
Glyn Daniels. 1962. *The Idea of Prehistory*. Penguin Books, Baltimore, Maryland.
Goodall, Jane 1986 *The Chimpanzees of Gombe: Patterns of Behavior*. Harvard University Press, Cambridge, Massachusetts
Goodall, Jane. 1987. A Plea for the Chimpanzees. *American Scientist* 75(6):574-577.
Goodman, Morris, W. J. Bailey, K. Hayaska, M.J. Stanhope, J. Slightom, J. Czelusniak. 1994. Molecular Evide4nce on Primate Phylogeny from DNA Sequences. *American Journal of Physical Anthropology* 94: 3-24.
Gorman, Chester. 1971. Hoabhinian and After: Subsistence Patterns in Southeast Asia During the Late Pleistocene and Early Recent Periods. *World Archaeology* 2: 300-320.
Gorman, Chester. 1977. *A Priori* Models and Thai Prehistory: Beginnings of Agriculture. In *Origins of Agriculture*, edited by C. A. Reed, pp. 334-335. Mouton, The Hague.
Goudie, Andrew. 1977. *Environmental Change*. Claredon Press, Oxford.
Goudie, Andrew. 1983. *Environmental Change*. Second Edition. Claredon Press, Oxford.
Goudie, Andrew. 1984. *The Nature of the Environment: An Advanced Physical Geography*. Blackwell, Oxford.
Gould, Stephen Jay. 1990. Down on the Farm: Review of From Foraging to Agriculture: The Levant at the End of the Ice Age by Donald O. Henry. *New York Review of Books* January 18, 1990, Pp. 26-27.
Graham, Russell W., Donald Lee Johnson and Marvin Kay. 1981. Kimmswick: A Clovis-Mastodon Association in Eastern Missouri. *Science* 213: 1115-1117.
Grayson, Donald K. 1983. *The Establishment of Human Antiquity*. Academic Press. New York.
Grayson, Donald K. 1984. Explaining Pleistocene Extinctions: Thoughts on the Structure of a Debate. In *Quaternary Extinctions*, edited by Paul S. Martin and Richard Klein, pp. 807-823. Tucson: The University of Arizona Press.
Grayson, Donald K. 1987. An Analysis of the Chronology of Late Pleistocene Mammalian Extinctions in North America." *Quaternary Research* 28: 281-289.
Green, W. Verney, Ulrich Achauer and Robert P. Meyer 1991. T Three-Dimensional Seismic Image of the Crust and Upper Mantle Beneath the Kenya Rift. *Nature* 354: 199-203.
Greenberg, Turner and Zegura, *Current Anthropology* 27: 477-497
Grigson, C. 1991. An African Origin for African Cattle? Some Archaeological Evidence. *African Archaeological Review*"9:119-144.
Grosswald, M. G. 1980. Late Weichselian Ice Sheet of Northern Eurasia. *Quaternary Research* 13:1-32.
Grove, J. M. 1979. The Glacial History of the Holocene. *Progress in Physical Geography* 3: 1-54.
Gruhn, Ruth. 1987. On the Settlement of the Americas, *Current Anthropology* 28: 363-364.
Grün, Rainer and Alan Thorne. 1997. Dating the Ngandong Humans. *Science* 276:1575-1576.
Guidon, Niede and G. Delibrias. 1986. Carbon-14 Dates Point to Man in the Americas 32,000 Years Ago. *Nature* 321: 769-771.
Haber, F.C. 1959a. Fossils and Early Cosmology. In *Forerunners of Darwin 1745-1859*, edited by Bentley Glass, Owsei Temkin, and William L. Strauss, Jr., pp. 3-29. John Hopkins University Press, Baltimore.
Haber, F.C. 1959b. Fossils and the Idea of a Process of Time in Natural History. In *Forerunners of Darwin 1745-1859*, edited by Bentley Glass, Owsei Temkin, and William L. Strauss, Jr., pp. 222-261. John Hopkins University Press, Baltimore.
Hahn, Joachim. 1972. Aurignacian Signs, Pendants and Art Objects in Central and Eastern Europe. *World Archaeology* 3: 252-266.
Hallam, Anthony. 1972. Continental Drift and the Fossil Record. *Scientific American* 227(5): 56-66.
Hallam, Anthony. 1987. End-Cretaceous Mass Extinction Event: Argument for Terrestrial Causation. *Science* 238:1237-1242.
Hallo, W. W. 1971. Mesopotamia and the Asiatic Near East. In *The Ancient Near East*, edited by W. W. Hallo and W. K. Simpson, pp. 3-183. Harcourt, Brace Jovanovich, Inc, New York.
Hammond, Mason. 1972. *The City in the Ancient World*. Harvard University Press, Cambridge.
Hance, William A. 1975. *The Geography of Modern Africa*. (2nd Edition). Columbia University Press, New York.
Hard, R. J. and Merrill, W. L. 1992. Mobile Agriculturalists and the Emergence of Sedentism: Perspectives from Northern Mexico. *American Anthropologist* 94(3): 601-620.
Harlan, Jack R. 1967. A Wild Wheat Harvest in Turkey. *Archaeology* 20: 197-201.
Harlan, Jack R. 1971. Agricultural Origins: Centers and Noncenters. *Science* 174: 468-474.
Harlan, Jack R. 1977. The Origins of Cereal Agriculture in the Old World. In *Origins of Agriculture*, edited by C. A. Reed, pp. 357-383. Mouton, The Hague.
Harlan, Jack R. 1989. The Tropical African Cereals. In *Foraging and Farming: The Evolution of Plant Exploitation*, edited by D. R. Harris and G. C. Hillman, pp. 335-343. Unwin Hyman, London.
Harlan, Jack R. 1992. *Indigenous African Agriculture*. In *The Origins of Agriculture: An International Perspective*. Edited by C. Wesley Cowan and Patty Jo Watson. Pp. 59-70. Smithsonian Institution Press, Washington.
Harlan, Jack R, J. M. J. De Wet, and A. Stemler. 1976. Plant Domestication and Indigenous African Agriculture. In *Origins of African Plant Domestication*, pp. 3-19. Mouton Publishers, The Hague.
Harlan, Jack R. and Daniel Zohary. 1966. Distribution of Wild Wheats and Barley. *Science* 153: 1075-1080.
Harris, Brown, Leakey, Walker and Leakey 1988
Harris, Edward C. 1979. The Laws of Archaeological

Stratigraphy. *World Archaeology* 11(1): 111-117.
Harris, Edward C. 1989. *The Principles of Archaeological Stratigraphy. Second Edition.* Academic Press, New York.
Harris, Marvin and Eric B. Ross. 1987. *Death, Sex, and Fertility: Population Regulation in Preindustrial and Developing Societies.* Columbia University Press, New York.
Harrison, G.A. and J.S. Weiner. 1964. Human Evolution in: *Human Biology*, pp. 3-98, edited by G.A. Harrison, J.S. Weiner, J.S. Tanner and N.A. Barnicot, Oxford University Press, New York.
Harrison, T. 1957. The Great Cave at Niah: a Preliminary Report on Bornean Prehistory. *Man* 57: 161-166.
Harrison, T. 1959. New Archaeological and Ethnological Results from Niah Cave, Sarawak. *Man* 59: 1-8.
Harrison, T. Mark, Peter Copeland, W. S. F. Kidd and An Yin. 1992. Raising Tibet. *Science* 256: 1663-1670.
Harrold, Francis B. 1980. A comparative Analysis of Eurasian Paleolithic Burials. *World Archaeology* 12(2): 195-211.
Harrold, Francis B. 1981. New Perspectives on the Châtelperronian. *Disputacío de Barcelona, Institut de Prehistória I Archqueologia, Monografies* LXII: 1-51.
Hassan, F. 1981. *Demographic Archaeology.* Academic Press, New York.
Haury, E.W. 1962. Recollections of a Dramatic Moment in Southwestern Archaeology. *Tree-Ring Bulletin* 24: 11-14.
Hawkes, J. G. 1989. The Domestication of American Roots and Tubers in the American Tropics. In*Foraging and Farming: The Evolution of Plant Exploitation*, edited by D. R. Harris and G. C. Hillman, pp. 481-503. Unwin Hyman, London.
Hawkes, Jaquetta. 1973. *The First Great Civilizations: Life in Mesopotamia, the Indus Valley, and Egypt.* Hutchinson, London.
Hawking, Stephen W. 1988. *A Brief History of Time: From the Big Bang to Black Holes.* Bantam Books, New York.
Haynes, C. Vance. 1969. The Earliest Americans. *Science* 166: 709-715.
Haynes, Vance. 1973. Calico Site: Artifacts or Geofacts? *Science* 181: 305-310.
Haynes, C. Vance. 1980. Paleo-Indian Charcoal from Meadowcroft Rockshelter: Is Contamination a Problem? *American Antiquity* 45: 582-587.
Haynes, C. Vance. 1982. Were Clovis Progenitors in Beringia? In the *Paleoecology of Beringia*, edited by David Hopkins, pp. 383-398. New York: Academic Press, Inc.
Haynes, C. Vance and E. Thomas Hemmings. 1968. Mammoth-Bone Shaft Wrench from Murray Springs, Arizona. *Science* 159: 186-187.
Heizer, R. F. 1962. The Background of Thomsen's Three Age System. *Technology and Culture* 3: 259-266.
Henry, Donald O. 1989. *From Foraging to Agriculture: The Levant at the End of the Ice Age.* University of Pennsylvania Press, Philadelphia.
Hester, James. 1972. *Blackwater Locality No. 1: A Stratified, Early Man Site in Eastern New Mexico.* Southern Methodist University, Fort Burgwin Research Center Publication No. 8.
Heusser, Calvin J. 1985. Quaternary Pollen Records from the Pacific Northwest Coast: Aleutians to the Orgeon-California Boundary. In *Pollen Records of Late-Quaternary North American Sediments*, edited by Vaughn M. Bryant and Richard G. Holloway, pp. 141-189. American Association of Stratigraphic Palynologists Foundation.
Hibbard, C. W. 1970. Pleistocene Mammalian Local Faunas from the Great Plains and Central Lowland Provinces of the United States. W. Dort and J.K. Jones (Editors), *Pleistocene and Recent Environments of the Central Great Plains.* Lawrence: Department of Geology, University of Kansas.
Higham, C. F. W. 1983. Prehistoric Rice Cultivation in Southeast Asia. *Scientific American* 250 (4): 138-146.
Higham, C. F. W. 1996. A Review of Archaeology in Mainland Southeast Asia. *Journal of Archaeological Research* 4(1): 3-49.
Hillman, Gordon C. and M. Stuart Davies. 1990. Measured Domestication Rates in Wild Wheats and Barley under Primitive Cultivation, and Their Archaeological Implications. *Journal of World Prehistory* 4(2): 157-222.
Hitch, Charles J. 1982. Dendrochronology and Serendipity. *American Scientist* 70: 300-305.
Hodder, Ian. 1992. *Theory and Practice in Archaeology.* Routledge, London.
Hodgett, G. A. 1972. *A Social and Economic History of Medieval Europe.* Harper Torchbooks, New York.
Holdaway, Simon and Richard Cosgrove. 1997. The Archaeological Attributes of Behaviour: Difference or Variability? *Endeavour* 21(2): 66-71.
Hole, Frank. 1989. A Two-part, Two-stage Model of Domestication. In *The Walking Larder: Patterns of Domestication, Pastoralism, and Predation*, Juliet Clutton-Brock, pp. 97-104. Unwin Hyman, London.
Hole, Frank, Kent V. Flannery, and James A. Neely. 1969. Prehistory and Human Ecology of the Deh Luran Plain. *University of Michigan Museum of Anthropology, Memoirs* 1.
Holland, Heinrich D. 1997. Evidence for Life on Earth More Than 3850 Million Years Ago. *Science* 275: 38-39.
Hooten, Ernest. 1949. *Up From the Apes.* MacMillan, New York.
Hopkins, David, editor. 1984. *Paleoecology of Beringia.* Academic Press New York.
Hopkins, David M. 1986. Sea-level History in Beringia During the past 250,000 Years. In *Beringia in the Cenozoic Era*, edited by V. L. Kontrimavichus, pp.3-29. Rotterdam: A.A. Balkema.
Horodyski, Robert J. and L. Paul Knauth 1994. Life on Land in the Precambrian. *Science* 263: 494-498.
Howard, Calvin D. 1974. The *Atlatl*: Function and

Performance. *American Antiquity* 39: 102-104.
Howell, F. Clark. 1957. Pleistocene Glacial Ecology and the Evolution of "Classic" Neanderthal Man. *Southwestern Journal of Anthropology* 8:377-410.
Howell, F. Clark. 1966. Observations on the Earlier Phases of the European Lower Paleolithic. *American Anthropologist* 68(2):111-140.
Howell, F. Clark. 1975. Neanderthal Man: Facts and Figures. In *Paleoanthropology*. Edited by R. Tuttle, pp. 389-407. Mouton Publishers, The Hague.
Howell, Nancy. 1979. *Demography of the Dobe !Kung*. Academic Press, New York.
Isaac, Erich. 1962. On the Domestication of Cattle. *Science* 137:195-204.
Hsu, K. J. 1983. *The Mediterranean Was a Desert*. Princeton University Press, Princton.
Hudson, M. and G. I. Barnes. 1991. Yoshinogari–a Yayoi Settlement in Northern Kyushu. *Mouumenta Nipponica* 46(2): 211-235.
Huelsenbeck, John P.and Bruce Rannala 1997. Phylogenetic Methods Come of Age: Testing Hypotheses in an Evolutionary Context. *Science* 276: 227-232.
Hugot, H. J. 1981. The Prehistory of the Sahara. In *General History of Africa* I, edited by J. Ki-zerbo, pp. 565-610.
Hunt, Charles B. 1974. *Natural Regions of the United States and Canada*. W. H. Freeman and Company, San Francisco.
Hutton, James. 1788. Theory of the Earth: Or an Investigation of the Laws Observable in the Composition, Dissolution, and Restoration of Land Upon the Globe. *Transactions of the Royal Society of Edinburgh* 1: 209-304.
Huxley, Julian. 1935. *Science and Social Needs*. Harper and Brothers, London.
Huxley, Julian. 1963. *Evolution: The Modern Synthesis*. Chatto and Winders, London.
Ikawa-Smith, F. and F. Kanjodori. 1992. Communal Cemeteries of the Late Jomon in Hokkaido. *Pacific Northeast Asia in Prehistory: Hunter-Fisher-Gatherers, Farmers and Sociopolitical Elites*. Edited by C. M. Aikens and S. N. Rhee". Pp.83-90. Washington State University Press, Pullman, Washington.
Ikawa-Smith, Fumiko. 1980. Current Issues in Japanese Archaeology. *American Scientist* 68: 134-145.
Ikeya, Motoji. 1986. Electron Spin Resonance. In *Dating and Age Determination of Biological Materials*, edited by Michael Zimmerman and J. Lawrence Angel, pp 59-125).
Irving, William N. and C. R. Harrington. 1973. Upper Pleistocene Radiocarbon-dated Artifacts from the Northern Yukon. *Science* 179: 335-340.
Irwin, Henry T. 1970. Paleo-Indian tool types in the Great Plains. *American Antiquity* 35(1): 24-34.
Irwin-Williams, Cynthia. 1967. Summary of Archaeological Evidence from the Valesquilla Region, Puebla, Mexico. In *Pleistocene Extinctions: The Search for a Cause*, edited by Paul S. Martin and Herbert E. Wright, pp. 337-347. New Haven: Yale University Press.
Isozaki, Yukio 1997. Permo-Triassic Boundary Superanoxia and Stratified Superocean: Records from Lost Deep Sea. *Science* 276: 235-238).
Izet, G. A., W. A. Cobban, J. D. Obradovich and M. J. Kunk.1993. The Manson Impact Structure: 40Ar/39Ar Age and its Distal Impact Ejecta in the Pierre Shale in Southeastern Dakota. *Science* 262: 729-732.
Jacobs. 1972 The Absolute Date of the Djetis at Modjokerto. *Antiquity* 47: 148-?
Jacobson, Jerome. 1979. Recent Developments in South Asian Prehistory and Protohistory. *Annual Review of Anthropology* 8: 467-502.
James, Edwin O. 1957. *Prehistoric Religion*. Frederick A. Praeger, New York.
Jarman, M. R. 1972. European Deer Economies and the Advent of the Neolithic. In *Papers in Economic Prehistory*, edited by Eric Higgs, pp. 125-148. Cambridge University Press, Cambridge.
Jeheskel, Shoshain and Pascal Tassy (editors). 1996. *The Proboscidea: Evolution and Paleoecology of Elephants and Their Relatives*. Oxford University Press, New York.
Jelinek, Arthur J. 1977. The Lower Paleolithic: Current Evidence and Interpretations. *Annual Review of Anthropology* 6: 11-32.
Jensen, Jorgen. 1982. *The Prehistory of Denmark*. Methuen, London.
Jett, Stephen C. 1983. Precolumbian Transoceanic Contacts. In *Ancient North Americans*, edited by Jesse D. Jennings, pp. 557-613. W. H. Freeman and Company, New York.
Jia, Lanpo and Weiwen Huang. 1991. The Paleolithic Culture of China. *Quaternary Science Reviews* 10(6): 519-521.
Jochim, Michael A. 1976. *Hunter-Gatherer Subsistence and Settlement: a Predictive Model*. Academic Press, New York.
Jochim, Michael. 1983. Paleolithic Cave Art in Perspective. In *Hunter-Gatherer Economy in Prehistory*, edited by G. N. Bailey, pp. 212-219. Cambridge University Press, Cambridge.
Johnson 1987. *Lubbock Lake: Late Quaternary Studies on the Southern High Plains*. Texas A&M Press, College Station, Texas.
Johnson, Samuel. 1990 [orig. 1775]. *A Journey to the Western Islands of Scotland*. The Folio Society, London.
Jowett, Benjamin (translator). 1943. *Politics of Aristotle*. Modern Library, New York.
Kahn, Patricia and Ann Gibbons. 1997. DNA from an Extinct Human. *Science* 277: 176-178.
Kano, Takayoshi. 1992. The Last ape: *Pygmy Chimpanzee Behavior and Ecology*. Stanford University Press, Stanford, California.
Kaplan, L. 1965. Archaeology and the Domestication in American *Phaseolus* (Beans). *Economic Botany* 19: 358-368.
Kasting, James F. and Ackerman, Thomas P. 1986. Climatic Consequences of Very High Carbon Dioxide

Levels in the Earth's Early Atmosphere. *Science* 234: pp.1383-1385.
Katz, S, H. Katz, M. I. Hediger, and I. A. Valleroy. 1974. Traditional Maize Processing Techniques in the New World: Traditional Alkali Processing Enhances the Nutritional Quality of Maize. *Science* 184: 765-773.
Kay, R.F., J.G. Fleagle, and E.L. Simon. 1981. A Revision of the Oligocene Apes of the Fayum Province, Egypt. *American Journal of Physical Anthropology* 55: 293-322.
Keene, Arthur S. 1981. *Prehistoric Foraging in a Temperate Forest: a Linear Programming Model*. Academic Press, New York.
Kelly, Robert L. and Lawrence C. Todd. 1988. Coming into the Country: Early Paleo-Indian Hunting and Mobility. *American Antiquity* 53(2): 231-244.
Kennedy, G. E. 1980. *Paleoanthropology*. McGraw-Hill, New York.
Kennett, James P. 1995. A Review of Polar Climatic Evolution During the Neogene, Based on the Marine Sediment Record. In *Paleoclimate and Evolution with Emphasis on Human Origins*. Edited by Elisabeth S. Vrba, Geoge H. Denton, Timothy C. Partridge, and Lloyd H. Burckle, pp.49-64. Yale University Press, New Haven and London.
Kennett, James P. and R. C. Thunell. 1975. Global Increase in Quaternary Explosive Volcanism. *Science* 187: 497-503.
Kenyon, Kathleen. 1979. *Archaeology in the Holy Land*. (4th Edition), Ernest Benn Limited, London.
Kerr , Richard A. 1993. The Greatest Extinction Gets Greater. *Science* 262: 1370-1371.
Kerr, Richard A. 1996. A Shocking View of the Permo-Triassic. *Science* 274: 1080.
Kershaw, A. P. 1978. Record of Last Interglacial-Glacial Cycle from Northeastern Queensland. *Nature* 272: 159-161.
King, Anthony. 1983. Terracotta Revetments, Figurines and Lamps. In *A Handbook of Roman Art*, edited by Martin Henig, pp. 191-204. Phaidon Press Limited, Oxford
Kislev, E. 1984. The Emergence of Wheat Agriculture. *Paléorient* 10(2): 61-70.
Klein, Richard G. 1969. The Mousterian of European Russia. *Proceedings of the Prehistoric Society* XXXV.
Klein, Richard G. 1974. Ice-Age Hunters of the Ukraine. *Scientific American* 220(6): 96-105.
Klein, Richard G. 1976. The Mammalian Fauna of the Klasies River Mouth Sites, Southern Cape Province, South Africa. *South African Archaeological Bulletin* 31:75-98.
Klein, Richard G. 1978. Stone Age Predation on Large African Bovids," *Journal of Archaeological Science* 5: 195-217.
Klein, Richard G. 1980. Late Pleistocene Hunters. In *The Cambridge Encyclopedia of Archaeology*, edited by A. Sheratt, pp. 87-96. Cambridge University Press, Cambridge.
Klein, Richard G. 1983. The Stone Age Prehistory of Southern Africa. *Annual Review of Anthropology* 12: 25-48.
Klein, Richard G. 1984a. The Large Mammals of Southern Africa. In *Southern African Prehistory and Paleoenvironments*, edited by Richard G. Klein, pp. 107-146. Rotterdam: A.A. Balkema.
Klein, Richard G. 1984b. Mammalian Extinctions and Stone Age People in Africa. In *Quaternary Extinctions*, edited by Paul S. Martin and Richard Klein. Tucson: University of Arizona Press.
Knoll, Andrew H. 1992. The Evolution of the Eucaryotes: A Geological Perspective. *Science* 256: 622-627.
Koebert, Christian, C. Wylie Poag, Wolf Uwe Reimold and Dion Brandt.1996. Impact Origin of the Chesapeake Bay Structure and the Source of the North American Tektites. *Science* 271: 1263-1266.
Koyama, S. 1978. Jomon Subsistence and Population. *Senri Ethnological Studies* 2: 1-65.
Kramer, Samuel N. 1959. *History Begins at Sumer*. Doubleday, New York.
Kroeber, A. L. and Clyde Kluckhohn. 1952. *Culture: A Critical Review of Concepts and Definitions*. Vintage Books, New York.
Kuhn, Steven L. 1995. *Mousterian Lithic Technology: An Ecological Perspective*. Princeton University Press, Princeton, New Jersey.
Kukla, G. J. 1977. Pleistocene Land Sea Correlations. I. Europe." *Earth Sciences Review* 13: 307-374.
Kukla, G.J., R.K. Mathews, and J.M. Mitchell. 1972. The End of the Present Interglacial. *Quaternary Research* 2: 261-269.
Kuman, Kathleen. 1986. Florisbad--New Investigations at a Middle Stone Age Hominid Site in South Africa. *Geoarchaeology* 1(2): 103-125.
Kurtén, Bjorn. 1968. *Pleistocene Mammals of Europe*. Aldine, Chicago.
Kurtén, Björn. 1972. *The Age of Mammals*. Columbia University Press, New York
Kurtén, Björn. 1980. *Pleistocene Mammals of North America*. Columbia University Press, New York.
Kutzbach, J. E. and Z. Liu. 1997. Response of the African Monsoon to Orbital Forcing and Ocean Feedbacks in the Middle Holocene. *Science* 278: 440-443.
Kuzmin, Yaroslav K. and Lyobov A. Orlova 1998. Radiocarbon Chronology of the Siberian Paleolithic. *Journal of World Prehistory* 12(1): 1-53. 1995
Kyte, Frank T., Lei Zhou, John T. Wasson. 1988. New Evidence on the Size and Possible Effects of a Late Pliocene Oceanic Asteroid Impact. *Science* 241: 63-65.
LaBarre, Weston. 1974. Morning Glory and Mother Goddess at Tepantitla. In *Mesoamerican Archaeology: New Approaches*, edited by Norman Hammond. Austin: University of Texas Press.
Lahr, M.M. 1996. *The Evolution of Modern Human Diversity*. Cambridge University Press, New York.
Laville, Henri, Jean-Philippe Rigaud, and James Sackett. 1980. *Rock Shelters of the Perigord: Geological Stratigraphy and Archaeological Succession*.

Academic Press, New York.
Lahrens, L. and Robson Bonnichen. 1976. Bone Foreshafts from a Clovis Burial in Southwestern Montana. *Science* 186: 147-149.
Lake, James A. 1988. Origin of the Eukaroytic Nucleus Determined by Rate-invariant Analysis of RNA Sequences. *Nature* 331:184-186.
Lamb, H. H. 1982. Reconstruction of the Course of Postglacial Climate over the World. In *Climate Change in Later Prehistory*, edited by A. F. Harding, pp. 11-32. University of Edinburgh Press, Edinburgh.
Lambrecht, J. B. 1964. Aspects of the Evolution and Ecology of Tsetse Flies and Trypanosomiasis in the Prehistoric African Environment. *Journal of African History* 5(1): 1-24.
Laville, Henri, Jean-Philippe Rigaud, and James Sackett 1980. *Rock Shelters of the Perigord: Geological Stratigraphy and Archaeological Succession*. Academic Press, New York.
Leakey, Louis S. B., Ruth D. Simpson, and Thomas Clements. 1968. Archaeological Excavations in the Calico Mountains, California. *Science* 160: 1022-1023.
Leakey, Louis B., P. V. Tobias, and J. R. Napier. 1964. A New Species of the Genus *Homo* from Olduvai Gorge. *Nature* 202: 5-7.
Leakey, Mary D. 1971. *Olduvai Gorge Volume III: Excavations in Beds I and II, 1960-1963*. Cambridge University Press, Cambridge.
Leakey, Meave G., C.S. Feibel, I. McDougall, C. Ward and A. Walker. 1998. New specimens and Confirmation of an Early Age for *Australopithecus anamensis*. *Nature* 393: 62-66.
Leakey, Richard E. F. and Alan Walker. 1976. *Australopithecus*, *Homo erectus*, and the Single-Species Hypothesis. *Nature* 261: 572-576.
Lee, David H., Kay Severin, Yohel Yokobayashi and M. Reza Ghadirii. 1997.Emergence of Symbiosis in Peptide Self-Replication Through a Hypercyclic Network. *Nature* 390: 591-594.
Lee, Richard B. 1965. *Subsistence Ecology of !Kung Bushman*. Unpublished Ph.D. Dissertation, Department of Anthropology, University of California at Berkeley, Berkeley, California.
Lee, Richard B. 1968. What Do Hunters Do for a Living or How to Make out on Scarce Resources, in Richard B. Lee and I. Devore, *Man the Hunter*. Aldine Publishing Company, Chicago.
Lee, Richard B. and Irven Devore, editors. 1968. *Man the Hunter*. Aldine Publishing Company, Chicago.
Lee, Richard B. Lee and Irven Devore. 1976. *Kalahari Hunter-Gatherers: Studies of the !Kung San and Their Neighbors*. Harvard University Press, Cambridge.
Leeds, Anthony. 1976. Village Settlements. In *Encyclopedia of Anthropology*, edited by David E. Hunter and Philip Whitten, p. 400. Harper and Row, Publishers, New York.
Lees, Susan H. and Daniel G. Bates. 1974. The Origins of Specialized Nomadic Pastoralism: a Systemic Model. *American Antiquity* 39(2): 187-193.
Legge, A. J. and P. Rowley-Conwy. 1987. Gazelle Hunting in Stone Age Syria. *Scientific American* 238: 88-95.
Lenski, Gerhard and Jean Lenski. 1970. Human Societies: An Introduction to Macrosociology. McGraw-Hill, New York.
Lenzer, Gertrud (editor). 1975. *Auguste Comte and Positivism: The Essential Writings*. Harper and Row, New York.
Leone, Mark. 1968. Neolithic Economic Autonomy and Social Distance. *Science* 162: 1150-1151.
Leroi-Gourhan, André and Arlette Leroi-Gourhan. 1965. Chronologie de Grottes d'Arcy-sur-Cure (Yonne). *Gallia-Préhistoire* 7:1-64.
Leveque, F. and Bernard Vandermeersch. 1981. Le Nandertalien de Saint-Cesaire. *La Recerche* 12: 242-244.
Levy, David H. 1995. *Impact Jupiter: The Crash of Comet Shoemaker-Levy 9*. Plenum Publishing, New York.
Lewin, Roger. 1984a DNA Reveals Surprises in Human Family Tree. *Science* 226: 1179-1182.
Lewin, Roger. 1984b. *Human Evolution: An Illustrated Introduction*. W. H. Freeman, New York.
Lewin, Roger. 1986a. Molecular Clocks Turn a Quarter Century. *Science* 239: 561-563.
Lewin, Roger. 1986b. New Fossil Upsets Human Family. *Science* 233:720-721.
Lewin, Roger. 1987a. The Earliest 'Humans' Were More like Apes. *Science* 236:1061-1063.
Lewin, Roger. 1987b. The First Americans Are Getting Younger. *Science* 238: 1230-1232.
Lewin, Roger. 1988a. Conflict Over DNA Results. *Science* 241: 1598-1600.
Lewin, Roger. 1988b. New Views Emerge on Hunters and Gatherers. *Science* 240: 1146-1148.
Li, chia-Wei, Jun-Yuan Chen and Tzu-en Hua. 1998. Precambmrian Sponges with Cellular Structures. *Science* 279: 879-882.
Libby, Leona Marshall. 1983. *Past Climates*. University of Texas Press, Austin.
Lieberman, P. 1976. Interactive Models for Evolution: Neural Mechanisms, Anatomy and Behavior. In Origins and Evolution of Language and Speech. Edited by S. Harnad, H.D. Steklis, and J. Lancaster. *Annals of the New York Academy of Sciences* 280: 660-672.
Lieberman, P. and E. S. Crelin. 1971. On the Speech of the Neanderthal Man. *Linguistic Enquiry* 11:203-222.
Linton, Ralph. 1955. *The Tree of Culture*. Alfred A. Knopf, Inc, New York.
Little, Michael A. 1989. Human Biology of African Pastoralists. *Yearbook of Physical Anthropology* 32: 215-247.
Livi-Bacci, Massimo. 1989. *A Concise History of World Population*. Blackwell Publishers, Inc., Malden, Massachusetts.
Livingstone, Frank B. 1958. Anthropological Implications of Sickle Cell Gene Distribution in West

Africa. *American Anthropologist* 60: 533-562.

Lourandos, Harry. 1987. Pleistocene Australia: Peopling a Continent. In *The Pleistocene Old World: Regional Perspectives*, edited by Olga Soffer, pp. 147-165. New York: Plenum Press

Loy, James D. 1970. Peri-menstrual Sexual Behavior among Rhesus Monkeys. *Folia Primatologica.* 13: 286-297.

Loy, James D. and Calvin B. Peters 1991. *Understanding Behavior: What Primate Studies Tell Us About Human Behavior.* Oxford University Press, New York.

Luchterhand, Kubet. 1978. Late Cenozoic Climate, Mammalian Evolutionary Patterns, and Middle Pleistocene Human Adaptation in Eastern Asia. In *Views of the Past: Essays in Old World Prehistory and Paleoanthropology*, edited by Leslie G. Freeman, pp ??-?? The Hague: Mouton Publishers.

Lundberg, John G., Antonio Machado-Allison, and Richard F. Kay. 1986. Miocene Characid Fishes from Colombia: Evolutionary Stasis and Extirpation. *Science* 234: 208-209.

Lundelius, Ernest L., Jr. 1976. Vertebrate Palaeontology of the Pleistocene: An Overview. In Ecology of the Pleistocene: A Symposium, edited by R. C.West and W.G. Haag, pp 45-59. *Geoscience and Man* XIII.

Lynch, Tom. 1978. The South American Paleo-Indians. In *Ancient Native Americans*, edited by Jesse D. Jennings, pp. 455-489. San Francisco: W.H. Freeman and Company.

MacNeish, Richard S. 1964. The Origins of New World Civilization. *Scientific American* ?

MacNeish, Richard S. 1978. *The Science of Archaeology?* Wadsworth Publishing Company, Belmont, California

MacNeish, Richard S. 1979. The Early Man Remains from Pikimachay Cave, Ayacucho Basin, Highland Peru, In *Pre-Llano Cultures of the Americas: Paradoxes and Possibilities*, edited by R. L. Humphrey and Dennis Stanford, pp. 1-48. Washington, D.C.: The Anthropological Society of Washington.

Maddin, Robert, James D. Muhly, and Tamara S. Wheeler. 1977. How the Iron Age Began. *Scientific American* 237(4): 122-131.

Maher, A. and D.J. Stevenson. 1988. Impact Frustration of the Orgin of Life. *Nature* 331:612-614.

Mania, D. 1991. The Zonal Division of the Lower Paleolithic Open-Air Site of Bilzingsleben. *Anthropologie* 29: 17-24.

Mania, D. and Mania U. 1988. Deliberate Engravings on Bone Artifacts by *Homo erectus*. *Rock Art Research* 5: 91-107.

Mann, Alan. 1976. Ecology of Early Man in the Old World. In Ecology of the Pleistocene: A Symposium, edited by R. C. West and W. G. Haag, pp 61-70. *Geoscience and Man* 13.

Mansouri, Mouhsin, Abderrahmane El Fouikar and Bernard Saint-Martin. 1996. Correlation Between 14C Ages and Aspartic Acid Racemization at the Upper Palaeolithic Site of the Abri Pataud (Dordogne, France). *Journal of Archaeological Science* 23: 803-809.

Marean, C.W. 1992. From Hunter to Herder: Large Mammmael Remains from the Hunter-Gatherer Occupation at Enkapune Ya Mune Rockshelter, Central Rift Valley, Kenya. *African Archaeological Review* 10: 65-127.

Marean, Curtis W. and Soo Yeun Kim. 1998. Mousterian Large-Mammal Remains from Kobeh Cave: Behavioral Implications for Neanderthals and Early Modern Humans. *Current Anthropology* 39 (Supplement): S79-S113.

Marquardt, William H. 1978. Advances in Archaeological Seriation. *Advances in Archaeological Method and Theory* 1: 257-314.

Marshack, Alexander. 1970. The Baton of Montgaudier. *Natural History* 79: 56-63.

Marshack, Alexander. 1972. *The Roots of Civilization: The Cognitive Beginnings of Man's First Art, Symbol and Notation.* McGraw-Hill, NY.

Marshack, Alexander. 1976. Some Implications of the Paleolithic Symbolic Evidence of the Origin of Language. *Current Anthropology* 17: 274-281.

Marshack, Alexander. 1977. The Meander as a System: The Analysis and Recognition of Iconographic Units in Upper Paleolithic Composition. In form in Indigenous Art: Schematisation in the Art of Aboriginal Australia and Prehistoric Europe, edited by Peter J. Ucko, pp 284-317. *Australian Institute of Aboriginal Studies, Prehistory and Material Culture Series* 13.

Marshack, Alexander. 1981. On Paleolithic Ochre and the Early Uses of Color and Symbol. *Current Anthropology* 22: 188-191.

Marshall, Fiona. 1989. Rethinking the Role of *Bos Indicus* in Sub-Saharan Africa. *Current Anthropology* 30(2): 235-240.

Marshall, Larry G. 1988. Land Mammals and the Great American Interchange. *American Scientist* 76(4): 380-388.

Marshall, Larry G. S. David Webb, J. John Sepkoski and David M.Raup.1982. Mammalian Evolution and the Great American Interchange. *Science* 215: 1351-1357.

Martin, P. S. 1984. Prehistoric Overkill: the Global Model. In *Quaternary Extinctions*, edited by Paul S. Martin and Richard Klein. Tucson: University of Arizona Press.

Martin, P.S., Robert S. Thompson, and Austin Long, "Shasta Ground Sloth Extinctions: a Test of the Blitzkrieg Model," In *Environments and Extinctions: Man in Late Glacial North America*, edited by Mead and Meltzer pp. 5-14. Orono: Center for the Study of Early Man, University of Maine at Orono.

Martin, Paul S. and H. E. Wright, editors. 1967. *Pleistocene Extinctions: The Search for a Cause.* New Haven: Yale University Press.

Martin, R. D. 1988. Several Steps Forward for Eocene Primates. *Nature* 331: 660-661.

Matson, Frederick R. 1965. *Ceramics and Man.* Aldine, Chicago.

Mayr, Ernst. 1972. The Nature of the Darwinian Revolution. *Science* 176: 981-989.

McBurney, C.B. M. 1967. *The Haua Fteah*. Cambridge University Press, Cambridge.

McBurney, Charles. 1973. From the Beginning of Man to 33,000 B.C. In *France Before the Romans*. Edited by Stuart Piggott, Glyn Daniel and Charles McBurney, pp. 9-29. Noyes Press, Park Ridge, New Jersey.

McEwan, Gordon F. and Dickson, D. Bruce. 1978. Valdivia, Jomon Fishermen, and the Nature of the North Pacific: Some Nautical Problems with Meggers, Evans, and Estradea's (1965) Transoceanic Contact Thesis. *American Antiquity* 43: 362-371.

McIntosh, S. K. and R. J. McIntosh. 1983. Current Directions in West African Prehistory. *Annual Review of Anthropology* 12: 215-258.

McKay, David S. Everett K. Gibson, Jr., Kathie Thomas-Keprta, Hojatollah Vali. 1997. Reply to Bradley, Harvey and McSween. *Nature* 390: 456-457.

McKay, David S., Everett K. Gibson, Jr., Kathie L. Thomas-Keprta, Hojatollah Valli, Christopher S. Romanek, Simon J. Clemett, Xavier D. F. Chillier, Claude R. Maechling, Richard N. Zare. 1996. Search for Past Life on Mars: Possible Relic Biogenic Activity in Martian Meteorite ALH84001. *Science* 273: 924-930.

McKern, Sharon S. and Thomas W. McKern. 1974. *Living Prehistory: An Introduction to Physical Anthropology and Archaeology*. Cummings Publishing Co., Menlo Park, CA.

Mcneill, William H. 1982. *The Pursuit of Power*. University of Chicago Press, Chicago.

Mead, Jim I. 1980. Is it Really That Old? A Comment about the Meadowcroft Rockshelter 'Overview.' *American Antiquity* 45: 579-582.

Meggers, Betty, Clifford Evans and Emilio Estrada. 1965. Early Formative Period of Coastal Ecuador: The Valdivia and Machalilla Phases. *Smithsonian Contributions to Anthropology* I: 1-34.

Meggers, Betty J. 1973. Some Problems of Cultural Adaptation in Amazonia, with Emphasis on the Pre-european Period. In *Tropical Forest Ecosystems in Africa and South America: a Comparative Review*, edited by B. J. Meggers, E. S. Ayensu, D. Duckworth, pp. 311-320. The Smithsonian Instituion Press, Washington, D.c.

Mellaart, James. 1975. *The Earliest Civilizations of the Near East*. Charles Scribner's Sons, New York.

Mellars, Paul A. 1973. The Character of the Middle-Upper Paleolithic Transition in Southwestern France. In *The Explanation of Cultural Change*, edited by Colin Renfrew, pp. 255-276. Duckworth, London.

Mellars, Paul A. 1974. The Paleolithic and Mesolithic. In *British Prehistory; a New Outline*, edited by Colin Renfrew, pp. 41-99. Noyes Press, Park Ridge, New Jersey.

Mellars, Paul. 1985. The Ecological Basis of Social Complexity in the Upper Paleolithic of Southwestern France. In *Prehistoric Hunter-Gatherers: The Emergence of Cultural Complexity*, edited by T. Douglas Price and James A. Brown, pp 271-297. Academic Press, New York.

Meltzer, David J. and Jim I. Mead. 1985. Dating Late Pleistocene Extinctions: Theoretical Issues, Analytical Bias, and Substantive Results. In *Environments and Extinctions: Man in Late Glacial North America*, edited by Mead and Meltzer. Orono: Center for the Study of Early Man, University of Maine at Orono.

Mercier, N., H. Valladas, G. Valladas, J-L. Reyss. A.Jelinek, L. Meignen and J-L. Joron. 1995. TL Dates of Burnt Flints from Jelinek's Excavations at Tabūn and Their Implications. *Journal of Archaeological Science* 22:495-509.

Michels, Joseph. 1973. *Dating Methods in Archaeology*. Seminar Press, New York.

Milburn, M. 1978. Towards and Absolute Chronology of Certain Saharan Rock Art. *Antiquity* LII: 135-136.

Miller, Richard. 1987. *Fact and Method: Explanation, Confirmation and Reality in the Natural and Social Sciences*. Princeton University Press, Princeton.

Milo, Richard G. I998. Evidence for Hominid Predation at Klasies River Mouth, South Africa, and its Implications for the Behavior of Early Modern Humans. *Journal of Archaeological Research* 25: 99-133.

Moffat, Anne Simon 1997. Teeth and Bone Tell Their Stories at Chicago Meeting. *Science* 278: 801-802

Moore, Andrew M. T. 1985. The Development of Neolithic Societies in the Near East. *Advances in World Archaeology* 4: 1-69.

Moore, Peter. 1987. A Thousand Years of Death. *New Scientist* 8 January 46-48.

Moores, E. M. 1995. A Tectonic Amalgamation. *Science* 268: 134-135.

Morgan, Jo, Mike Warner and the Chicxulub Working Group, John Brittan, Richard Buffler, Antonio Camargo, Gail Christeson, Paul Denton, Alan Hilldebrand, Richard Hobbs, Hamish Macintyre, Graeme Mackenzie, Peter Maguire, Luis Marin, Yosio Nakamura, Mark Pilkington, Virgil Sharpton, Dave Snyder, Gerardo Suarez, and Alberto Trejo. 1997. Size and Morphology of the Chicxulub Impact Crater. *Nature* 390: 472-476.

Morgan, Paul. 1991. A Deep Look at African Rifting. *Nature* 354: 188-189

Morrell, Virginia. 1997. Fossilized Hatchling Heats Up the Bird-Dinosaur Debate. *Science* 276: 1501.

Movius, Hallam L., Jr. 1944. Early Man and Pleistocene Stratigraphy in Southern and Eastern Asia. *Papers of the Peabody Museum, Harvard University, Volume* 19.

Movius, Hallam L., Jr. 1949. The Lower Paleolithic Cultures of Southern and Eastern Asia. *Transactions of the American Philosophical Society* 38(4): 329-420.

Movius, Hallam L., Jr. 1966. The Hearths of the Upper Perigordian and the Aurignacian Horizons at Abri pataud, Les Eyzies (Dordogne) and Their Possible Significance. *American Anthropologist* 68: 296-325.

Movius, Hallam L., Jr. 1974. The Abri Pataud Program

of the French Upper Paleolithic in Retrospect. In *Archaeological Researches in Retrospect*, edited by Gordon R. Willey, pp. 87-116. Cambridge University Press, Cambridge.

Movius, Hallam L., Jr., ed. 1975. Excavation of the Abri Pataud, Les Eyzies (Dordogne). *American School of Prehistoric Research, Peabody Museum of Archaeology and Ethnology, Harvard University, Bulletin* 30.

Mueller-Wille, C. S. and D. Bruce Dickson. 1991. An Examination of Some Models of Late Pleistocene Society in Southwestern Europe. In *Perspectives on the Past*, edited by G. A. Clark. University of Pennsylvannia Press, Philadeslphia.

Muller, Richard A. and Gordon J. MacDonald. 1997. Glacial Cycles and Astronomical Forcing. *Science* 277: 215-218.

Mumford, Lewis. 1967. *The Myth of the Machine: Technics and Human Development*. Harcourt Brace Jovanovich, New York.

Murdock, George Peter. 1959. *Africa: Its Peoples and Their Culture History*. McGraw- Hill, New York.

Murra, John V. 1985a "El Archipièlago Vertical" Revisited. In *Andean Ecology and Civilzation: an Interdisciplinary Perspective on Andean Ecological Complementarity*, edited by Shozo Masuda, Izumi Shimada, and Craig Morris, pp. 3-13. University of Tokyo Press, Tokyo.

Murra, John V. 1985b The Limits and Limitations of the "Vertical Archipelago" in the Andes. In *Andean Ecology and Civilzation: an Interdisciplinary Perspective on Andean Ecological Complementarity*, edited by Shozo Masuda, Izumi Shimada, and Craig Morris, pp. 15-20. University of Tokyo Press, Tokyo.

Noli, Dieter and Graham Avery. 1988. Protein Poisoning and Coastal Subsistence. *Journal of Archaeological Science* 15: 395-401.

Murray, Jacqueline. 1970. *The First European Agriculture: a Study of the Osteological and Botanical Evidence Until 2,000 B.C.* Edinburgh University Press, Edinburgh.

Naroll, Raoul and Frada Naroll. 1973. *Main Currents in Cultural Anthropology*. Appleton-Century-Crofts, New York.

Nelso, Sarah Milledge. 1993. *The Archaeology of Korea*. Cambridge University Press, New York.

Nelson, D.E., R.E. Morlan, J.S. Vogel, J.R. Southon and C.R. Harrington. 1986. New Dates on Northern Yukon Artifacts: Holocene Not Upper Pleistocene. *Science* 232: 749-751.

Netboy, Anthony. 1968. *The Atlantic Salmon: A Vanishing Species?* Houghton Mifflin, Boston.

Newcomer, M. H. 1975. "Punch Technique" and Upper Paleolithic Blades. In *Lithic Technology: Making and Using Stone Tools*, edited by Earl Swanson, pp. 97-102, Mouton Publishers, The Hague.

Ninkovich, D. and Lloyd H. Burkle. 1978. Absolute Age of the Base of the Hominid-Bearing Beds in Eastern Java. *Nature* 275: 306-7.

Nurse, G. T., J. S. Weiner and Trefor Jenkins. 1985. *The Peoples of Southern Africa and Their Affinities*. Oxford University Press, New York.

Nussbaum, Martha. 1986. Sex in the Head. *New York Review of Books* 33 (20): 49-52.

O'Connell, James F. and Jim Allen. 199-. When Did Humans First Arrive in Greater Australia and Why Is It Important to Know? *Evolutionary Anthropology*

Oakley, Kenneth P. 1957. *Man the Toolmaker*. The University of Chicago Press, Chicago.

Oates, David and Joan Oates. 1976. Early Irrigation Agriculture in Mesopotamia. In *Problems in Economic and Social Archaeology*, edited by G. and De G. Sieveking, I. H. Longworth and K. E. Wilson, pp. 109-135. Duckworth, London.

Oates, Joan. 1972. Prehistoric Settlement Patterns in Mesopotamia. In *Man, Settlement and Urbanism*, edited by P. J. Ucko, Ruth Tringham, and G. W. Dimbleby, pp. 299-310. Schenkman, Cambridge, Massachusetts.

Oates, Whitney J., editor. 1940. The Stoic and Epicurean Philosophers: The Complete Extant Writings of Epicurus, Epictetus, Lucretius and Marcus Aurelius. Random House, New York.

Oiby, Robert. 1982. Mendel, Gregor 1822-1884. In *Makers of 19th Century Culture: 1800-1914*, edited by Justin Wintle , pp. 417-418. Routledge and Kegan Paul, London.

Oliver, 1982. In the *Paleoecology of Beringia*, edited by David Hopkins, pp. Academic Press, New York.

Olsen, John W. 1987. Recent Developments in the Upper Pleistocene Prehistory of China." In *The Pleistocene Old World; Regional Perspectives*, edited by Olga Soffer, pp. 135-146. New York: Plenum Press.

Olsen, John W. and Sari Miller-Antonio. 1992. *The Paleolithic in Southern California*. Asian Perspectives 31(2): 129-160.

Olsen, P.E., N.H. Shubin and M.H. Anders. 1987. New Early Jurassic Tetrapod Assemblages Constrain Triassic-Jurassic Extinction Event. *Science* 237:1025-1029.

Olsen, Stanley. 1985. *Origins of the Domesticated Dog*. Tucson: University of Arizona Press.

Olson, Storrs L. and D. Tab Rasmussen. 1986. Paleoenvironment of the Earliest Hominoids: New Evidence from the Oliogocene Avifauna of Egypt. *Science* 233:1202-1204.

Oppo, D.W., J.F. McManus and J.L. Cullen. 1998. Abrupt Climate Events 500,000 to 340,000 Years Ago: Evidence from Subpolar North Atlantic Sediments. *Science*.279: 1335-1338.

Orme, Bryony. 1977. The Advantages of Agriculture. In *Hunters, Gatherers and First Farmers Beyond Europe*, edited by J. V. S. Magaw, pp. 41-49. Leicester University Press, Leicester, England.

Orr, Philip C. 1978. *The Prehistory of Santa Rosa Island*. Santa Barbara Museum of Natural History.

Orton, Clive, Paul Tyers and AlanVince. 1993. *Pottery in Archaeology*. Cambridge University Press, Cambridge.

Osborn, Alan J. 1988. Limitations of the Diffusionist Approach: Evolutionary Ecology and Shell-tempered Ceramics. In *The Transfer and Transformation of Ideas and Material Culture*, edited by Peter J. Hugill and D. Bruce Dickson, pp. 23-44. Texas A&M University Press, College Station.

Oswalt, Wendell H. 1976. *An Anthropological Analysis of Food-Getting Technology*. Wiley-Interscience, New York

Pachur, H-J. and S. Kröplin. 1987. Wadi Howar: Paleoclimatic Evidence from an Extinct River System in the Southeastern Sahara. *Science* 237: 298-300.

Parker, S.P. 1995. *World Geographical Encyclopedia, Volume 1: Africa*. McGraw-Hill, New York.

Parson, Jeffrey R. 1976. The Role of *Chinampa* Agriculture in the Food Supply of Aztec Tenochtitlan. In *Culture Change and Continuity*, edited by Charles Cleland, pp. 233-257. Academic Press, New York.

Parsons, James J. and William M. Denevan. 1967. Precolumbian Ridged Fields. *Scientific American* 217(10): 92-110.

Pasini, Giancarlo and Maria Luisa Colalongo. 1997. The Pliocene-Pleistocene Boundary-Stratotype at Vrica, Italy. In *The Pleistocene Boundary and the Beginning of the Quaternary*. Edited by John A. Van Courvering. Pp. 15-45. Cambridge University Press, New York.

Pavlish, L. A. and P. J. Sheppard. 1983. Thermoluminescent Determination of Paleo-Indian Heat Treatment in Ontario, Canada. *American Antiquity* 48(4): 793-799.

Pavlish, L.A. and E.B. Banning. 1980. Revolutionary Developments in Carbon-14 Dating. *American Antiquity* 45 (2): 290-295

Peltier, W. Richard. 1994. Ice Age Paleotopography. *Science* 265: 195-201.

Pennisi, Elisabeth and Wade Roush. 1997. Developing a New View of Evolution. *Science* 277: 34-37.

Penny, David. 1988. What Was the First Living Cell? *Nature* 331:11-112.

Petrie, Sir Flinders. 1901. Diospolis Parva. *Egyptian Exploration Fund, Memoir* 20.

Peyrony, Denis. 1933. Les Industries Aurignaciennes Dons le Bassin de La Vézère. *Bulletin Société Préhistorique Française* 30: 543-559.

Phillipson, David W. 1984. Early Food-production in Central and Southern Africa. In J. Desmond Clark and Steven A. Brandt, Pp. 272-280. *From Hunters to Farmers: The Causes and Consequences of Food Production in Africa*. The University of California Press, Berkeley.

Phillipson, David W. 1985. *African Archaeology*. Cambridge: Cambridge University Press

Pickersgill, Barbara. 1989. Cytological and Genetical Evidence on the Domestication and Diffusion of Crops Within the Americas. In *Foraging and Farming: The Evolution of Plant Exploitation*, edited by D. R. Harris and G. C. Hillman, pp. 426-439. Unwin Hyman, London.

Pielou, E.C. 1991. *After the Ice Age: The Return of Life to Glaciated North America*. University of Chicago Press, Chicago.

Piggot, Stuart,. 1968. The Beginning of Wheeled Transport. *Scientific American.*

Pilbeam, David. 1984. The Descent of Hominoids and Hominids. *Scientific American* 250(3): 84-96.

Piperno, Dolores R. 1988. *Phytolith Analysis: An Archaeological and Geological Perspective.* Academic Press, New York.

Piperno, Dolores R. and Deborah M. Pearsall. 1998. *The Origins of Agriculture in the Lowland Neotropics*. Academic Press, New York.

Poirier, Frank E. 1977. *In Search of Ourselves*. (Second ed.). Burgess Publishing Company, Minneapolis.

Pokines, James T.1998. Experimental Replication and Use of Cantabrian Lower Magdalenian Antler Projectile Points. *Journal of Archaeological Science* 25: 875-886.

Pokras, Edward M. and Alan C. Mix. 1985. Eolian Evidence for Spatial Variability of Late Quaternary Climates in Tropical Africa. *Quaternary Research* 24: 137-149.

Pomerol, Charles. 1982. *The Cenozoic Era: Tertiary and Quaternary*. John Wiley and Sons, New York.

Pope, G. G. 1985. Taxonomy, Dating and Palaeoenvironment: The Palaeoecology of the Early Far Eastern Hominids. *Modern Quaternary Research in Southeast Asia* 9:65-80.

Pope, Geoffrey G. 1989. Bamboo and Human Evolution. *Natural History* 10: 49-56.

Popenoe, Hugh. 1986. African Soils: Opportunities and Constraints. In *Food in Sub-Saharan Africa*, edited by A. Hansen and D. E. McMillan, pp. 169-206. Lynne Rienner Publishers, Inc, Boulder.

Prasad, K. N. 1983. Was *Ramapithecus* a Tool-User? *Journal of Human Evolution* 11: 101-104.

Price, Cynthia R. 1979. 19th Century Ceramics in the Eastern Ozark Border Region. *Center for Archaeological Research, Southwest Missouri State University, Monograph Series* 1.

Price, T. Douglas and James A. Brown. 1985. Aspects of Hunter-gatherer Complexity. In *Prehistoric Hunter-Gatherers: The Emergence of Cultural Complexity*, edited by T. D. Price and J. A. Brown. Academic Press, New York.

Pryor, Frederic L. 1982. The Invention of the Plow. *Comparative Studies in History and Society* 27: 727-743.

Pryor, Frederic L. 1986. The Adoption of Agriculture: Some Theoretical and Empirical Evidence. *American Anthropologist* 88(4): 879-897.

Pryor, Frederic L. and Steven B. Mauer. 1982. On Induced Economic Change in Precapitalist Societies. *Journal of Development Economics* 10(3): 325-353.

Pusey, Anne, Jennifer Williams and Jane Goodall 1997. The Influence of Dominance Rank on the Reproductive Success of Female Chimpanzees. *Science* 277: 828-831

Pyramarn, Kosum. 1989. New Evidence on Plant Exploitation and Environment During the Hoabinhian (Late Stone Age) from Bao Kao Caves, Thailand. In

Foraging and Farming: The Evolution of Plant Exploitation, edited by D. R. Harris and G. C. Hillman, pp. 282-291. Unwin Hyman, London.

Quechon, G.. 1976. Le Sépultures des Hommes du Paléolithique Supérior. In *La Préhistoire Française, I: Civilisations Paléolithques et Mésolithiques*, edited by H. de Lumley, pp. 728-733. CNRS, Paris.

Ramsköld, L. and Hou Xianguang 1991. New Early Cambrian Animal and Onychophoran affinities of Enigmatic Metazoans. *Nature* 351: 225-228.

Rank and Arensberg. 1987. Kebara 2 Neanderthal Pelvis. *American Journal of Physical Anthropology* 73: 227-

Rathje, William and Michael B. Schiffer. 1980. *Archaeology*. Harcourt Brace Jovanovich, New York.

Raup, D and J. Sepkoski 1982.Mass Extinction in the Marine Fossil Record. *Science* 215: 1501-1503.

Reed, Charles A. 1969. The Pattern of Animal Domestication in the Prehistoric Near East. In *The Domestication and Exploitation of Plants and Animals*, edited by Peter J. Ucko and G. W. Dimbleby, pp. 361-380. Aldine Publishing Company, Chicago.

Renfrew, Colin. 1973. *Before Civilization: The Radiocarbon Revolution and Prehistoric Europe*. Cambridge University Press, New York.

Renfrew, Colin. 1987. *Archaeology and Language: The Puzzle of Indo-European Origins*. Cambridge University Press, New York.

Renne, P.R., W.D. Sharp, A. L. Deino, G. Orsi, L. Civetta. 1997. 40Ar/39Ar Dating into the Historical Realm: Calibration Against Pliny the Younger. *Science* 277: 1279-1280.

Retallack, G.J. and Carolyn R. Feakes. 1987. Trace Fossil Evidence for Late Ordovician Animals on Land. *Science* 225:61-63.

Rich,Thomas H., Patricia Vickers-Rich, Andrew Constantine, Timothy F. Flannery, Lesley Kool, Nicholas van Klaveren. 1997. A Tribosphenic Mammal from the Mesozoic of Australia. *Science* 278: 1438-1442.

Rieseberg, L.H. and Seller, G. J. 1990. Molecular Evidence and the Origin and Development of the Domesticated Sunflower (*Helanthus annus* Asteraceae"). *Economic Botany* 44 (Suppl.): 79-91.

Rightmire, G. P. 1981. Patterns in the Evolution of *Homo erectus*. *Paleobiology* 7:241-246.

Rindos, David. 1980. Symbiosis, Instability, and the Origins and Apread of Agriculture: a New Model. *Current Anthropology* 21(6): 751-772.

Rindos, David. 1984. *The Origins of Agriculture: an Evolutionary Perspective*. Academic Press, New York.

Roberts, Neil. 1989. *The Holocene: An Environmental History*. Basil Blackwell, New York.

Robertshaw, P. 1991. On Agricultural Beginnings in Kenya: Retrospect and Prospects. In *Origins and Development of Agriculture in East Africa: The Ethnosystems Approach to the Study of Early Food Production in Kenya*, edited by R. E. Leakey and L. J. Slikkerveer, pp 7-15, Studies in Technology and Social Change, No. 19, Iowa State University, Ames.

Robertshaw, P. 1995. The Last 200,000 Years (Or Thereabouts) in Eastern Africa: Recent Archaeological Research. *Journal of Archaeological Research* 3: 55-86.

Robertshaw, P. and D. Collett. 1983. A New Framework for the Study of Early pastoral Communities in East Africa. *Journal of African History* 24: 289-302.

Robertshaw, Peter. 1991. Gogo Falls, Excavation at a Complex Archaeological Site East of Lake Victoria. Azania 26: 63-195.

Robinson, J. T. 1965. *Homo "habilis"* and the Australopithecines. *Nature* 205:121-124.

Rogers, Richard A. 1986. Spurred End Scrapers as Diagnostic North American Paleo-Indian Artifacts: A Distributional Analysis on Stream Terraces. *American Antiquity* 51(2): 338-341.

Rolland, Nicolas and Harold L. Dibble. 1990. A New Synthesis of Middle Paleolithic Variability. *American Antiquity* 55(3): 480-499.

Roosevelt, A.C., M. Lima da Costa, C. Lopes Machado, M. Michab, N. Mercier, H. Valladas, J. Feathers, W. Barnett, M. Imazio da Silveira, A. Henderson, J. Silva, B. Chernoff, D. S. Reese, J.A. Holman, N. Toth, K. Schick. 1996. Paleo-Indian Cave Dwellers in the Amazon: The Peopling of the Americas. *Science* 272: 373-384.

Rothwell, G. W. and S. E. Scheckler. 1988. Biology of Ancestral Gymnosperms. In *Origin and Evolution of the Gymnosperms*, edited by C. B. Beck pp. 85-135. Columbia University Press, New York.

Roubet, C. 1978. Une economie pastorale, pre-agricaole en Algerie orientale: Le Neolithique de tradition capsienne. *L'Anthropologie* 82: 583-586.

Rowe, John Howland. 1962. Worsaae's Law and the Use of Grave Lots for Archaeological Dating. *American Antiquity* 28:129-137.

Rowe, John Howland. 1963. Stratigraphy and Seriation. *American Antiquity* 26: 324-330.

Ruben, John N., Terry D. Jones, Nicholas R. Geist and W. Jaap Hillenius 1997. Lung Structure and Ventilation in Theropod Dinosaurs and Early Birds. *Science* 278: 1267-1270.

Ruddiman, W.F. and J.-C. Duplessy. 1983. Conference on the Last Deglaciation: Timing and Mechanisms. *Quaternary Research* 23: 1-17.

Rudwick, Martin J. S. 1985. *The Meaning of Fossils: Episodes in the History of Palaeontology*. The University of Chicago Press, Chicago.

Runnels, Curtis and Tjeerd H. Van Andel. 1988. Trade and the Origins of Agriculture in the Eastern Mediterranean. *Journal of Mediterranean Archaeology* 1(1): 83-109.

Russell, Bertrand. 1948. *Human Knowledge: Its Scope and Limits*. Simon and Schuster, New York.

Russell, Donald E. 1975. Paleoecology of the Paleocene-Eocene Transition in Europe. In Approaches to Primate Paleobiology. *Contributions to Primatology* 5:28-61.

Ryder, Michael L. 198?. The Evolution of the Fleece. *Scientific American* Vol?: 112-119.

Sackett, James and Jean Gaussen. 1976. Upper Paleolithic Habitation Structures in the Sud-ouest of France. *IXth Congress of the International Union of Prehistoric and Protohistoric Sciences, Colloquim XIII: Les Structures d'Habitat au Paleolithic Superieur*, pp. 55-83.

Sadr, Karim. 1998. The First Herds at the Cape of Good Hope. *African Archaeological Review* 15(2): 101-132.

Sahlins, Marshall. 1972. *Stone Age Economics*. The University of Chiciago Press, Chicago.

Saldarriaga, Juan G. and Darrell C. West. 1986. Holocene Fires in the Northern Amazon Basin. *Quaternary Research* 26: 358-366.

Sankalia, H.D. 1978. The Early Paleolithic in India and Pakistan. In *Early Paleolithic in South and East Asia*, edited by Fumiko Ikawa-Smith, pp. 97-127. The Hague: Mouton Publishers.

Sanz, José, Luis M. Chiaape, Bernardino P. Pérez-Moreno, José J. Moratalla, Francisco Hernández-Carrasquilla, Angela D. Buscalioni, Francisco Ortega, Francisco J. Poyato-Ariza, Diego Rasskin-Gutman, Xavier Martínez-Delclòs. 1997. A Nestling Bird from the Lower Cretaceous of Spain: Implications for Avian Skull and Neck Evolution. *Science* 276: 1543-1546.

Sarich, Vincent and V. M. Wilson 1967. Rates of Albumin Evolution in Primates. *Proceedings of the National Academy of Science* 58:142-148.

Sauer, Carl O. 1952. *Agricultural Origins and Dispersals*. American Geographical Society, New York.

Sauer, Carl. 1944. A Geographic Sketch of Early Man in America, *Geographical Review* XXXIV (4).

Schaller, George 1963. *The Mountain Gorilla: Ecology and Behavior*. University of Chicago Press, Chicago.

Schaller, George B. 1972. Are You Running With Me, Hominid? *Natural History* LXXXI(3):60-69.

Schaller, George. 1977. *Mountain Monarchs, Wild Sheep and Goats of the Himalayas*. University of Chicago Press, Chicago.

Schiffer, Michael B. 1987. *Formation Processes of the Archaeological Record*. University of New Mexico Press, Albuquerque.

Schmandt-Besserat, Denise. 1980. Ochre in Prehistory, 300,000 years of the use of Iron Ores as Pigments. In *The Coming of the Age of Iron*, edited by T. A. Wertime and J. D. Muhly, pp. 127-150. Yale University Press, New Haven.

Schoell, M., S. Schouten, J. S. Sinninghe Damsté, J. W. de Leeuw, R. E. Summons. 1994. A Molecular Organic Carbon Isotope Record of Miocene Climate Changes. *Science* 263: 1122-1125.

Schopf, William. 1976. How Old Are the Eukaryotes? *Science* 193:47-49.

Schopf, William and Bonnie M. Packer. 1987. Early Archean (3.3 billion to 3.5 billion-year-old) Microfossils from Warrawoona Group, Australia. *Science* 237:70-773.

Semaw, S., P. Renne, J. W. K. Harris, C. S. Felbel, R. L. Bernor, N. Fesseha and K. Mowbray. 1997. 2.5 Million Year Old Stone Tools from Gona, Ethiopia. *Nature* 385:333-336.

Semenov, S. A. 1964. *Prehistoric Technology*. Barnes & Noble, New York.

Senshui, Zhang. 1985. The Early Palaeolithic of China. In *Palaeoanthropology and Palaeolithic Archaeology in the People's Republic of China*, edited by Wu Rukang and John W. Olsen, pp. 147-186. New York: Academic Press.

Service, Elman R. 1979. *The Hunters*. (2nd Edition) Prentice-hall, Inc, New York.

Shackleton, M. L. 1975. *Archaeological Sediments: A Survey of Analytical Methods*. Halsted, New York.

Shanks, Michael and Christopher Tilley. 1992. *Re-Constructing Archaeology: Theory and Practice. Second Edition*. Routledge, London.

Sharer, Robert J. and Wendy Ashmore 1993. *Archaeology: Discovering Our Past. Second Edition*. Mayfield Publishing, Mountain View, California.

Shaw, Thurstan. 1977. Hunters, Gatherers and First Farmers in West Africa. In *Hunters, Gatherers and First Farmers Beyond Europe*, edited by J. V. S. Megaw, pp. 69-126. Leicester University Press, Leicester.

Sheratt, A. G. 1983. The Secondary Exploitation of Animals in the Old World. *World Archaeology* 15: 90-104.

Sheratt, A. G. 1984. Social Evolution: Europe in the Later Neolithic and Copper Ages. In *European Social Evolution*, edited by J. Bintliff, Pp. 123-134. University of Bradford, Bradford, England.

Sherwood, Morgan B. 1969. Genesis, Evolution, and Geology in America Before Darwin: The Dana-Lewis Controversy, 1856-1857. In *Toward a History of Geology*, edited by Cecil J. Schneer, pp 305-316. MIT Press, Cambridge, Mass.

Shimkin, Edith M. 1978. The Upper Paleolithic in North-Central Eurasia. In *Views of the Past*, edited by Leslie G. Freeman, pp. 193-315. The Hague: Mouton Publishers.

Shotton, F. W. 1977. *British Quaternary Studies*. Clarendon Press, Oxford.

Shutler, Richard and Mary Elizabeth Shutler. 1975. *Oceanic Prehistory*. Cummings, Menlo Park, Calif.

Sieveking, Ann. 1976. Settlement Patterns of the Late Magdalenian in the Central Pyrenees. In *Problems in Economic and Social Archaeology*, edited by G. de G. Sieveking, I. H. Longworth, and K. E. Wilson, pp 583-603. Duckworth, London.

Simmons, I. G, G. W. Dimbleby and Caroline Grigson. 1981. The Mesolithic. In *The Environment in British Prehistory*, edited by I. G. Simmons and M. J. Tooley, pp. 82-124. Cornell University Press, Ithaca, New York.

Simon, E. L. 1985. African Origin, Characteristics and Context of Earliest Higher Primates. In *Hominid Evolution; Past, Present, and Future*. Edited by Philip V. Tobias, pp. 101-106. Alan R. Liss Inc.

New York.

Simons, E. L., P. Andrews, and D. R. Pilbeam. 1978. Cenozoic Apes. In *Evolution of African Mammals*. Edited by Vincent J. Maglio and H.B.E. Cooke, pp. 120-146. Harvard University Press, Cambridge.

Simpson, George Gaylord. 1944. *Tempo and Mode in Evolution*. Columbia University Press, New York.

Singer, Ronald and John Wymer. 1982. *The Middle Stone Age at Klasies River Mouth in South Africa*, Chicago: The University of Chicago Press.

Smith, A.B. 1980. The Neolithic Tradition In the Sahara. In *The Sahara und the Nile*, edited by M. A. J. Williams and H. Faure, pp. 451-65. Rotterdam.

Smith, Andrew B. 1986. Review Article: Cattle Domestication in North Africa. *The African Archaeological Review* 4: 197-203.

Smith, Bruce D. 1989. Origins of Agriculture in Eastern North America. *Science* 246: 1566-1571.

Smith, Fred H. 1984. Fossil Hominids from the Upper Pleistocene of Central Europe and the Origin of Modern Europeans. In *The Origins of Modern Humans*, edited by Fred Smith and Frank Spencer, pp. 137-209. Alan R. Liss, Inc, New York.

Smith, Philip E. L. 1964. The Solutrean Culture. *Scientific American* 211(2): 86-94.

Smith, Philip E. L. 1965. Some Solutrean Problems and Suggestions for Future Research. *Disputación Provincial de Barcelona, Instituduo de Prehistória y Arqeuologia. Mongrafias* 18: 398-408.

Smith, Philip E. L. 1973. Some Thoughts on Variations Among Certain Solutrean Artifacts. In *Estudios Dedicados al Professor Dr. Luis Pericot*, edited by J. Malvquer de Motes, pp. 67-75. Instituto de Arqueologia y Prehistoria.

Smith, Philip E. L. 1982. The Late Paleolithic and Epi-Paleolithic of Northern Africa. *Cambridge History of Africa, Volume 1: From Earliest Times to 500 B.C.* Cambridge University Press, Cambridge.

Smith, Vernon. 1975. The Economics of the Primitive Hunter Culture, Pleistocene Extinctions and the Rise of Agriculture. *Journal of Political Economy* Lxxxiii: 727-756.

Soffer, Olga. 1985. *The Upper Paleolithic of the Central Russian Plain*. Academic Press, New York.

Soffer, Olga. 1985. *The Upper Paleolithic of the Central Russian Plain*. Academic Press, New York.

Sonneville-Bordes, Denise. 1973. The Upper Paleolithic: Ca. 33,000-10,000 B. C. In *France Before the Romans*, edited by S. Piggot, G. Daniel, and Charles McBurney, pp 30-60. Noyes Press, Park Ridge, New Jersey.

Speth, J. D. and K. A. Spielmann. 1983. Energy Source, Protein Metabolism, and Hunter-Gatherer Subsistence Strategies. *Journal of Anthropological Archaeology* 2: 1-31.

Spiess, Arthur E. 1979. *Reindeer and Caribou Hunters: An Archaeological Study*. Academic Press, New York.

Sponsel, Leslie E. 1986. Amazon Ecology and Adaptation. *Annual Review of Anthropology* 15: 67-97.

Stahl, A. 1993. Intensification in the West African Late Stone Age: a View from Central Ghana. In *The Archaeology of Africa: Food, Metals, and Towns*. Edited by T. Shaw, P. Sinclair, B. Andah, and A. Okpoko. Pp. 261-273. Routledge, London.

Stahl, Ann B. 1989. Plant-Food Processing: Implications for Dietary Quality. In *Foraging and Farming: The Evolution of Plant Exploitation*, edited by D. R. Harris and G. C. Hillman, pp. 171-194. Unwin Hyman, London.

Stanley, Steven M. 1987. *Extinction*. Scientific American Books, Inc., New York.

Stanton, W. R. 1969. Some Domesticated Lower Plants in South-East Asian Food Technology. In *The Domestication and Exploitation of Plants and Animals*, edited by Peter J. Ucko and G. W. Dimbleby, pp. 463-469. Aldine Publishing Company, Chicago.

Stebbins, G. Ledyard, F.J. Ayala. 1984. The Evolution of Darwinism *Scientific American* 253 (1): 72-82.

Stein, Julie K. 1987. Deposits for Archaeologists. *Advances in Archaeological Method and Theory* 11: 337-395.

Stern, John and Randall Susman. 1983. The Locomotor Anatomy of *Australopithecus afarensis*. *American Journal of Physical Anthropology*. 60: 279-317.

Steward, Julian. 1938. *Basin-Plateau Aboriginal Sociopolitical Groups*. Smithsonian Institution. Bureau of American Ethnology, Bulletin 120.

Straus, L. G. 1977. Thoughts on Solutran Concave Base Point Distribution. *Lithic Technology* 6: 32-35.

Straus, L. G. 1979. Mesolithic Adaptations along the Coast of Northern Spain. *Quaternaria* 21: 305-327

Straus, L. G. 1985. Stone Age Prehistory of Northern Spain. *Journal of Anthropological Research* 38: 75-96.

Straus, L. G. 1995. The Upper Paleolithic of Europe: An Overview. *Evolutionary Anthropology* 4: 4-16.

Straus, L. G, Geoffrey A. Clark, Jesus Altuna, and Jesus A. Ortea. 1980. Ice-age Subsistence in Northern Spain. *Scientific American* 242(6): 120-129.

Stringer, C. B. 1974. Population Relationships of Later Pleistocene Hominids: A Multi variate Study of Available Crania. *Journal of Human Evolution* 11: 431-438.

Stringer, C. B. 1982. Towards a Solution of the Neanderthal Problem. *Journal of Human Evolution* 11: 431-438.

Stringer, C. B. and P. Andrews. 1988. Genetic and Fossil Evidence for the Origin of Modern Humans. *Science* 239:1261-1268.

Stringer, C. B., J. J. Hublin, and B. Vandermeersch. 1984. The Origins of Anatomically Modern Humans in Western Europe. In *The Origins of Modern Humans*. Edited by Fred Smith and Frank Spencer, pp. 51-135. Alan R. Liss, Inc., New York.

Stuiver, M. 1978. Radiocarbon Timescale Tested Against Magnetic and Other Dating Methods. *Nature* 273: 271-274.

Stuiver, M. 1982. A High-Precision Calibration of the A.D. Radiocarbon Time Scale. *Radiocarbon* 24: 1-

26.
Stuiver, M., C. J. Heusser, and I.C. Yang. 1978. North American Glacial History Extended to 75,000 Years Ago. *Science* 200: 16-21.
Stuiver, M. and P. J. Reimer. 1993. Extended ^{14}C Data Base and Revised Calib 3.0 ^{14}C Age Calibration Program. *Radiocarbon* 35: 215-230.
Suess, R. E. 1980. Bristlecone-Pine Calibrations of the Radiocarbon Time-Scale, 5200 B.C. to the Present. In *Radiocarbon Variations and Absolute Chronology*, edited by I. U. Olsson, pp. 303-309. Nobel Symposium 12. Wiley Interscience, New York.
Sugiyama, Yukimaru. 1997. Social Tradition and the Use of Tool-Composites by Wild Chimpanzees. *Evolutionary Anthropology* 6(1): 23-27.
Sussman, Robert. 1972. Child Transport, Family Size, and the Increase in Human Population Size During the Neolithic. *Current Anthropology* 13: 258-267.
Sutton, J.E.G. 1974. The Aquatic Civilization of Middle Africa. *Journal of African History* 15: 527-546.
Sutton, J.E.G. 1977. The African Aqualithic *Antiquity* 51:25-34.
Suwa, Gen, Berhane Asfaw, Yonas Beyene, Tim D. White, Shlgehiro Katoh, Shinji Nagaoka, Hideo Nakaya, Kazuhiro Uzawa, Paul Renne and Giday Wolde Gabriel. 1997. The First Skull of *Australopithecus boisei*. *Nature* 389: 489-492.
Swisher, C.C., III, W.J. Rink, S.C.Antón, H.P. Schwarcz, G.H. Curtis, A. Suprijo, Widiasmoro. 1996. Latest *Homo erectus* of Java: Potential Contemporaneity with *Homo sapiens* in Southeast Asia. *Science* 274: 1870-1874.
Taranik, Linda L., 1977. The Beauronnian at Solvieux: An Upper Paleolithic Industry from Southwestern France. Unpublished Ph.D. dissertation, Department of Anthropology, University of California, Los Angeles.
Tattersall, Ian, Eric Delson, and John Van Couvering (Editors). 1988. Pliocene. *Encyclopedia of Human Evolution and Prehistory*. Pp. 467-469. Garland Publishing, New York.
Tauber, H. 1970. The Scandinavian Varve Chronology and C-14 Dating. In *Radiocarbon Variation and Absolute Chronology*, pp. 173-196, edited by I. U. Olsson. John Wiley, New York
Taylor, R.E., L.A. Payen, C.A. Prior, P.J. Slota, Jr., R. Gillespie, J.A.J. Gowlett, R.E.M. Hedges, A.J.T. Jull, T.H. Zabel, D.J. Donahue, and R. Burger. 1985. Major Revisions in the Pleistocene Age Assignments for North American Human Skeletons by C-14 Accelerator Mass Spectrometry: None Older than 11,000 C-14 Years B. P. *American Antiquity* 50(1): 136-140.
Teilhard de Chardin, Pierre. 1964. *The Future of Man*. Harper and Row, New York.
Thompson, L. G, E. Mosley-Thompson, W. Dansgaard, P. M. Grootes. 1986. The Little Ice Age as Recorded in the Stratigraphy of the Tropical Quelccaya Ice Cap. *Science* 234: 361-364.
Thunell, Robert C. and P. Graham Mortyn. 1995. Glacial Climate Instability in the Northeast Pacific Ocean. *Nature* 376: 504-506.
Tobias, P. V. 1965. Early Man in East Africa. *Science* 149:22-23.
Tobias, P. V. 1981. The Emergence of Man in Africa and Beyond. *Philosophical Transactions of the Royal Society, London B* 292: 43-56.
Tolestoy, Paul. 1978. Western Mesoamerica: A.D. 900-1520. In *Chronologies in New World Archaeology*, edited by Taylor, R.E. and Clement W. Meighan, pp. 241-284. Academic Press, New York.
Trewartha, Glenn Thomas. 1981. *Earth's Problem Climates. Second Edition*. University of Wisconsin Press, Madison, Wisconsin.
Trigger, Bruce G. 1989. *A History of Archaeological Thought*. Cambridge University Press, Cambridge.
Trikaus, Erik. 1989. Comment. *Current Anthropology* 30(2): 183-184.
Trinkaus, Erik and William W. Howells 1979. The Neanderthals. *Scientific American* 241(6):118-133.
Trompette, Roland. 1994. *Geology of Western Gondwana (2000-500 Ma). Pan-African-Brasilano Aggregation of South America and Africa*. Translated by Albert V. Carozzi. Balkema, Brookfield, Vermont.
Tsukada, M. 1983. Vegetation and Climate During the Last Glacial Maximum in Japan. *Quaternary Research* 19: 212-235.
Turnbaugh, William A. 1973. Toward an Explantion of the Broadpoint Dispersal in Eastern North American Prehistory. *Journal of Anthropological Research* 31(1): 51-68.
Turner, Christy G. 1979. Dental Anthropological Indications of Agriculture among the Jomon People of Central Japan. *American Journal of Physical Anthropology* 51: 619-636.
Turner, Christy G. 1987. Telltale Teeth; from the Mouths of Prehistoric Indians Comes Evidence Tracing Their Northeast Asian Ancestry. *Natural History* 96(1): 6-10.
Valladas, H., J. Reyss, J. L. Joron, G. Vallada, O. Bar-Josef, B. Vandermeersch. 1988. Thermoluminescence Dating of Mousterian "Proto-Cro-Magnon" Remains from Israel and the Origin of Modern Man. *Nature* 331: 614-617.
Vallois, H. 1961. The Social Life of Early Man: The Evidence of the Skeletons. In The Social Life of Early man, edited by Sherwood L. Washburn. *Viking Fund Publications in Anthropology* 31: 214-235.
van der Hammen, T. 1974. The Pleistocene Changes of Vegetation and Climate in Tropical South America. *Journal of Biogeography* 1: 3-26.
van der Hammen, T., T.A. Wjimstra, and W.H. Zagwijn. 1971. The Floral Record of the Late Cenozoic. In *Late Cenozoic Glacial Ages*, edited by Richard Flint and K. K. Turekian, pp. 51-135. New Haven: Yale University Press.
van der Hammern, T. A. Wijmstra, W. H. Zagwijn. 1971. The Floral Record of the Late Cenozoic of Europe. In *The Late Cenozoic Glacial Ages*, edited by Richard Flint and K. K. Turekian, Pp. 391-424. Yale University Press, New Haven.

Van Devender, T. R., J.L. Betancourt, and M. Wimberly. 1984. *Quaternary Research* 22: 344.

Van Devender, T. R. and W. G. Spaulding. 1979. *Science* 204: 701.

Van Zinderen Bakker, Sr, E. M. and J. Maley. 1979. Late Quaternary Paleoenvironments of the Sahara Region. *Palaeoecology of Africa* 11: 83-104.

Vandermeersch, Bernard. 1976. Le Sépultures Néandertaliennes. In *La Préhistoire Française, I: Civilisations Paléolithiques et Mésolithiques*, edited by H. de Lumley, pp 725-727. CNRS, Paris.

Voder, Anne D. 1997. Back to the Future: a Synthesis of Strepsirrhine Systematics. *Evolutionary Anthropology* 6(1): 11-22.

Vogel, Gretchen 1997. Searching for the Living Relics of the Cell's Early Days. *Science* 277: 1604.

Waal, Frans de 1982. *Chimpanzee Politics: Power and Sex Among the Apes*. Harper and Row, New York.

Wagner, Philip L. 1960. *The Human Use of the Earth.* Collier-Macmillan Ltd., London.

Walker, Alan and Richard E. F. Leakey. 1978. The Hominids of East Turkana. *Scientific American* 239(2): 54-66.

Washburn, S. L. and V. Avis 1958. Evolution of Human Behavior. In *Behavior and Evolution*. Edited by A. Roe and G. G. Simpson, pp. 421-436. Yale University Press, New Haven.

Washburn, Sherwood L. 1961. *The Social Life of Early Man*. Aldine, Chicago.

Washburn, Sherwood L. and Irven DeVore. 1961b. The Social Life of Baboons. *Scientific American* 204(6): 62-71.

Washburn, Sherwood L. and Irven DeVore. 1961a. Social Behavior of Baboons and Early Man. In *The Social Life of Early Man*, edited by Sherwood L. Washburn, pp 91-105. Aldine, Chicago.

Waters, Michael R. 1985. Late Quaternary Alluvial Stratigraphy of Whitewater Draw, Arizona: Implications for Regional Corrlation of Fluvial Deposits in the American Southwest. Geology 13: 705-708.

Waters, Michael R. 1986. *The Geoarchaeology of Whitewater Draw, Arizona*. Anthropological Papers of the University of Arizona, No. 45. University of Arizona Press, Tucson.

Waters, M.R., S. L. Forman, J. M. Pierson. 1997. Diring Yuriakh: A Lower Paleolithic Site in Central Siberia. *Science* 275: 1281-1294.

Waters, Michael R., Steven L. Forman, and James M. Pierson 1997. Diring Yuriakh: A Lower Paleolithic Site in Central Siberia. *Science* 275: 1281-1284.

Waters, Michael. 1986a. The Sulphur Spring Stage and its Place in New World Prehistory. *Quaternary Research* 25: 251-256.

Waters, Michael. 1986b. The Geoarchaeology of Whitewater Draw, Arizona. *Anthropological Papers of the University of Arizona* 45.

Waters, Tom. 1992. Greetings From Pangea. *Discover* 13(2): 38-43.

Watts, W. A. 1983. Vegetational History of the Eastern United States 25,000 to 10,000 Years Ago. In *Late-Quaternary Environments of the United States, Volume I, The Late Pleistocene*, edited by S. C. Porter, pp. 294-310. Longmans, London.

Webb, David S., Richard C. Hulbert and W. David Lambert. 1995. Climatic Implications of Large-Herbivore Distributions in the Miocene of North America. In *Paleoclimate and Evolution with Emphasis on Human Origins*. Edited by Elisabeth S. Vrba, Geoge H. Denton, Timothy C. Partridge, and Lloyd H. Burckle, pp. 91-108. Yale University Press, New Haven and London.

Webb, S. David, Jerald T. Milanich, Roger Alexon, and James S. Dunbar. 1984. A *Bison antiquus* Kill Site, Wacissa River, Jefferson County, Florida. *American Antiquity* 49(2): 384-392.

Wedel, Waldo. 1961. *Prehistoric Man on the Great Plains*. Norman: University of Oklahoma Press.

Weinberg, Steven. 1977. *The First Three Minutes*. Basic Books, New York.

Wendorf, Fred and James Hester. 1962. Early Man's Utilization of the Great Plains Environment. American Antiquity 28: 159-71.

Wendorf Fred and Romuald Schild. 1994. Are the Early Holocene Cattle in the Eastern Sahara Domestic or Wild? *Evolutionary Anthropology* 3(4): 118-128.

West, R. G. 1977. *Pleistocene Geology and Biology*. Second Edition. Longman, New York.

West, Robert Cooper and John Augelli. 1989. *Middle America, Its Lands and Peoples. Third Edition*. Prentice Hall, Englewood Cliffs, New Jersey.

Williams, Douglas F., R. C. Thunell, E. Tappa, D. Rio and I. Raffi. 1988. Chronology of the Pleistocene Oxygen Isotope Record: 0-1.88 M.Y.B.P. *Palaeogeography, Palaeoclimatology, Palaeoecology* 64: 221-240.

Wheat, Joe Ben. 1967. A Paleo-Indian Bison Kill. *Scientific American* 216(1): 44-52.

Wheat, Joe Ben. 1972. *The Olsen-Chubbock Site: A Paleo-Indian Bison Kill*. Society for American Archaeology, Memoir No. 26.

White, Randall. 1980. The Upper Paleolithic Occupation of the Périgord: A Topographic Approach to Subsistence and Settlement. Unpublished Ph.D. dissertation, University of Toronto, Toronto, Ontario.

White, Randall. 1982. Rethinking the Middle/Upper Paleolithic Transition. *Current Anthropology* 23: 169-191.

White, T. D., Suwa, G., and Asfaw, B. 1994. *Australopithecus ramidus*, A New Species of Early Hominid from Aramis, Ethiopia. *Nature* 371:306-312.

Whittington, S.L. and Dyke, B. 1984. Simulating Overkill: Experiments with the Mosimann and Martin Model. In *Quaternary Extinctions*, edited by Paul S. Martin and Richard Klein, pp. 451-465. Tucson: The University of Arizona Press.

Wilkenson, Paul F. 1972. Current Experimental Domestication and its Relevance to Prehistory. In *Papers in Economic Prehistory*, edited by E. S. Higgs, pp. 107-118. Cambridge University Press,

Cambridge.
Willey, Gordon R. 1966. *An Introduction to American Archaeology. Volume 1: North and Middle America*. Prentice-Hall, Englewood Cliffs, New Jersey.
Willey, Gordon R. and Phillip Phillips. 1958. *Method and Theory in American Archaeology*. The University of Chicago Press, Chicago.
Wilmsen, Edwin. 1974. *Lindenmeier: A Pleistocene Hunting Society*. Chicago: Harper and Row, Publishers
Wilson, A. T. 1981. Isotope Evidence for Past Climatic and Environmental Change. In *Climate and History*, edited by R.I. Rutberg and T.K. Rabb, pp. 215-232. Princeton University Press, Princeton, N.J.
Woilland, G.M. and W.G. Mook. 1966 Carbon-14 Dates at Grand Pile: Correlation of Land and Sea Chronologies. *Science* 215: 159-161.
Woillard, G. M. 1978. Grande Pile Bog: Continuous Pollen Record for the Last 140,000 Years. *Quaternary Research* 9: 1-21.
Wolpoff, Milford H. 1979. The Krapina Dental Remains. *American Journal of Physical Anthropology* 50(1): 67-114.
Wolpoff, Milford H. 1980. *Paloeanthropology*. Alfred A. Knopf, New York.
Wolpoff, Milford H. 1981 Cranial Capacity Estimates for Olduvai Hominid 7. *American Journal of Physical Anthropology* 56: 297-304.
Wrangham, Richard W. 1997. Subtle, Secret Female Chimpanzees. *Science* 277: 774-775
Wrangham, Richard W and Dale Peterson. 1997. *Demonic Males: Apes and the Origins of Human Violence*. Houghton Mifflin, New York.
Wreschner, E. E. 1980. Red Ochre and Human Evolution: A Case for Discussion. *Current Anthropology* 21: 631-644.
Wright, H. E, Jr. 1987. Synthesis; The Land South of the Ice Sheets. In *North America and Adjacent Oceans During the Last Deglaciation*, edited by W. F. Ruddiman and H. E. Wright, Jr., pp 479-488. The Geological Society of America.
Wright, H. E, Jr. 1976. The Dynamic Nature of Holocene Vegetation: a Problem in Paleoclimatology, Biogeography, and Stratigraphic Nomenclature. *Quaternary Research* 6: 581-596.
Wright, H. T. 1981. Appendix: The Southern Margins of Sumer: Archaeological Survey of the Area of Eridu and Ur. In *Heartland of Cities: Surveys of Ancient Settlement and Land Use on the Central Floodplain of the Euphrates*, by Robert McC. Adams, pp. 295-345. University of Chicago Press, Chicago.
Wuethrich, Bernice. 1997. Will Fossil From Down Under Upend Mammal Evolution? *Science* 278: 1401-1402.
Wymer, John. 1982. *The Paleolithic Age*. New York: St. Martin's Press.
Wynn, Thomas. 1979. The Intelligence of Later Acheulean Hominids. *Man* (ns) 14:371-391.
Xinzhi, Wu and Wang Linghong. 1985. Chronology in Chinese Paleoanthropology. In *Palaeoanthropology and Palaeolithic Archaeology in the People's Republic of China*, edited by Wu Rukang and John W. Olsen, pp. 29-51. New York: Academic Press.
Yamagata., M. 1992. The Shakado Figurines and Middle Jomon Ritual in the Kofu Basin. *Japanese Journal of Religious Studies* 19(2-3): 129-38.
Yarnell, Richard A. 1964. Aboriginal Relationships Between Culture and Plant Life in the Upper Great Lakes Region. *Anthropological Papers, Museum of Anthropology, University of Michigan* 23.
Yellen, John. 1977. Archaeological Approaches to the Present: Models for Reconstructing the Past. Academic Press, New York.
Yen, Douglas E. 1977. Hoabinhian Horticulture: the Evidence and Questions from Northwest Thailand. In *From Sunda to Sahel*, edited by J. Allen, J. Golson and Philip Jones. Academic Press, New York.
Yi, Seon-bok and G. A. Clark. 1983. Observations on the Lower Paleolithic of Northeast Asia. *Current Anthropology* 24(2): 184-202.
Zebrowski, Ernest, Jr. 1997. *Perils of a Restless Planet: Scientific Perspectives on Natural Disasters*. Cambridge University Press, New York.
Zeuner, Frederick E. 1963. *A History of Domesticated Animals*. Harper and Row, Publishers, New York.
Zhao, Zhijun. 1998. The Middle Yangtze Region in China is One Place Where Rice Was Domesticated: Phytolith Evidence From the Diaotonghuan Cave, Northern Jiangxi. *Antiquity* 72: 685-697.
Zhimin, A. 1989. Prehistoric Agriculture in China. In *Foraging and Farming: The Evolution of Plant Exploitation*, edited by D. R. Harris and G. C. Hillman, pp. 643-649. Unwin Hyman, London.
Zhongland, Qiu. 1985. The Middle Palaeolithic of China. In *Palaeoanthropology and Palaeolithic Archaeology in the People's Republic of China*, edited by Wu Rukang and John W. Olsen, pp. 187-210. New York: Academic Press.
Zohary, Daniel. 1989. Domestication of the Southwest Asian Neolithic Crop Assemblage of Cereals, Pulses, and Flax: the Evidence from Living Plants. In *Foraging and Farming: The Evolution of Plant Exploitation*, edited by D. R. Harris and G. C. Hillman, pp. 358-373. Unwin Hyman, London.
Zukerkandl, Emile and Linus Pauling. 1965. Molecules as Documents of Evolutionary History. *Journal of Theoretical Biology* 8: 357-366.
Zvelebil, Mark, editor. 1986. *Hunters in Transition*. Cambridge University Press, Cambridge.